YEARBOOK OF CELL AND TISSUE TRANSPLANTATION 1996-1997

Yearbook of Cell and Tissue Transplantation 1996-1997

Edited by
ROBERT P. LANZA and WILLIAM L. CHICK
BioHybrid Technonologies, Inc., Shewsbury, Massachusetts

Kluwer Academic Publishers
Dordrecht / Boston / London

A C.I.P. Catalogue record for this book is available from
the Library of Congress.

ISBN 0-7923-3844-8

Published by Kluwer Academic Publishers,
P.O. Box 17, 3300 AA Dordrecht, The Netherlands.

Kluwer Academic Publishers incorporates
the publishing programmes of
D. Reidel, Martinus Nijhoff, Dr W. Junk and MTP Press.

Sold and distributed in the U.S.A. and Canada
by Kluwer Academic Publishers,
101 Philip Drive, Norwell, MA 02061, U.S.A.

In all other countries, sold and distributed
by Kluwer Academic Publishers Group,
P.O. Box 322, 3300 AH Dordrecht, The Netherlands.

Printed on acid-free paper

Printed in the Netherlands

To Bonnie Baines Chick
For her friendship, love and devotion

Table of Contents

Foreword

Thomas E. Starzl, M. D., Ph. D. Dr. Starzl is Professor of Surgery and Director of the Pittsburgh Transplantation Institute. He is one of the pioneers who ushered in the modern era of transplantation. Together with Nobel laureate Joseph Murray and colleagues, Dr. Starzl used immunosuppressive agents along with superb surgical skills to open up the field of transplantation of kidney, heart and liver.

Yearbooks have always had a special place in the scientific literature. The justification for these compendia is greater than ever as we approach the third millennium, overwhelmed with information. The all but unmanageable avalanche of yearly publications has not spared tissue transplantation and especially the relatively new field of cell transplantion. These special fields are evolving at a rapid pace that began to gather momentum along with whole organ transplantation more than 30 years ago. The cumulative progress in engraftment of tissue has been relatively even. In contrast, the concept of successfully engrafting single cells or cell clumps was virtually secret until recently, except for bone marrow transplantation. This changed in the last decade. When in 1992 the Cell Transplantation Society was formed, a special discipline based on developments with virtually every kind of cell already was buttressed by a large literature for each.

Those working in tissue and cell transplantation have a common language that is shared by workers in seemingly unrelated medical and non-medical fields. As it has turned out, the strongest interdisciplinary linkage and even frank overlap may well prove to be with workers in whole organ transplantation. In 1992, evidence was obtained that the ubiquitous traffic of nonparenchymal cells of bone marrow origin (the so-called passenger leukocytes of these organs) and the survival of these migrant leukocytes peripherally was the long sought after explanation of organ "acceptance", and the key to the controlled induction of tolerance. Organ transplantation had turned out to be merely cell transplantation in artful disguise. Thus, it is appropriate that the first chapter of the 1996 yearbook should cover the topic of the stem cells that are necessary for maintenance of chimerism. The last chapter is a wrap-up of chimerism, tolerance, and how to produce them.

Many of the chapters in between touch peripherally on these two topics, either by explaining how cell renewal (or long survival of mature cells) can be achieved in tissue and cell grafts, or else by indicating why this is not a therapeutically important objective. No major topic in the global field is left uncovered in the many chapters, including the application of molecular techniques. The result will be a feast for those already well informed, and a life raft for those who are not.

Preparation of the yearbook has been a labor of love for the editors, Robert Lanza and William Chick, and apparently also for the contributors who uniformly are from the highest rank of an unusually dedicated and heterogenous professional group. There is a need for this yearbook and this need has been met with a product that is relevant, current, and easy to read.

Thomas E. Starzl, M.D., Ph. D.
Professor of Surgery
Director of the Pittsburgh Transplantation Institute

Preface

Cell and tissue transplantation is one of the most exciting and rapidly expanding areas in medicine. This first edition of the Yearbook of Cell and Tissue Transplantation summarizes the latest advances in this revolutionary field, including developments in tissue engineering and transplantation of hybrid organs and tissues, while reviewing those data which, while not new, add to the usefulness of this Yearbook as both a text and a reference. The potential therapeutic applications for cell and tissue transplantation are enormous. Although a single volume cannot cover all of the cells and tissues which might reasonably be transplanted to treat human disease, several of the most promising and important of these are included.

Section One, the transplantation of "**Bone Marrow and Stem Cells**", contains chapters on one of the most successful forms of cell transplantation being performed in clinical practice. Bone marrow and blood cell transplants are increasingly used to treat patients with diverse diseases, to reconstitute the bone marrow after either intensive chemotherapy and/or irradiation for leukemia, breast cancer, or to replace defective stem cell function in aplastic anemia or congenital immunodeficiency disorders. Since 1964, over 45,000 persons have received allogeneic bone marrow transplants. Widescale use of autotransplants only began more recently, although the number of annual procedures now exceeds 8,000, with over 30,000 transplants since 1980.

Section Two encompasses the transplantation of **cartilage, bone and muscle**. Bone and cartilage have been transplanted for almost a century. Today, there are nearly one hundred bone banks in the United States alone. Banked bone is widely used in spinal fusions, fracture treatment, dental applications, and joint replacement revisions, including application in bone loss engendered by osteolysis or stress shielding. Extensive experience with cartilage transplants has also been reported for joint replacement, for use in traumatic defects, and in skeletal reconstruction following resection of tumors. The transplantation of isolated chondrocytes is now also being studied for the treatment of condylar cartilage defects such as the resurfacing of joints for arthritis.

A variety of skeletal muscle transplantation procedures are practiced clinically. The mature differentiated tissue is commonly transplanted, either as minced muscle fragments or as standard free muscle grafts which can be surgically connected with the host through a vascular anastomosis. Surgical techniques can also be employed to enhance the degree of reinnervation and consequently the quality of the muscle grafts. Considerable attention has also been given in recent years to myoblast transfer as a means of treating diseased skeletal muscle, in particular, for introducing normal genes into diseased dystrophic muscle that is lacking the protein, dystrophin. Normal myoblasts are cultured *in vitro* and then injected into dystrophic muscles. The myoblasts then fuse with either intact or regenerated dystrophic muscle fibers, thus forming a chimeric syncytium containing the product of the dystropin gene.

Section Three, "**Chromaffin Cells**", details the use of these cells as a source of neuroactive agents for transplantation in the central nervous system. Chromaffin cells produce a rich diversity of biologically active substances, including neurotransmitters, neuropeptides and trophic factors. To date, most of the work has focused on the alleviation of Parkinsonian symptoms, although adrenal medullary transplants have also been used for other applications, including chronic pain, memory loss, Huntington's disease, and psychological depression.

Section Four, "**Corneal Transplantation**", reviews developments in both clinical and experimental corneal transplantation. Corneal transplantation is a widely used, highly successful treatment for many diseases of the cornea which cause blindness. Between 50,000 and 100,000 corneal graft procedures are performed worldwide each year. Common indications for keratoplasty vary depending on geographic location and local patterns of practice; these include keratoconus, bullous keratopathy (corneal decompensation associated with removal of a cataract and intraocular lens implantation), corneal dystrophy, corneal ulcers, herpetic keratitis, and corneal scars and opacities resulting from the sequelae of trauma and ocular infection.

Sections Six through Eleven include chapters on the transplantation of several other cell types, including **fetal tissue**, **germ cells**, **hepatocytes**, **pancreatic islets**, **neural tissue**, **pituitary cells**, and **retinal cells**. Although presently more experimental than use of corneal grafts or bone marrow, the transplantation of these cells and tissues appear likely to play a major role in medicine in the future. Diabetes mellitus, for example, afflicts an estimated 100 million people worldwide. An ideal treatment for this disease would be the transplantation of the islets of Langerhans. Restoration of normal glucose metabolism has already been achieved in several patients with Type I diabetes by the transplantation of human islets. Successful islet grafts should therefore also prevent or retard the development of the complications associated with the disease, which are believed to result from hyperglycemia.

Section Twelve focuses on the transplantation of **skin and endothelial cells**. The ability to control and accelerate various aspects of cutaneous wound healing has played an important role in improving functional and cosmetic outcomes following severe burn injuries and skin ulcers. Much progress has been made to optimize the development of laboratory-grown temporary and permanent skin replacements. The transplantation of endothelial cells on the lumenal surface of vascular implants has also been evaluated in both preclinical animal trials and human trials. With the development of gene therapy methodologies, endothelial cell transplantation is now moving beyond prosthetic devices and studies are ongoing to establish the use of these cells in other anatomical sites for gene therapy.

Section Thirteen, "**Tissue Engineering/Hybrid Organs and Tissues**", contains chapters on the utilization and transplantation of living cells to restore and reconstruct tissues and organs damaged by disease, accident, or congenital abnormalities and defects. Great strides have been made in this emerging discipline during the past decade. New approaches promise to provide solutions to tissue creation and repair, and include replacements for deficient endocrine and exocrine glands, such as the pancreas and the liver; cardiovascular replacement including small arteries, veins, coronary and peripheral stents; orthopedic replacements including cartilage, bone, and muscle; replacements of various tubular structures such as the trachea and the intestines; and skin and filler tissues for the treatment of burns, skin ulcers, deep wounds, and other injuries.

Finally Section Fourteen explores the use of **bone marrow transplantation for tolerance induction**. Hematopoietic reconstitution of cytoablated mice, via transplantation of bone marrow from an allogeneic donor, has been found to result in the production of chimeric animals which possess varying degrees of both host and donor derived hematopoietic cells. The recent demonstration that chimerism can be achieved in non-cytoblated animals via multiple infusions of donor bone marrow has renewed interest in potential applications of this approach in humans.

In view of the important and rapid changes occurring in the area of cell and tissue transplantation, a new edition of the *Yearbook of Cell and Tissue Transplantation* will appear periodically. The Editors would welcome any comments or suggestions for future editions. We would also like to express our gratitude and appreciation to our many associates and colleagues who as experts in their fields have contributed to this volume.

Robert P. Lanza
William L. Chick

Section I:

Bone Marrow and Stem Cells

R. P. Lanza and W. L. Chick (eds.), Yearbook of Cell and Tissue Transplantation 1996/1997, 3–12.

Chapter 1

Bone marrow transplantation

James O. Armitage and Michael R. Bishop
Section of Oncology/Hematology, University of Nebraska Medical Center, Omaha, NE 68198, U.S.A.

Bone marrow transplantation

Bone marrow transplantation (BMT) involves the intravenous infusion of hematopoietic progenitor cells to re-establish marrow function in a patient with a damaged or defective bone marrow. The term bone marrow transplantation has become somewhat of an anachroism since hematopoietic stem cells may be collected from the peripheral blood or the umbilical cord and successfully transplanted [1, 2]. The beginnings of modern bone marrow transplantation may be traced to work showing that rodents could be protected against lethal hematopoietic injury by the intravenous infusion of bone marrow [3]. The subsequent identification of transplantation antigens (i.e. the HLA system in man) and the development of cryobiology to allow reproducibly successful freezing and thawing of hematopoietic cells laid the ground work for the difficult and time consuming clinical trials that have brought allogeneic and autologous bone marrow transplantation to their present states. The first successful allogeneic transplants in the late 1960s [4]. Autologous bone marrow transplantation was first successfully employed for the treatment of patients with lymphoma in the late 1970's and today the annual number of autologous transplants has surpassed that of allogeneic transplants [5].

Allogeneic and syngeneic bone marrow transplantation

Allogeneic bone marrow transplantation involves the transfer of marrow between individuals. In the special case where the donor and recipient are genetically identical – i.e. identical twins – the correct term is syngeneic transplantation. For patients without a twin, an HLA-matched sibling donor is preferable for an allogeneic bone marrow transplantation. Because of the relatively small average size of American families, only 30 percent of Americans actually have an HLA-identical sibling. For patients not fortunate enough to have an HLA-matched sibling donor, but who may benefit from an allogeneic bone marrow transplantation, an unrelated but closely HLA-matched person willing to donate bone marrow may be used [6, 7]. The extremely large number of possible HLA-phenotypes (i.e. the theoretical number of possibilities are greater than the total world population) make finding an unrelated donor a difficult undertaking. Fortunately, certain HLA-phenotypes occur more frequently than would be expected in persons with a similar genetic background. For example, it has been estimated that a registry of 200,000 persons of European ancestry would allow another person of European ancestry a 40 to 50 percent chance of finding an HLA-matched donor [8, 9]. Because of the imprecision of traditional serologic studies, HLA typing at the molecular level using oligonucleotide probes has been widely adopted in identifying unrelated donors.

An alternate approach is to identify related individuals who share some but not all HLA antigens [10]. Current results of transplantation with HLA mismatched related donors suggest that mismatching gives less good results than with perfectly matched related donors, and the results become worse with higher degrees of mismatching.

Once a donor has been identified, high doses of chemotherapy and/or radiotherapy are necessary to provide a sufficient degree of immunosuppression to avoid destruction of the allograft by residual immunologically active cells in the host and to destroy any residual cancer cells in the patient. Most preparative regimens for bone marrow transplantation use some combination of radiotherapy, alkylating agents, etoposide, and cytarabine. In addition to the anticipated problems with severe, prolonged myelosuppression, allogeneic BMT is associated with several unique complications, in particular, graft-versus-host disease. Graft-versus-host Disease (GVHD) describes an illness that is predominantly manifested by symptoms and signs referable to the skin, gastrointestinal system, and liver, and the severity of the condition is graded based on the involvement of these organs (Table 1) [11]. The pathogenesis of this condition involves immunologically competent cells in the graft targeting antigens on the cells of the transplant recipient and producing the syndrome. GVHD is referred to as acute (i.e. occurring in the first 1–2 months after an allogeneic BMT) or chronic (i.e. developing at least 2–3 months after an allogeneic BMT). Patients undergoing allogeneic BMT typically receive prophylactic immunosuppression for acute GVHD, including cyclosporine, methotrexate, corticosteroids, and T-cell depletion of the graft [12–16]. Even so the majority of adults develop some degree of acute GVHD following allogeneic bone marrow transplantation. Treatments for established, severe acute GVHD include high doses of corticosteroids, antithymocyte globulin, and monoclonal antibodies [17–19].

Chronic GVHD shares certain clinical characteristics with other immunologic disorders such as scleroderma [20]. The strongest predictive factors in its development are the occurrence of acute GVHD and increasing patient age [21, 22]. Untreated, the condition is often fatal. Treatment utilizing prednisone, cyclosporine and/or thalidomide has improved the long-term outlook for these patients [23, 24]. Major adverse prognostic factors for patients who develop chronic GVHD are thrombocytopenia, progressive presentation, and elevated bilirubin [25, 26].

There is evidence that GVHD is also accompanied by a graft-versus-malignancy effect. It includes the lower relapse rate in leukemia patients who develop GVHD [27, 28] and the higher leukemia relapse rate observed in patients who receive T-cell depleted allografts or transplants from identical twins where GVHD would not be expected [16, 29]. However, separating the dangerous effects of GVHD from the presumed graft-versus-malignancy effect has not been simple. Routine infusions of donor buffy coat cells lead only to a higher incidence of severe GVHD [29, 30].

Autologous bone marrow transplantation

Autologous BMT involves re-establishing hematopoietic cell function in patients after high-dose therapy for cancer. The hematopoietic stem cells used for transplantation come from the patients themselves and may be derived from the bone marrow or peripheral blood. Autologous BMT differs from allogeneic BMT in a variety of ways (Table 2). Autologous BMT can be performed in older patients, probably because of the absence of GVHD as a major complication.

A major concern with autologous bone marrow transplantation is the reinfusion of viable tumor cells. Numerous methods, including *in vitro* treatment with chemotherapeutic agents or monoclonal antibodies plus complement, have been developed to remove contaminating tumor cells from the graft, a process often referred to as "purging" [12, 31–37]. Retrospective analyses have suggested that purging leads to a reduced relapse rate in patients with acute myeloid leukemia and B-cell non-Hodgkin's lymphomas [31, 32]. Tumor cells can be cultured from histologically negative bone marrow in some patients with lymphoma, leukemia, and breast cancer, and patients with positive cultures have a poorer outlook [33–40]. However, relapses often occur at sites of previously known disease, raising the question of treatment resistance rather than reinfusion of tumor cells as the cause of relapse [32]. This question is not likely to be settled until gene transfer experiments make it clear exactly which patients are likely to relapse because of tumor cells in the graft [41].

Complications of bone marrow transplantation

In addition to the anticipated problems with severe, prolonged myelosuppression, bone marrow transplantation is associated with several unique complications. A major cause of death following allogeneic bone marrow transplantation is idiopathic interstitial pneumonia [42]. When it occurs in the second month or later after bone marrow transplantation, this disorder is most frequently caused by cytomegalovirus. Patients at highest risk are those with severe GVHD [43]. The use of cytomegalovirus negative blood products in patients who are cytomegalovirus antibody negative, administration of intravenous immune globulin, and the prophylactic administration of acyclovir or gancyclovir to patients at high risk have all reduced the incidence of this potentially lethal complication [44–48]. Interstitial

Table 1. Classification of patients with acute graft-versus-host disease

	Extent of organ injury		
Level	Skin	Liver	Intestinal tract
1	Maculopapular rash < 25% of body surface	Bilirubin 2–3 mg/100 ml	> 500 ml diarrhea/day
2	Maculopapular rash 25–50% body surface	Bilirubin 3–6 mg/100 ml	> 1000 ml diarrhea/day
3	Generalized erythro-derma	Bilirubin 6–15 mg/100 ml	> 1500 ml diarrhea/day
4	Generalized erythro-derma with bullous formation & desquamation	Bilirubin > 15 mg/100 ml	Severe abdom-inal pain, with or without ileus

Clinical grading

	Level of injury		
Grade*	Skin	Liver	Intestinal tract
I	1–2	0	0
II*	1–3	1	1
III	2–3	2–3	2–3
IC+	2–4	2–4	2–4

* Grade II or higher requires skin plus either liver of intestinal involvement or both.
+ Requires extreme decrease in performance status.

pnuemonia may occur after autologous BMT, butfar less frequently than after allogeneic BMT [49].

Veno-occlusive disease (VOD) of the liver has been frequently reported following both allogeneic and autologous BMT. The diagnosis of VOD is made when two of the three major symptoms (i.e. jaundice, tender hepatomegaly, and ascites or unexplained weight gain) are present, and the condition is frequently fatal [50]. Therapy has generally been unsatisfactory; however, a recent report of the use of thrombolytic therapy was encouraging [51].

Clinical results with bone marrow transplantation

Non-malignant diseases

Allogeneic bone marrow transplantation leads to disease-free survival in more than 50 percent of patients with severe aplastic anemia [52]. Allogeneic bone marrow transplantation is effective in all forms of aplastic anemia including that following paroxysmal nocturnal hemoglobinuria [53]. However, patients with Fanconi anemia pose special problems and require modified preparative regimens because of increased treatment-related toxicity [54].

The effectiveness of allogeneic BMT in patients with thalassemia has been well documented [55]. The largest reported series included 222 patients and used busulfan and cyclophosphamide rather than total body irradiation and cyclophosphamide as a preparative reg-

Table 2. Comparison of allogeneic and autologous bone marrow transplantation

	Allogeneic	Autologous
Oldest age to which applicable	40–55 years	60–70 years
Major problem in finding a donor	finding a closely HLA-matched sibling or willing unrelated donor	ability to collect sufficient numbers of hematopoietic progenitor cells uncontaminated by tumor cells
Most important complication	graft-vs-host disease	relapse of original disease
Anti-cancer effect of infused cells	proved or suspected in a number of malignancies	probably no – but data regarding cyclosporine-induced GVL effect or possible anti-lymphoma effect of peripheral stem cells are interesting
Applicable to non-malignant disorders	potentially curative in both genetic and immunologic disorders	not until gene therapy becomes practical or the potential to induce immunologic alterations becomes clearer

imen in an attempt to avoid long-term complications [56]. Event-free survival after one year was 75 percent. Sickle cell anemia represents another serious hemoglobinopathy that is potentially curable with allogeneicBMT. The available literature suggest that allogeneic bone marrow transplantation can establish normal hematopoiesis and alleviate the symptoms of the disorder [57].

Allogeneic BMT was first reported to be successful in children with life-threatening immunodeficiency disorders [58, 59]. Allogeneic BMT can replace the defective stem cells in severe combined immunodeficiency, Wiskott-Aldrich syndrome, and Chediak-Higashi syndrome [60–62]. When an HLA identical sibling is available for a child recognized to have any of these lethal disorders, allogeneic BMT is the treatment of choice. Unfortunately, in many cases an HLA matched sibling donor is not available. In such cases, both the use of T-cell depleted marrow from an HLA non-identical but related donor or HLA-matched unrelated individuals have been successful [63, 64]. Unfortunately, patients receiving T-cell depleted grafts seem at especially high risk to develop Epstein-Barr virus associated B-cell lymphoproliferative disorders [65].

Allogeneic BMT has been used successfully to treat a number of genetic disorders including the reversal of infantile malignant osteopetrosis, Gaucher's disease, infantile metachromatic leukodystrophy, and x-linked adrenoleukodystrophy [66–69].

Malignant diseases (Table 3)

Some patients with acute myeloid leukemia can be cured with allogeneicBMT, even when treated for end-stage, refractory leukemia, although the cure rate is very low [70]. Long-term survival and apparent cure of 20 to 40 percent has been achieved in patients treated in second or subsequent complete remission and cure rates of 40 to 70 percent have been reported in patients transplanted in first complete remission [71–73]. Because some patients can be cured with standard chemotherapy regimens without bone marrow transplantation, withholding bone marrow transplantation until the first sign of treatment failure has been an increasingly popular strategy [74]. Attempts to compare allogeneic with autologous bone marrow transplantation in patients with acute myeloid leukemia have generally reached the conclusion that autologous transplantation is associated with a lower treatment-related mortality but a higher leukemia relapse rate [75]. Patients with myelodyspastic syndrome, when an HLA identical sibling can be identified, can obtain

Table 3. The application of bone marrow transplantation in selected malignancies

Malignancy	Preferred of transplant	BMT is potentially curative in advanced disease	"Standard" therapy
AML, ALL	Probably allogeneic	Yes	Yes
CML	Allogeneic	Yes	Yes
CLL	Probably allogeneic	Uncertain	No
Multiple myeloma	Controversial	Uncertain	No
Histologically aggressive NHL	Probably autologous	Yes	Yes
Histologically indolent NHL	Probably autologous	Uncertain	Yes
Hodgkin's disease	Probably autologous	Yes	Yes
Neuroblastoma	Uncertain	Probably	Yes
Breast cancer	Autologous	Uncertain	No
Testicular cancer	Autologous	Yes	Yes

long-term, disease-free survival and potentially cure with allogeneic BMT [76, 77].

The results of standard therapy in children with acute lymphocytic leukemia (ALL) are sufficiently good using intensive chemotherapy regimens that BMT as part of the primary therapy should probably be performed only in special situations, such as in patients with Philadelphia chromosome positive ALL where the cure rate with standard therapy is very low [78]. Children who fail to be cured with their primary chemotherapy regimen, particularly those who relapse early and patients who relapse in the first six months after achieving an initial remission are candidates for allogeneic BMT.

Allogeneic BMT done in first complete remission in adults has been reported to produce long-term, disease-free survival in 40 to 70 percent of patients [79, 80]. However, one retrospective comparison of two large data bases suggested no advantage of autologous BMT over an effective chemotherapy regimen [81]. As with children, BMT is indicated in first remission for patients whose leukemia is Philadelphia chromosome positive, and, probably, in second remission for patients with a short initial remission [78].

Allogeneic bone marrow transplantation from an HLA-matched sibling donor has become the treatment of choice for patients of an appropriate age with stable phase chronic myelogenous leukemia [82–84]. Patients seem to do better when transplanted within the first year after diagnosis and when they receive hydroxyurea rather than busulfan as their initial therapy for the leukemia [82, 85].

Chronic lymphocytic leukemia usually occurs in elderly people. However, when younger patients develop the disease, allogeneic and autologous BMT have both been employed [86–88]. Both approaches have produced leukemia-free survival in more than 50 percent of treated patients with brief follow up. It will take many years to document the curability of this disorder with BMT because of the long natural history of this usually indolent hematologic malignancy.

Both allogeneic and autologous BMT have been performed for multiple myeloma, since standard chemotherapy regimens rarely, if ever, cure the disease. The largest series of patients reported with allogeneic bone marrow transplantation found a 43 percent complete remission rate with 50 percent of the complete responders alive and relapse free 48 months after transplant [89]. Autologous BMT has frequently been performed for multiple myeloma [90–92]. The complete remission rate is lower than with allogeneic BMT, but progression-free survivals at one year is similar to due to a lower treatment related mortality as compared to allogeneic BMT [90].

Allogeneic, syngeneic, and autologous bone marrow transplantation have all been reported to yield long-term, disease-free survival and apparent cure for patients with intermediate and high-grade non-

Hodgkins lymphomas (NHL) [93–95]. The disease-free survival with autologous or allogeneic BMT appears to be comparable. Patients with relapsed lymphoma are more likely to be cured if their tumor remains sensitive to chemotherapy; however, patients who fail to achieve an initial complete remission, but do not have other adverse prognostic factors, can also achieve long-term, disease-free survival [94, 95].

For patients with low grade NHL autologous BMT using either purged bone marrow, or peripheral blood stem cells, can result in disease-free survivals of 40 to 60 percent with median follow up periods of approximately 3 years [96–98]. However, the late relapses seen in this illness with conventional therapy make very long follow-up necessary to document the curative potential of this approach. The encouraging results seen in relapsed patients has led to early trials of the use of autologous bone marrow transplantation as part of the primary therapy of patients with low grade lymphoma [99].

High-dose therapy and autologous or allogeneic BMT has been widely performed in patients with recurrent Hodgkin's disease [100–103]. Because of the lower treatment related mortality, autologous transplantation is preferred by most, but not all, investigators [104]. One controlled trial of autologous transplantation found a superior event-free survival when compared to the same chemotherapy drugs given at lower doses [105].

Autologous BMT has been used to allow the administration high doses of chemotherapy to patients with advanced neuroblastoma. When incorporated into the treatment for Stage IV disease, disease-free survival as high as 40 percent at 2 years has been seen [106, 107].

Breast cancer is an increasingly common indication for autologous BMT. Autologous BMT can produce complete responses in a higher proportion of patients than is seen with traditional doses of chemotherapy [108, 109]. The results in a large number of patients have shown that disease-free survival at 2 to 5 years is seen in 10 to 30 percent of patients with chemotherapy sensitive, Stage IV disease [108–110]. If longer follow-up shows that the survival with transplantation is superior to conventional therapy in controlled trials, this treatment will likely become a standard approach to young women with chemotherapy sensitive metastatic disease. Encouraging results of incorporation of autologous bone marrow transplantation into the primary therapy of patients who present with poor prognostic factors has led to this treatment being tested in an adjuvant setting in controlled trials on a national basis [111].

In patients with testicular carcinoma who fail to be cured with platinum based chemotherapy regimens, autologous bone marrow transplantation has resulted in disease-free survival of 10 to 20 percent at 2 years in these patients [112–113]. These results are similar to those reported in the treatment of lymphoma, and suggest that results would be superior if the treatment was incorporated earlier in the course of the disease.

Future of bone marrow transplantation

Malignant diseases are likely to remain the major indication for bone marrow transplantation in the foreseeable future. The development of new, more effective chemotherapeutic agents may make transplantation unnecessary, and the reinfusion of hematopoietic stem cells may be made unnecessary in certain situations by more effective hematopoietic growth factors. On the other hand, better selection of patients and decreased acute morbidity and mortality associated with BMT, may increase its use as part of standard therapy for various malignancies. If the transduction of genetic vector can be improved, the indications for BMT will expand far beyond the boundaries of treating malignancies. It appears that BMT is just in its infancy and will hopefully grow and mature in the years to come.

References

1. Kessinger A, Armitage JO and Landmark JD (1988) Autologous peripheral hematopoietic stem cell transplantation restores hematopoietic function following marrow ablative therapy. Blood 71: 723–727.
2. Gluckman E, Broxmeyer HE, Auerbach AD *et al.* (1989) Hematopoietic reconstitution in a patient with Fanconi's anemia by means of umbilical blood from an HLA-identical sibling. N Engl J Med 321: 1174.
3. Lorenz E, Uphoff DE, Reid TR and Shelton E (1951) Modification of irradiation injury in mice and guinea pigs by bone marrow injections. J Natl Cancer Inst 12: 197–201.
4. Thomas ED, Storb R, Clift RA *et al.* (1975) Bone marrow transplantation. N Engl J Med 292: 832–843 and 895–902.
5. Appelbaum FR, Herzig GP, Ziegler JC *et al.* (1978) Successful engraftment of cryopreserved autologous bone marrow in patients with malignant lymphoma. Blood 52: 85–95.
6. Hansen JA, Clift RA, Thomas ED, Buckner CD, Storb R and Giblett ER (1980) Transplantation of marrow from an unrelated donor to a patient with acute leukemia. N Engl J Med 303: 565–567.

7. Kernan NA, Bartsch G, Ash RC *et al.* (1993) Analysis of 462 transplantations from unrelated donors facilitated by the National Marrow Donor Program. N Engl J Med 328: 593–602.
8. Gahrton G (1991) Bone marrow transplantation with unrelated volunteer donors. Eur J Cancer 27: 1537–1539.
9. Beatty PG, Dahlberg S, Mickelson EM *et al.* (1988) Probability of finding HLA-matched unrelated marrow donors. Transplantation 45: 714–718.
10. Beatty PG, Clift RA, Mickelson EM *et al.* (1985) Marrow transplantation from related donors other than HLA-identical siblings. N Engl J Med 313: 765–771.
11. Thomas ED, Storb R, Clift RA *et al.* (1975) Bone-marrow transplantation. N Engl J Med 292: 895–902.
12. Morecki S, Margel S and Slavin S (1988) Removal of breast cancer cells by soybean agglutinin in an experimental model for purging human marrow. Cancer Res 48: 4575–4577.
13. Storb R, Deeg HJ, Whitehead J *et al.* (1986) Methotrexate and cyclosporine compared with cyclosporine alone for prophylaxis of acute graft versus host disease after marrow transplantation for leukemia. N Engl J Med 324: 729–735.
14. Ramsay NK, Kersey JH, Robison LL *et al.* (1982) A randomized study of the prevention of acute graft-versus-host disease. N Engl J Med 306: 392–397.
15. Storb R, Deeg HJ, Pepe M *et al.* (1989) Methotrexate and cyclosporine versus cyclosporine alone for prophylaxis of graft-versus-host disease in patients given HLA-identical marrow grafts for leukemia: Long-term follow-up of a controlled trial. Blood 73: 1729–1734.
16. Mitsuyasu RT, Champlin RE, Gale RP *et al.* (1986) Treatment of donor bone marrow with monoclonal anti-T-cell antibody and complement for the prevention of graft-versus-host disease. Annals Intern Med 105: 20–26.
17. Martin PJ, Schoch G, Fisher L *et al.* (1990) A retrospective analysis of therapy for acute graft-versus-host disease: initial treatment. Blood 76: 1464–1472.
18. Kennedy MS, Deeg HJ, Storb R *et al.* (1985) Treatment of acute graft-versus-host disease after allogeneic marrow transplantation. Randomized study comparing corticosteroids and cyclosporine. Am J Med 78: 978–983.
19. Herve P, Wijdenes J, Bergrat JP *et al.* (1990) Treatment of corticosteroid resistant acute graft-versus-host disease by in vivo administration of anti-interleukin-2 receptor monoclonal antibody (B-B10). Blood 75: 1017-1023.
20. Shulman HM, Sullivan KM, Weiden PL *et al.* (1980) Chronic graft-versus-host syndrome in man. A long-term clinicopathologic study of 20 Seattle patients. Am J Med 69: 204–217.
21. Atkinson K, Horowitz MM, Gale RP *et al.* (1990) Risk factors for chronic graft-versus-host disease after HLA-identical sibling bone marrow transplantation. Blood 75: 2459–2464.
22. Storb R, Prentice RL, Sullivan KM *et al.* (1983) Predictive factors in chronic graft-versus-host disease in patients with aplastic anemia treated by marrow transplantation from HLA-identical siblings. Annals Intern Med 98: 461–466.
23. Sullivan KM, Witherspoon RP, Storb R *et al.* (1988) Alternating-day cyclosporine and prednisone for treatment of high-risk chronic graft-versus-host disease. Blood 72: 555–561.
24. Vogelsang GB, Farmer ER, Hess AD *et al.* (1992) Thalidomide for the treatment of chronic graft-versus-host disease. N Engl J Med 326: 1055–1058.
25. Sullivan KM, Witherspoon RP, Storb R *et al.* (1988) Prednisone and azathioprine compared with prednisone and placebo for treatment of chronic graft-v-host disease: prognostic influence of prolonged thrombocytopenia after allogeneic marrow transplantation. Blood 72: 546–554.
26. Wingard JR, Piantadosi S, Vogelsange GB *et al.* (1989) Predictors of death from chronic graft-versus-host disease after bone marrow transplantation. Blood 74: 1428–1435.
27. Weiden PL, Flournoy N, Thomas ED *et al.* (1979) Antileukemic effect of graft-versus-host disease in human recipients of allogeneic-marrow grafts. N Engl J Med 300: 1068–1073.
28. Weiden PL, Sullivan KM, Flournoy N *et al.* (1981) Antileukemic effect of chronic graft-versus-host disease. N Engl J Med 304: 1529–1533.
29. Horowitz MM, Gale RP, Sondel PM *et al.* (1990) Graft-versus-leukemia reactions after bone marrow transplantation. Blood 75: 555–562.
30. Sullivan KM, Storb R, Buckner CD *et al.* (1989) Graft-versus-host as adoptive immunotherapy in patients with advanced hematologic neoplasms. N Engl J Med 320: 828–834.
31. Gorin NE, Aegerter P, Auvert B *et al.* (1990) Autologous bone marrow transplantation for acute myelocytic leukemia in first remission: A European survey of the role of marrow purging. Blood 75: 1606–1614.
32. Gribben JG, Freedman AS, Neuberg D *et al.* (1991) Immunologic purging of marrow assessed by PCR before autologous bone marrow transplantation for B-cell lymphoma. N Engl J Med 325: 1525–1533.
33. Atzpodien J, Gulati SC, Strife A and Clarkson BD (1987) Photoradiation models for the clinical ex vivo treatment of autologous bone marrow grafts. Blood 70: 484–489.
34. Roy DC, Griffin JD, Belvin M, Blattler WA, Lambert JM and Ritz J (1991) Anti-MY9-Blocked-Ricin: An immunotoxin for selective targeting of acute myeloid leukemia cells. Blood 77: 2404–2412.
35. Chang J, Coutinho L, Morgenstern G *et al.* (1986) Reconstitution of haemopoietic system with autologous marrow taken during relapse of acute myeloblastic leukaemia and grown in long-term culture. The Lancet 1: 294–295.
36. Shpall EJ, Stemmer SM, Johnston CF *et al.* (1992) Purging of autologous bone marrow transplantation: The protection and selection of the hematopoietic progenitor cell. J Hematotherapy 1: 45–54.
37. Philip I, Philip T, Favrot M *et al.* (1984) Establishment of lymphomatous cell lines from bone marrow samples from patients with Burkitt's lymphoma. JNCI 73: 835–840.
38. Sharp JG, Joshi SS, Armitage JO *et al.* (1992) Significance of detection of occult non-Hodgkin's lymphoma in histologically uninvolved bone marrow by a culture technique. Blood 79: 1074–1080.
39. Estrov Z, Grunbrger T, Dube ID, Wange Y-P, Freedman MH *et al.* (1986) Detection of residual acute lymphoblastic leukemia cells in cultures of bone marrow obtained during remission. N Engl J Med 315: 538–541.
40. Sharp JG, Mann SL, Kessinger A, Joshi SS, Crouse DA and Weisenburger DD (1987) Detection of occult breast cancer cells in cultured pretransplantation bone marrow. In: Dicke KA, Spitzer G and Jagannath S (eds) Autologous Bone Marrow Transplantation, pp 497–502. The University of Texas Publ.
41. Brenner MK, Rill DR, Moen RC *et al.* (1993) Gene-marking to trace origin of relapse after autologous bone-marrow transplantation. The Lancet 341: 85–86.
42. Wingard JR, Mellits ED, Sostrin MB *et al.* (1988) Interstitial pneumonitis after allogeneic bone marrow transplantation.

Nine-year experience at a single institution. Medicine 67: 175–186.
43. Miller W, Flynn P, McCullough J *et al.* (1986) Cytomegalovirus infection after bone marrow transplantation: an association with acute graft-v-host disease. Blood 67: 1162–1167.
44. Bowden RA, Sayers M, Flournoy N *et al.* (1986) Cytomegalovirus immune globulin and seronegative blood products to prevent primary cytomegalovirus infection after marrow transplantation. N Engl J Med 314: 1006–1010.
45. Winston DJ, Ho WG, Lin C-H *et al.* (1987) Intravenous immune globulin for prevention of cytomegalovirus infection and interstitial pneumonia after bone marrow transplantation. Annals Intern Med 106: 12–18.
46. Meyers JD, Leszczyzynski J, Zaia JA *et al.* (1983) Prevention of cytomegalovirus infection by cytomegalovirus immune globulin after marrow transplantation. Annals Intern Med 98: 442–6.
47. Meyers JD, Reed ED, Shepp DH *et al.* (1988) Acyclovir for prevention of cytomegalovirus infection and disease after allogeneic marrow transplantation. N Engl J Med 318: 70–75.
48. Goodrich JM, Mori M, Gleves CA *et al.* (1991) Early treatment with ganciclovir to prevent cytomegalovirus disease after allogeneic bone marrow transplantation. N Engl J Med 325: 1601–1607.
49. Wingard JR, Yen-Hung Chen D, Burns WH *et al.* (1988) Cytomegalovirus-infection after autologous bone marrow transplantation with comparison to infection after allogeneic bone marrow transplantation. Blood 71: 1432–437.
50. Shulman HM and Hinterberger W (1992) Hepatic veno-occlusive disease – liver toxicity syndrome after bone marrow transplantation. Bone Marrow Transplantation 10: 197–214.
51. Bearman SI, Shuhart MC, Hinds MS and McDonald GB (1992) Recombinant human tissue plasminogen activator for the treatment of established severe venocclusive disease of the liver after bone marrow transplantation. Blood 80: 2458–2462.
52. Gluckman E, Horowitz MM, Champlin RE *et al.* (1992) Bone marrow transplantation for severe aplastic anemia: Influence of conditioning and graft-versus-host disease prophylaxis regimens on outcome. Blood 79: 269-275.
53. Szer J, Deeg J, Witherspoon RP *et al.* (1984) Long-term survival after marrow transplantation for paroxysmal nocturnal hemoglobinuria with aplastic anemia. Annals Intern Med 101: 193–195.
54. Flower MED, Doney KC, Storb R *et al.* (1992) Marrow transplantation for Fanconi anemia with or without leukemic transformation: an update of the Seattle experience. Bone Marrow Transpl 9: 167–173.
55. Thomas ED, Buckner CD, Sanders JE *et al.* (1982) Marrow transplantation for thalassanemia. Lancet 2: 227–229.
56. Lucarelli G, Galimberti M, Polchi P *et al.* (1990) Bone marrow transplantation in patients with thalassemia. N Engl J Med 322: 417–421.
57. Vermylen C, Fernandez Robles E, Ninane J and Cornu G (1988) Bone marrow transplantation in five children with sickle cell anaemia. The Lancet 1: 1427–1428.
58. Gatti RA, Meuwissen HJ, Allen HD *et al.* (1968) Immunological reconstitution of sex-linked lymphopenic immunological deficiency. Lancet 2: 1366–1369.
59. Bach FH, Albertini RJ, Anderson JL, Joop P and Bortin MM (1968) Bone marrow transplantation in a patient with the Wiskott-Aldrich syndrome. Lancet 2: 1364–1366.
60. Bortin MM and Rim AA (1977) Severe combined immunodeficiency disease. Characterization of the disease and results of transplantation. JAMA 238: 591–600.
61. Parkman R, Rappeport J, Geha R *et al.* (1977) Complete correction of the Wiskott-Aldrich syndrome by allogeneic bone marrow transplantation. N Engl J Med 298: 921–927.
62. Kazmierowski JA, Elin RJ, Reynolds HY, Durbin WA and Wolff SM (1976) Chediak-Higashi syndrome: reversal of increased susceptibility to infection by bone marrow transplantation. Blood 47: 555–559.
63. O'Reilly RJ, Keever CA, Small TN and Brochstein J (1989) The use of HLA-non-identical T-cell-depleted marrow transplants for correction of severe combinated immunodeficiency disease. Immunodefic Rev 1: 273.
64. O'Reilly RJ, Dupont B, Pahwa S *et al.* (1977) Reconstitution in severe combined immunodeficiency by transplantation of marrow from an unrelated donor. N Engl J Med 297: 1311–1318.
65. Shearer WT, Ritz J, Finegold MJ *et al.* (1985) Epstein-Barr virus-associated B-cell proliferations of diverse clonal origins after bone marrow transplantation in a 12-year-old patient with severe combined immunodeficiency. N Engl J Med 312: 1151–1159.
66. Coccia PF, Krivit W, Cervenka J *et al.* (1980) Successful bone-marrow transplantation for infantile malignant osteopetrosis. N Engl J Med 302: 701–708.
67. Rappeport JM and Ginns EI (1984) Bone-marrow transplantation in severe Gaucher's disease. N Engl J Med 311: 84–88.
68. Krivit W, Shapiro E and Kennedy W (1990) Treatment of late infantile metachromatic leukocystrophy by bone marrow transplantation. N Engl J Med 322: 28–32.
69. Auborg P, Blanche S, Jambauqe' I *et al.* (1990) Reversal of early neurologic and neuroradiologic manifestations of X-linked adrenoleukodystrophyby bone marrow transplantation. N Engl J Med 322: 1860–1866.
70. Thomas ED, Buckner CD, Banaji M *et al.* (1977) One hundred patients with acute leukemia treated by chemotherapy, total body irradiation, and allogeneic marrow transplantation. Blood 49: 511–533.
71. Thomas ED, Buckner CD, Clift RA *et al.* (1979) Marrow transplantation for acute nonlymphoblastic leukemia in first remission. N Engl J Med 301: 597–599.
72. Blume KG, Beutler E, Bross KJ *et al.* (1980) Bone-marrow ablation and allogeneic marrow transplantation in acute leukemia. N Engl J Med 302: 1041–1046.
73. Santos GW, Tutschka PJ, Brookmeyer R *et al.* (1983) Marrow transplantation for acute nonlymphocytic leukemia after treatment with busulfan and cyclophosphamide. N Engl J Med 309: 1347–1353.
74. Clift RA, Buckner CD, Appelbaum FR *et al.* (1992) Allogeneic marrow transplantation during untreated first relapse of acute myeloid leukemia. J Clin Oncol 10: 1723–1729.
75. Lowenberg B, Verdonck LJ, Dekker AW *et al.* (1990) Autologous bone marrow transplantation in acute myeloid leukemia in first remission: Results of a Dutch prospective study. J Clin Oncol 8: 287–294.
76. Appelbaum FR, Storb R, Rambert RE *et al.* (1987) Treatment of preleukemic syndromes with marrow transplantation. Blood 69: 92–96.
77. O'Donnell MR, Nademanee AP, Snyder DS *et al.* (1987) Bone marrow transplantation for myelodysplastic and myeloproliferative syndromes. J Clin Oncol 5: 1822–1826.

78. Barrett AJ, Horowitz MM, Ash RC *et al.* (1992) Bone marrow transplantation for Philadelphia chromosome-positive acute lymphoblastic leukemia. Blood 79: 3067–3070.
79. Doney K, Buckner CD, Kopecky KJ *et al.* (1987) Marrow transplantation for patients with acute lymphoblastic leukemia in first marrow remission. Bone Marrow Transplant 2: 355–363.
80. Chao NJ, Forman SJ, Schmidt GM *et al.* (1991) Allogeneic bone marrow transplantation for high-risk acute lymphoblastic leukemia during first complete remission. Blood 78: 1923–1927.
81. Horowitz MM, Messerer D, Hoelzer D *et al.* (1991) Chemotherapy compared with bone marrow transplantation for adults with acute lymphoblastic leukemia in first remission. Annals Intern Med 115: 13–18.
82. Fefer A, Cheever MA, Thomas ED *et al.* (1979) Disappearance of Ph^1-positive cells in four patients with chronic granulocytic leukemia after chemotherapy, irradiation and marrow transplantation from an identical twin. N Engl J Med 300: 333–337.
83. Thomas ED, Clift RA, Fefer A *et al.* (1986) Marrow transplantation for the treatment of chronic myelogenous leukemia. Annals Intern Med 104: 155–163.
84. Goldman JM, Apperley JF, Jones L *et al.* (1986) Bone marrow transplantation for patients with chronic myeloid leukemia. N Engl J Med 314: 202–207.
85. Goldman JM, McGlave P, Szydlo P *et al.* (1990) Impact of disease duration and prior treatment on outcome of bone marrow transplants for chronic myelogenous leukemia (CML). Blood 80 (Suppl 1): 170a.
86. Michallet M, Corront B, Hollard D *et al.* (1991) Allogeneic bone marrow transplantation in chronic lymphocytic leukemia: 17 cases. Report from the EBMTG. Bone Marrow Transplant 7: 275–279.
87. Rabinowe SN, Soiffer RJ, Gribben JG *et al.* (1992) Autologous and allogeneic bone marrow transplantation (BMT) for patients with binet stage B and C B-cell chronic lymphocytic leukemia (B-CLL). Blood 80: 170a.
88. Khouri I, Thomas M, Andersson B, Deisseroth A, Keating M and Champlin R (1992) Purged autologous bone marrow transplantation for chronic lymphocytic leukemia: Preliminary results. Blood 80: 66a.
89. Gahrton G, Tura S, Ljungman P *et al.* (1991) Allogeneic bone marrow transplantation in multiple myeloma. N Engl J Med 325: 1267–1273.
90. Jagannath S, Vesole DH, Glenn L, Crowley J and Barlogie B (1992) Low-risk intensive therapy for multiple myeloma with combined autologous bone marrow and blood stem cell support. Blood 80: 1666–1672.
91. Reiffers J, Marit G and Boiron JM (1989) Autologous blood stem cell transplantation in high-risk multiple myeloma. Br J Haematol 72: 296–297.
92. Anderson KC, Barut BA, Ritz J *et al.* (1991) Monoclonal antibody-purged autologous bone marrow transplantation therapy for multiple myeloma. Blood 77: 712–220.
93. Phillips GL, Herzig RH, Lazarus HM *et al.* (1984) Treatment of resistant malignant lymphoma with cyclophosphamide, total body irradiation, and transplantation of cryopreserved autologous marrow. N Engl J Med 310: 1557–1561.
94. Philip T, Armitage JO, Spitzer G *et al.* (1987) High-dose therapy and autologous bone marrow transplantation after failure of conventional chemotherapy in adults with intermediate-grade of high-grade non-Hodgkins's lymphoma. N Engl J Med 316:1493–1498.
95. Vose JM, Bierman PJ, Anderson JR *et al.* (1992) High-dose chemotherapy with hematopoietic stem cell rescue for non-Hodgkin's lymphoma (NHL): evaluation of event-free survival based on histologic subtype and rescue product. Proc Amer Soc Clin Oncol 11: 318.
96. Freedman AS, Ritz J, Neuberg D *et al.* (1991) Autologous bone marrow transplantation in 69 patients with a history of low-grade B-cell non-Hodgkin's lymphoma. Blood 77: 2524–2529.
97. Rohatiner AZS, Price CGA, Amott SJ *et al.* (1991) Ablative therapy with autologous bone marrow transplantation as consolidation of remission in patients with follicular lymphoma. In: Dicke KA, Armitage JO and Dicke-Evinger MJ (eds) Autologous Bone Marrow Transplantation V, Proceedings of the Fifth International Symposium, p. 465. Omaha: University of Nebraska Medical Center Press.
98. Bierman P, Vose J, Armitage J *et al.* (1992) High dose therapy followed by autologous hematopoietic rescue for follicular low grade non-Hodgkin's lymphoma (NHL). Proc Am Soc Clin Oncol 11: 317.
99. Freedman A and Nadler L (1992) Bone marrow transplantation in low-grade non-Hodgkin's lymphoma. Issues in Hematol, Oncol, Immunol 2(3): 33–38.
100. Jagannath S, Dicke KA, Armitage JO *et al.* (1986) High-dose cyclophosphamide, carmustine, and etoposide and autologous bone marrow transplantation for relapsed Hodgkin's disease. Annals Intern Med 104: 163–168.
101. Carella AM, Congiu AM, Gaozza E *et al.* (1988) High-dose chemotherapy with autologous bone marrow transplantation in 50 advanced resistant Hodgkin's disease patients: An Italian study group report. J Clin Oncol 6: 1411–1416.
102. Gribben JG, Linch DC, Singer CRJ, McMillan AK, Jarrett M and Goldstone AH (1989) Successful treatment of refractory Hodgkin's disease by high-dose combination chemotherapy and autologous bone marrow transplantation. Blood 73: 340–344.
103. Armitage JO, Bierman PJ, Vose JM *et al.* (1991) Autologous bone marrow transplantation for patients with relapsed Hodgkin's disease. Amer J Med 91: 605–611.
104. Jones RJ, Piantadosi S, Mann RB *et al.* (1990) High-dose cytotoxic therapy and bone marrow transplantation for relapsed Hodgkin's disease. J Clin Oncol 8: 527–537.
105. Linch DC, Winfield D, Goldstone AH *et al.* (1993) Dose intensification with autologous bone-marrow transplantation in relapsed and resistant Hodgkin's disease: results of a BNLI randomized trial. The Lancet 341: 1051–1054.
106. Philip T, Zucker JM, Bernard JL *et al.* (1991) Improved survival at 2 and 5 years in the LMCE1 unselected group of 72 children with stage IV neuroblastoma older than 1 year of age at diagnosis: Is cure possible in a small subgroup? J Clin Oncol 9: 1037–1044.
107. Pole JG, Casper J, Elfenbein G *et al.* (1991) High-dose chemoradiotherapy supported by marrow infusions for advanced neuroblastoma: A pediatric oncology group study. J Clin Oncol 9: 152–158.
108. Peters WP, Shpall EJ, Jones RB *et al.* (1988) High-dose combination alkylating agents with bone marrow support as initial treatment for metastatic breast cancer. J Clin Oncol 6: 1368–1376.
109. Antman K, Ayash L, Elias A *et al.* (1992) A phase II study of high-dose cyclophosphamide, thiotepa, and carboplatin with autologous marrow support in women with measurable advanced breast cancer responding to standard-dose therapy. J Clin Oncol 10: 102–110.

110. Williams SF, Gilewski T, Mick R and Bitran JD (1992) High-dose consolidation therapy with autologous stem-cell rescue in stage IV breast cancer: Follow-up report. J Clin Oncol 10: 1743–1747.
111. Peters WP (1992) Evolving concepts in dose-intensive chemotherapy for node-positive breast cancer. Adv In Oncol 8: 17–25.
112. Broun ER, Nichols CR, Kneebone P *et al.* (1992) Long-term outcome of patients with relapsed and refractory germ cell tumors treated with high-dose chemotherapy and autologous bone marrow rescue. Annals Intern Med 117: 124–128.
113. Nichols CR, Andersen J, Lazarus HM *et al.* (1992) High-dose carboplatin and etoposide with autologous bone marrow transplantation in refractory germ cell cancer: An eastern cooperative oncology group protocol. J Clin Oncol 10: 558–563.

R. P. Lanza and W. L. Chick (eds.), Yearbook of Cell and Tissue Transplantation 1996/1997, 13–17.

Chapter 2

Circulating hematopoietic stem cell transplantation

Anne Kessinger
Section of Oncology, Department of Internal Medicine, University of Nebraska Medial Center, NE 68198, U.S.A.

Autologous grafts

Applications

In 1986, six separate centers independently reported that infusion of autologous hematopoietic progenitors collected from peripheral blood to patients who had received high-dose therapy for a variety of malignancies was followed by complete and sustained hematopoietic recovery [1–6]. Since then, the number of autologous peripheral stem cell transplants (PSCT) has increased yearly. This increase, which is at the expense of the number of autologous bone marrow transplantation (ABMT) procedures performed, has occurred as a result of certain characteristics unique to peripheral stem cell collection and transplant.

Patients who are ineligible for ABMT because of marrow abnormalities (e.g. hypocellularity or metastatic disease) and those with characteristics that make marrow harvesting difficult (e.g. obesity, pelvic skeletal metastases or impenetrable bones) can receive high-dose therapy and PSCT. Administration of cytokines following myelosuppressive chemotherapy or cytokines alone increases the number of circulating progenitors thereby facilitating their collection. Transplantation of these mobilized cells results in very rapid hematopoietic recovery. In contrast to ABMT, mobilized PSCT offers the advantages of graft product harvesting with apheresis procedures that do not require anesthesia or penetration of bones with aspiration needles and also offers earlier hematopoietic reconstitution [7]. PSCT research has been very active since the inception of the technique, and several important papers have appeared in the literature in the past year.

Occult tumor cell contamination

Although a difference in the potential of occult metastatic tumor cells in marrow and of occult tumor cells circulating in blood has been postulated [8, 9], the possibility that occult tumor cells may be included in bone marrow or peripheral stem cell graft products has been, and continues to be, a liability of hematopoietic stem cell autografting. Using gene marking techniques, occult neuroblastoma cells present in bone marrow autografts have clearly been shown to contribute to clinical malignant relapse following high dose therapy and autotransplantation [10]. The clonogenic potential of occult tumor cells in the circulation is as yet unknown.

For some diseases (e.g. breast cancer and lymphoma), occult tumor cells have been found less often or in fewer numbers in autologous blood than in marrow [11–13]. Chemotherapy and/or cytokine mobilized peripheral stem cell (PSC) collections from advanced breast cancer patients contain fewer detectable tumor cells than do autologous marrow harvests [14]. However, occult tumor cells identified both in marrow and mobilized blood have demonstrated the capability of growth *in vitro* [11, 12, 15]. Very recently, three studies were reported which suggest that techniques routinely employed to mobilize stem cells to the circulation may cause increased numbers of both progenitor cells and tumor cells in the blood stream. The first indication came from a study by Brugger et al. [16] where chemotherapy plus granulocyte colony-stimulating factor (G-CSF) was administered to patients with newly diagnosed advanced breast or lung cancer. The purpose of drug administration was anti-cancer therapy and not mobilization for PSC collection, but the peripheral blood was monitored for a progenitor mobilization effect and also for the presence of tumor cells in the circulation. The appearance of or an increase in numbers of circulating occult tumor cells was documented, and the timing of tumor cell mobilization differed depending on whether overt tumor was present in the bone marrow of these patients. For the great majority of patients, the mobilized tumor cells disappeared from the circulation within 15 days of initiation of therapy and before blood stem/progenitor cell collections would customarily begin. The second report considered a different population; patients with metastatic breast cancer that was responding to standard doses of chemotherapy. These patients received chemotherapy and granulocyte-macrophage colony-stimulating factor (GM-CSF) for the expressed purpose of mobilizing PSC for collection. No evidence of increased numbers of tumor cells in the collections, which were performed 15 days after the initiation of mobilizing therapy, was found [14]. In contrast, the third study by Vora et al. [17] showed that one of three patients with multiple myeloma who received G-CSF for PSC mobilization had an increased number of malignant cells in the PSC collection. Even though the mobilized tumor cells in each of these studies were detected with techniques incapable of assessing functional capabilities, the concern nonetheless arises that some patients could have an increased risk of tumor relapse following a mobilized PSCT because mobilized tumor cells may have been present in the graft product. Prospective studies of the potential risks of infusing mobilized tumor cells are not yet available, but one report has described a retrospective analysis of two groups of low-grade lymphoma patients who received either mobilized or non-mobilized PSCT following similar high dose therapy [18]. No difference in progression-free survival was found between the two groups, suggesting that either mobilized low-grade lymphoma cells were not present in the mobilized graft products or, if present, they were either not capable or less capable of restoring disease than any occult non-mobilized lymphoma cells in either the mobilized or non-mobilized graft products. Whether mobilized tumor cells are likely to be present in a mobilized peripheral blood stem cell harvest, whether the timing of tumor cell mobilization varies from one mobilization technique to another, whether mobilized tumor cells have similar characteristics of occult tumor cells in blood or in bone marrow and whether such cells have malignant potential after infusion are all issues to be resolved. In addition, the clinical significance of reinfusing small numbers of malignant cells may vary by disease. Resolution of these points will be important in order to make the best decisions regarding the controversial issues of the need for purging mobilized graft products prior to transplant.

Anti-tumor cytotoxic properties of the peripheral stem cell autograft product

The majority of cells collected in a peripheral blood stem/progenitor cell (PSC) apheresis product are T cells and cells with NK activity [19]. Many more of these "accessory" cells are present in a PSC autograft product than in an autologous marrow harvest [20]. Anti-tumor cytotoxicity of mononuclear cells in cytokine-mobilized PSC collections has been quantitated. PSC apheresis products from normal volunteers contained more cellular cytotoxic activity against K562 target cells (an NK sensitive human erythromyeloid leukemic cell line) than did products from patients with malignancies, but the cellular cytotoxicity against Raji target cells (an NK-resistant human B-cell lymphoma cell line) was similar in patient and normal volunteer products [19]. These cytotoxic cells might have activity against occult tumor cells in a PSC harvest and, after transplantation, might influence malignant cells remaining after high dose therapy. If so, perhaps *ex vivo* manipulation of blood stem cell harvests to elevate the cytotoxicity levels, or *in vivo* cytokine mobilization of accessory cells with high levels of cytotoxicity along with progenitor cell mobilization, will prove valuable in the future.

Immune reconstitution following high dose therapy and autografting

Since the apheresis protocols used to collect blood stem cells are similar to those used to collect circulating lymphocytes, the fact that PSC autograft products contain more immunocompetent lymphocytes than do autologous bone marrow harvests is not surprising [19]. This difference in the two graft products prompted the prediction that immune recovery would be different if blood stem cells rather than marrow cells were infused after high dose therapy [21]. Recently, immune reconstitution following PSCT has been compared with immune recovery after ABMT, and possible advantages to patients who received PSCT were observed [22].

The immune status of cancer patients at the time of diagnosis was found to be functionally inferior when compared to immune measurements after high dose therapy and PSCT [23]. Not only may T cells recover faster following PSCT than ABMT [24], but antigen presenting dendritic cells and their precursors have recently been identified in blood [25]. Their presence in a PSC graft product may facilitate recovery of immune functions and help better explain the differences in immune recovery following PSCT versus ABMT.

Allogeneic grafts

Background

Just as 1986 has been considered the inaugural year of modern clinical autologous PSCT, 1995 is a good candidate for the year PSC were first accepted as a reasonable product for allogeneic transplantation. Allogeneic peripheral stem cell transplantation (allo-PSCT) has been slower to develop than autologous PSCT because of concerns unique to hematopoietic allografting. To date, an assay for the pluripotent stem cell capable of sustained self-renewal has not been described, so that the presence of these cells in the circulation cannot be confirmed. Therefore, allo-PSCT could result in eventual aplasia if such cells were not included in the allograft. A concern also arose regarding the number of T cells in PSC products, a log more than the number in allogeneic bone marrow graft products [20]. In pre-clinical models, the incidence of graft-versus-host disease (GVHD) was reported to increase with infusion of a larger number of T cells [26] and depletion of T cells in clinical allogeneic marrow harvests has been associated with less graft versus host disease [27]. In addition, infusion of excess numbers of T cells with allogeneic bone marrow transplant (allo-BMT) has been reported to increase the incidence of chronic GVHD [28]. A third concern centered around the use of mobilization therapies for normal donors. Clearly, chemotherapy could not be used for the purpose and, until recently, experience with mobilizing cytokines most often used for patients, G-CSF and granulocyte-macrophage colony-stimulating factor (GM-CSF), was minimal. In fact, the potential for long term toxicity from administration of G-CSF or GM-CSF is still not well characterized. The use of non-mobilized PSC for allografting was considered unattractive because of the larger number of apheresis procedures that would be required to collect an adequate graft product.

Initial experiences

The first clinical allo-PSCT was reported in 1989 [29]. The transplant itself took place in late 1987 before cytokines were known to be useful mobilizing agents. The donor declined to donate marrow but agreed to undergo multiple apheresis procedures to produce an allograft product for his HLA-identical sibling. The collections were partially depleted of T cells, resulting in a product with approximately the same number of T cells as would be present in a marrow allograft. More than the number of apheresis procedures considered adequate to collect an adequate graft product were performed, to assure as much as possible that engraftment would occur. Engraftment of donor cells was documented, but the patient died 32 days later of an infection, thereby ending the opportunity to determine whether the engraftment of donor cells would be sustained.

The next report of allo-PSCT appeared in 1993, when engraftment failed to occur following two consecutive marrow infusions from an HLA-matched sibling donor to a patient who had received marrow-ablative therapy for acute lymphocytic leukemia [28]. Rather than harvesting marrow a third time from this donor, investigators administered G-CSF to the donor and collected mobilized PSC. Infusion of these cells resulted in rapid hematopoietic restoration of donor origin. While the PSC infusion was likely responsible, the fact that two marrow infusions from the same donor had occurred earlier clouded the precise origin of the sustained engraftment which followed.

Later in 1993, another report of a successful allo-PSCT appeared [31]. In this instance, the only suitable HLA-matched related donor had an unacceptable anesthetic risk and donated G-CSF-mobilized allogeneic PSC rather than marrow. Engraftment followed promptly after infusion of the cells which continued at the time of the report. Severe acute GVHD was not encountered. This successful experience demonstrated feasibility and provided an impetus to evaluate the clinical usefulness of the procedure.

Clinical trials

In the spring of 1995, three separate reports simultaneously appeared in the literature describing successful allo-PSCT for a total of 25 patients [32–34]. The early follow-up of these patients showed that acute GVHD was certainly no more likely, and perhaps less likely, to occur than after allo-BMT. No serious untoward effects from administration of G-CSF to normal donors were described, and platelet recovery in peripheral blood appeared to be faster than would be expected after allo-BMT. The longest follow-up in these series of patients was 489 days. No patient developed subsequent unexplained aplasia, and for all but one of the patients tested, hematopoiesis was of donor cell origin. One patient was discovered to have chimerism at the time of malignant relapse. Continued follow-up will be required to determine if cells capable of continuous self-renewal were transplanted and if the incidence and severity of chronic GVHD is different than following allo-BMT. Nonetheless, the results available thus far suggest that allo-PSCT may provide some advantages to certain patient populations in need of allogeneic hematopoietic transplantation.

Future directions

Hematopoietic stem and progenitor cells collected from either the marrow or the peripheral blood appear to have similar functional characteristics. These cells retrieved from either source are clearly capable of restoring and maintaining hematopoiesis in clinical circumstances. Manipulating these hematopoietic cells with cytokines can improve their functional usefulness whether they are in marrow or peripheral blood. The more striking differences between cells harvested from the marrow or from the blood for grafting likely reside in the other (accessory) cell populations present in the collections. The value, both positive and negative, of these accessory cells in the transplant setting is just beginning to be appreciated. Some accessory cells may have more usefulness if they are allogeneic rather than autologous (e.g. the mediators of the graft versus leukemia effect). As was so elegantly pointed out by Goldman [35], the eventual issue may not be whether peripheral stem cell transplants are better or worse than marrow transplants, but rather what combination of allogenic and/or autologous hematopoietic and accessory cells derived from the blood and/or the marrow, manipulated or not, will offer the optimal hematopoietic and therapeutic advantage to the patient. The risks and benefits of each population of accessory cells need to be defined in order to resolve that issue. When those definitions are known, engineering the quintessential graft product for an individual patient can become routine.

References

1. Korbling M, Dorken B, Ho A *et al.* (1986) Autologous transplantation of blood-derived hematopoietic stem cells after myeloablative therapy in a patient with Burkitt's lymphoma. Blood 67: 529.
2. Reiffers J, Bernard P, David B *et al.* (1986) Successful autologous transplantation with peripheral blood hemopoietic cells in a patient with acute leukemia. Exp Hematol 14: 312.
3. Bell AJ, Figes A, Oscier DG *et al.* (1986) Peripheral blood stem cell autografting. Lancet 1: 1027.
4. Tilly H, Bastit D, Lucet JC *et al.* (1986) Haemopoietic reconstitution after autologous peripheral blood stem cell transplantation in acute leukaemia. Lancet 2: 154.
5. Castaigne S, Calvo F, Douay L *et al.* (1986) Successful haematopoietic reconstitution using autologous peripheral blood mononucleated cells in a patient with acute promyelocytic leukaemia. Br J Haematol 62: 209.
6. Kessinger A, Armitage JO, Landmark JD *et al.* (1986) Reconstitution of human hematopoietic function with autologous cryopreserved circulating stem cells. Exp Hematol 14: 192.
7. Pettengell R, Morgenstern GR, Woll PJ *et al.* (1993) Peripheral blood progenitor cell transplantation in lymphoma and leukemia using a single apheresis. Blood 82: 3770–3777.
8. Mansi JL, Berger U, McDonnell T *et al.* (1989) The fate of bone marrow micrometastases in patients with primary breast cancer. J Clin Oncol 7: 445–449.
9. Sharp JG and Crouse DA (in press) Minimal disease: detection and significance In: Armitage JO and Antman KH (eds) High-dose cancer therapy: pharmacology, hematopoietins and stem cells. 2nd ed. Baltimore: Williams & Wilkins.
10. Rill DR, Santana VM, Roberts WM *et al.* (1994) Direct demonstration that autologous bone marrow transplantation for solid tumors can return a multiplicity of tumorigenic cells. Blood 84: 380–383.
11. Ross AA, Cooper BW, Lazarus HM *et al.* (1993) Detection and viability of tumor cells in peripheral blood stem cell collections from breast cancer patients using immunocytochemical and clonogenic assay techniques. Blood 82: 2605–2610.
12. Sharp JG, Kessinger A, Pirruccello SJ *et al.* (1991) Frequency of detection of suspected lymphoma cells in peripheral blood stem cell collections. In: Dicke KA, Armitage JO and Dicke-Evinger MJ (eds) Autologous bone marrow transplantation. Proceedings of the fifth international symposium, pp 801–810. Omaha: University of Nebraska Medical Center.
13. Langlands K, Craig JIO, Parker AC *et al.* (1990) Molecular determination of minimal residual disease in peripheral blood stem cell harvests. Bone Marrow Transplantation 5 (suppl 1): 64.
14. Passos-Coelho JL, Ross AA, Moss TJ *et al.* (1995) Absence of breast cancer cells in a single-day peripheral blood progenitor cell collection after priming with cyclophosphamide and granulocyte-macrophage colony-stimulating factor. Blood 85: 1138–1143.
15. Sharp JG, Kessinger A, Armitage JO *et al.* (1993) Influence of minimal tumor contamination of hematopoietic harvests on clinical outcome of patients undergoing high dose therapy and transplantation. In: Dicke KA and Keating A (eds) Autologous Bone Marrow Transplantation. Proceedings of the Sixth International Symposium, pp 223–226. Arlington, Texas: Cancer Treatment Research Fund.

16. Brugger W, Bross KJ, Glatt M *et al.* (1994) Mobilization of tumor cells and hematopoietic progenitor cells into peripheral blood of patients with solid tumors. Blood 83:636–640.
17. Vora AJ, Tho CH, Peel J *et al.* (1994) Use of granulocyte colony-stimulating factor (G-CSF) for mobilizing peripheral blood stem cells: risk of mobilizing clonal myeloma cells in patients with bone marrow infiltration. Br J Haematol 86: 180–182.
18. Kessinger A, Anderson J, Bierman P *et al.* (1994) Mobilized versus non-mobilized peripheral stem cell transplantation after high-dose therapy for low grade non-Hodgkin lymphoma: effect on progression free survival. Blood 84 (suppl 1): 394a.
19. Verbik DJ, Jackson JD, Pirruccello SJ *et al.* (1995) Functional and phenotypic characterization of human peripheral blood stem cell harvests: a comparative analysis of cells from consecutive collections. Blood 85: 1964–1970.
20. Weaver CH, Longin K, Buckner CD *et al.* (1994) Lymphocyte content in peripheral blood mononuclear cells collected after the administration of recombinant human granulocyte colony-stimulating factor. Bone Marrow Transplantation 13: 411–415.
21. Zander AR and Cockerill K (1987) Autologous transplantation with circulating hemopoietic stem cell. J Clin Apheresis 3: 191–201.
22. Roberts MM, To LB, Gillis D *et al.* (1993) Immune reconstitution following peripheral blood stem cell transplantation, autologous bone marrow transplantation and allogeneic bone marrow transplantation. Bone Marrow Transplantation 12: 469–475.
23. Scambia G, Panici PB, Pierelli L *et al.* (1993) Immunological reconstitution after high-dose chemotherapy and autologous blood stem cell transplantation for advanced ovarian cancer, Eur J Cancer 29A: 1518–1522.
24. Gordy C, Perry G, Thomas M *et al.* (1994) Immune function and phenotype of peripheral blood and bone marrow stem cell products and peripheral blood reconstitution. J Cell Biochem 18B: 100.
25. Thomas R and Lipsky PE (1994) Human peripheral blood dendritic cell subsets. Isolation and characterization of precursor and mature antigen-presenting cells. J Immunol 153: 4016–4028.
26. Owens AH and Santos GW (1986) The induction of graft versus host disease in mice treated with cyclophosphamide. J Exp Med 128: 277.
27. Kernan NA, Collins NH, Juliano L *et al.* (1986) Clonable T lymphocytes in T cell-depleted bone marrow transplants correlate with development of graft-v-host disease. Blood 68: 770.
28. Storb R, Etzioni R, Anasette C *et al.* (1994) Cyclophosphamide combined with antithymocyte globulin in preparation for allogeneic marrow transplants in patients with aplastic anemia. Blood 84: 941.
29. Kessinger A, Smith DM, Strandjord SE *et al.* (1989) Allogeneic transplantation of blood-derived, T-cell depleted hematopoietic stem cells after myeloablative treatment in a patient with acute lymphoblastic leukemia. Bone Marrow Transplantation 4: 643–646.
30. Dreger P, Suttorp M, Haferlach T *et al.* (1993) Allogeneic granulocyte colony-stimulating factor-mobilized peripheral blood progenitor cells for treatment of engraftment failure after bone marrow transplantation. Blood 81: 1404.
31. Russel NH, Hunter A, Rogers S *et al.* (1993) Peripheral blood stem cells as an alternative to marrow for allogeneic transplantation. Lancet 341: 1482.
32. Bensinger WI, Weaver CH, Appelbaum FR *et al.* (1995) Transplantation of allogeneic peripheral blood stem cells mobilized by recombinant human granulocyte colony-stimulating factor. Blood 85: 1655.
33. Korbling M, Przepiorka D, Huh YO *et al.* (1995) Allogeneic blood stem cell transplantation for refractory leukemia and lymphoma: potential advantage of blood over marrow autografts. Blood 85: 1659.
34. Schmitz N, Dreger P, Suttorp M *et al.* (1995) Primary transplantation of allogeneic peripheral blood progenitor cells mobilized by filgrastim (granulocyte colony stimulating factor). Blood 85: 1666.
35. Goldman J (1995) Peripheral blood stem cells for allografting. Blood 85: 1413.

R. P. Lanza and W. L. Chick (eds.), Yearbook of Cell and Tissue Transplantation 1996/1997, 19–34.

Chapter 3

Current status of allogeneic and autologous blood and marrow transplantation: Report from the IBMTR and ABMTR – North America

Philip A. Rowlings, Jakob R. Passweg, James O. Armitage, Robert P. Gale, Kathleen A. Sobocinski, John P. Klein, Mei-Jie Zhang and Mary M. Horowitz
IBMTR and ABMTR Statistical Center, Health Policy Institute, Medical College of Wisconsin, Milwaukee, WI 53226, U.S.A.

The International Bone Marrow Transplant Registry (IBMTR) has collected data from over 320 institutions performing allogeneic bone marrow transplants, worldwide since 1972. This database includes information for about 40% of all allogeneic transplants performed between 1964 and 1994. The Autologous Blood and Marrow Transplant Registry – North America (ABMTR) began collecting data on transplants using autologous bone marrow and/or blood cells (autotransplants) performed in North America in 1991. More than 130 autotransplant centers now contribute data to the ABMTR. The ABMTR database includes information for about 50% of autotransplants in North America performed between 1989 and 1994. Recent use and outcome of allogeneic and autologous transplants are reviewed.

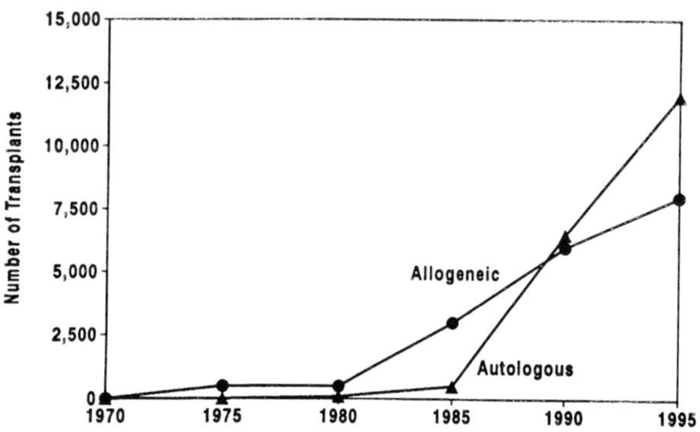

Fig. 1. Annual numbers of allogeneic and autologous transplants, worldwide.

Increasing use of blood and marrow transplants

Bone marrow and blood transplants are increasingly used to treat patients with diverse diseases (Fig. 1, Tables 1 and 2). The results of an IBMTR survey indicated more than 5,000 allogeneic transplants were performed in 1990 with a 15% annual increase over the preceding five years [1]. Over 45,000 persons received allogeneic transplants between 1964 and 1994. Widespread use of autotransplants began more recently and annual numbers now exceed 8,000. Use of autotransplants is increasing by about 20% per year, with over 30,000 recipients since 1980.

Indications for allogeneic and autologous transplants

The indications for allogeneic and autologous transplants are presented in Tables 1 and 2. The most common indications differ according to source of hematopoietic stem cells. Seventy-six percent of allogeneic transplants are for leukemia or preleukemia: 25% for chronic myelogenous leukemia (CML), 26% for acute myelogenous leukemia (AML), 21% for acute lymphoblastic leukemia (ALL), 3% for myelodysplas-

Table 1. Indications for allogeneic bone marrow transplants between 1964 and 1994, reported to the IBMTR

Disease	Number of patients
Chronic myelogenous leukemia	5757
Acute myelogenous leukemia	5921
Acute lymphoblastic leukemia	4770
Non-Hodgkin lymphoma	1285
Hodgkin disease	186
Multiple myeloma	313
Other malignancies[a]	802
Aplastic anemia	2811
Immunodeficiency states	520
Other[b]	405
Total	22,770

[a] Includes acute undifferentiated leukemia, other leukemia, myelodysplastic syndrome, myeloproliferative disease, and unclassified malignancies.
[b] Includes hemoglobinopathies, lysosomal storage diseases, and other unclassified diseases.

Table 2. Indications for autologous bone marrow or peripheral blood stem cell transplants between 1989 and 1994, reported to the ABMTR

Disease	Number of patients
Chronic myelogenous leukemia	160
Acute myelogenous leukemia	1516
Acute lymphoblastic leukemia	398
Non-Hodgkin lymphoma	4098
Hodgkin disease	2275
Breast cancer	4563
Multiple myeloma	671
Neuroblastoma	430
Testicular cancer	266
Ovarian cancer	282
Intracranial neoplasms	231
Other[a]	755
Total	15,645

[a] Includes acute undifferentiated leukemia, myelodysplastic syndrome, chronic lymphocytic leukemia, melanoma, small cell lung cancer, Ewing's sarcoma and other unclassified malignancies.

tic syndromes and 1% for other leukemias. Eight percent are for other cancers including non-Hodgkin lymphoma (NHL) (6%), multiple myeloma (1%), and Hodgkin disease (< 1%). The remainder are for aplastic anemia (12%), immune deficiencies (2%), inherited disorders of metabolism (1%) and other non-malignant disorders. Autotransplants are used to treat cancer, although autotransplant gene therapy to treat immune deficiencies is now under investigation. The most common autotransplant indications during the past five years were breast cancer (29%), NHL (26%), Hodgkin disease (15%), AML (10%), multiple myeloma (4%), ALL (3%), and neuroblastoma (3%), with 10% for a variety of other cancers. The dramatic increase in autotransplants for breast cancer has been the most strking and controversial change in autotransplant use. In 1989, about 16% of autotransplants in North America were for breast cancer while in 1994, 40% were for breast cancer.

Treatment failure in allogeneic and autologous transplants

Causes of failure after allogeneic and autologous transplants differ. Allogeneic transplants are associated with relatively high (25–50%) risks of transplant-related mortality from graft-versus-host disease (GVHD), infection and liver toxicity [2]. Most of these complications result from major or minor histocompatibility differences between donor and host. In contrast, transplant-related mortality is generally < 10% after autotransplants but relapse is frequent. Higher relapse rates after autologous versus allogeneic transplants may result from re-infusion of tumor cells with the autograft and/or absence of an immune-mediated graft-versus-tumor effect associated with allografts [3–5].

An IBMTR study of HLA-identical sibling transplants for leukemia demonstrated a decrease in transplant-related mortality during the 1980's, from 38 to 31% in early leukemia, 47 to 41% in intermediate leukemia and 68 to 46% in advanced leukemia [6]. Possible reasons include altered radiation schedules, use of blood products screened for cytomegalovirus, improved antiviral therapy, and improved GVHD prophylaxis. There was no significant change in the risk of posttransplant relapse during this time.

Increasing use of blood cells alone for transplant

Traditionally cells for transplant were obtained from the bone marrow space only. In the 1980s it was shown that cells harvested from the peripheral blood during recovery from chemotherapy could also be used for transplant [7]. It was subsequently shown that large numbers of these cells could be obtained after administration of hematopoietic growth factors,

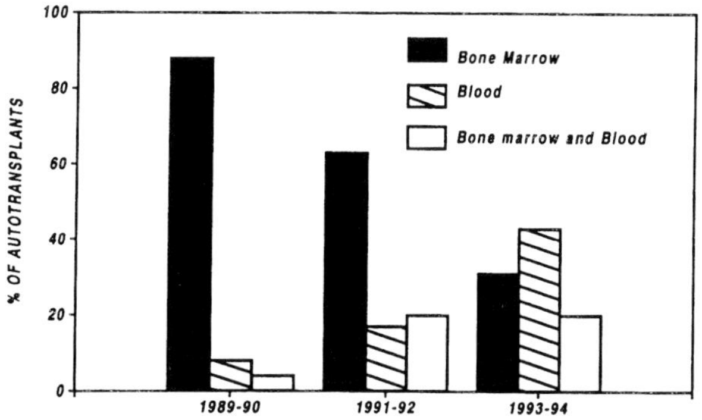

Fig. 2. Stem cell source for autotransplants in North America.

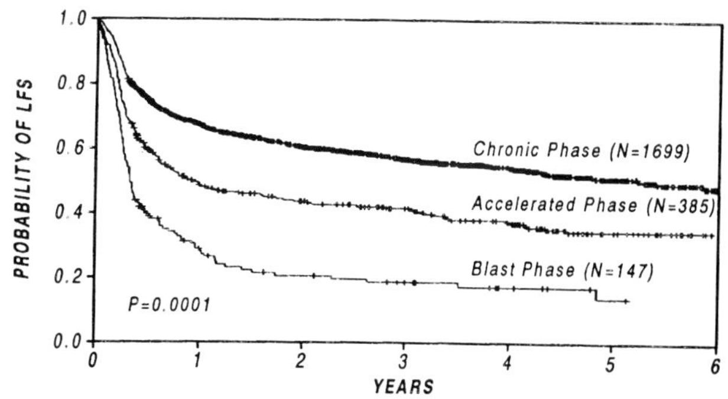

Fig. 3. Probability of leukemia-free survival after HLA-identical sibling bone marrow transplants for chronic myelogenous leukemia, 1987–1994.

such as G-CSF [8]. These cells have the advantage of providing a more rapid recovery of platelet count posttransplant. Hematopoietic cells collected from the blood were initially used to supplement bone marrow cells. However, blood alone is now the commonest source of hematopoietic stem cells for autotransplants (Fig. 2).

Use of blood cells for allogeneic transplants is less common. There are some concerns about administering growth factors to normal donors, although an increasing amount of data suggests that these agents are safe. Additionally, cells collected from the blood include large numbers of T-lymphocytes which have the potential to induce GVHD. Nevertheless, several recent trials indicate that allografts using blood rather than bone marrow are feasible (9–11) and this approach is rapidly gaining acceptance.

Results of blood and marrow transplants

Chronic myelogenous leukemia (CML)

Allogeneic transplant remains the only known cure for CML [12]. Among 2231 recipients of HLA-identical sibling transplants done between 1987 and 1994, reported to the IBMTR, three-year actuarial probabilities of relapse (95% confidence interval) were 13 ± 2% for transplants done in first chronic phase, 26 ± 6% in accelerated phase, and 58 ± 11% in blast phase. Three-year probabilities of leukemia-free survival (LFS) were 57 ± 3%, 41 ± 5% and 18 ± 7%, respectively (Fig. 3).

Results of HLA-identical sibling transplants done in chronic phase are best in younger patients, those treated with hydroxyurea rather than busulfan pretransplant and in those transplanted earlier rather than later after diagnosis [13, 14]. Disease-related features predicting response to conventional therapy such as spleen size, % myeloblasts and hemoglobin do not appear to affect transplant outcome.

Immune-mediated antileukemia effects (often termed graft-versus-leukemia or GVL) are important in preventing relapse after HLA-identical sibling transplants for CML [5, 15]. This is evidenced by lower rates of relapse in patients with GVHD versus those without and high rates of relapse after identical twin transplants and transplants using marrow depleted of T-lymphocytes prior to infusion. This GVL effect is now purposefully induced in patients relapsing after allogeneic transplant by infusion of donor leukocytes without additional cytotoxic therapy, with subsequent disappearance of both hematologic and cytogenetic evidence of CML [16–18].

Autotransplants are infrequently performed in CML due to difficulties in eradicating Philadelphia (Ph^1)-chromosome positive cells from the bone marrow or blood cells. To avoid re-infusion of CML, cells are often obtained following intensive chemotherapy or interferon treatment when all or some cells are Ph^1-chromosome negative. Some centers attempt to remove CML cells from the graft by *in vitro* culture or drug treatment. Patients receiving autotransplants may become partially or completely Ph^1-chromosome negative, but cytogenetic responses are generally incomplete and transient. Three-year LFS is less than 5%. A recent survey of autotransplants done at several centers suggests that survival, though not leukemia-free survival, may be prolonged after autotransplants [19].

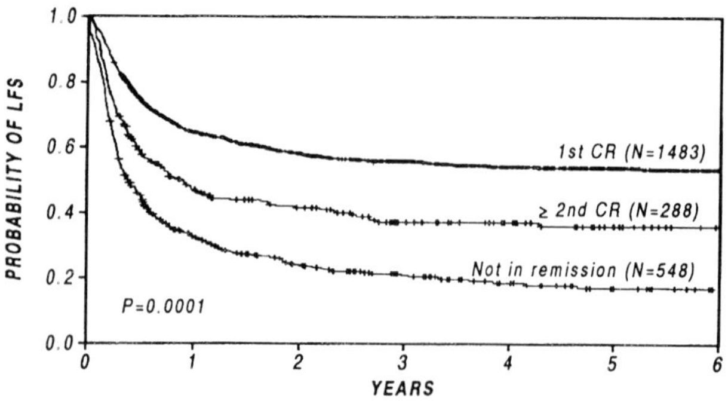

Fig. 4. Probability of leukemia-free survival after HLA-identical sibling bone marrow transplants for acute myelogenous leukemia, 1987–1994.

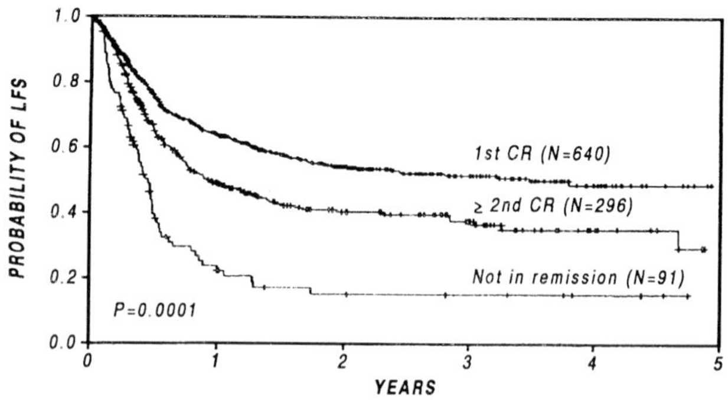

Fig. 5. Probability of leukemia-free survival after autotransplants for acute myelogenous leukemia, 1987–1994.

Acute myelogenous leukemia

Results of allogeneic transplants for AML correlate with disease stage as in CML. Among 2319 recipients of HLA-identical sibling transplants done between 1987 and 1994, reported to the IBMTR, three-year probabilities of relapse were 23 ± 3% for transplants done in first remission, 39 ± 8% in second or subsequent remission and 63 ± 6% in relapse. Three-year probabilities of LFS were 56 ± 3%, 37 ± 6% and 20 ± 4%, respectively (Fig. 4).

In contrast to CML, disease-related factors predicting outcome of conventional chemotherapy for AML also predict outcome of HLA-identical sibling transplants in first remission [20]. Patients with high leukocyte counts at diagnosis, French American British classification of M4 or M5 subtypes, and certain chromosome abnormalities have lower LFS.

As in CML, GVL appears to be important in preventing relapse after allogeneic transplants for AML. Relapse rates after identical twin transplants where there is no GVL effect are about 60%, which is similar to that reported with conventional chemotherapy [5]. In contrast to CML, T-cell depletion does not eliminate the antileukemia effect of allografts for AML [15].

Autotransplants for AML use cells obtained during remission and, in some cases, treated with drugs or monoclonal antibodies *in vitro* (purging) to remove residual leukemia cells. Among 1027 recipients of autotransplants for AML between 1989 and 1994, reported to the ABMTR, three-year probabilities of relapse were 42 ± 5% for transplants done in first remission, 56 ± 7% in second or subsequent remission and 81 ± 10% in relapse. Three-year probabilities of LFS were 52 ± 5%, 38 ± 6% and 15 ± 9%, respectively (Fig. 5).

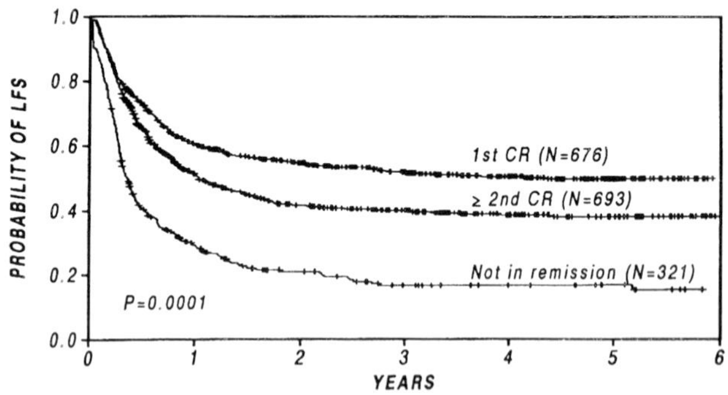

Fig. 6. Probability of leukemia-free survival after HLA-identical sibling bone marrow transplants for acute lymphoblastic leukemia, 1987–1994.

Acute lymphoblastic leukemia

In contrast to AML and CML, most patients with ALL are cured with conventional chemotherapy. Therefore, transplants are generally reserved for patients failing conventional therapy, i.e., in relapse or second or subsequent remission [21], or patients in first remission with prognostic factors predicting a high risk of failure with conventional therapy. Such factors are older age, high leukocyte count at diagnosis, Ph^1 and other chromosome abnormalities and difficulty obtaining a first remission; it is controversial whether transplants increase LFS in these patients [22, 23].

Among 1690 recipients of HLA-identical sibling transplants between 1987 and 1994, reported to the IBMTR, three-year probabilities of relapse were 27 ± 4% for transplants done in first remission, 44 ± 5% in second or subsequent remission, and 70 ± 7% in relapse. Three-year probabilities of LFS were 51 ± 4%, 40 ± 4% and 16 ± 5%, respectively (Fig. 6).

GVL does not appear to be as important in ALL as in CML and AML. Although GVHD decreases relapse risk, relapse rates are not markedly increased after

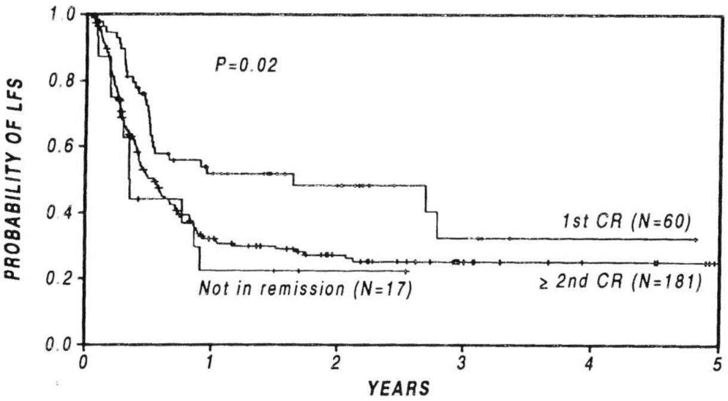

Fig. 7. Probability of leukemia-free survival after autotransplants for acute lymphoblastic leukemia, 1987–1994.

identical twin or T-cell depleted transplants compared to non T-cell depleted HLA-identical sibling transplants [5, 15].

Most autotransplants for ALL use cells treated with drugs or monoclonal antibodies *in vitro* to remove leukemia cells. Among 258 recipients of autotransplants for ALL between 1989 and 1994, reported to the ABMTR, three-year probabilities of relapse were 59 ± 22% for transplants done in first remission, 72 ± 8% in second or subsequent remission, and 75 ± 24% in relapse. Three-year probabilities of LFS were 32 ± 21%, 25 ± 7% and 22 ± 22%, respectively (Fig. 7).

Non-Hodgkin lymphoma (NHL)

Only about 10% of transplants for NHL use allogeneic donors, the remainder being autotransplants. The three-year probability of survival after 612 HLA-identical sibling transplants for non-Hodgkin lymphoma between 1987 and 1994, reported to the IBMTR, is 35 ± 4%.

Most autotransplants for NHL are for disease refractory to initial therapy or relapsing after an initial response [24]. Complete remission rates are about 25% in patients with primary refractory disease, 75% in those in relapse with disease still responsive to chemotherapy and 45% in those in relapse resistant to chemotherapy. Among 445 autotransplants for intermediate grade and immunoblastic (aggressive) NHL between 1989 and 1994, reported to the ABMTR, three-year survival was 30 ± 15% for 150 patients with induction failure (primary refractory disease) and 37 ± 11% for 173 in first relapse. Adverse prognostic factors for disease-free survival after autotransplants for NHL include tumor mass ≥ 10 cm at the time of transplant, > 3 prior chemotherapy regimens; elevated serum LDH and use of bone marrow versus blood as the autologous stem cell source. There are recent reports of patients with high-risk aggressive NHL receiving autotransplants in first complete remission with disease-free survival rates of 60–80% [25]. Among 39 autotransplants for aggressive NHL in first remission between 1989 and 1994, reported to the ABMTR, three-year survival was 67 ± 19%.

Most treatment failures after autotransplants for NHL are due to relapse. Relapse may result from lymphoma cells remaining in the recipient, infusion of tumor cells, or both. Most relapses occur in prior sites of bulky disease, suggesting that residual lymphoma in the recipient is the predominant cause of relapse. A similar conclusion is suggested by studies reporting similar relapse rates after autologous, syngeneic and allogeneic transplants. However, some investigators report lower relapse rates in recipients of allo- versus autotransplant recipients [26, 27]. This may reflect relapse from lymphoma cells in autografts or an immune-mediated antilymphoma effect of allografts, similar to GVL.

Hodgkin disease

As in ALL, most patients with Hodgkin disease are cured with conventional chemotherapy. However, in the 20–30% who fail conventional therapy, transplant offers a potential option for salvage therapy. Allogeneic transplants offer several potential advantages over autotransplants. There is no risk of malignant cells in the graft, no prior exposure of hematopoietic stem cells to drugs and/or radiation and possible immune-mediated graft-versus-tumor effect. However, relatively few allogeneic transplants for Hodgkin disease have been done. Among 108 HLA-identical sibling transplants for Hodgkin disease reported to the IBMTR, the three-year probability of disease-free survival is 16 ± 7%. Autotransplants for Hodgkin disease are more common. Among 393 autotransplants between 1989 and 1994, reported to the ABMTR, three-year probability of survival was 51 ± 16% for 110 patients never achieving first remission, 56 ± 10% for 191 in first relapse and 57 ± 25% for 92 in second remission.

Multiple myeloma

Relatively few allogeneic transplants are done for multiple myeloma, largely because most affected persons are older than 50 years and thus have a high risk of transplant-related mortality. A review from the European Bone Marrow Transplant Group of 90 recipients

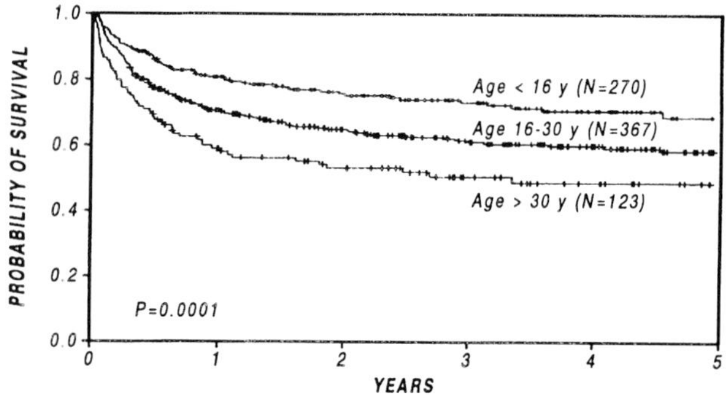

Fig. 8. Probability of survival after HLA-identical sibling bone marrow transplants for severe aplastic anemia by age, 1987–1994.

of HLA-identical sibling transplants indicates six-year disease-free survival of about 40% [28].

Increasing numbers of patients receive autologous transplants for multiple myeloma [29]. Median age of 578 autotransplant recipients reported to the ABMTR was 49 years with a range of 16 to 71 years. Among patients transplanted for progressive or advanced disease, complete remission is achieved in about 30%; two-year progression-free survival is about 20%. Among those transplanted for responsive disease, the complete remission rate is about 40% and two-year progression-free survival about 40%. Prognostic factors suggested to be of importance include age, performance score, and β_2-microglobulin [30].

Severe aplastic anemia

Bone marrow transplant is the treatment of choice for young patients with aplastic anemia who have an HLA-identical sibling. The three-year probability of survival after 760 HLA-identical sibling transplants between 1987 and 1994, reported to the IBMTR, was 73 ± 6% for patients less than 16 years of age, 61 ± 6% for those between 16 and 30 years and 50 ± 10% for patients over 30 years of age (Fig. 8).

Alternative allogeneic donors

Most allogeneic transplants use HLA-identical sibling donors. However, since only about 30% of transplant candidates will have such a donor, there is increasing interest in using alternative donors including HLA-mismatched related donors and unrelated donors [31–33]. These now account for 12% and 20% of all allogeneic transplants, respectively.

Transplants from alternative donors are associated with significantly increased risks of graft failure and GVHD resulting in increased transplant-related mortality. Relapse risks are either similar or decreased compared to HLA-identical sibling transplants. Disease-free survival is decreased. Although some centers report similar outcome with HLA-identical sibling and one-antigen mismatched related donors, in general, transplant outcome correlates inversely with degree of donor-recipient HLA-incompatibility. Among patients transplanted for acute leukemia in first remission or CML in first chronic phase and reported to the IBMTR, two-year probabilities of LFS are about 35% with a one-antigen-mismatched related donor, about 20% with a two-antigen mismatched related donor, about 30% with a fully matched unrelated donor and about 25% with a one-antigen mismatched unrelated donor.

Summary

Bone marrow and blood cell transplants are increasingly used treatments for cancer and disorders of bone marrow function. This is due to use of autotransplants to treat common diseases such as breast cancer. Other significant changes in the past five years are the increasing use of blood cells alone for autotransplants and unrelated bone marrow donors for allogeneic transplants. The IBMTR and ABMTR provide a unique resource for studying the role of allogeneic and autologous bone marrow and blood cell transplants.

Acknowledgements

We thank Melodee Nugent for the help with data analysis and Lisa Schneider for manuscript preparation. Supported by Public Health Service Grant PO1-CA-40053 from the National Cancer Institute, the National Institute of Allergy and Infectious Diseases, and the National Heart, Lung and Blood Institute, and Contract No. CP-21161 from the National Cancer Institute of the U.S. Department of Health and Human Services; the U.S. Army Medical Research and Development Command; and grants from Activated Cell Therapy; Alpha Therapeutic Corporation; Amgen, Inc.; Applied Immune Sciences; Armour Pharmaceutical Company; Astra Pharmaceutical; Baxter Healthcare Corporation; Bayer Corporation; Biogen; Lynde and Harry Bradley Foundation; Bristol-Myers Squibb Company; Frank G. Brotz Family Foundation; Burroughs-Wellcome Company; Cancer Center, Medical College of Wisconsin;

Caremark, Inc.; CellPro, Inc; Center for Advanced Studies in Leukemia; Cigna HealthCare; COBE BCT Inc.; Charles E. Culpeper Foundation; Eleanor Naylor Dana Charitable Trust; Deborah J. Dearholt Memorial Fund; Eppley Foundation for Research; Genentech, Inc.; Glaxo Pharmaceutical; Hewlett-Packard Company; Immunex Corporation; Janssen Pharmaceutica; Kabi Pharmacia; Kettering Family Foundation; Kirin Brewery Company; Robert J. Kleberg, Jr. and Helen C. Kleberg Foundation; Herbert H. Kohl Charities; Lederle Laboratories; Eli Lilly Company Foundation; Nada and Herbert P. Mahler Charities; Marion Merrell Dow, Inc.; Milstein Family Foundation; Milwaukee Foundation/Elsa Schoeneich Research Fund; Samuel Roberts Noble Foundation; Ortho Biotech Corporation; John Oster Family Foundation; Elsa U. Pardee Foundation; Jane and Lloyd Pettit Foundation; Pharmacia; Quadra Logic Technologies; RGK Foundation; Roche Laboratories; Roerig Division of Pfizer Pharmaceuticals; Sandoz Oncology; Schering-Plough International; Walter Schroeder Foundation; Stackner Family Foundation; Starr Foundation; StemCell Technologies; Joan and Jack Stein Charities; SyStemix; Therakos; Upjohn Company; and Wyeth-Ayerst Laboratories.

Table 3. Centers reporting data to the IBMTR and ABMTR

Institute	City	Country	IBMTR	ABMTR
British Hospital of Buenos Aires	Buenos Aires	Argentina	X	
Institutos Medicos Antartida	Buenos Aires	Argentina	X	
Navy Hospital Pedro Mallo	Buenos Aires	Argentina	X	X
Alexander Fleming Institute	Buenos Aires	Argentina		X
Hospital Privado De Oncologia	Buenos Aires	Argentina		X
Centro De Internacion e Investigation	Buenos Aires	Argentina		X
Hospital Privado de Cordoba	Cordoba	Argentina	X	X
Inst. de Transplante de Medula Osea	Gonnet	Argentina	X	
Hanson Center for Cancer Research	Adelaide	Australia	X	
Royal Children's Hospital Brisbane	Brisbane	Australia	X	
Royal Brisbane Hospital (Adults)	Brisbane	Australia	X	
Royal Prince Alfred Hospital	Camperdown	Australia	X	X
Royal Alexandra Hosp. for Children	Camperdown	Australia	X	
St. Vincent's Hospital	Darlinghurst	Australia	X	
Royal Hobart Hospital	Hobart	Australia	X	
Royal Melbourne Hospital	Parkville	Australia	X	
Royal Children's Hospital	Parkville	Australia	X	
Royal Perth Hospital	Perth	Australia	X	
Alfred Hospital	Prahran	Australia	X	
Prince of Wales Children's Hospital	Randwick	Australia	X	
Royal North Shore Hospital	St. Leonards	Australia	X	
Westmead Hospital	Westmead	Australia	X	
Queen Elizabeth Hospital	Woodville	Australia	X	
Donauspital/SM2-OST	Vienna	Austria		X
Univ. Klinik fur Innere Medizin I	Vienna	Austria	X	
St. Anna Children's Hospital	Vienna	Austria	X	
AZ Sint-Jan	Brugge	Belgium	X	
Cliniques Universitaires Saint-Luc	Bruxelles	Belgium	X	
Children's University Hospital	Bruxelles	Belgium	X	
University Hospital Antwerp	Edegem	Belgium	X	
Universiteits Ziekenhuis Gasthuisberg	Leuven	Belgium	X	
Universitaire De Liège	Liège	Belgium	X	
Hospital de Clinicas	Curitiba	Brazil	X	X
Universidade de Sao Paulo	Ribeirao Preto	Brazil	X	
Instituto Nacional de Cancer	Rio de Janeiro	Brazil	X	
Santa Casa Medical School	Sao Paulo	Brazil	X	
Instituto do Coracao – INCOR	São Paulo	Brazil	X	
University of Calgary	Calgary	Canada	X	X
Alberta Children's Hospital	Calgary	Canada	X	X
Victoria General Hospital	Halifax	Canada	X	X
Chedoke-McMaster Hospitals	Hamilton	Canada	X	
University Hospital	London	Canada	X	
Jewish General Hospital	Montreal	Canada		X

Table 3. Continued

Institute	City	Country	IBMTR	ABMTR
Montreal Children's Hospital	Montreal	Canada	X	X
Sacre Coeur Hospital	Montreal	Canada		X
Royal Victoria Hospital	Montreal	Canada	X	X
Ottawa General Hospital	Ottawa	Canada		X
Hôpital du Saint-Sacrement	Quebec City	Canada		X
NE Ontario Regional Cancer Centre	Sudbury	Canada		X
Toronto General Hospital	Toronto	Canada		X
Hospital for Sick Children	Toronto	Canada	X	X
Princess Margaret Hospital	Toronto	Canada	X	
British Columbia's Children's Hospital	Vancouver	Canada	X	X
Vancouver General Hospital	Vancouver	Canada	X	X
Manitoba Cancer Treatment Center	Winnipeg	Canada	X	X
Hospital Militar	Santiago	Chile	X	
Bei Tai Ping Lu Hospital	Beijing	China	X	
Beijing Medical University	Beijing	China	X	
Lanzhou General Hospital	Lanzhou	China	X	
Institute of Hematology, CAMS	Tianjin	China	X	
Klin. Unutrasnje bolesti KBC-Rebro	Zagreb	Croatia	X	
Hermanos Arneijeiras Hospital	Habana	Cuba	X	X
Institute Hematologia E. Immunologia	Havana	Cuba	X	
Faculty Hospital	Pilsen	Czech Republic	X	
University Hospital Motol	Prague	Czech Republic	X	
Inst. of Hem./Blood Transfusion	Prague	Czech Republic	X	
Rigshospitalet	Copenhagen	Denmark	X	
NCI Cairo University	Cairo	Egypt	X	
Helsinki University Central Hospital	Helsinki	Finland	X	
Turku University	Turku	Finland	X	
Angers University	Angers	France	X	
Ctr. Hospitalier Univ. Bésançon	Bésançon	France	X	
Universitaire de Caen	Caen	France	X	
Hôpital Nord	Etienne	France	X	
Hospital A. Michallon	Grenoble	France	X	
Centre Hospitalier Régional De Lille	Lille	France	X	
Hôpital Edouard Herriot	Lyon	France	X	
Hôpital Debrousse	Lyon	France	X	
Institut J. Paoli I. Calmettes	Marseille	France	X	
Hôtel Dieu de Paris	Paris	France	X	
Hôpital Saint-Louis	Paris	France	X	
Hôpital Saint-Antoine	Paris	France	X	
Hôpital Robert Debre	Paris	France	X	
De l'Hôpital Cochin	Paris	France	X	
Hôpital des Enfants Malades	Paris	France	X	
Groupe Hospitalier du Hau Leveque	Pessac	France	X	
Hôpital Jean Bernard	Poitiers	France	X	
Centre Henri Becquerel	Rouen	France	X	
Hospital Regional de Toulouse	Toulouse	France	X	

Table 3. Continued

Institute	City	Country	IBMTR	ABMTR
Hôpital de Purpan	Toulouse	France	X	
Groupe Hôpital Paul-Brousse	Villejuif	France	X	
Universitätsklinikum Rudolf Virchow	Berlin	Germany	X	
Heinrich-Heine Universität	Düsseldorf	Germany	X	
University of Hamburg	Hamburg	Germany	X	
Med. Hochschule Hannover Ped	Hannover	Germany	X	
Medizinische Hochschule Hannover	Hannover	Germany	X	
Christian-Albrechts-Universität	Kiel	Germany	X	
University of Leipzig	Leipzig	Germany	X	
Universität München	Münich	Germany	X	
Universitäts Kinderklinik	Münich	Germany	X	
Children's Hospital	Tübingen	Germany	X	
Medizinische Universitätsklinik	Tübingen	Germany	X	
Universität Ulm	Ulm/Donau	Germany	X	
Evangelismos Hospital	Athens	Greece	X	
Queen Mary Hospital	Hong Kong	Hong Kong	X	
Prince of Wales Hosp.-Chinese Univ.	Hong Kong	Hong Kong	X	
National Institute of Haematology	Budapest	Hungary	X	
Tata Memorial Hospital	Bombay	India	X	
Institute Rotary Cancer Hospital	New Delhi	India	X	
Medical Science Univ. of Tehran	Tehran	Iran	X	
St. James's Hospital	Dublin	Ireland	X	
Hadassah University Hospital	Jerusalem	Israel	X	
Università Di Bologna	Bologna	Italy	X	
S. Orsola University Hospital	Bologna	Italy	X	
Spedali Civili – Brescia	Brescia	Italy	X	
Ospedale San Martino	Genoa	Italy	X	
Ospedale Cervello	Palermo	Italy	X	
Ospedale di Pesaro	Pesaro	Italy	X	
Ospedale Civile	Pescara	Italy	X	
University of Rome	Rome	Italy	X	
Università degli Studi, La Sapienza	Rome	Italy	X	
Ospedale S. Camillo	Rome	Italy	X	
University of Torino	Torino	Italy	X	
Di Midollo Osseo Ospedale Molinette	Torino	Italy	X	
Chiba University School of Medicine	Chiba	Japan	X	
Hyogo College of Medicine	Hyogo	Japan	X	
Tokai University School of Medicine	Kanagawa	Japan	X	
Kanazawa Univ. School of Medicine	Kanazawa-shi	Japan	X	
Nagoya Second Red Cross Hospital	Nagoya	Japan	X	
Center For Adult Diseases	Osaka	Japan	X	
Jichi Medical School	Tochigi-ken	Japan	X	
University of Tokyo	Tokyo	Japan	X	
Keio University	Tokyo	Japan	X	
Nihon University	Tokyo	Japan	X	
Kanagawa Cancer Center	Yokohama	Japan	X	

Table 3. Continued

Institute	City	Country	IBMTR	ABMTR
Jordan University Hospital	Amman	Jordan	X	
Catholic University Medical College	Seoul	Korea	X	
Asan Medical Center	Seoul	Korea	X	
University of Malaya	Kuala Lumpur	Malaysia	X	
Centro de Hem. y Medicina Interna	Puebla	Mexico		X
Inst. Nacional de Cancerologia	San Fernando	Mexico		X
Leiden University Hospital (Pediatrics)	Leiden	Netherlands	X	
Academic Hospital Maastricht	Maastricht	Netherlands	X	
University of Nijmegen	Nijmegen	Netherlands	X	
Dr. Daniel Den Hoed Cancer Center	Rotterdam	Netherlands	X	
University Hospital Utrecht	Utrecht	Netherlands	X	
Auckland Medical School	Auckland	New Zealand	X	
Auckland Hospital	Auckland	New Zealand	X	
Starship Children's Health	Auckland	New Zealand	X	
Christchurch Hospital	Christchurch	New Zealand	X	
Wellington School of Medicine	Wellington	New Zealand	X	
Silesian Medical Academy	Katowice	Poland	X	
I Klinika Chorob Dzieci	Poznan	Poland	X	
Academy of Medicine	Poznan	Poland	X	
Postgraduate Medical Center	Warsaw	Poland	X	
Instituto Portugues de Oncologia	Lisbon	Portugal	X	
Instituto Portugues de Oncologia	Porto	Portugal	X	X
Clinical Hospital Number 6	Moscow	Russia	X	
Petrov Res. Inst. of Oncology	St. Petersburg	Russia	X	X
Riyadh Armed Forces Hospital	Riyadh	Saudi Arabia	X	
King Faisal Spec. Hosp. & Research	Riyadh	Saudi Arabia	X	
University Hospital	Bratislava	Slovak Rep.	X	
Univ. of Cape Town Medical School	Cape Town	South Africa	X	
Univ. of Witwatersrand Med. School	Johannesburg	South Africa	X	
University of Barcelona	Barcelona	Spain	X	
Hospital de la Santa Creui Sant Pau	Barcelona	Spain	X	
Hospital General Vall d'Hebron	Barcelona	Spain	X	
Hospital Infantil Vall d'Hebron	Barcelona	Spain	X	
Hospital Reina Sofia	Cordoba	Spain	X	
Hospital Ntra Sra Del Pino	Islas Canarias	Spain	X	
Hospital Ramon y Cajal	Madrid	Spain	X	
Hospital de la Princesa	Madrid	Spain	X	
Hospital Puerta de Hierro	Madrid	Spain	X	
Hospital Regional "Carlos Haya"	Malaga	Spain	X	
Hospital Marques De Valdecilla	Santander	Spain	X	
Hospital La Fe	Valencia	Spain	X	
University of Goteborg	Goteborg	Sweden	X	
Huddinge Hospital	Huddinge	Sweden	X	
University of Lund	Lund	Sweden	X	
University Hospital Uppsala	Uppsala	Sweden	X	
Kantonsspital	Basel	Switzerland	X	

Table 3. Continued

Institute	City	Country	IBMTR	ABMTR
Kantonsspital Zurich	Zurich	Switzerland	X	
Kinderspital Zurich	Zurich	Switzerland	X	
National Yang-Ming Medical College	Taipei	Taiwan	X	
National Taiwan Univ. Hospital (Peds)	Taipei	Taiwan	X	
National Taiwan Univ. Hosp. Int Med	Taipei	Taiwan	X	
Gulhane Military Medical Academy	Ankara	Turkey	X	X
Ankara University Medical School	Ankara	Turkey	X	
SSK Tepecik Teaching Hospital	Izmir	Turkey	X	
Birmingham Heartlands Hospital	Birmingham	United Kingdom	X	
Queen Elizabeth Medical Centre	Birmingham	United Kingdom	X	
University Hospital of Wales	Cardiff	United Kingdom	X	
Royal Infirmary of Edinburgh	Edinburgh	United Kingdom	X	
Glasgow Royal Infirmary	Glasgow	United Kingdom	X	
Royal Hospital for Sick Children	Glasgow	United Kingdom	X	
St. James's University Hospital	Leeds	United Kingdom	X	
Royal London Hospital Whitechapel	London	United Kingdom	X	
Great Ormond St. Hosp. for Children	London	United Kingdom	X	
St. George's Hospital Medical School	London	United Kingdom	X	
Royal Free Hospital	London	United Kingdom	X	
Royal Postgraduate Med. School	London	United Kingdom	X	
Charing Cross Hosp & Westminster	London	United Kingdom	X	
London Clinic	London	United Kingdom	X	
Royal Marsden Hospital	London	United Kingdom	X	
Royal Victoria Infirmary	Newcastle	United Kingdom	X	
City Hospital	Nottingham	United Kingdom	X	
John Radcliffe Hosp./Univ. of Oxford	Oxford	United Kingdom	X	
Presbyterian Health Care Services	Albuquerque	United States		X
C.S. Mott Children's Hospital	Ann Arbor	United States	X	X
Emory Clinic	Atlanta	United States	X	X
Emory University	Atlanta	United States		X
Northside Hospital	Atlanta	United States		X
Johns Hopkins Hospital	Baltimore	United States	X	X
University of Maryland Cancer Center	Baltimore	United States		X
Alta Bates Hospital	Berkeley	United States	X	X
University of Alabama at Birmingham	Birmingham	United States	X	X
Dana-Farber Cancer Institute	Boston	United States		X
Massachusetts General Hospital	Boston	United States	X	
Montefiore Medical Center	Bronx	United States		X
Roswell Park Cancer Institute	Buffalo	United States	X	X
Univ. of North Carolina Chapel Hill	Chapel Hill	United States		X
Medical University of South Carolina	Charleston	United States	X	X
Univ. of Virginia Medical Center	Charlottesville	United States	X	X
University of Chicago Medical Center	Chicago	United States		X
Children's Memorial Hospital	Chicago	United States	X	X
Rush Presbyterian/St. Luke's Med. Ctr.	Chicago	United States	X	X
Wyler Children's Hospital	Chicago	United States	X	
Michael Reese Hosp. and Medical Ctr	Chicago	United States	X	
Jewish Hospital of Cincinnati	Cincinnati	United States	X	X

Table 3. Continued

Institute	City	Country	IBMTR	ABMTR
University Hospital Cincinnati	Cincinnati	United States	X	X
Children's Hospital Medical Center	Cincinnati	United States	X	X
Rainbow Babies & Children's Hospital	Cleveland	United States	X	X
Cleveland Clinic Foundation	Cleveland	United States	X	X
Case Western Reserve Univ. Hospital	Cleveland	United States	X	X
Univ. of South Carolina	Columbia	United States	X	X
Ohio State University Hospital	Columbus	United States		X
Children's Hospital	Columbus	United States	X	
Baylor University Medical Center	Dallas	United States	X	X
Children's Medical Center Dallas	Dallas	United States	X	X
Univ. of Colorado Health Sci. Ctr.	Denver	United States	X	
Presbyterian St. Luke's Hospital	Denver	United States	X	X
Wayne State University	Detroit	United States		X
City of Hope National Medical Center	Duarte	United States	X	X
Duke University Medical Center	Durham	United States	X	X
Univ. of Connecticut Health Center	Farmington	United States		X
Harris Methodist Oncology Program	Fort Worth	United States	X	X
Cook-Fort Worth Children's Med. Cen.	Fort Worth	United States	X	X
Univ. of Florida (Adult) – JHMHC	Gainesville	United States	X	
University of Florida, Shands Hospital	Gainesville	United States	X	X
East Carolina Univ. School of Med.	Greenville	United States		X
Hackensack Medical Center	Hackensack	United States		X
Hinsdale Hem-Oncology Associates	Hinsdale	United States		X
St. Francis Medical Center	Honolulu	United States	X	X
St. Joseph's Hospital Medical Center	Houston	United States	X	
Texas Children's Hospital	Houston	United States	X	
Baylor College of Medicine	Houston	United States	X	X
M.D. Anderson Cancer Center	Houston	United States	X	X
Methodist Hospital of Indiana	Indianapolis	United States	X	X
St. Vincent Hosp. & Health Care Ctr.	Indianapolis	United States		X
Indiana Univ. Hosp. & Outpatient Ctr.	Indianapolis	United States	X	X
Mayo Clinic Jacksonville	Jacksonville	United States	X	X
University of Kansas	Kansas City	United States	X	X
Children's Mercy Hospital	Kansas City	United States		X
Univ. of Tennessee Medical Center	Knoxville	United States	X	X
Scripps Clinic & Research Fdn	La Jolla	United States	X	
Wilford Hall USAF Medical Center	Lackland AFB	United States	X	
Dartmouth-Hitchcock Medical Center	Lebanon	United States		X
Univ. of Kentucky Medical Center	Lexington	United States	X	X
Arkansas Cancer Research Center	Little Rock	United States	X	X
Children's Hospital of Los Angeles	Los Angeles	United States	X	
USC/Norris Cancer Hospital	Los Angeles	United States		X
UCLA Center for Health Sciences	Los Angeles	United States	X	X
Kaiser Permanente of Southern Calif.	Los Angeles	United States	X	X
UCLA- Center for Health Sci. Ped	Los Angeles	United States	X	
James Graham Brown Cancer Center	Louisville	United States	X	X

Table 3. Continued

Institute	City	Country	IBMTR	ABMTR
University of Wisconsin	Madison	United States	X	X
North Shore University Hospital	Manhasset	United States		X
Marshfield Clinic	Marshfield	United States	X	
Loyola University Medical Center	Maywood	United States	X	X
Methodist Hospital Central	Memphis	United States		X
Response Technologies	Memphis	United States		X
St. Jude Children's Research Hospital	Memphis	United States	X	X
Miami Children's Hospital	Miami	United States	X	X
Baptist Hospital of Miami	Miami	United States		X
St. Luke's Medical Center	Milwaukee	United States		X
John L. Doyne Hospital	Milwaukee	United States	X	X
Abbott Northwestern Hospital	Minneapolis	United States		X
University of Minnesota	Minneapolis	United States	X	X
West Virginia University	Morgantown	United States	X	X
Vanderbilt University Medical Center	Nashville	United States		X
Louisana State Medical Center	New Orleans	United States	X	
Louisiana State Univ. Children's Hosp.	New Orleans	United States	X	X
Tulane University Medical Center	New Orleans	United States		X
New York Hosp. Cornell Med. Center	New York	United States		X
Mem. Sloan-Kettering Cancer Center	New York	United States	X	X
Columbia University	New York	United States		X
Mt. Sinai Medical Center	New York	United States	X	X
Medical Center of Delaware	Newark	United States	X	X
Hoag Cancer Center	Newport Beach	United States		X
Univ. of Oklahoma Health Sciences	Oklahoma	United States	X	X
Bishop Clarkson Memorial Hospital	Omaha	United States	X	
Univ. of Nebraska Medical Center	Omaha	United States	X	X
Children's Hospital of Orange County	Orange	United States	X	X
Saint Joseph Hospital	Orange	United States	X	X
Lutheran General Hospital	Park Ridge	United States		X
St. Joseph's Hospital and Medical Ctr.	Paterson	United States	X	
Hematology Associates	Peoria	United States		X
Children's Hospital of Philadelphia	Philadelphia	United States	X	X
Temple Univ. Comp. Cancer Center	Philadelphia	United States	X	X
University of Pennsylvania Hospital	Philadelphia	United States		X
Hahnemann University Hospital	Philadelphia	United States	X	X
Children's Hospital of Pittsburgh	Pittsburgh	United States	X	X
Montefiore University Hospital	Pittsburgh	United States	X	X
Western Pennsylvania Cancer Institute	Pittsburgh	United States	X	
The Cancer Center of Boston	Plymouth	United States		X
St. Charles & John T. Mather Hosp.	Pt. Jefferson Stat.	United States		X
Oregon Health Sciences Univ.	Portland	United States	X	X
Oregon Health Sciences Univ. Peds	Portland	United States	X	
Roger Williams Medical Center	Providence	United States	X	X
Cancer & Blood Inst. of the Desert	Rancho Mirage	United States		X
Washoe Medical Center	Reno	United States	X	X

Table 3. Continued

Institute	City	Country	IBMTR	ABMTR
Mayo Clinic	Rochester	United States	X	X
Strong Memorial Hospital	Rochester	United States	X	
University of Rochester	Rochester	United States		X
Sutter Memorial Hospital	Sacramento	United States	X	X
Univ. of Calif. Davis Cancer Center	Sacramento	United States		X
LDS Hospital	Salt Lake City	United States	X	X
University of Utah School of Medicine	Salt Lake City	United States	X	X
Univ. of Texas Health Sciences Ctr.	San Antonio	United States	X	X
Children's Hospital San Diego	San Diego	United States		X
University of CA, San Diego	San Diego	United States		X
Univ. of CA, San Francisco Med. Ctr.	San Francisco	United States		X
Univ. of CA, San Francisco Pediatrics	San Francisco	United States	X	X
Mayo Clinic Scottsdale	Scottsdale	United States	X	X
LSU Medical Center-Shreveport	Shreveport	United States	X	X
Memorial Medical Center	Springfield	United States		X
Cardinal Glennon Children's Hospital	St. Louis	United States	X	X
St. Louis Children's Hospital	St. Louis	United States	X	X
St. Louis University Medical Center	St. Louis	United States		X
Meth. Hosp./Nicollet Cancer Center	St. Louis Park	United States		X
All Children's Hospital	St. Petersburg	United States	X	X
Stanford University Hospital	Stanford	United States	X	X
SUNY-Health Science Center	Syracuse	United States		X
H. Lee Moffitt Cancer Center	Tampa	United States	X	X
Arizona Cancer Center	Tucson	United States	X	X
St. Francis Hospital	Tulsa	United States		X
New York Medical College	Valhalla	United States		X
Children's National Medical Center	Washington	United States	X	
Georgetown Univ. Medical Center	Washington	United States	X	
George Washington Univ. Med. Ctr.	Washington	United States	X	X
Walter Reed Army Medical Center	Washington	United States		X
Westlake Comprehensive Cancer Ctr.	Westlake Village	United States		X
St. Francis Hospital	Wichita	United States	X	X
Wake Forest University	Winston-Salem	United States	X	X
British Hosp. & Faculty of Medicine	Montevideo	Uruguay	X	X
Hospital Central de Valencia	Valencia	Venezuela	X	

References

1. Bortin MM, Horowitz MM and Rimm AA (1992) Increasing utilization of allogeneic bone marrow transplantation. III. Results of the 1988–1990 survey. Ann Intern Med 116: 505–12.
2. Bortin MM, Ringdén O, Horowitz MM *et al.* (1989) Temporal relationships between the major complications of bone marrow transplantation for leukemia. Bone Marrow Transplant 4: 339–44.
3. Brenner MK, Rill DR, Moen RC *et al.* (1993) Gene-marking to trace origin of relapse after autologous bone marrow transplant. Lancet 341: 85–86.
4. Kersey JH, Weisdorf D, Nesbit ME *et al.* (1987) Comparison of autologous and allogeneic bone marrow transplantation for treatment of high risk refractory acute lymphoblastic leukemia. N Engl J Med 317: 461–67.
5. Gale RP, Horowitz MM, Ash RC *et al.* (1994) Identical-twin bone marrow transplants for leukemia. Ann Intern Med 120: 646–52.
6. Bortin MM, Horowitz MM, Gale RP *et al.* (1992) Changing trends in bone marrow transplantation for leukemia in the 1980's. JAMA 268: 607–12.
7. Juttner CA, To LB, Haylock DN *et al.* (1985) Circulating autologous stem cells collected in very early remission from acute non-lymphoblastic leukaemia produce prompt but incomplete haemopoietic reconstitution after high dose melphalan or supralethal chemoradiotherapy. Br J Haematol 61: 739–45.
8. Sheridan WP, Begley CG, Juttner CA *et al.* (1992) Effect of peripheral blood progenitor cells mobilized by filgrastim (G-CSF) on platelet recovery after high dose chemotherapy. Lancet 339: 640–44.
9. Körbling M, Przepiorka D, Huh YO *et al.* (1995) Allogeneic blood stem cell transplantation for refractory leukemia and lymphoma: Potential advantage of blood over marrow allografts. Blood 85: 1659–65.
10. Bensinger WI, Weaver CH, Appelbaum FR *et al.* (1995) Transplantation of allogeneic peripheral blood stem cells mobilized by recombinant human granulocyte colony-stimulating factor. Blood 85: 1655–58.
11. Schmitz N, Dreger P, Suttorp M *et al.* (1995) Primary transplantation of allogeneic peripheral blood progenitor cells mobilized by filgrastim (granulocyte colony-stimulating factor). Blood 85: 1666–72.
12. Thomas ED and Clift RA (1989) Indications for marrow transplantation in chronic myelogenous leukemia. Blood 73: 861–64.
13. Thomas ED, Clift RA, Fefer A *et al.* (1986) Marrow transplantation for the treatment of chronic myelogenous leukemia. Ann Intern Med 104: 155–63.
14. Goldman JM, Szydlo R, Horowitz MM *et al.*, for the Advisory Committee of the IBMTR (1993) Choice of pretransplant treatment and timing of transplants for chronic myelogenous leukemia in chronic phase. Blood 82: 2235–38.
15. Horowitz MM, Gale RP, Sondel PM *et al.* (1990) Graft-versus-leukemia reactions after bone marrow transplantation. Blood 75: 555–62.
16. Kolb HJ, Mittermuller J, Clemm C *et al.* (1990) Donor leukocyte transfusions for treatment of recurrent chronic myelogenous leukemia in marrow transplant patients. Blood 76: 2462–65.
17. Porter DL, Roth MS, McGarigle C *et al.* (1994) Induction of graft-versus-host disease as immunotherapy for relapsed chronic myeloid leukemia. New Engl J Med 330: 100–6.
18. Drobyski WR, Keever CA, Roth MS *et al.* (1993) Salvage immunotherapy using donor leukocyte infusions as treatment for relapsed chronic myelogenous leukemia after allogeneic bone marrow transplantation: efficacy and toxicity of a defined T-cell dose. Blood 82: 2310–18.
19. McGlave PB, De Fabritiis P, Deisseroth A *et al.* (1994) Autologous transplants for chronic myelogenous leukaemia: results from eight transplant groups. Lancet 343: 1486–88.
20. Gale RP, Horowitz MM, Biggs JC *et al.* (1989) Transplant or chemotherapy in acute myelogenous leukaemia. Lancet 1: 1119–22.
21. Barrett AJ, Horowitz MM, Pollock BH *et al.* (1994) Bone marrow transplants from HLA-identical siblings as compared with chemotherapy for children with acute lymphoblastic leukemia in a second remission. N Engl J Med 331: 1253–58.
22. Chao NJ and Blume KG (1990) Bone marrow transplantation: What is the question? [editorial]. Ann Intern Med 113: 340–41.
23. Horowitz MM, Messerer D, Hoelzer D *et al.* (1991) Comparison of chemotherapy and bone marrow transplantation for adults with acute lymphoblastic leukemia in first remission. Ann Intern Med 115: 13–8.
24. Takvorian T, Canellos GP, Ritz J *et al.* (1987) Prolonged disease-free survival after autologous bone marrow transplantation in patients with non-Hodgkin's lymphoma with a poor prognosis. N Engl J Med 316: 1499–1505.
25. Haioun C, Lepage E, Gisselbrecht C *et al.* (1994) Comparison of autologous bone marrow transplantation with sequential chemotherapy for intermediate-grade and high-grade non-Hodgkins lymphoma in first complete remission – a study of 464 patients. J Clin Oncol 12: 2543–51.
26. Gulati SC, Shank B, Black P *et al.* (1988) Autologous bone marrow transplantation for patients with poor-prognosis lymphoma. J Clin Oncol 6: 1303–13.
27. Jones RJ, Ambinder RF, Piantadosi S *et al.* (1991) Evidence of a graft-versus-lymphoma effect associated with allogeneic bone marrow transplantation. Blood 77: 649–53.
28. Gahrton G, Tura S, Ljungman P *et al.* (1991) Allogeneic bone marrow transplantation in multiple myeloma. N Engl J Med 325: 1267–73.
29. Jagannath S, Vesole DH, Glenn L *et al.* (1992) Low-risk intensive therapy for multiple myeloma with combined autologous bone marrow and blood stem cell support. Blood 80: 1666–72.
30. Jagannath S, Barlogie B, Dicke KA *et al.* (1990) Autologous bone marrow transplantation in multiple myeloma: Identification of prognostic factors. Blood 76: 1860–66.
31. Ash RC, Horowitz MM, Gale RP *et al.* (1991) Bone marrow transplantation from related donors other than HLA-identical siblings: Effect of T-cell depletion. Bone Marrow Transplant 7: 443–52.
32. Ash RC, Casper JT, Chitambar C *et al.* (1990) Successful allogeneic marrow transplantation from closely HLA-matched unrelated donors using T-cell depletion. N Engl J Med 322: 485–94.
33. Kernan NA, Bartsch G, Ash RC *et al.* (1993) Analysis of 462 transplantations from unrelated donors facilitated by the National Marrow Donor Program. N Engl J Med 328: 593-602.

Section II:

Cartilage, Bone and Muscle

R. P. Lanza and W. L. Chick (eds.), Yearbook of Cell and Tissue Transplantation 1996/1997, 37–40.

Chapter 4

Bone and cartilage transplantation *

William W. Tomford and Henry J. Mankin
Department of Orthopaedic Surgery, Massachusetts General Hospital, Boston, MA 02114, U.S.A.

Bone and cartilage have been transplanted as tissues for almost one hundred years. During much of this time, these tissues were seldom modified following removal from the donor and transplantation into the recipient. Advances were made in surgical technique as well as in methods of storage of the tissues, but few advances were made in manipulating the tissues in order to improve incorporation or function. In the past few years, biochemical advances have resulted in modification of these tissues which has produced a promising future in their transplantation. This paper will review bone and cartilage transplantation through historical, current, and future perspectives.

Bone transplantation

Historical

Allograft bone has been used by surgeons for over one hundred years. The first human bone allograft transplant was performed by MacEwen in 1881 [1]. Alexis Carrell is credited with early studies on the storage of musculoskeletal tissues [2] and developed the idea of transplanting cadaver bone. Following Carrell's report, Albee, working in New York, transplanted refrigerated autogenous bone as well as cadaver bone in spinal fusions [3]. Gallie used boiled bone for fusions in patients with Pott's disease (tuberculous involvement of the spine) [4]. Haas recommended the use of frozen bone following studies of viability of bone in which he determined that frozen bone was superior to boiled bone. [5] Orell used "*os purum*" or xenografts purified by chemical and mechanical processes and "*os novum*" which consisted of "*os purum*" that was stored under the periosteum of the recipient's tibia for three months prior to transplantation [6].

In the 1940s, bone grafting and the use of allografts began to increase. In 1942, citing Carell and Albee as the basis for his investigations, Inclan reported from Havana the first true bone bank [7]. Inclan was one of the early bankers of cadaver bone and reported good or excellent results on 75% of his cases in which allografts were used in spinal fusions and hip fractures. In 1947, Bush reported his ideas on the storage and use of allograft bone [8]. These included the use of homogenous bone grafts in 126 operations in 104 patients. All of these patients received deep frozen bone, and 91 of the 104 patients had successful spinal fusions. In 1947, Philip D. Wilson, working at the Hospital for Special Surgery in New York, reported the first femoral head bone bank for the use of femoral heads in spine fusions [9]. Hault reported the use of bones stored deep frozen at the Karolinska Institute in Stockholm, Sweden [10]. In 1950, the Navy Tissue Bank in Bethesda, Maryland, under the direction of George Hyatt, began the storage of bone by lyophilization [11]. The Navy Tissue Bank supplied several thousand tissue grafts to military and civilian surgeons during its thirty years of existence.

The next major advance in bone transplantation was the use of massive frozen bone allografts. This technique was developed by Dr. Frank Parrish [12] following original work by Lexer [13]. In the 1970s, Mankin began a program of transplantation of frozen long bone allografts in limb-sparing procedures for the treatment of bone cancer [14]. Mankin's efforts stimulated the development of hospital bone banks throughout the

* Work supported in part by NIH Grant AR-21896.

United States which were the forerunners of the large regional banks that exist today.

Current

Currently there are nearly one hundred bone banks in the United States. Many of these bone banks are members of the American Association of Tissue Banks, an organization developed in the mid 1970s to serve as a forum for the discussion of tissue banking including standards, guidelines, and a voluntary inspection and accreditation program (AATB, 1350 Beverly Road, Suite 220-A, McLean, Virginia 22101).

Banked bone is in popular use in spinal fusions, fracture treatment, dental applications, and joint replacement revisions. The last application is particularly important in bone loss engendered by osteolysis or stress shielding. Bone replacement by the use of bone allografts as freeze-dried chips or cortical struts has provided solutions of very difficult bone reconstruction problems.

The most important development in bone transplantation in the past few years has been the isolation and experimental application of bone cytokines or growth factors. Originally postulated by Urist [15], further progress has included the work of Glowacki [16] and others [17–19]. Bone graft factors are now entering the marketplace in studies which attempt to develop and confirm a commercially feasible application. Proven experimentally to be useful in the stimulation of bone formation in animal models, the method and success of application in humans remains to be determined.

Future

In the future, bone growth factors will be available to the clinician either as isolated factors to be used with autogenous bone grafts or in combination with banked bone or synthetic substitutes such as bovine collagen or even devices. The specific application of bone growth factors has produced controversy on the type and extent of tissure formation that occurs. Bone growth in these circumstances is endochondral with the formation of cartilage prior to the formation of bone. Control of the factors in the conversion of cartilage to bone is as yet an incompletely solved problem, but once this is resolved, current methods of bone healing, incorporation and fixation may be radically changed.

Cartilage transplantation

Historical

Lexer was the first to publish a large series of fresh cartilage allografts [13], although Tuffier had previously performed such transplants [20]. Lexer's osteochondral grafts were used for replacement of the most distal or proximal 1–2 cm of a long bone with the articular surface. He noted that he preferred to implant grafts "while the tissue is of body temperature". Lexer frequently replaced both sides of a diarthrodial joint because he was treating septic arthritis due to bacterial infections which usually destroyed all joint cartilage. Lexer found in long term follow-up of his cartilage transplants that collapse of the subchondral bone allowed degeneration of the articular cartilage. He believed this problem was due to revascularization of the bony part of the osteochondral graft which led to fatigue fractures and failure of the underlying support of the articular surface. He therefore advised his patients to avoid weight bearing for up to two years following surgery. Lexer also reported the use of stored cadaver cartilage grafts, but he believed that fresh cartilage transplants were superior.

Entin subsequently published an interesting study of fresh autogenous whole joint transplants in humans [21]. He noted that articular cartilage remained relatively normal for about five months after transplanting metatarso-phalangeal joints taken from the foot to replace damaged joints in the hand. The entire joint was transplanted using midshaft osteosyntheses on each side of the original joint. Entin found that cartilage destruction depended mostly on the function of the joint after replacement. In non-functioning joints, where scar tissue prevented motion, the cartilage lasted longer. He concluded that abnormal mechanical stresses were the major factor in transplanted cartilage destruction.

In an extensive series on small osteochondral wafer grafts, Gross reported the use of articular cartilage transplants which included up to 3 mm of subchondral bone. In several reports on these types of transplants, Gross found that many survived for several years following their use in traumatic defects, but he did not find that osteoarthritis or osteonecrosis were good indications for these transplants. [22–24]. Gross also noted that chondrocytes remained viable several years after the transplant [25]. Others who have used fresh grafts include Garrett who has transplanted fresh grafts for condylar defects [26]. He used a specially sized cutting bit to remove similar sized grafts from the donor and recipient. Meyers has also published articles in which he found that small fresh osteochondral grafts in the hip and knee were useful for prolonging the need for joint replacement [27,28].

Frozen cartilage allografts were pioneered by Ottolenghi and Parrish and developed by Mankin, all of whom used osteochondral grafts in skeletal reconstruction following replacement of tumors. Ottolenghi noted that the subchondral bone of frozen grafts was subject to collapse just as Lexer had noted this problem in the results of his massive fresh grafts [29]. In spite of the slow deterioration of the articular cartilage, Ottolenghi noted that the joint function frequently remained excellent, a finding that had been previously noted by Lexer [29]. In follow-up of his long bone osteoarticular allografts, Parrish noted degenerative changes at the joint in most cases, including upper extremity transplants [12]. He found the cartilage space well preserved in one case with an eleven year follow-up, but complete loss was noted in another at four years. Parrish believed that degenerative changes could be expected in almost all cartilage grafts.

Extensive experience with massive osteochondral allograft transplants has been reported by Mankin [30]. Mankin has used cryopreserved (10% DMSO) osteochondral allografts. In a study on Mankin's grafts, Waber found that deterioration occurred in about 10% of the grafts after they had been in for a minimum of three years [31]. Waber noted that subchondral collapse as a result of metaphyseal fractures was the major cause of deterioration and was most prevalent in weight bearing joints.

Current

Tissue engineering methods are now being reported which use cartilage cells to fill defects as appeared in a recent report from Sweden [32]. Patients had full thickness cartilage defects that ranged up to 6.5 cm^2. Chondrocytes obtained from a normal area of the knee joint during arthroscopy are isolated and cultured in a laboratory and then reinjected into the area of the defect. The defect is covered by a periosteal flap taken from the medial proximal tibia and sutured around the periphery of the defect, under which the chondrocytes are transplanted. The authors noted that two years after transplantation, fourteen of sixteen patients with femoral condylar transplants had good to excellent results, but approximately thirty-six months after transplantation, only two of seven patients with transplants into the patella had good to excellent results. The authors concluded that the operation is not indicated for patellae but may be indicated for selected femoral condylar cartilage defects.

Future

Future work in cartilage transplants will in part be concentrated on autogenous transplants. One of the more exciting areas will be in the use of cartilage growth factors. Cytokines are now known which will induce cartilage formation as a part of the cascade of bone formation. These chemicals no doubt will be applied in cartilage transplants to provide a biologic therapeutic approach to cartilage loss.

References

1. Macewen W (1881) Observations concerning transplantation of bone. Proc R Soc London 32: 232–247.
2. Carrell A (1912) The preservation of tissues and its application in surgery. JAMA 59: 523–527.
3. Albee FH (1915) The fundamental principles involved in the use of the bone graft in surgery. Am J Med Sci 149: 313–325.
4. Gallie WE (1918) The use of boiled bone in operative surgery. Am J Orthop Surg 16: 373–383.
5. Haas SL (1923) A study of the stability of bone after removal from the body. Arch Surg 7: 213–226.
6. Orell S (1937) Surgical bone grafting with "os purum", "os novum" and boiled bone. J Bone and Joint Surg 19: 873–885.
7. Inclan A (1942) The use of preserved bone graft in orthopaedic surgery. J Bone and Joint Surg 24: 81–96.
8. Bush LF (1947) The use of homogenous bone grafts: A preliminary report on the bone bank. J Bone and Joint Surg 29: 620–628.
9. Wilson PD (1947) Experience with a bone bank. Ann Surg 126: 932-946.
10. Hault L (1950) Some experiences with a bone bank. Acta Orthop Scand 19: 476–480.
11. Hyatt GW, Turner TC, Bassett CAL, Pate JW and Sawyer PN (1952) New methods for preserving bone, skin, and blood vessels. Postgrad Med 12: 239–254.
12. Parrish FF (1973) Allograft replacement of all or part of the end of a long bone following excision of a tumor: report of 21 cases. J Bone and Joint Surg 55A: 1–22.
13. Lexer E (1925) Joint transplantations and arthroplasty. Surg Gynecol Obstet 40: 782–809.
14. Mankin HJ, Gebhardt MC and Tomford WW (1987) The use of frozen cadaveric allografts in the management of patients with bone tumors of the extremities. Ortho Clin N Am 18: 275–289.
15. Urist MR (1965) Bone formation by autoinduction. Science 150: 893-899.
16. Glowacki J, Kaban LB, Murray JE, Folkman J and Mulliken JB (1981) Application of the bone induction principle of induced osteogenesis for craniofacial defects. Lancet 1: 959–963.
17. Wozney JM, Rosen V, Celeste AJ, Mitsock LM, Whitters MJ, Kriz RW, Hewick RM and Wang EA (1988) Novel regulators

of bone formation: molecular clones and activities. Science 242: 1528–1534.

18. Yasko AW, Lane JM, Fellinger EJ, Rosen V, Wozney JM and Wang EA (1992) The healing of segmental bone defects induced by recombinant human bone morphogenetic protein (rhBMP-2). A radiographic, histological, and biomechanical study in rats. J Bone and Joint Surg 74-A: 659–670.
19. Cook SD, Wolfe MW, Salkeld SL and Rueger DC (1995) Effect of recombinent osteogenic protein-1 on healing of segmental defects in non-human primates. J Bone and Joint Surg 77A: 734–750.
20. Tuffier T (1913) Sur les graffes osteo-articulaires. Bull Mem Soc Chir Paris 39: 1078–1096.
21. Entin MA, Alger JR and Barrett RM (1962) Experimental and clinical transplantation of autogenous whole joints. J Bone and Joint Surg 44A: 1518–1536.
22. Gross AE, Silverstein EA, Falk J *et al.* (1975) The allotransplantation of partial joints in the treatment of osteoarthritis in the knee. Clin Orthop 108: 7–14.
23. McDermott AGP, Langer F, Prizker KPH *et al.* (1985) Fresh small-fragment osteochondral allografts: long term follow-up study on first 100 cases. Clin Orthop 197: 96–102.
24. Oakeshott RD, Farine I, Prizker KPH *et al.* (1988) A clinical and histologic analysis of failed fresh osteochondral allografts. Clin Orthop 233: 283–294.
25. Czitrom AA, Keating S and Gross AE (1990) The viability of articular cartilage in fresh osteochondral allografts after clinical transplantation. J Bone and Joint Surg 72A: 574–581.
26. Garrett JC (1986) Treatment of osteochondritis dissecans of the distal femur with fresh osteochondral allografts. Arthroscopy 2: 222–226.
27. Meyers MH (1985) Resurfacing of the femoral head with fresh osteochondral allografts. Clin Orthop 197: 111–115.
28. Convery FR, Meyers MH and Akeson WH (1991) Fresh osteochondral allografting of the femoral condyle. Clin Orthop 273: 139–145.
29. Ottolenghi CE (1972) Massive osteo- and osteo-articular bone grafts: technique and results of 62 cases. Clin Orthop 87: 156–164.
30. Mankin HJ, Fogelson FS and Thrasher AZ (1976) Massive resection and allograft transplantation in the treatment of malignant bone tumors. New England J Med 294: 1247–1255.
31. Waber BA, Tomford WW and Mankin HJ (1989) Long term results of osteoarticular allografts in weight bearing joints. In: Aebi M and Regazzoni P (eds) Bone Transplantation, pp 275–283. Berlin: Springer-Verlag.
32. Brittberg M, Lindahl A, Nilsson A, Ohlsson C, Isaksson O and Peterson L (1994) Treatment of deep cartilage defects in the knee with autologous chondrocyte transplantation. New England J Med 331: 889–895.

R. P. Lanza and W. L. Chick (eds.), Yearbook of Cell and Tissue Transplantation 1996/1997, 41–51.

Chapter 5

Chondrocytes

Stanislaw Moskalewski
Department of Histology and Embryology, Institute of Biostructure, Warsaw Medical School, Warsaw, 02–004 Poland

Chondrocytes liberated from the intercellular substance are able to produce and deposit matrix components both in a suitable tissue culture system and after transplantation. Cartilage reformed by transplanted chondrocytes displays basic similarity to cartilage which served as the source of cells. Thus, chondrocyte transplantation may be of some interest in studies on cartilage morphogenesis. Moreover, it is hopefully expected that isolated chondrocytes may be used for treatment of some cartilage defects, e.g. resurfacing of joints [1]. This chapter reviews problems encountered during chondrocyte isolation, observations on the reformation of cartilage by various types of transplanted chondrocytes, rejection of cartilage produced by allogeneic chondrocytes and trials directed at clinical application of chondrocyte grafts.

Isolation of chondrocytes

Chondrocytes from fetal or mature but young cartilage can be liberated by enzymatic digestion usually within 3–4 hours. Crude collagenase supplemented with DNase and α-tosyl-lysine chloromethyl ketone (TLCK), gives good results [2] and seems to be less harmful to cells than trypsin [3] or trypsin and collagenase [4] used in earlier studies. DNase is needed to digest DNA released from injured cells [5], and TLCK, by inhibiting the activity of clostripain present in crude collagenase preparations, decreases cell damage [6]. Recently, a special cartilage digestion medium (low Mg^{2+} with added K^{+} and TLCK) improving both tissue digestion rate and viable cell yield was developed [7].

Cell yield considerably depends on the age of the cartilage. Cartilage matrix from older individuals frequently contains areas of degeneration [8–10] and resists digestion [9,11]. Moreover, cell density in cartilage decreases rapidly with age [12]. In older costal [13], nasal septal [14] or auricular [9,15], but not articular [13] cartilage chondrocytes acquire large lipid droplets and are easy to damage during isolation [9]. Incidentally, the large lipid droplets may be removed from cytochalasin B treated chondrocytes by high speed centrifugation [16]. In an attempt to "rejuvenate" rib cartilage degenerated in older subjects [8] the cartilage was subperichondrially removed in adult dogs and new cartilage allowed to regenerate from the perichondrium [11]. Unfortunately, the cells isolated from the latter only occasionally produced matrix in intramuscular transplants [11].

During enzymatic digestion some chondrocytes may be more susceptible to destruction than others, or some areas of cartilage may resist digestion. Therefore, in the resulting cell suspension the share of chondrocytes from various regions of cartilage may not correspond to that existing *in situ*. Moreover, isolated chondrocytes may be contaminated by other cell types, originating from the surrounding tissues difficult to remove during dissection of cartilage [17].

The articular-epiphyseal cartilage complex [18] from newborn animals, growth plate cartilage from older, but immature ones and shivers of articular cartilage from mature individuals are frequently used as a source of chondrocytes. For isolation of human chondrocytes the costal cartilage obtained from children undergoing surgical correction of pectus excavatum may be used [19]. It is usually possible to obtain

1×10^7 cells from articular-epiphyseal complexes of all limbs of 3-4-day-old rat or $1–4\times10^8$ from a shoulder of a newborn calf [19]. Adult human articular cartilage yielded only $1\times10^5 - 5\times10^5$ chondrocytes from 1 g of tissue [20]. In contrast, 1 g of articular cartilage from immature rabbits yielded, under optimal conditions, 2.8×10^7 viable cells [7]. Isolated chondrocytes can be easily preserved by freezing [21].

Transplantation of chondrocytes

Syngeneic, allogeneic and xenogeneic chondrocytes are able to produce cartilage after transplantation. The presence of a similar number of sex chromatin bodies in chondrocytes in intact female cartilage and in cartilage formed by female chondrocytes grafted into males helped to establish that the cartilage in transplants was formed by the injected cells and not by induction at the site of transplantation [14]. In the histoincompatible systems cartilage formed by transplanted chondrocytes is slowly rejected. Chondrocyte transplantation was used in morphogenetic and immunological studies as well as in attemps to find clinical application for such transplants, most frequently for repair of articular cartilage defects [22].

Morphogenetic studies

One week after intramuscular transplantation chondrocytes isolated from lapine nasal septal (hyaline) cartilage formed islands of cartilage with scanty amount of matrix and cells with vesicular nuclei [14,23]. In four-week-old transplants cartilage nodules were surrounded by a perichondrium-like layer, contained a few rows of flattened cells at the periphery and rounded cells in the center. The latter contained large lipid droplets and flattened nuclei displaced to the periphery. Thus, the general appearance of cartilage formed in transplants was similar to that of nasal septal cartilage *in situ*.

Chondrocytes isolated from the articular-epiphyseal complex of guinea pigs [24], mice [25,26] and rats [27] and transplanted intramuscularly into closely related or syngeneic animals form cartilage nodules in which hypertrophy of chondrocytes, matrix calcification and endochondral ossification take place converting transplants into ossicles (Fig. 1a). Similar results have been obtained with lapine growth plate chondrocytes transplanted subcutaneously into nude mice [28]. Nevertheless, distribution of cells in transplants is irregular, hypertrophied chondrocytes appear at random in various areas of the nodules, and ossicles contain only small islands of cartilage without any structure which could correspond to the growth plate normally individualizing from the articular-epiphyseal complex *in situ* [29]. Attempts at reconstruction of physeal regions with cultured chondrocytes in an inbred strain of mice, were also unsuccessful since the transplanted chondrocytes did not produce growth [30]. In contrast to articular-epiphyseal cartilage chondrocyte grafts, transplants of isolated rabbit or dog articular chondrocytes into nude mice resulted in the formation of hyaline cartilaginous nodules, without matrix calcification and ossification [31]. Moreover, cartilage produced by isolated murine rib chondrocytes in intramuscular transplants contained hypertrophied chondrocytes and calcified matrix in the central part, similarly as in rib cartilage *in situ* [32]. Thus, the processess occurring in reformed cartilage are basically similar to those in cartilages from which the transplanted chondrocytes originated.

The site of chondrocyte implantation may affect development of grafts. In intrahepatic transplants of cells from the articular-epiphyseal complex, hypertrophy of chondrocytes occurred faster than in control intramuscular transplants [33]. Epiphyseal or rib chondrocytes in intrarenal transplants reformed cartilage, and afterwards most probably transdifferentiated into bone cells without involvement of endochondral ossification [32]. Still unexplained remains the observation that cartilage formed by human chondrocytes from the articular-epiphyseal complex, and transplanted into a hamster cheek pouch (an immunologically privileged site [34]), differs from cartilage produced by similar chondrocytes from hamster, rat or rabbit [35,36]. In the former, chondrocytes did not hypertrophy and the matrix remained non-calcified, while the latter contained numerous hypertrophic cells and strongly calcified matrix. Thus, either development of human cartilage was inhibited by some immunological factor or human chondrocytes do not respond to some hamster hormones.

Intramuscularly transplanted chondrocytes from rabbit auricular (elastic) cartilage produced in intramuscular transplants nodules of cartilage with morphological appearance strongly dependent on the age of the cartilage donor [15,37]. Chondrocytes in auricular cartilage during the first two weeks of postnatal life are rather small but during further development the centrally located cells considerably enlarge. Cartilage formed by chondrocytes from animals up to two

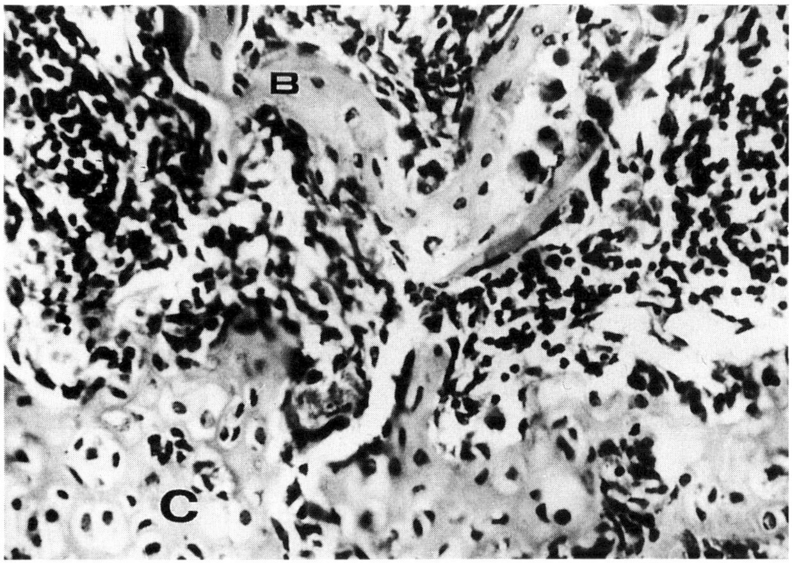

Fig. 1a.

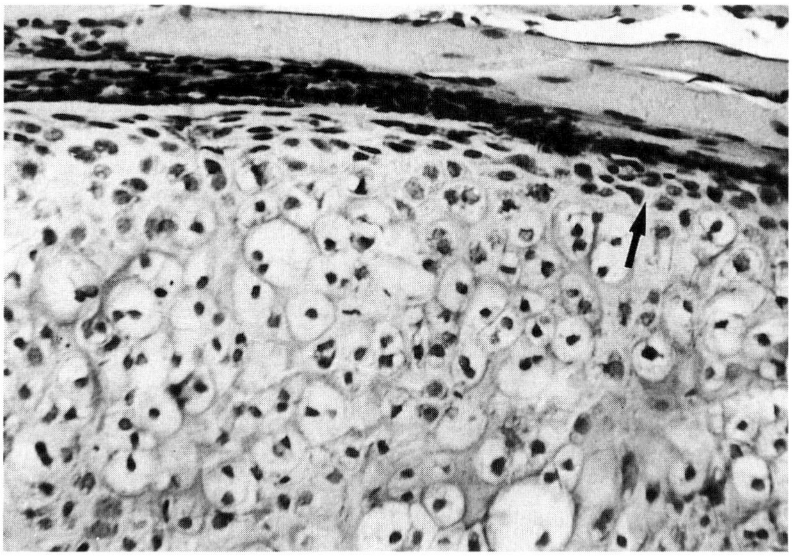

Fig. 1b.

Figs. 1(a)–(b) Cartilage formed by syngeneic (a) and allogeneic (b) chondrocytes isolated from an articular-epiphyseal complex of young rats four weeks after intramuscular transplantation. a – Cartilage (C) is partially resorbed and substituted by bone (B). b – Chondrocytes are irregularly arranged and no signs of ossification are present. Infiltrating cells accumulate at the border between cartilage and muscle. Some cells (arrow) penetrate into the matrix. (Hematoxylin-eosin, original magnification × 160).

weeks of age produced cartilage with regular arrangement of cells and with a set of elastic fibers similar to that in control cartilage of appropriate age (Fig. 2). Chondrocytes from older animals formed cartilage in which small and large cells were randomly distributed. Moreover, elastic fibers were reduced in number and irregularly arranged [15]. Nodules formed by chondrocytes from mature animals contained only a small amount of elatin [37,38]. Therefore, it seems that only the chondrocytes from an elastic cartilage of young animals have the ability to reform original tissue.

Fate of cartilage formed by allogeneic chondrocytes

Transplanted fragments of allogeneic cartilage usually remain unrejected owing to sequestration of chondrocytes from the immune system by the matrix [39–41]. Nevertheless, chondrocytes are endowed with both the Class I and Class II major histocompatibility complex (MHC) antigens, as well as tissue-specific antigens [41–45]. They are also able to express some weak histocompatibility antigens [25]. Since isolated allogeneic chondrocytes lack matrix protection immediately after transplantation, they immunize the host and the reformed cartilage is slowly resorbed by infiltrating cells [14,23,46]. Infiltrations already accumulate around one- two-week-old transplants [26,46]. The border of cartilage in such transplants is uneven and the infiltrating cells are in close contact with the matrix. They penetrate into the matrix and are frequently present in lacunae devoid of chondrocytes or containing remnants thereof. Numerous chondrocytes at the periphery of cartilage are dead judging by the lack if nuclear stainability. In comparison with syngeneic transplants the peripheral region of cartilage is strongly eosinophilic due to the loss of proteoglycans [23,25,26]. In cartilage produced by transplanted allogeneic chondrocytes from the articular-epiphyseal complex, hypertrophy of chondrocytes and matrix calcification take place but bone induction is not observed [25,26] (Fig. 1b). Bone formation, however, may occur in older transplants after destruction of non-calcified parts of cartilage and subsidence of local immunological response [25]. Lymphocytes predominate among the infiltrating cells, although macrophages, granulocytes and occasionally giant cells of foreign body type are also present [23,25,26]. Analysis of infiltrates with monoclonal antibodies demonstrated the presence of numerous Class II MHC- and CD4 positive cells, as well as CD8-positive lymphocytes (Romaniuk *et al.*, submitted).

Since transplanted chondrocytes produce intercellular substance rather rapidly, it seemed feasible to prevent rejection of reformed cartilage by short-term immunosuppression during the critical period of their exposure to the immune system of the host. However, cortisone acetate, cyclophosphamide, procarbazine (PCH) and antithymocyte serum (ATS) administered alone did not prevent rejection of cartilage formed by transplanted allogeneic chondrocytes from the articular-epiphyseal complex in mice [26]. Endochondral ossification was also absent. Nevertheless, immunosuppression consisting of combined treatment with ATS and PCH strongly inhibited immune response, completely prevented infiltrate formation and allowed endochondral ossification similar to that in the syngeneic control [26]. A similar study with isolated rib chondrocytes, demonstrated that a combination of ATS and PCH was superior to cyclosporine which did not significantly improve survival of cartilage formed by allogeneic chondrocytes [47].

Potential clinical applications of chondrocyte transplants

Chondrocyte transplants could be used either to modify the shape of non-cartilaginous tissues or to repair damaged cartilage. These applications, however, are still at the experimental stage. Intrascleral introduction of chondrocytes in rabbits was tried as a potential treatment for the detachment of the retina [48]. Auto- or allogeneic chondrocytes were injected into vocal cords in dogs to test whether such transplants could be used to fill the unilaterally paralysed vocal cord [49]. Cartilage formed by autogeneic chondrocytes survived and seemed to fulfil its function but the allogeneic one was slowly resorbed [49]. Furthermore, autogeneic auticular chondrocytes expanded in culture and suspended in alginate, a biodegradable polymer, were successfully used for treatment of vesicouretheral reflux created in mini-pigs [50,51].

The main efforts in clinically-oriented experimental chondrocyte transplantation were, however, directed towards the tempting goal of resurfacing destroyed articular surface. An extensive review concerning this particular application of chondrocyte transplants has already been published [52]. It remains to be seen, however, whether chondrocyte transplants will ever be able to complete with artificial joint prostheses.

The majority of studies of this type was done either in rabbits [53–60] or chickens [61–66]. Only a few studies were undertaken in guinea pigs [67], inbred rats [68,69] or horses [70]. There is, however, also a report describing successful treatment of traumatic lesions of knee articular cartilage by autologous chondrocyte transplants in man [71].

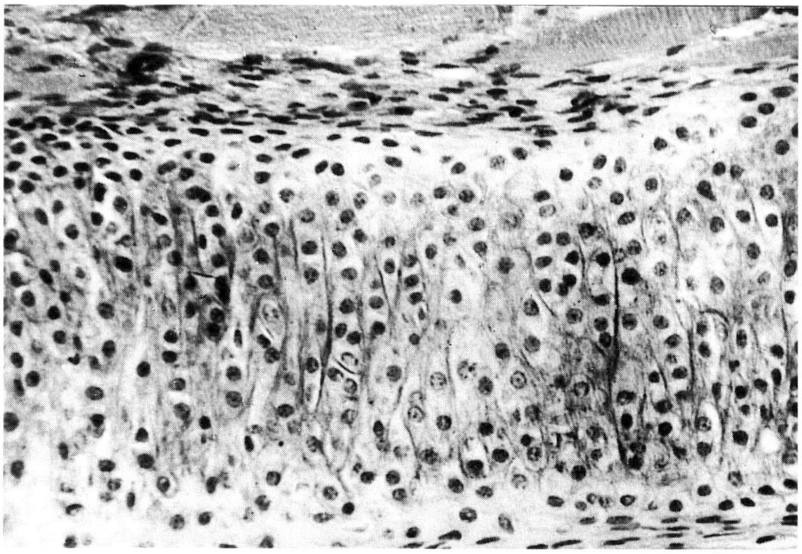

Fig. 2. Elastic cartilage produced by chondrocytes isolated from auricular cartilage of one-day-old rabbit two weeks after intramuscular transplantation. Regular arrangement of cells and thin elastic fibers are seen. (Orcein-hematoxylin, original magnification × 160).

In the usual experimental design a defect is created in the articular cartilage and the underlying bone. Control defects without chondrocyte transplants either did not show regenerating cartilage [54] or contained only small islands of chondrocytes surrounded by matrix and probably originating from the subchondral bone marrow [55,56,59]. In the older works chondrocytes were pipetted into a defect [53–57]. In more recent papers chondrocytes were introduced into a defect in collagen gel [59], fibrin [60,70], high-molecular weight hyaluronic acid [63, 64, 66, 68] or in alginate [50]. The concentration of chondrocytes in the medium is usually within the range of $0.5–4\times10^7$ cells/ml [51,63,68]. The depth of the drill hole might influence results since a deep hole may cause a bleeding from the bone marrow tending to wash out the transplanted cells and a shallow one may lead to washing out of the cells by the synovial fluid [56].

Allogeneic articular chondrocytes from young rabbits, pipetted into articular cartilage defects yielded irregular areas of newly formed cartilage surrounded by fibrous tissue [53], and some inflammatory cells [54,56,57]. Transplants of rabbit autogeneic articular chondrocytes, propagated in culture for three weeks, resulted in partial or complete healing of articular cartilage defects [58]. Frozen allogeneic articular chondrocytes suspended in collagen gel also repaired joint cartilage defects forming a smooth articular surface [59]. These observations were extended to 24 weeks and microscopic examination revealed that the reformed tissue was similar to the surrounding host cartilage. There was neither histological evidence of immunological rejection nor blastic transformation of recipients blood mononuclear cells by the transplanted chondrocytes [59]. Chondrocytes from the growth plate of young rabbits produced cartilage which completely filled the defects, without an evident inflammatory reaction, but the percentage of successful grafts was rather low [54,56]. Interestingly, endochondral ossification occurred in the base of the grafts up to the level of the original subchondral bone [54]. In another study, isolated growth plate chondrocytes from young rabbits were allowed to form a cohesive disk in ten days lasting culture subsequently grafted into the articular cartilage defect [60]. The disks displayed well formed matrix, but infiltrating cells accumulated at their margins. After twelve weeks the grafts were usually replaced by fibrous tissue. Recipients of the transplants developed cytotoxic lymphocytes against donor chrondrocytes according to the chromium.release cytotoxicity test [60].

In another animal model, full-thickness defects in articular cartilage of roosters were repaired with alogeneic embryonic chick chondrocytes from the articular-epiphyseal cartilage complex [61]. Chondrocytes were cultures first in agar and afterwards in a spinner bottle. After fourteen days of culture the cells were embedded in a fibrin clot and transplanted into a defect. Proliferation of chondrocytes and intense matrix deposition were observed in early transplants. Within eight weeks the defects were completely filled with hyaline cartilage without formation of fibrous tissue at the interface. Six months after transplantation two zones could be distinquished in the implant i.e. the superficial articular cartilage zone and the deep zone with hypertrophic chondrocytes and bone formation [61]. This was similar to the development of analogous transplants in rabbits [54]. Embryonic chondrocytes implanted into the articular surface defects in 3-year-old recipients in high-molecular weight hyaluronic acid stimulated endochondral ossification much faster than in 4-month-old animals and caused an increase in density of metaphyseal bone [62–64].

Embryonal chick chondrocytes were also used to heal the defects in articular cartilage of adult guinea pigs [67]. Cartilage with cells similar to typical avian chondrocytes appeared in transplants and more than half of the defects was partially or completely filled after an observation period lasting several months. Interestingly, these xenogeneic transplants did not evoke accumulation of inflammatory cells [67].

Extensive cartilage defects (12 mm) were resurfaced in horses with fibrin-embedded allogeneic chondrocytes from a foal [70]. Four or eight months after grafting cartilage was present in the middle and deep zone of the grafted defects. No signs of immunological reaction were observed. In control defects which were left empty after surgery only fibrous tissue was present [70].

Repair of traumatic full-thickness defects of the knee articular cartilage was undertaken in man as a preventive measure against osteoarthritis. As a source of chondrocytes served cartilage obtained through an arthroscope from a minor load-bearing area on the femoral condyle of the damaged knee. Isolated chondrocytes were cultures for two-three weeks and introduced into the cartilage defect covered with a periosteal flap. In the majority of patients (23 in total) the results were good-to-excellent [71].

Observations on the difference in behavior of transplanted chondrocytes from the articular-epiphyseal cartilage complex above and under the tidemark emerging from studies with rabbit [54] and avian [62–64] allogeneic chondrocytes were further confirmed and extended to syngeneic transplants in inbred rats [68]. Above the tidemark, tissue filling the defect retained a structure a similar to the surrounding articular cartilage, while under the tidemark, hypertrophy of chondrocytes and matrix calcification, followed by invasion of cartilage by blood vessels and ossification took place [68] (Fig. 3). In contrast, in similar intramuscular transplants hypertrophy of chondrocytes occurred at random in various areas of reformed cartilage and nearly the whole cartilage was converted into ossicle [68]. The lack of hypertrophy above the tidemark could depend on the influence of the local environment including nourishment by the synovial fluid [61]. Alternatively, since cell suspension from the articular-epiphyseal complex contained both articular and growth plate chondrocytes their segregation into two zones soon after transplantation cannot be exluded [68]. These findings emerging from transplantation studies may be important for understanding the relationship between normal articular cartilage and joint environment.

Another point deserving comment is the considerable difference in the immunological response towards chondrocytes grafted into articular surface defects in various experimental systems. The differences in the case of transplants made in rabbits could depend on the more or less close genetic relationship between donors and recipients used in various laboratories [53–60]. Similarly, lack of an evident immunological response towards transplanted chondrocytes in chickens [61–63] could be caused by the close relationship between donors and recipients. But how to explain lack of response of guinea pigs to chicken chondrocytes [67] or horses [70] to allogeneic chondrocytes? Moreover, in one work, both syngeneic and allogeneic articular chondrocyte transplants into joint cartilage defects in rats had a similar success rate after long periods of observation [69], while in another study cartilage formed by similarly transplanted allogeneic chondrocytes was slowly resorbed by infiltrating cells (Fig. 4a, b) (Hyc *et al.*, in preparation). Thus, immunological reactions evoked by allogeneic chondrocytes transplanted into articular cartilage defects and factors determining their intensity deserve further study.

As evident from the above summary, both articular and growth plate chondrocytes produce cartilage when transplanted into joint surface defects. However, the serious drawback of chondrocytes as material for clinical use is the lack of an easily available source of cells for autogeneic and even for allogeneic transplants. Cartilage for chondrocyte isolation may be obtained only in special cases, e.g. from the uninvolved area of cartilage during arthroscopy of the injured knee [71] or from nasal septum cartilage removed during surgery

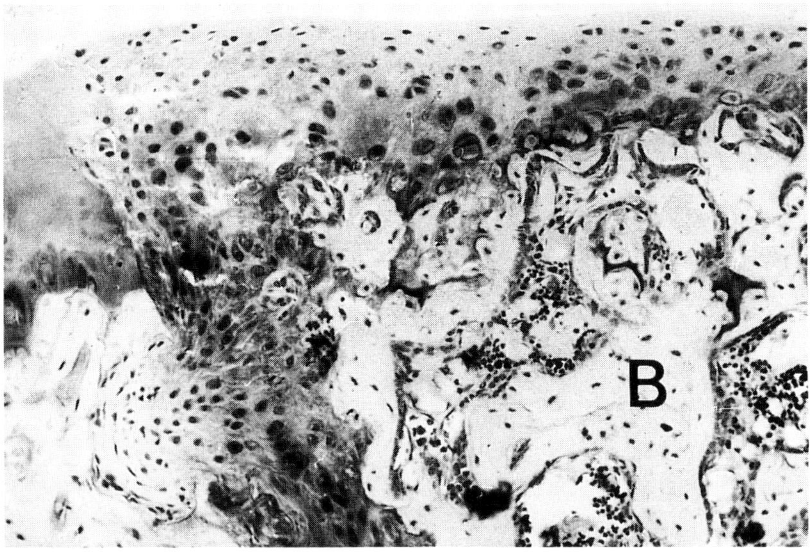

Fig. 3. Eight-week-old syngeneic transplant of chondrocytes from an articular-epiphyseal complex of young rats into a defect in articular cartilage. Cartilage contains small chondrocytes above the tidemark and is ossified in the deeper part. At left a necrotic fragment of an original articular cartilage damaged during preparation of the defect is seen. B – bone. (Toluidine blue, original magnification × 160).

[72]. Chondrocytes for allogeneic transplants could be obtained from costal cartilage of young children [19] or from fetal material but in many countries the latter is prevented by legal restrictions. Moreover, rejection of allogeneic chondrocyte transplants must be taken into account. A good solution would be to propagate articular chondrocytes in cell culture similarly as it is done with keratinocytes from severely burned patients [73]. Unfortunately, the characteristic phenotype of chondrocytes expressing collagen type II and cartilage specific proteoglycans is gradually lost during serial monolayer culture and replaced by a complex collagen phenotype consisting primarily of type I collagen and a low level of proteoglycan synthesis [74–78]. Dedifferentiated chondrocytes may, however, regain their original phenotype in three-dimensional culture systems either in suspension or in gel (agarose) culture [79,80,81].

There are also several reports describing formation of cartilage implants by chondrocytes cultured on synthetic biodegradable polymers [82–86]. Chondrocytes seeded into a porous polyglycol acid scaffold produced 3-D cartilage both *in vitro* and *in vivo* [82]. Similarly chondrocytes laden into porous collagen disks proliferated, and produced matrix within the disks [83]. Furthermore, to produce cartilage implants of predetermined shape, synthetic biopholymers of various configuration were seeded with bovine articular chondrocytes and implanted into nude mice. Cartilage formed in implants had approximately the same shape and dimensions as the original construct. Thus, using this method a cartilage with desired shape could be produced for use in plastic and reconstructive surgery [84–86].

Several attempts have also been undertaken to heal joint surface defects by transplantation of periosteum or perichondrium fragments [87,88] or presumable chondrocyte precursors derived from periosteum or bone marrow [89]. The cells present in all types of these transplants usually differentiated into chondrocytes and produced cartilage matrix. There was no apparent difference between the cartilage produced by precursor cells obtained from the bone marrow and from the periosteum [89]. Another approach was to stimulate chondrogenesis in mesenchymal stem cells obtained from muscle biopsies or bone marrow and cultured in the presence of bone morphogeneic protein [90–93]. Autologous transplants of these cells in hyaluronic acid or scaffold made of poly-lactic acid or carbon fibers and implanted into articular cartilage defects in rabbits or goats usually produced cartilage. The principal technical obstacle was the lack of adhesion of the implants to the surrounding bone [90–93].

The use of autogeneic precursors would eliminate the need for much less accessible differentiated chon-

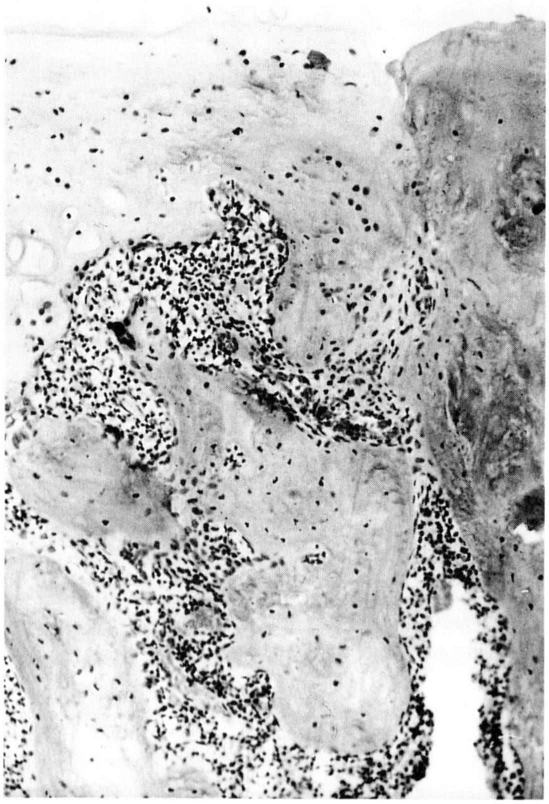

Fig. 4a

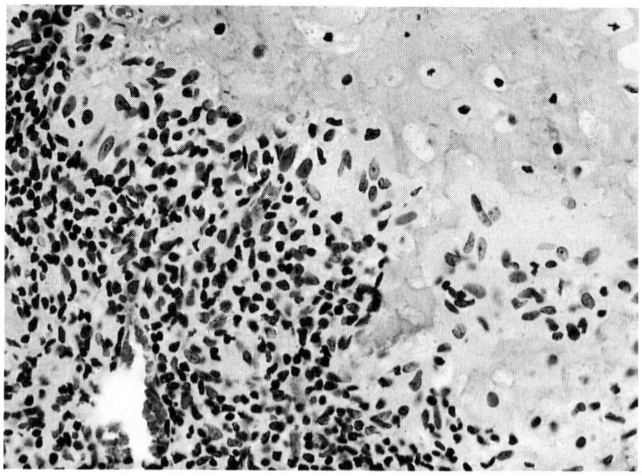

Fig. 4b

Figs. 4(a)–(b). Eight-week-old allogeneic transplant of chondrocytes from an articular-epiphyseal complex of young rats into a defect in articular cartilage. a – Cartilage is invaded and partially destroyed by infiltrating cells. At right host articular cartilage and bone partially damaged during preparation of the defect are present. b – Infiltrating cells within the cartilage matrix (Hematoxylin-eosin, original magnification: a × 160, b × 400).

drocytes and allow to avoid immunological complications inherent to allogeneic transplants. Nevertheless, it is uncertain, whether the quality of matrix produced by chondrocyte precursors differentiating within the articular cartilage defect is sufficient to fulfil the demands of long-term joint function. Twenty-four weeks after transplantation of chondrocyte precursors into such a defect, the tissue produced by them was less stiff and more compliant than the intact cartilage [89]. It must be particularly borne in mind that collagen fibers in articular cartilage have a complex arrangement adjusted to weight-bearing [94–97]. Chondrocytes or their precursors transplanted into the joint surface defect may be able to produce matrix, but fail to reconstruct the original arrangement of collagen fibers. If so, reformed cartilage may be prone to degeneration, and thus be of limited benefit to the patient. Accordingly, the word of caution expressed by Mankin [98] in his comment on clinical use of chondrocyte grafts may be well founded.

Acknowledgement

Supported by the State Committee for Scientific Research (KBN) grant No 4 0325 91 01.

References

1. Editorial. Cartilage transplants brings hope to arthritis. New Scientist 1988; 117: 38.
2. Malejczyk J, Kaminski MJ, Malejczyk M, *et al.* Natural cell-mediated cytotoxic activity against isolated chondrocytes in the mouse. Clin Exp Immunol 1985; 59: 110–16.
3. Moscona AA. Cell suspensions from organ rudiments of chick embryos. Exp Cell Res 1952; 535–39.

4. Kawiak J, Moskalewski S, Darzynkiewicz Z. Isolation of chondrocytes from calf cartilage. Exp Cell Res 1965; 39: 59–68.
5. Steinberg MS. "EMC": its nature, origin and function in cell aggregation. Exp Cell Res 1963; 30: 257–79.
6. Hefley T, Cushing J, Brand JS. Enzymatic isolation of cells from bone; Cytotoxic enzymes of bacterial collagenase. Am J Physiol 1981; 240: C234–38.
7. Glade MJ, Kanware YS, Hefley TJ. Enzymatic isolation of chondrocytes from immature rabbit articular cartilage and maintenance of phenotypic expression in culture. J. Bone Miner Res 1991; 6: 217–26.
8. Dearden LC, Bonucci E, Cuicchio M. An investigation of ageing in human costal cartilage. Cell Tissue Res 1974; 152: 305–37.
9. Madsen K, Moskalewski S, Mark von der K, *et al.* Synthesis of proteoglycans, collagen, and elastin by cultures of rabbit auricular chondrocytes-relation to age of the donor. Dev Biol 1983; 96: 63–73.
10. Quintarelli G, Dellovo MC. Age changes in the localization and distribution of glycosaminoglycans in human hyaline cartilage 1966; 7: 141–67.
11. Moskalewski S, Bator S. Regenerating rib cartilage tentatively used as a source of chondrocytes for transplantation. Arch Immunol Ther Exp 1985; 33: 685–92.
12. Stockwell RA. The cell density of human articular and costal cartilage. J Anat 1967; 101: 753–63.
13. Collins DH, Ghadially FN, Meachim G. Intra-cellular lipids of cartilage. Ann Rheum Dis 1965; 24: 123–35.
14. Moskalewski S, Kawiak J, Cartilage formation after transplantation of isolated chondrocytes. Transplantation 1965; 3: 737–47.
15. Moskalewski S, Rybicka E. The influence of the degree of maturation of donor tissue on the reconstruction of elastic cartilage by isolated chondrocytes. Acta Anat 1977; 97: 231–40.
16. Moskalewski S. Removal of large lipid droplets from cytochalasin B treated chondrocytes by high speed centrifugation. Cell Biol Int Rep 1985; 9: 613–618.
17. Bryan J. Studies on clonal cartilage strains I. Effect of contaminant non-cartilage cells. Exp Cell Res 1968; 52: 319–26.
18. Van Sickle DC, Kincaid SA. The articular-epiphyseal cartilage complex. Proc Am Assoc Vet Anat 1983; 120: 14.
19. Klagsbrun M. Large-scale preparation of chondrocytes. Methods in Enzymol 1979; 58: 560–64.
20. Manning WK, Bonner WM Jr. Isolation and culture of chondrocytes from human adult articular cartilage. Arthritis Rheum 1967; 10: 235–9.
21. Smith UA. Survival of frozen chondrocytes isolated from cartilage of adult mammals. Nature 1965; 205: 782–84.
22. Moskalewski S. Transplantation of isolated chondrocytes. Clin Orthop Rel Res 1991; 272: 16–20.
23. Moskalewski S, Kawiak J, Rymaszewska T. Local cellular response evoked by cartilage formed after auto- and allogeneic transplantation of isolated chondrocytes. Transplantation 1966; 4: 572–81.
24. Thyberg J, Moskalewski S. Bone formation in cartilage produced by transplanted epiphyseal chondrocytes. Cell Tissue Res 1979; 204: 77–94.
25. Ksiazek T, Moskalewski S. Studies on bone formation by cartilage reconstructed by isolated epiphyseal chondrocytes, transplanted syngeneically or across known histocompatibility barriers in mice. Clin Orthop Rel Res 1983; 172: 233–42.
26. Malejczyk J, Moskalewski S. Effect of immunosuppression on survival and growth of cartilage produced by transplanted allogeneic epiphyseal chondrocytes. Chin Orthop Rel Res 1988; 232: 292–303.
27. Moskalewski S, Hyc A, Grzela T *et al.* Differences in cartilage formed intramuscularly or in joint surface defects by syngeneic rat chondrocytes isolated from the articular-epiphyseal cartilage complex. Cell Transpl 1993; 2: 467–73.
28. Wright GC Jr, Miller F, Sokoloff L. Induction of bone by xenografts of rabbit growth plate chondrocytes in the nude mouse. Calc Tissue Int 1985; 37: 250–256.
29. Hunziker EB. Biology of the growth plate. Pathol Immunopathol Res 1988; 81: 1–12.
30. Barr SJ, Zaleske DJ, Mankin HJ. Physeal replacement with cultured chondrocytes of varying developmental time: failure to reconstruct a functional or structural physis. J. Orthop Res 1993; 11: 10–19.
31. Lipman JM, McDevitt CA, Sokoloff L. Xenografts of articular chondrocytes in the nude mouse. Calcif Tissue Int 1983; 35: 767–772.
32. Moskalewski S, Malejczyk J, Bone formation following intrarenal transplantation of isolated murine chondrocytes: chrondrocyte-bone cell transdifferentiation? Development 1989; 107: 473–80.
33. Moskalewski S, Hyc A, Kupinska U *et al.* Influence of liver environment on the maturation of isolated epiphyseal chondrocyte transplants. Folia Histochem Cytobiol 1993; 31: 15–22.
34. Billingham RE. Concerning the laboratory career of *Mesocricetus auratus* with special reference to transplantation. Fed Proc 1978; 37: 2024–27.
35. Moskalewski S, Kaminski MJ. Cartilage formation by isolated human chondrocytes transplanted into the hamster cheek pouch. Ann Immunol 1970; 2: 21–6.
36. Kaminski MJ, Kaminska G, Moskalewski S. Species differences in the ability of isolated epiphyseal chondrocytes to hypertrophy after transplantation into the wall of the Syrian hamster cheek pouch. Folia Biol (Krakow), 1980; 28: 27–36.
37. Kawiak J, Moskalewski S, Hinek A. Reconstruction of the elastic cartilage by isolated chondrocytes in autogeneic transplants. Acta Anat (Basel) 1970; 76: 530–44.
38. Thyberg J, Hinek A: Fine structure of rabbit ear chondrocytes *in vitro* and after transplantation. Cell Tissue Res 1977; 180: 341–356.
39. Bacsich P, Wyburn GM. The significance of the mucoprotein content on the survival of homografts of cartilage and cornea. Proc R Soc Edin Sect B 1947; 62: 321–327.
40. Heyner S. The antigenicity of cartilage grafts. Surg Gynec Obst 1973; 1336: 298–305.
41. Malseed ZM, Heyner S. Antigenic profile of the rat chondrocytes. Arthritis Rheum 1976; 19: 223–31.
42. Tiku ML, Liu S, Weaver CW *et al.* Class II histocompatibility antigen-mediated immunologic function of normal articular chondrocytes. J. Immunol; 1985; 135: 2923–28.
43. Malejczyk J, Romaniuk A. Reactivity of normal rat epiphyseal chondrocytes with monoclonal antibodies recognizing different leucocyte markers. Clin Exp Immunol 1989; 75: 477–80.
44. Lance EM, Kimura LH, Manibog CN. The expression of major histocompatibility antigens on human articular chondrocytes. Clin Orthop Rel Res 1993; 291: 266–82.
45. Bang H, Mollenhauer J, Schulmeister A *et al.* Isolation and characterization of a cartilage-specific membrane antigen (CH65): comparison with cytokeratins and heat-shock proteins. Immunology 1994; 81: 322–29.
46. Heyner S, The significance of the intercellular matrix in the survival of cartilage allografts. Transplantation 1969; 8: 666–677.

47. Malejczyk J, Osiecka A, Hyc A, *et al.* Effect of immunosuppression on rejection of cartilage formed by transplanted allogeneic rib chondrocytes in mice. Clin Orthop Rel Res 1991; 269: 266–73.
48. Ptasinska-Urbanska M, Mokalewski S, Kaminski M *et al.* Intrascleral introduction of isolated allogeneic chondrocytes capable of cartilage reformation in rabbits. Possible procedure in treatment of detachment of retina. Exp Eye Res 1977; 24: 214–247.
49. Bator S. Medial shifting of the canine vocal cord by injection of isolated chondrocytes. Arch Otorhinolaryngol 1985; 242: 19–25.
50. Atala A, Coma LG, Kim W *et al.* Injectable alginate seeded with chondrocytes as a potential treatment for vesicoureteral reflux. J Urol 1993; 150: 745–747.
51. Atala A, Kim W, Paige KT *et al.* Endoscopic treatment of vesicoureteral reflux with a chondrocyte-alginate suspension. J. Urol 1994; 152: 641–3.
52. Nevo Z, Robinson D, Halperin N. The use of grafts composed of cultured cells for repair and regeneration of cartilage and bone. In: Hall B.K. ed. Bone. Boca Raton Ann Arbor London: CRC Press, 1992; 5: 123–52.
53. Chesterman PJ, Smith AU. Homotransplantation of articular cartilage and isolated chondrocytes. J. Bone Joint Sutg 1968; 50B: 184–197.
54. Bentley G, Greer RB III. Homotransplantation of isolated epiphyseal and articular cartilage chondrocytes into joint surfaces of rabbits. Nature 1971; 230: 385–88.
55. Green WT Jr. Articular cartilage repair: Behaviour of rabbit chondrocytes during tissue culture and subsequent allografting. Clin Orthop Rel Res 1977; 124: 237–250.
56. Bentley G, Smith AU, Mulkerjhee R. Isolated epiphyseal chondrocyte allografts into joint surface – An experimental study in rabbits. Ann Rheum Dis 1978; 37: 449–58.
57. Aston JE, Bently G. Repair of articular surfaces by allografts of articular and growth plate cartilage. J Bone Joint Surg 1986: 68B: 29–35.
58. Grande DA, Singh IJ, Pugh J. Healing of experimentally produced lesions in articular cartilage following chondrocyte transplantation. Anat Rec 1987; 218: 142–148.
59. Wakitani S, Kimura T, Hirooka A *et al.* Repair of rabbit articular surface with allograft chondrocytes embedded in collagen gel. J Bone Joint Surg 1989; 71B: 74–80.
60. Kawabe N, Yoshinao M. The repair of full-thickness articular cartilage defects. Clin Orthop Rel Res 1991; 268: 279–293.
61. Itay S, Abramovici A, Nevo Z. Use of cultured embryonal chick epiphyseal chondrocytes as grafts for defects in chick articular cartilage. Clin orthop Rel Res 1986; 220: 284–303.
62. Robinson D, Halperin N, Nevo Z. Fate of allogeneic embryonal chick chondrocytes implanted orthotopically as determined by the host's age. Mech Ageing Dev 1989; 50: 71–80.
63. Robinson D, Halperin N, Nevo Z Regenerating hyaline cartilage in articular defects of old chickens using implants of embryonal chick chondrocytes embedded in a natural delivery substance. Calcif Tissue Int 1990; 46: 246–53.
64. Robinson D, Halperin N, Nevo Z. Bone density in old chicken's metaphyses, as affected by local trauma and chondrocyte implantation. Bull Hospit Joint Dis 1993; 53: 83–87.
65. Robinson D, Halperin N, Nevo Z. Use of cultured chondrocytes as implants for repairing cartilage defects. In: Maroudas A, Kuettner K, eds. Methods in Cartilage Research. London San Diego New York: Academic Press, 1990; 327–35.
66. Nevo Z, Robinson D, Halperin N *et al.* Culturing chondrocytes for implantation. In: Maroudas A, Kuettner K, eds. Methods in Cartilage Research. London San Diego New York: Academic Press, 1990; 98–104.
67. Robinson D, Halperin N, Novo Z. Long-term follow-up of the fate of xenogeneic transplants of chondrocytes implanted into joint surfaces. Transplantation 1991; 52: 380–83.
68. Moskalewski S, Hyc A, Grzela T *et al.* Differences in cartilage formed intramuscularly or in joint surface defects by syngeneic rat chondrocytes isolated from the articular-epiphyseal cartilage complex. Cell Transpl. 1993; 2: 467–73.
69. Noguchi T, Oka MM, Fujino M *et al.* Repair of osteochondral defects with grafts of cultured chondrocytes. Clin Orthop Rel Res 1994; 302: 251–58.
70. Hendrickson DA, Nixon AJ, Grande DA *et al.* Chondrocyte-fibrin matrix transplants for resurfacing extensive articular cartilage defects. J Orthop Res 1994; 12: 485–97.
71. Brittberg M, Lindahl A, Nilsson A *et al.* Treatment of deep cartilage defects in the knee with autologous chondrocyte transplantation. N Engl J Med 1994; 331: 889–95.
72. Bujia J, Pitzke P, Wilmes E *et al.* Culture and cryopreservation of chondrocytes from human cartilage: relevance for cartilage allografting in orolaryngology. J Otorhinolaryngol Rel Spec 1992; 54: 80–84.
73. Faure M, Mauduit G, Schmitt D *et al.* Growth and differentiation of human epidermal cultures used as auto- and allografts in humans. Br J Dermatol 1987; 116: 161–70.
74. Benya PD, Padilla SR, Nimni ME. Independent regulation of collagen types by chondrocytes during the loss of differentiated function in culture. Cell; 1978: 1313–21.
75. Okayama M, Pacifici M, Holtzer H. Differences among sulfated proteoglycans synthesized in nonchondrogenic cells, presumptive chondroblasts, and chondroblasts. Proc Natl Acad Sci USA 1976; 73: 3224–28.
76. Schiltz JR, Mayne R, Holtzer H. The synthesis of collagen and glycosaminoglycans by dedifferentiated chondroblasts in culture. Cell Differ 1973; 1: 97–108.
77. Duchéne M, Sobel ME, Müller PK. Levels of collagen mRNA in dedifferentiating chondrocytes. Exp Cell Res 1982; 142: 317–24.
78. Oegema TR Jr, Thompson RC Jr. Characterization of a hyaluronic acid-dermatan sulfate proteoglycan complex from dedifferentiated chondrocyte cultures. J Biol Chem 1981; 256: 1015–22.
79. Benya PD, Shaffer JD. Dedifferentiated chondrocytes reexpress the differentiated collagen phenotype when cultured in agarose gels. Cell; 1982: 215–24.
80. Aulthouse AL, Beck M, Griffey E *et al.* Expression of the human chondrocyte phenotype *in vitro* Cell Dev Biol 1989; 25: 659–668.
81. Solursh M. Formation of cartilage *in vitro*. J Cell Biochem 1991; 45: 258–60.
82. Freed LE, Marquis JC, Emmanual J *et al.* Neocartilage formation *in vitro* and *in vivo* using cells cultured on synthetic biodegradable polymers. J Biomed Mater Res 1993; 27: 11–23.
83. Nixon AJ, Sams AE, Lust G *et al.* Temporal matrix synthesis and histological features of a chondrocyte-laden porous collagen cartilage analoque. Am J Vet Res 1993; 54: 349–56.
84. Vacanti CA, Langer R, Schloo R *et al.* Synthetic polymers seeded with chondrocytes provide a template for new cartilage formation. Plast Reconstr Surg 1991; 88: 753–759.
85. Kim WS, Vacanti JP, Cima L *et al.* Cartilage engineered in predetermined shapes emploing cell transplantation on biodegradable polymers. Plast Reconstr Surg 1994; 94: 233–37.

86. Puelacher WC, Kim SW, Vacanti JP *et al.* Tissue-engineered growth of cartilage: the effect of varying the concentration of chondrocytes seeded onto synthetic polymer matrices Int J Oral Maxillofac Surg 1994; 23: 49–53.
87. Ritsilä VA, Santavirta S, Alhopuro S *et al.* Periosteal and perichondral grafting in reconstructive surgery. Clin Orthop Rel Res 1994; 302: 259–65.
88. O'Driscoll SW, Keeley FW, Salter RB. Durability of regenerated articular cartilage produced by free autogenous periosteal grafts in major full-thickness defects in joint surfaces under the influence of continuous passive motion. J Bone Joint Surg 1988; 70A: 595–606.
89. Wakitani S, Goto T, Pineda SJ *et al.* Mesenchymal cell-based repair of large, full-thickness defects of articular cartilage. J Bone Joint Surg 1994; 76A: 579–592.
90. Robinson D, Efrat M, Mendes DG *et al. In vitro* construction of cells-containing implants for articular cartilage regeneration. In: van den Berg WB, van der Kraan PM, van Lent PLEM, eds. Agents Action Supplements. Joint Destruction in Arthritis and Osteoarthritis. Basel: Birkhaüser Verlag, 1993; 39: 231–35.
91. Robinson D, Halperin N, Efrat M *et al.* Implants composed of carbon fiber mesh and bone marrow derived, chondrocyte-enriched cultures for joint surface reconstruction. Bull Hosp Joint Dis 1993; 53: 75–82.
92. Nevo Z, Robinson D, Efrat M *et al.* Cartilaginous implants containing cultured cells. Orthop Int Ed 1993; 1: 441–46.
93. Robinson D, Efrat M, Halperin N *et al.* The mesenchymal stem cell in orthopaedics. Orthop Int Ed 1993; 1: 448–53.
94. Benninghoff A. Form und Bau der Gelenkknorpel in ihren Beziehungen zu Funktion. II Teil. Der Aufbau des Gelenkknorpels in seinen Beziehungen zu Funktion. Zsch Zelfforsch Mikroskop Anat 1925; 2: 783–862.
95. Weiss CH, Rosenberg L, Helfet AJ. An ultrastructural study of normal young adult human articular cartilage. J Bone Joint Surg 1968; 59A: 663–74.
96. Clark JM. The organisation of collagen fibrils in the superficial zones of articular cartilage. J Anat 1990; 171: 117–30.
97. Poole CA, Flint MH, Beaumont BW. Morphological and functional interrelationships of articular cartilage matrices. J Anat 1984; 138: 113–38.
98. Mankin HJ. Chondrocyte transplantation – one answer to an old question. N Engl J Med 994; 331: 940–41.

R. P. Lanza and W. L. Chick (eds.), Yearbook of Cell and Tissue Transplantation 1996/1997, 53–59.

Chapter 6

Myoblast transplantation

Terence Partridge
Muscle Cell Biology Group, MRC Clinical Sciences Centre, Royal Postgraduate Medical School, Hammersmith Hospital, Du Cane Road, London W12 0NN, U.K.

Skeletal muscle transplantation is practised in two quite different ways. Plastic surgeons commonly strive to transplant the mature differentiated tissue, attempting to minimize damage to the structure of the muscle fibres by restoring vascular and neural connections as quickly as possible [1]. The second method, and the one with which this article is concerned, involves the transplantation of precursors of skeletal muscle into sites in which they are able to follow their differentiative pathway to the formation of mature muscle tissue [2]. In this case, the need for re-establishment of vascularization and innervation is less urgent, for myogenic precursors are less sensitive to anoxia than mature muscle fibres. This was demonstrated by the observations of Studitsky [3] and Carlson [4] that when skeletal muscle is transplanted, intact or minced, without deliberate re-attachment of the vascular supply, then the mature muscle fibres undergo necrosis to be replaced, as the tissue eventually becomes spontaneously invaded by the vasculature, by new fibres formed from progeny of the surviving myogenic precursors resident in the muscle.

Therapeutic myoblast transplantation

Our interest in grafting muscle in the form of its precursors arose initially from the desire to devise a therapy for primary genetic diseases of skeletal muscle, notably Duchenne muscular dystrophy (DMD), by transplanting genetically normal myogenic cells into diseased muscles[5, 6, 7, 8, 9]. The underlying rationale for this approach was that the normal myogenic cells would participate in the repair, regeneration and remodelling of diseased, degenerating muscle fibres and thus gain entry for their normal nuclei into the recipient's muscle fibres. Once within the fibre, the normal nucleus, now a myonucleus, would express its normal genes, including that which is defective in the myonuclei of the recipient, and thus compensate for the biochemical lesion consequent upon the genetic defect. However, this practical goal was preceded by some years by the use of myoblast transplantation for study of the basic cell biology of muscle lineages [10, 11], and this remains an important objective of the use of this technique, for effective therapy for muscle disease will surely require a better understanding of the cell biology of this tissue than we presently enjoy.

Animal studies

To examine the feasibility of myoblast transplantation required the use of an animal model of a genetic myopathy on which the principle could be established. The first such model available, the dy/dy dystrophic mouse, was extensively used by Law and his colleagues [6, 7, 12, 13] for this purpose. Some difficulties in the interpretation of these studies arise from the fact that the genetic/biochemical defect was not known at the time and that simple surgical manipulation could produce functional improvements in dy/dy muscles which are difficult to distinguish from improvements induced by genetic mechanisms [2]. This animal has become more interesting with the recent demonstration that the primary defect lies in the gene for the laminin M-chain [14–16], a constituent of the basement membrane of the muscle fibre, and that this animal is a model of some of the 'congenital muscular dystrophies' in children. But it has not been shown whether this protein is synthesized by the muscle fibre or by interstitial cells, leaving open the question as to whether a graft of myoblasts or fibroblasts would be the more appropriate to correct the defect.

The alternative approach, in the face of the lack of a suitable animal model of a genetic primary myopathy at the time, was to demonstrate that grafts of myogenic cells could be used to introduce new genes into muscle fibres of the recipient [17–19]. When the mdx mouse became established as a homologue of DMD, it was a straightforward matter to use the same techniques to introduce the missing protein, namely dystrophin, into the muscle fibres of this animal [20, 21]. Subsequently, a series of studies has demonstrated that muscle fibres, when converted to being dystrophin-positive in this way, persist in the long term and are stable relative to unconverted dystrophin-negative fibres [22–24].

Human studies

Preliminary trials of myoblast transplantation in DMD patients were announced by 5 groups; 3 in the U.S.A. and 2 in Canada, in a meeting organized by the Muscular Dystrophy Association of America in June 1989 [25], soon – too soon in the view of many – after the initial demonstration of its feasibility in mdx mice. Other trials have been started since and are currently being reported.

Markers of success

Curiously few of the lessons generated by the mouse myoblast transplantation studies were taken into account in designing the equivalent experiments in man. It was, for example, evident that the existence of 'revertant' muscle fibres would cause a problem. These are occasional single fibres or small groups of fibres found sporadically in the muscles of mdx mouse [26] and in many DMD patients [27, 28] in which the genetic defect appears to have been accomodated in some way such that they are able to synthesize dystrophin. As such they present the difficulty that they are difficult to distinguish from small numbers of normal muscle fibres which might be expected to be generated from injected normal myoblasts. In the mouse, it was possible to use other markers as independent confirmation of the contribution of injected myoblasts to the recipient muscle, but no such markers were available in man and much of the data suggesting low-level success of myoblast transplantation in DMD patients is equivocal. In some instances the claims of success are appropriately muted by this uncertainty [29], others seem undaunted by it and opt for a positive interpretation [30].

Because revertant muscle fibres cannot contain those parts of the normal dystrophin gene which have been lost by deletion in the dystrophin genes of many the DMD patients, it has been possible to establish the survival of injected myogenic cells within recipient muscles by demonstrating the presence of mRNA derived from the deleted region; this can only have been transcribed from full-length genomic DNA of the normal donor. Of the two instances where this marker has been used, one failed to detect any evidence of surviving donor muscle [31], the second detected low levels of donor dystrophin mRNA [32]. In the second case, where this marker was positive, it showed no particular association with the small numbers of dystrophin-positive muscle fibres in the biopsy, so the causal link between the two phenomena is uncertain.

Probably the best experimental criterion of primary success for myoblast transplantation in man is induction of the synthesis of full length dystrophin in the recipient muscle. Better than detection of the presence of mRNA, within the treated muscle, containing transcript of exons missing from the patient's genome, would be the detection of dystrophin containing the parts of the molecule encoded by these exons. This has become possible because of the production of monoclonal antibodies specific for particular exons of dystrophin (A. Burghes and Glen Morris: personal communications) and current studies are being conducted with these tools.

Immune rejection

A second problem which had long been evident in myoblast transplantation experiments in the mouse was that of immune rejection: more than trace amounts of donor-derived muscle were produced only when the hosts were matched at the Major Histocompatibility Complex (MHC) locus and were tolerized to the host strain at birth [33–36]. Furthermore, the best results were always obtained when immunodeficient, nu/nu host animals were used [19, 20]. With the exception of grafts between identical twins, which are of no interest for genetic disease, such methods have no true counterparts in human transplantation. Matching of the Major Histocompatibility antigens can sometimes be achieved with sibling donors. But reactions against mismatched minor antigens are likely to be exacerbated by such matching at the major locus, for response to minor antigens is MHC-restricted (i.e. it requires the MHC of the antigen-bearing tissue to be matched with that of the responding immune system in order to trigger an immune response).

More recently, following the disappointing results with human myoblast transplantation, extensive studies have been conducted on the immunological constraints on myoblast transplantation in animals. These have shown that grafts mismatched at the MHC are usually rapidly rejected but that grafting between matched host and donor are successful [37–39]. However with some combinations of MHC mismatch, rejection may be chronic [35]. In mice, immunosuppression with cyclosporin A is adequate to suppress rejection of MHC-mismatched grafts but, on withdrawal of this immunosuppressant, rejection recommences and there is evidence of a 'bystander effect' involving damage to muscle of the recipient in addition to that derived from the mismatched myoblasts [40]. Such results should engender proper caution in the conduct of experiments in man. Experiments in mice have also been used in the search for more effective immunosuppressants than cyclosporin A, such as rapamycin [41], to permit the survival of mismatched grafts – FK506 having been identified as the most effective so far in supporting survival of donor muscle [42].

Much discussion has been devoted to the possibility that the very objective of myoblast transplantation, dystrophin production from the normal gene of donor myonuclei, might act as a minor antigen in Duchenne boys grafted with normal myogenic cells. The natural occurrence of 'revertant' muscle fibres in many DMD boys means that the majority of the dystrophin molecule is not a truly foreign antigen and may well have induced tolerance in these instances. However, the prospect remains of an immune response against the portion of dystrophin encoded by exons which have been lost in the patients. The second point of debate is, given the potential for an immune response against dystrophin or a portion of it, whether this immune response could be cytotoxic to dystrophin-positive muscle fibres. As an intracellular antigen, dystrophin would not be perceptible within healthy muscle fibres to circulating antibody against it, but dystrophin-derived peptides could be presented to cytotoxic T-lymphocytes in the clefts of Class I histocompatibility antigens and thus excite a cytolytic attack by these cells. Muscle fibres fall into the rare category of cells which do not routinely express Class I histocompatibility antigens (or perhaps express them only at very low levels) but do express them conspicuously in a variety of inflammatory conditions [43–45]. Notably, they have been found to express Class I MHC in DMD boys in sites of myoblast transplantation [31]. In order to prime a T-lymphocyte response however, the dystrophin must first be presented by an antigen-presenting cell in the clefts of Class II histocompatibility antigens, and mature muscle fibres in which dystrophin is expressed have not been found to express Class II antigens. Equally, myoblasts, which can express Class II antigens under stimulation by lymphokines, do not express dystrophin. It remains possible however, that dystrophin in damaged muscle fibres could be processed and presented to the immune system by antigen-presenting cells infiltrating the lesions.

An extra layer of complexity, has been added to this question by the demonstration that dystrophin is associated with complex of proteins and glycoproteins which traverse the membrane and which are absent, or much reduced in amount, when dystrophin is absent [46]. It seems however, that the absence of all of these dystrophin-associated proteins is the result of their lack of stabilization rather than lack of production [47, 48] and therefore that they may be present in sufficient amounts in DMD boys to be tolerogenic.

Empirically, it has been found that fully MHC-matched grafts in DMD boys, without further immunosuppression, do, in some instances excite an immune response which is specific in tissue culture for the muscle cells of the donor of the myoblast graft [49]; in one instance the serum of the graft recipient being specifically cytotoxic to cultured muscle of the myoblast donor.

In most cases, the donor and recipient were partially matched at the Major Histocompatibility locus and were immunosuppressed either with cyclosporin A [32, 50], or with cyclophosphamide [31]. No overt sign of rejection was observed in any of these studies but evidence of survival was absent [31], minimal [32] or at best equivocal [30]. In addition, there is the complication that Cyclosporin A itself appears to produce a degree of amelioration of muscle function in Duchenne patients [51] which must be allowed for in any functional test of efficacy of myoblast transplantation.

Evaluation of success

While the difficulties which were foreseeable by extrapolation from the mdx mouse experiments had 'spiked' our expectations of the outcome of the human trials with a degree of caution, it did not prepare us for the almost complete lack of evidence of any significant survival of transplanted myogenic cells in muscles of DMD patients. So far, the only support of the efficacy of myoblast transplantation in DMD comes from the reports of Law's group which have concentrated on the evaluation of functional improvement in treated muscles [30, 50, 52–54]. By and large, this work is inadequately controlled from an experimental point of view, particularly as regards the beneficial effects of Cyclosporin A and placebo effects. In defence of this conduct, the case that it is unethical to perform unilateral treatment of ambulatory muscles, and to undertake potentially damaging sham injections on muscles of the control side, has undoubted force. Nonetheless, the resulting loss of information is an inevitable and unfortunate consequence of this and, in the absence of corroborative muscle biopsy data, the outcome of such studies must be viewed tentatively.

The ethical question, in fact, has presented major problems for the proper design and conduct of all of the studies conducted so far. All but one have been restricted to the use of children who were in the major phase of clinical decline of muscle function, largely on the basis that this would facilitate detection of any functional benefit which might accrue from the treatment, but also because it is relatively easy to obtain permission to treat clinically affected children. It is quite natural for clinicians to concern themselves mainly with alleviation of the disease symptoms and for the patient this is undoubtedly the only acceptable end-point. But, in this case, such a bias away from the primary biochemical criteria of success has been detrimental to the gathering of useful information. Concentration on clinical and functional effects has quite clearly distorted experimental design in such a way as to lessen the chances of finding positive results; by waiting for sufficient time to make valid measurements of functional improvement the early events after myoblast transplantation have been missed and, with them, the likelihood of discovering what has gone wrong. But for the constraint of the functional/clinical imperative, it would have been better to have tested the therapy on younger children whose muscles had not yet accumulated the irreversible fibrotic and fatty changes which seem to accompany function deterioration and clinical decline. On the face of it, it seems unlikely that such secondary pathological changes could be reversed by replacing dystrophin within the surviving muscle fibres, indeed the loss of normal muscle structure might be predicted to diminish the efficacy of the therapeutic mechanisms of myoblast transplantation. This caveat against the use of older children for initial trials holds still greater force when it comes to the testing of purely genetic therapies, where the restoration and reconstitution of damaged tissue is not part of the therapeutic strategy and replacement of the missing gene would not, of itself, be expected to reverse secondary and tertiary degenerative pathological changes.

Efficiency

Until recently, little attempt has been made to measure the efficiency of myoblast transplantation. Relevant parameters are: survival of transplanted myogenic cells, proliferation of the surviving cells, dispersal of the cells from the site of injection, and the total amount of muscle derived from a given number of cells.

We have attempted to make some of these measurements in the mdx mouse in order to generate a quantitative basis for comparison of different regimes and for making absolute estimates of the amount of effort required to achieve a given effect [55]. From these studies, we estimate that, under the most favourable conditions we can arrange, approximately 10 mg of genetically normal muscle is produced in an mdx mouse muscle by injection of half a million muscle cells (to put this in context, it would require 10^5 such injections to produce a kilogram of normal muscle in a DMD boy). This amount of muscle in the mouse would contain approximately 10^6 muscle nuclei, implying at least one cell division by some of the injected cells. However, our investigations of the *in vivo* fate of injected myogenic cells reveal a surprisingly complex process. By injecting cells derived from male donors, and also labelled with ^3H or ^{14}C thymidine, into muscles of female recipient mice we have been able to follow the survival of the injected cells (persistence of radiolabelled thymidine) and the total yield of donor derived cells (amount of Y chromosome) and from these two, to estimate the net number of cell divisions undergone by the surviving fraction of cells. To our surprise we found that less than 1% of the injected fraction of myogenic cells survives for more than a few days and that this small fraction undergoes at least one cell-division per day to achieve the number of nuclei required to form the measured yield of muscle over a period of 7–10 days. This combination of massive cell death in the majority of the cell population with contemporaneous proliferation by a small surviving fraction suggests that the latter is a distinct sub-population with some of the properties of stem cells. This picture of a stem cell sub-population within mature skeletal muscle, supports our earlier demonstration of a minor population of myogenic cells which were passaged through a series of mouse muscles, forming normal skeletal muscle fibres and undifferentiated, extractable myogenic cells at each passage [24]. The existence of such a stem-cell-like population within skeletal muscle of mature animals is of great importance for the development of improved methods of myoblast transplantation and gene therapy. Interestingly, evidence of uncommitted cells lying in non-muscle tissue but with a potential to differentiate into skeletal muscle under appropriate conditions has come from a number of sources in recent years [56, 57], Caplan (personal communication), and could provide a source of material for myoblast transplantation.

Prospects

When the genetic cause of Duchenne muscular dystrophy was discovered, myoblast transplantation was the only potential genetic therapy available for consideration [2]. Since this time, it has been shown that genes can be inserted into skeletal muscle by a number of means; direct injection of DNA [58], infection of myogenic cells with replication-defective recombinant retroviral vectors [59, 60] and infection of immature muscle fibres with replication deficient recombinant adenoviral vectors [61]. Of these possibilities, only myoblast transplantation has been subjected to trials in man, an exercise which has shown it to be safe when it is ineffective. This conclusion points to an absurdity in the notion of separating safety from efficacy in an area of research where the dangers inherent in the method can only be adequately revealed when its prime objective, namely significant levels of genetic conversion, has been successfully achieved.

Despite the lack of success to date, it is too early to abandon myoblast transplantation as a method. First, as described above, its failure, as conducted in man, is attributable to a number of identified problems which need to be eliminated before the principle as a whole is written off. Second, myoblast transplantation is still the only approach to restoration of muscle architecture in muscles damaged by chronic degeneration generated by lack of dystrophin; purely genetic therapies would have no direct effect on the reversal of secondary and tertiary pathological changes in affected muscle and would need to be applied at an early stage in the disease, before irreversible pathological damage had accrued.

In addition, myoblast transplantation has potential uses outside the field of genetic myopathies, skeletal muscle having been shown to a suitable tissue for the synthesis and secretion of a number of pharmacologically active peptides and proteins [62–67]. Since these products have a systemic function, this use does not require the wide dispersal of the genetically altered muscle: because of the absence of such a major obstacle [2], this option may be the first to bear therapeutic fruit.

Finally, myoblast transplantation is one of the most versatile tools we have for the investigation of muscle cell biology; helping to bridge the gap between, on the one hand, *in vitro* studies in which we can observe closely but interpret with difficulty, and, on the other hand, the less artefact-ridden but imprecise observations we can gain from *in vivo* experiments.

References

1. Frey M and Giovanoli P (eds) (1995) 4th International Muscle Symposium: Proceedings. Zurich: Klinik für Hand, Plastische und Wiederherstellungschirugie, Universitätsspital Zürich, 1995.
2. Partridge TA (1991) Invited review: myoblast transfer: a possible therapy for inherited myopathies? Muscle Nerve 14: 197–212.
3. Studitsky AN (1964) Free auto- and homografts of muscle tissue in experimental animals. Ann N Y Acad Sci 120: 789–801.
4. Carlson BM (1973) The regeneration of skeletal muscle: a review. Am J Anat 137: 119–150.
5. Bateson RG, Woodrow DF and Sloper JC (1967) Circulating cell as a source of myoblasts in regenerating injured mammalian skeletal muscle. Nature 213: 1035–1036.
6. Law P (1982) Beneficial effects of transplanting normal limb-bud mesenchyme into dystrophic mouse muscles. Muscle Nerve 5: 619–627.
7. Law PK and Yap JL (1979) New Muscle Transplant method produces normal twitch tension in dystrophic Muscle. Muscle Nerve 2: 356–363.
8. Partridge TA, Grounds M and Sloper JC (1978) Evidence of fusion between host and donor myoblasts in skeletal muscle grafts. Nature 273: 306–308.
9. Partridge TA (1982) Cellular interactions in the development and maintenance of skeletal muscle. In: Bellairs R, Curtis A and Dunn G (eds) The Social Abilities of Cells, pp 555–581. Cambridge: Cambridge University Press.

10. Lipton BH and Schultz E (1979) Developmental fate of skeletal muscle satellite cells. Science 205: 1292–1294.
11. Jones PH (1979) Implantation of cultured regenerate muscle cells into adult rat muscle. Exp Neurol 66: 602–610.
12. Law PK, Goodwin TG and Wang MG (1988) Normal myoblast injections provide genetic treatment for murine dystrophy. Muscle Nerve 11: 525–533.
13. Law PK, Goodwin TG and Li HJ (1988) Histoincompatible myoblast injection improves muscle structure and function of dystrophic mice. Transplant Proc 20: 1114–1119.
14. Arahata K, Hayashi YK, Koga R *et al.* (1993) Laminin in animal models for muscular dystrophy: defect of Laminin M in skeletal and cardiac muscles and peripheral nerve of the homozygous dystrophic dy/dy mouse. Proc Japan Acad 69: 259–264.
15. Sunada Y, Bernier SM and Kozak CA, *et al.* (1994) Deficiency of merosin in dystrophic *dy* mice and genetic linkage of laminin M chain gene to *dy* locus. J Biol Chem 269: 13729–13732.
16. Xu H, Christmas P and Wu XR *et al.* (1994) Defective basement membrane and lack of M-laminin in the dystrophic *dy/dy* mouse. Proc Nat Acad Sci U S A 91: 5572–5576.
17. Watt DJ, Lambert K, Morgan JE, *et al.* (1982) Incorporation of donor muscle precursor cells into an area of muscle regeneration in the host mouse. J Neurol Sci 57: 319–331.
18. Watt DJ, Morgan JE and Partridge TA (1984) Use of mononuclear precursor cells to insert allogeneic genes into growing mouse muscles. Muscle Nerve 7: 741–750.
19. Morgan JE, Watt DJ, Sloper JC *et al.* (1988) Partial correction of an inherited biochemical defect of skeletal muscle by grafts of normal muscle precursor cells. J Neurol Sci 86: 137–147.
20. Partridge TA, Morgan JE, Coulton GR *et al.* (1989) Conversion of mdx myofibres from dystrophin-negative to -positive by injection of normal myoblasts. Nature 337: 176–9.
21. Karpati G, Pouliot Y, Zubrzycka GE *et al.* (1989) Dystrophin is expressed in mdx skeletal muscle fibers after normal myoblast implantation. Am J Pathol 135: 27–32.
22. Morgan JE, Pagel CN, Sherratt T *et al.* (1993) Long-term persistence and migration of myogenic cells injected into pre-irradiated muscles of mdx mice. J Neurol Sci 115: 191–200.
23. Morgan JE, Hoffman EP and Partridge TA (1990) Normal myogenic cells from newborn mice restore normal histology to degenerating muscles of the mdx mouse. J Cell Biol 111: 2437–2449.
24. Morgan JE, Beauchamp JR, Pagel CN *et al.* (1994) Myogenic Cell Lines Derived from Transgenic Mice Carrying a Thermolabile T Antigen: A Model System for the Derivation of Tissue-Specific and Mutation-specific Cell Lines. Dev Biol 162: 1–13.
25. Griggs RC and Karpati G (eds) (1990) Myoblast Transfer Therapy. New York: Plenum Press.
26. Hoffman EP, Morgan JE, Watkins SC, *et al.* (1990) Somatic reversion/suppression of the mouse mdx phenotype in vivo. J Neurol Sci 99: 9–25.
27. Shimizu T, Matsumura K, Hashimoto K *et al.* (1988) A monoclonal antibody against a synthetic polypeptide fragment of dystrophin (amino acid sequence from position 215—264). Proc Japan Acad 64: 205–208.
28. Nicholson LVB, Davison K, Johnson MA *et al.* (1989) Dystrophin in skeletal muscle: II. Immunoreactivity in patients with Xp21 muscular dystrophy. J Neurol Sci 94: 137–146.
29. Huard J, Bouchard JP, Roy R *et al.* (1992) Human myoblast transplantation: preliminary results of 4 cases. Muscle Nerve 15: 550–60.
30. Law PK, Bertorini TE, Goodwin TG *et al.* (1990) Dystrophin production induced by myoblast transfer therapy in Duchenne muscular dystrophy. Lancet 336: 114–5.
31. Karpati G, Ajdukovic D, Arnold D *et al.* (1993) Myoblast transfer in Duchenne muscular dystrophy. Ann Neurol 34: 8–17.
32. Gussoni E, Pavlath GK, Lanctot AM *et al.* (1992) Normal dystrophin transcripts detected in Duchenne muscular dystrophy patients after myoblast transplantation. Nature 356: 435–8.
33. Watt DJ (1982) Factors which affect the fusion of allogeneic muscle precursors *in vivo*. Neuropath Appl Neurobiol 8: 137–147.
34. Grounds MD, Partridge TA and Sloper JC (1980) The contribution of exogenous cells to regenerating skeletal muscle: an isoenzyme study of muscle allografts in mice. J Pathol 132: 325–341.
35. Watt DJ, Morgan JE and Partridge TA (1992) Allografts of muscle precursor cells persist in the non-tolerized host. Neuromusc Disord 1: 345–355.
36. Morgan JE, Coulton GR and Partridge TA (1987) Muscle precursor cells invade and repopulate freeze-killed skeletal muscles. Journal of Muscle Research and Cell Motility 8: 386–396.
37. Labrecque C, Roy R and Tremblay JP (1992) Immune reactions after myoblast transplantation in mouse muscles. Transplant Proc 24: 2889–92.
38. Guérette B, Asselin I and Vilquin J-T *et al.* (1995) Lymphocyte infiltration following allo- and xenomyoblast transplantation in mdx mice. Muscle Nerve 18: 39–51.
39. Irintchev A, Zweyer M and Wernig A (1995) Cellular and molecular reactions in mouse muscles after myoblast implantation. J Neurocytol 24: 319–331.
40. Wernig A, Irintchev A and Lange G (1995) Functional effects of myoblast implantation into histoincompatible mice with or without immunosuppression. J Physiol 484: 493–504.
41. Vilquin J-T, Asselin I, Guérette B *et al.* (1995) Successful myoblast transplantation in mdx mice using Rapamycin. Transplantation 59: 422–449.
42. Kinoshita I, Vilquin J-T, Guerette B *et al.* (1994) Very efficient myoblast allotransplantation in mice under FK506 immunosuppression. Muscle Nerve 17: 1407–1415.
43. Appleyard ST, Dunn MJ, Dubowitz V *et al.* (1985) Increased expression of HLA ABC class I antigens by muscle fibres in Duchenne muscular dystrophy, inflammatory myopathy, and other neuromuscular disorders. Lancet i: 361–363.
44. Karpati G, Pouilot Y and Carpenter S (1988) Expression of immunoreactive major histocompatibility complex products in human skeletal muscles. Ann Neurol 23: 64–72.
45. Emslie-Smith AM, Arahata K and Engel AG (1989) Major histocompatibility complex class 1 antigen expression, immunolocalization of interferon subtypes and T cell-mediated cytotoxicity in myopathies. Hum Pathol 20: 224–231.
46. Ervasti JM, Ohlendieck K, Kahl SD *et al.* (1990) Deficiency of a glycoprotein component of the dystrophin complex in dystrophic muscle. Nature 345: 315–9.
47. Matsumura K, Lee CC, Caskey CT *et al.* (1993) Restoration of dystrophin-associated proteins in skeletal muscle of mdx mice transgenic for dystrophin gene. Febs Lett 320: 276–80.
48. Matsumura K and Campbell KP (1993) Deficiency of dystrophin-associated proteins: a common mechanism leading to muscle cell necrosis in severe childhood muscular dystrophies. Neuromusc Disord 3: 109–18.
49. Huard J, Roy R, Bouchard JP *et al.* (1992) Human myoblast transplantation between immunohistocompatible donors and

recipients produces immune reactions. Transplant Proc 24: 3049–51.
50. Law PK, Goodwin TG, Fang Q *et al.* (1991) Long-term improvement in muscle function, structure and biochemistry following myoblast transfer in DMD. Acta Cardiomiol 3: 281–301.
51. Sharma KR, Mynhier MA and Miller RG (1993) Cyclosporine increases muscular force generation in Duchenne muscular dystrophy. Neurobiology 43: 527–532.
52. Law PK, Goodwin TG, Fang QW *et al.* (1991) Myoblast transfer therapy for Duchenne muscular dystrophy. Acta Paediatr Jpn 33: 206–15.
53. Law PK (1992) Myoblast transplantation [letter]. Science 257: 1329–30.
54. Law PK, Goodwin TG, Fang Q *et al.* (1993) Cell Transplantation as an experimental treatment for Duchenne muscular dystrophy. Cell Transplant 2: 485–505.
55. Beauchamp JR, Morgan JE, Pagel CN *et al.* (1994) Quantitative studies of the efficacy of myoblast transplantation. Muscle Nerve Suppl. 1: S261.
56. Young HE, Mancini ML, Wright RP *et al.* (1995) Mesenchymal stem cells reside withi the connective tissues of many organs. Dev Dynam 202: 137–144.
57. Gibson AJ, Karasinski J, Relvas J *et al.* (1995) Dermal fibroblasts convert to a myogenic lineage in mdx mouse muscle. J Cell Sci 108: 207–214.
58. Acsadi G, Dickson G, Love DR *et al.* (1991) Human dystrophin expression in mdx mice after intramuscular injection of DNA constructs. Nature 352: 815–8.
59. Dunckley MG, Love DR, Davies KE *et al.* (1992) Retroviral-mediated transfer of a dystrophin minigene into mdx mouse myoblasts in vitro. Febs Lett 296: 128–34.
60. Dunkley MG, Wells DJ, Walsh FS *et al.* (1993) Direct retroviral-mediated transfer of a dystrophin minigene into *mdx* mouse muscle *in vivo*. Hum Mol Genet 2: 717–723.
61. Ragot T, Vincent N, Chafey P *et al.* (1993) Efficient adenovirus-mediated transfer of a human minidystrophin gene to skeletal muscle of mdx mice. Nature 361: 647–50.
62. Barr E and Leiden JM (1991) Systemic delivery of recombinant proteins by genetically modified myoblasts. Science 254: 1507–9.
63. Dhawan J, Pan LC, Pavlath GK *et al.* (1991) Systemic delivery of human growth hormone by injection of genetically engineered myoblasts. Science 254: 1509–12.
64. Jiao S, Gurevich V and Wolff JA (1993) Long term correction of rat model of Parkinson<prime>s disease by gene therapy. Nature 362: 450–455.
65. Yao SN, Wilson JM, Nabel EG *et al.* (1991) Expression of human factor IX in rat capillary endothelial cells: toward somatic gene therapy for hemophilia B. Proc Natl Acad Sci U S A 88: 8101–5.
66. Yao S-N, Smith KJ and Kurachi K (1994) Primary myoblast-mediated gene transfer: persistent expression of human factor IX in mice. Gene Therapy 2: 99–107.
67. Zhu N, Liggit D, Liu Y *et al.* (1993) Systemic gene expression after intravenous DNA delivery into adult mice. Science 261: 209–211.

R. P. Lanza and W. L. Chick (eds.), Yearbook of Cell and Tissue Transplantation 1996/1997, 61–67.

Chapter 7

Skeletal muscle transplantation

Bruce M. Carlson
Department of Anatomy and Cell Biology, University of Michigan, Ann Arbor, MI 48109, U.S.A.

Skeletal muscle is a tissue adapted for the generation of mechanical force, and a principal objective for the clinical transplantation of skeletal muscle is to restore force generation to a region in which the resident skeletal muscle has for some reason (e.g. denervation, trauma, pathology) lost is functional capacity (Frey and Freilinger, 1986; Freilinger and Deutinger, 1992; Frey and Giovanoli, 1995). In the research setting, muscle transplantation represents a convenient tool for producing the degeneration and regeneration of muscle fibers (Carlson *et al.*, 1979; Carlson and Faulkner, 1983). Transplantation has an advantage over many other muscle regeneration models in that whole muscle physiology can be readily performed on the regenerates. In addition, transplantation of muscles into inbred hosts can be used to investigate the effects of different host environments on the process of muscle regeneration.

Specific techniques and applications of muscle transplantation are discussed below. Because some of the transplantation techniques developed in parallel between research and clinical applications, certain assumptions and terminologies differ. An important one to note is the term "free graft". In most basic research reports, a free graft is defined as a muscle that has been completely removed from its bed, replaced into a new or its original site and allowed to become spontaneously revascularized and reinnervated. In contemporary clinical reports, free graft is typically used to describe a graft which is surgically connected with the host through a vascular anastomosis. This distinction is important because in a non-vascularized graft most of the original muscle fibers degenerate and subsequently regenerate, whereas in a vascularized graft most of the original muscle fibers survive.

Techniques of muscle transplantation

A variety of muscle transplantation techniques have been devised for specific purposes. Some of the more important ones are listed below.

Minced muscle regeneration
The first successful method for transplanting muscle consisted of mincing a muscle into one mm^3 fragments and placing the mince into an orthotopic or a heterotopic site. Originally done to prove the validity of a Lysenkoist theory (Studitsky, 1949), this technique nevertheless provided one of the earliest models by which an entire mammalian muscle could be regenerated, although the quality of mature regenerates is typically severely compromised by excessive fibrosis (Studitsky, 1959; Carlson, 1968). The minced muscle technique (autografts) has proven to be useful in studies of morphogenetic control of regenerating muscle (Carlson, 1972). Allografts of minced muscle between normal and dystrophic mice were used in early tests of the myogenic vs. the neurogenic hypotheses of muscular dystrophy (Salafsky, 1971), and grafts of minced muscle from rats into mice have been used to investigate the myogenic capacity of muscle in different states of ischemia (Phillips *et al.*, 1987).

In minced muscle regeneration, the muscle fragments constituting the original mince fall into a state of ischemic necrosis. After 2–3 days blood vessels begin to grow into the graft. With the initiation of revascularization, large numbers of macrophages enter the damaged muscle fibers and phagocytize the cytoplasmic contents. Satellite cells, located beneath the persisting basal laminae of the muscle fiber fragments, proliferate and fuse into myotubes, precursors of the cross-striated muscle fibers that ultimately repopulate the original mince (Snow, 1977). A centripetal gradient of muscle fiber breakdown and regeneration accompanies the ingrowth of the vasculature. Initially, the orientation of the regenerating muscle fibers within the mince is random, but as the regenerate becomes functional, the regenerating muscle fibers become oriented parallel to lines of tension within the regenerate (Carlson, 1972).

Standard free muscle grafts

The first successful whole muscle grafting technique consisted of removing a small skeletal muscle entirely from its bed by severing the tendons and all vascular and neural connections. The muscle is placed into its original or a heterotopic site by suturing the proximal and distal tendon stump to mechanically appropriate sites in the host. No attempt is made to reconnect vascular or nerve supply. Revascularization and reinnervation are accomplished through spontaneous outgrowth and regeneration. The fundamental processes and topography of muscle fiber breakdown and regeneration are similar to those occurring in a minced muscle graft, except that a thin rim of surviving intact muscle fibers (2–5% of the total) persists at the periphery of the graft. The cellular development of a free muscle graft is summarized in Fig. 1.

Standard free muscle grafts have been successfully carried out in mice (Roberts and McGeachie, 1990), rats (Carlson and Gutmann, 1972a, b), and cats (Faulkner *et al.*, 1980). Free grafts of the palmaris longus muscle in monkeys become filled with a large core of dense connective tissue surrounded by a rim of regenerated muscle fibers (Markley *et al.*, 1978).

Free non-vascularized grafts of muscles into the paralyzed human face were first introduced by Thompson (1974). He predenervated the muscles to be grafted for several weeks because at the time it was felt that the success of a non-vascularized free muscle graft was due to the survival of the original muscle fibers and that predenervation enhanced the ability of muscle fibers in the graft to withstand the initial period of ischemia that follows grafting. Subsequent animal research demonstrated that even when all original muscle fibers in the graft were destroyed by Marcaine, recovery of the grafts was not impaired (Carlson, 1976).

Early muscle fiber regeneration in standard grafts is vigorous, but the quality of stabilized standard grafts is quite variable, principally due to inconsistencies in the degree of reinnervation. Several techniques have been employed to enhance the degree of reinnervation and consequently the quality of the grafts.

Nerve-implant and nerve-intact grafts

Nerve-implant and nerve-intact muscle grafts are variations on the standard grafting technique designed to improve the final result by facilitating the reinnervation of the regenerating muscle fibers. Both types of grafts are similar to standard grafts in that the proximal and distal tendons are severed and the blood supply is interrupted, with no attempt made to anastomose vessels at the time of transplantation.

In a nerve-implant graft the proximal nerve stump is simply pulled into the substance of the grafted muscle. The nerve stump may or may not be held in place by a suture. The nerve-implant procedure ensures that regenerating axons will enter the muscle graft, but it does not provide the regenerating axons with performed neural channels into which they can grow. Nerve-implant grafts typically become restored to a greater degree of mass and contractile function than are standard grafts (Segal *et al.*, 1986).

A nerve-intact graft is one in which the muscle is completely removed from its bed and then replaced, but the nerves leading to the muscles are not severed. The muscle fibers within the graft degenerate and regenerate as in a standard graft. Axons within the ischemic graft die back, but they quickly regenerate and are well distributed throughout the graft because regeneration occurs within basal laminae of the Schwann cell sheaths. Nerve-intact grafts in rats typically become restored to the same mass and from 85–95% of the maximum tetanic force of the original muscle (Carlson *et al.*, 1981). Studies on motor units in the rat EDL muscle have shown the presence of nearly normal numbers in nerve-intact grafts as compared with a nearly 40% reduction in motor unit numbers in standard grafts (Côté and Faulkner, 1986).

Another technique designed to improve the quality of free muscle grafts through the facilitation of reinnervation is the approximation of host and graft nerve segments through epineurial repair. This allows axons regenerating from the host access to endoneurial channels leading into the graft. Preliminary results to date support that in grafts of normal muscle, epineurial repair produces stabilized regenerates that produce greater force than nerve-implant grafts but less than nerve-intact grafts (Gu, unpublished).

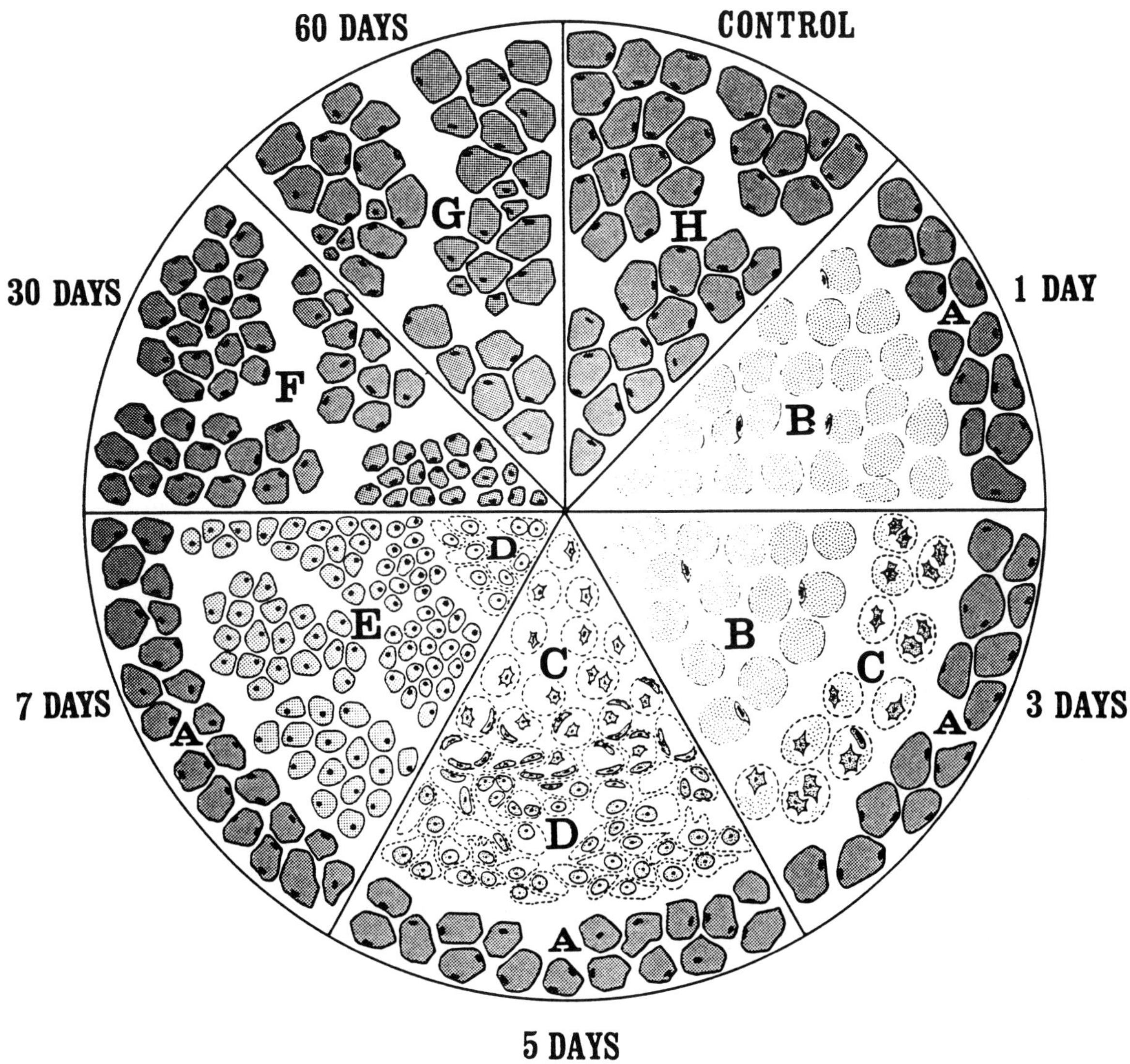

Fig. 1. Summary of the development of a free graft of the rat extensor digitorum longus muscle. Each wedge represents the histological appearance of a cross-section of the graft at the day indicated. Letters refer to zones of cellular activity. A – peripheral rim of surviving muscle fibers; B – central zone of muscle fibers in a state of ischemic necrosis; C – zone of macrophage-mediated muscle fiber degeneration and myoblastic activity. This also represents the region of furthest ingrowth of the vasculature; D – zone of early regenerating myotubes; E – zone of maturing myotubes; F – maturing cross-striated muscle fibers; G – stable regenerated muscle fibers with central nuclei; H – normal muscle fibers. From Carlson *et al.* (1979).

Vascular-anastomosed muscle grafts

Experience in the laboratory showed that muscles with a mass greater than 3 gm could not be successfully freely grafted if they had to rely upon spontaneous revascularization. Commonly free grafts of larger muscles undergo liquifiying necrosis within days after grafting or, if the muscle persists, the muscle fibers become to a large extent replaced by dense connective tissue. The advent of microvascular surgery brought about a profound change in the practice of muscle transplantation in the clinical treatment of muscular deficits (Freilinger and Deutinger, 1992). With current techniques it is possible to transplant large muscles in humans to restore function to limbs that have been severely traumatized or which have undergone massive ischemic events, such as Volkmanns ischemic contracture. Vascular anastomotic techniques also allow muscles to be transplanted for reconstructive purposes (e.g.

myocutaneous flaps) (Freilinger and Deutinger, 1992). With vascular-anastomosed grafts of large muscles, simultaneous epineurial repair of nerves is routinely performed.

Vascular-anastomosed grafts differ greatly from the other types of muscle grafts discussed above in that the intention is to allow survival of the muscle fibers. Nevertheless, scattered pockets of muscle fiber breakdown and regeneration occur even in vascular-anastomosed grafts, but the amount is insufficient to produce a significant functional deficit (Guelinckx *et al.*, 1992).

Myoblast transfer (cellular transplantation)

Great attention has been given in recent years to myoblast transfer as a means of tissue engineering for diseased skeletal muscle. This technique has been the subject of numerous reviews (e.g. Morgan and Watt, 1993) and will only be briefly summarized here.

Myoblast transfer was devised as a means of introducing normal genes into diseased muscle, in particular dystrophic muscle that is lacking the protein, dystrophin. The essence of the technique is to culture large numbers of myoblasts *in vitro* and then inject the normal myoblasts into dystrophic muscles. The myoblasts then fuse with either intact or regenerating dystrophic muscle fibers, thus forming a chimeric syncytium containing both normal (graft) and dystrophic (host) nuclei. The normal nuclei contribute the product of the dystrophin gene to the chimeric muscle fibers. The most significant practical question relating to this technique is whether or not an adequate distribution of dystrophin can be accomplished at both the intracellular level and at the level of the entire muscle.

Myoblast transfer has been shown to improve the functional mass of specific injected muscles in dystrophin-deficient (mdx) mice (Karpati *et al.*, 1989; Partridge *et al.*, 1989). Several clinical trials have been attempted, but with equivocal results (Law *et al.*, 1991; Tremblay *et al.*, 1991).

Reintegration of a muscle graft with the host

Like any transplant, a muscle graft (in clinical practice always an autograft) is completely removed from the body of the host and must become reintegrated with the host after transplantation. Reintegration must be accomplished in three areas – mechanical reintegration, revascularization and reinnervation.

In order for a muscle graft to function properly, it must be positioned in a manner that permits optimal mechanical function. In the laboratory, this means suturing the tendons of a transplanted limb muscle so that the muscle is fixed at a slight tension upon grafting. Clinically, especially in cases of muscle grafting to the face, it is common to reoperate upon the patient several months after transplantation to adjust the tension of the muscle. In minced muscle regeneration, both the gross form and internal architecture of the muscle reflect the mechanical environment into which the mince is placed (Carlson, 1972), and studies on whole muscle transplantation suggest that regenerating muscle fibers are more likely to undergo architectural adaptation than are intact muscle fibers. It has been known for years that muscle fibers can adapt to longitudinal tensile forces by adding or subtracting sarcomeres (Goldspink, 1980).

As was mentioned earlier, revascularization is the key to initial survival of muscle graft. In rats, satellite cells do not survive ischemia for more than 4 hours (Phillips *et al.*, 1987; Schultz *et al.*, 1988). A prolonged ischemic phase is eliminated in vascular anastomosed grafts, but in standard free muscle grafts most of the interior of the graft undergoes prolonged ischemia until blood vessels have grown into the area. In small rat muscles the center of the graft may remain ischemic for several days, whereas in grafted muscles in cats ischemic regions persist for as long as 7–8 weeks (Mufti *et al.*, 1977). Yet when these areas become revascularized, macrophage-mediated muscle fiber degeneration and regeneration take place. It is postulated that myogenic cells accompany the blood vessels that grow toward the center of the graft and that these cells repopulate the basal laminae left behind by the degenerating ischemic muscle fibers.

Once a muscle graft is revascularized, whether through surgical anastomosis or through spontaneous means, it contains a large population of non-innervated muscle fibers. If these muscle fibers do not become reinnervated, the graft remains functionally useless and progressive atrophy sets in. Although a variety of techniques have been devised to obtain the most effective reinnervation of a transplanted muscle, producing effective reinnervation remains one of the most important problems in muscle transplantation. Although it is common to utilize existing nerve channels and to attempt to direct regenerating nerve fibers back to the original zone of motor endplates, experimental work has shown that substantial reinnervation can be obtained in rat muscles from which the original motor endplate zone was surgically removed (Womble, 1986). In addition to motor innervation, muscles also have a sensory apparatus in the form of muscle spindles and Golgi tendon organs. That muscle spindles can regenerate has been demonstrated many times (Carlson and Gutmann, 1975; Milburn, 1976; Rogers and Carlson, 1981). The innervation pattern of normal muscle spindles is very complex, and typically regenerated muscle spindles are only incompletely reinnervated (Rogers, 1982). The regeneration of Golgi tendon organs has been reported in young rats (Bulyakova, 1988), but little is yet known about the process.

Limiting factors in muscle transplantation

The successful transplantation of skeletal muscle is subject to several constraints. For free non-vascular anastomosed grafts, overall size is the principal limiting factor. Spontaneous vascular ingrowth is not sufficient in either amount or rate to supply the entire cross-section of the muscle before irreversible fibrotic changes occur in the core of the muscle. Depending upon the species and type of muscle, this limit ranges from a mass of 1–3 gm. Innervation of a muscle graft is another limiting factor, and to a certain extent this is also related to the size of the muscle along with the availability of pre-existing neural pathways. In rodents, successful reinnervation can be obtained by the growth of regenerating axons through the connective tissue elements of the muscle, but this is progressively less likely as the cross-sectional area of the muscle increases. For larger muscles, epineurial repair gives regenerating host axons access to nerve channels in the graft, but as with any form of nerve repair there is no assurance that all sectors of the muscle will become reinnervated. In large vascular-anastomosed grafts, even of the nerve-intact variety, there is typically a maximum force deficit of approximately 20% that remains unexplained by any known mechanism. Limiting factors in vascular-anastomosed grafts have recently been reviewed by Faulkner *et al.* (1994).

The host environment represents another set of factors that can limit the success of muscle grafts. For whole muscle allografts, which are rarely attempted, classical cellular rejection occurs, although it is important to note that vigorous muscle fiber regeneration can occur in the face of a strong cellular immune response (Carlson, 1970; Gulati, 1985; Phillips *et al.*, 1987). Cyclosporin A treatment of host rats effectively eliminates the muscle allograft rejection response in rats (Gulati and Zalewski, 1982). Immune rejection is a much more important consideration in myoblast transfer therapy, where in the case of muscular dystrophy it is necessary to obtain myoblasts from donors possessing functional dystrophin genes (see Irintchev and Wernig, 1995 for details).

In rats muscle regenerates less well in streptozotocin-induced diabetic than in normal rats (Gulati and Swamy, 1991) and in old than in young rats (Carlson and Faulkner, 1989). However, in both of these cases, the quality of regeneration in the grafts can be greatly improved by grafting diabetic or old muscles into normal or young inbred hosts. Such experiments illustrate the importance of the overall host environment on the quality of muscle regeneration.

Clinical applications of muscle transplantation

The first effective clinical application of skeletal muscle transplantation was reported by Thompson in 1974. In the era preceding microsurgery, muscle transplantation in humans was limited to small muscles. Most applications of the standard grafting technique involved the correction of symptoms associated with facial palsy or with anal or urethral sphincter repair (Freilinger *et al.*, 1981). In all of these cases the generation of sustained force rather than great strength was required of the muscle grafts. In retrospect, it is now suspected that many of the human muscle grafts performed during this era were effective because of a static sling effect rather than through active contraction.

Currently, virtually all clinical muscle transplantation involves both vascular anastomosis and epineurial repair. In cases of unilateral facial palsy, it is common practice first to perform cross-face nerve grafts in order to lead facial nerve axons from branches in the healthy side of the face to the paralyzed side. This phase typically takes 5–7 months (Anderl, 1977). When regenerating axons have reached the end of the nerve graft, a muscle transplant is then performed (Freilinger 1975). The muscle graft, usually performed to elevate the corner of a drooping mouth, is anastomosed to a local blood vessel and also to the distal end of the cross-face nerve graft. The reason for delaying the muscle grafting phase of this operation is that potentially irreversible denervation atrophy of the muscle graft could occur if the muscle graft is put in place at the same time as the cross-face nerve graft. The reason for grafting muscles in the paralyzed face is that after 6–12 months of denervation, atrophied muscles respond very poorly to reinnervation (Bateman, 1962; Anerl, 1977). The reasons for this remain obscure.

The grafting of large muscles into traumatized limbs is case-specific, but in all cases the operation involves both vascular anastomosis and nerve repair (Schenck, 1978; Manktalow and Zucker, 1992). Although the results of such operations are not always aesthetically pleasing, transplanted limb muscles can function well enough to allow some patients to return to manual labor.

Summary

Skeletal muscle transplantation is a technique that has proven to be a useful model for studying whole muscle regeneration in laboratory animals. With the improvement of microsurgical techniques, the successful to autotransplantation of muscles become a reality for the treatment of certain neuromuscular deficits. Much remains to be learned about the integration of a transplanted muscle with the host and about host effects on transplanted muscles.

Acknowledgements

Original research reported in this review has been supported by NIH grants PO1 DE07687 and PO1 AG10821.

References

1. Anderl H (1977) Cross-face nerve grafting – up to 12 months of seventh nerve disruption. In: Rubin LR (ed.) Reanimation of the Paralyzed Face, pp 241–277. St. Louis: Mosby.
2. Bateman JE (1962) Trauma to Nerves in Limbs. Philadelphia: Saunders.
3. Bulyakova NV (1988) Sensory nerve endings in minced muscle of young and old rats. In: Hník PT, Vejsada R and Zelená J (eds) Mechanoreceptors: Development, Structure, and Function, pp 127–130. New York: Plenum Press.
4. Carlson BM (1968) Regeneration of the completely excised gastrocnemius muscle in the frog and rat from minced muscle fragments. J Morph 125: 447–472.
5. Carlson BM (1970) Regeneration of the rat gastrocnemius muscle from sibling and non-sibling fragments. Am J Anat 128: 21–32.
6. Carlson BM (1972) The Regeneration of Minced Muscles. Monographs in Developmental Biology, Vol. 4. Basel: S. Karger AG, 130 pp.
7. Carlson BM and Gutmann E (1975a) Regeneration in free grafts of normal and denervated muscles in the rat: Morphology and histochemistry. Anat Rec 183: 47–61.
8. Carlson BM and Gutmann E (1975b) Regeneration in grafts of normal and denervated rat muscles. Contractile properties. Pflügers Arch 353: 227–239.
9. Carlson BM, Hansen-Smith FM and Magon DK (1979) The life history of a free muscle graft. In: Mauro A (ed.) Muscle Regeneration, pp 493–507. New York: Raven Press.
10. Carlson BM, Hník P, Tuček S, Vejsada R, Bader DM and Faulkner JA (1981) Comparison between grafts with intact nerves and standard free grafts of the rat extensor digitorum longus muscle. Physiol Bohemoslovaca 30: 505–513.
11. Carlson BM and Faulkner JA (1983) The regeneration of skeletal muscle fibers following injury. Med and Sci in Sports and Exercise 15: 187–198.
12. Carlson BM and Faulkner JA (1989) Muscle transplantation between young and old rats: Age of host determines functional recovery. Am J Physiol 256 (Cell Physiol 25): C1262–1266.
13. Côté C and Faulkner JA (1986) Characteristics of motor units in muscles of rats grafted with nerves intact. Am J Physiol 250 (Cell Physiol 19): C828–C833.
14. Faulkner JA, Carlson BM and Kadhiresan VA (1994) Review: Whole skeletal muscle transplantation: Mechanisms responsible for functional data. Biotech Bioengin 43: 757–763.
15. Faulkner JA, Niemeyer JH, Maxwell LC and White TP (1980) Contractile properties of transplanted extensor digitorum longus muscles of cats. Am J Physiol 238: C120–C126.
16. Freilinger G (1975) A new technique to correct facial paralysis. Plast & Reconstr Surg 1: 44–48.
17. Freilinger G, Holle J and Carlson BM (eds) (1981) Muscle Transplantation. Vienna: Springer-Verlag, 3ll pp.
18. Freilinger G and Deutinger M (eds) (1992) Proceedings of the Third Vienna Muscle Symposium. Austria: Blackwell-MZV, 403 pp.
19. Frey M and Freilinger G (eds) (1986) Proceedings of the 2nd Vienna Muscle Symposium. Austria: Facultas-Verlag, 386 pp.
20. Frey M and Giovanoli P (eds) (1995) Proceedings of the 4th International Muscle Symposium. Zürich, Switzerland: Klinik für Hand-, Plastische und Wiederherstellungschirurgie, Universitätsspital, 285 pp.
21. Goldspink B (1980) Growth of muscle. In: Goldspink DF (ed.) Development and Specialization of Skeletal Muscle, pp 19–25. Cambridge: Cambridge University Press.
22. Guelinckx PJ, Carlson BM and Faulkner JA (1992) Morphological characteristics of muscles grafted in rabbits with neurovascular repair. J Reconstruc Microsurg 8: 481–483.
23. Gulati AK and Zalewski AA (1982) Muscle allograft survival after cyclosporin A immunosuppresion. Exp Neurol 77: 378–385.
24. Gulati AK (1985) Basement membrane component changes in skeletal muscle transplants undergoing regeneration or rejection. J Cell Biochem 27: 337–346.
25. Gulati AK and Swamy MS (1991) Regeneration of skeletal muscle in streptozotocin-induced diabetic rats. Anat Rec 229: 298–304.
26. Irintchev A and Wernig A (1995) Formation of muscle fibers and tumorgenesis after implantation of myogenic cells. In: Proceedings of the 4th International Muscle Symposium. Zürich, Switzerland: Klinik für Hand-, Plastische und Wiederherstellungschirurgie, Universitätsspital, pp 17–19.
27. Karpati G, Pouilot Y, Zubrzycka-Gaarn E *et al.* (1989b) Dystrophin is expressed in mdx skeletal muscle fibres after normal myoblast transplantation. Am J Pathol 135: 27–32.
28. Law P, Goodwin T, Fang Q *et al.* (1991) Pioneering development of myoblast transfer therapy. In: Angelini C, Danieli GA and Fontanari D (eds) Muscular Dystrophy Research: From Molecular Diagnosis Towards Therapy. Amsterdam, New York, Oxford: Excerpta Medica, pp 109–116.

29. Manktelow RT and Zuker RM (1992) Extremity reconstruction with functioning muscle transplantation – factors affecting functional return. In: Freilinger G and Deutinger M (eds) Proceedings of the Third Vienna Muscle Symposium. Austria: Blackwell-MZV, pp 234–243.
30. Markley JM, Faulkner JA and Carlson BM (1978) Regeneration of skeletal muscle after grafting in monkeys. Plastic Reconstr Surg 62: 415–422.
31. Milburn A (1976) The effect of the local anaesthetic bupivacaine on the muscle spindle of rat. J Neurocytol 5: 425–446.
32. Morgan JE and Watt DJ (1993) Myoblast transplantation in inherited myopathies. In: Partridge T (ed.) Molecular Cell Biology of Muscular Dystrophy. London: Chapman and Hall, pp 303–331.
33. Mufti SA, Carlson BM, Maxwell LC and Faulkner JA (1977) The free grafting of entire limb muscles in the cat: morphology. Anat Rec 188: 417–430.
34. Partridge TA, Morgan JE, Coulton GR *et al.* (1989) Conversion of mdx myofibres from dystrophin-negative to -positive by injection of normal myoblasts. Nature 337: 176–179.
35. Phillips GD, Lu D, Mitashov VI and Carlson BM (1987) Survival of myogenic cells in freely grafted rat rectus femoris and extensor digitorum longus muscles. Am J Anat 180: 365–372.
36. Roberts P and McGeachie JK (1990) Endothelial cell activation during angiogenesis in freely transplanted skeletal muscles in mice and its relationship to the onset of myogenesis. J Anat 169: 197–207.
37. Rogers SL and Carlson BM (1981) A quantitative assessment of muscle spindle formation in reinnervated and non-reinnervated grafts of the rat extensor digitorum longus muscle. Neuroscience 6: 87–94.
38. Rogers SL (1982) Muscle spindle formation and differentiation in regenerating rat muscle grafts. Devel Biol 94: 265–283.
39. Salafsky B (1971) Functional studies of regenerated muscles from normal and dystrophic mice. Nature (Lond.) 229: 270–273.
40. Schenck R (1978) Rectus femoris muscle and component skin transplantation by microneurovascular anastomoses for avulsion of fore-arm. J Hand Surg 3: 60.
41. Schultz E, Albright DJ, Jaryszak DL and David TL (1988) Survival of satellite cells in whole muscle transplants. Anat Rec 222: 12–17.
42. Segal SS, White TP and Faulkner JA (1986) Architecture, composition, and contractile properties of rat soleus muscle grafts. Am J Physiol I: C474–C479.
43. Snow MH (1977) Myogenic cell formation in regenerating rat skeletal muscle injured by mincing. II. An autoradiographic study. Anat Rec 188: 200–218.
44. Studitsky AN (1949) Regeneration of the biceps muscles in birds (Russian). Doklady Akad Nauk SSSR 44: 391–394.
45. Studitsky AN (1959) Experimental Surgery of Muscles (Russian). Izdatel Akad Nauk SSSR, Moscow.
46. Thompson N (1974) A review of autogenous skeletal muscle grafts and their clinical applications. Clin Plast Surg 1: 349–403.
47. Tremblay JP, Roy R, Bouchard JP *et al.* (1991) Human myoblast transplantation. In: Angelini C, Danielli GA and Fontanari D (eds) Muscular Dystrophy Research: From Molecular Diagnosis Towards Therapy. Amsterdam, New York, Oxford: Excerpta Medica, pp 123–130.
48. Womble MD (1986) The clustering of acetylcholine receptors and formation of neuromuscular junctions in regenerating mammalian muscle grafts. Am J Anat 176: 191–205.

Section III:

Chromaffin Cells

R. P. Lanza and W. L. Chick (eds.), Yearbook of Cell and Tissue Transplantation 1996/1997, 71–89.

Chapter 8

Chromaffin cell transplants in the CNS: Basic and clinical update

Jacqueline Sagen
Department of Anatomy and Cell Biology, University of Illinois at Chicago, Chicago, IL 60612, U.S.A.

The potential usefulness of chromaffin cells as a source of neuroactive agents for transplantation in the CNS is based on several promising features, including the diversity of biologically active neurotransmitters, neuropeptides, and trophic factors produced by the cells, their apparent neurochemical and morphological plasticity, and the ability to utilize cells derived from adult donors. The majority of studies using adrenal medullary transplants have been attempted for the alleviation of Parkinsonian symptoms, both in animal models and clinical studies (for review, see [1–3]). However, the general outcome from this work has been disappointing, most likely due to poor graft survival, suboptimal levels of dopamine production, or the failure of the grafts to re-establish appropriate host neural circuitry. More recent studies utilizing trophic factor support, co-grafts with trophic factor producing cells, and novel tissue preparation or cellular isolation methods may improve the outcome and feasibility of adrenal medullary transplantation approaches. In addition to Parkinson's disease, adrenal medullary transplants have been used for other applications, including chronic pain, memory, Huntington's disease, and depression models. The focus of this review is on recent developments in the use of adrenal medullary chromaffin cells as a graft source for disorders of the CNS.

Advantages of chromaffin cells as transplant donors: Neurochemistry and plasticity

Phenotypic plasticity

The rich diversity of neuroactive agents produced and their apparent plasticity and adaptability to novel environments make chromaffin cells attractive candidates for transplantation into the CNS. In addition, a strong motivation for using these cells as a graft source is the ability to transplant cells from adult donors, in contrast to CNS sources which require fetal donor material for survival in the adult brain. Even in the adult, chromaffin cells retain a good deal of phenotypic and neurochemical plasticity in response to environmental factors such as physiological stimulation, neural and hormonal influences, and growth factors [4–7]. For example, a well-studied phenomenon is the influence of trophic factors, particularly nerve growth factor (NGF) on enhancement of process outgrowth from chromaffin cells in culture [5, 7–9]. Thus, NGF promotes chromaffin cell transformation into a neuronal phenotype with similarities to sympathetic neurons. However, this transformation can be inhibited, and the cuboidal chromaffin cell morphology maintained in the presence of glucocorticoids, which simulate the normal environment of adrenal medullary cells in the *in situ* adrenal gland in proximity with the corticosteroid secreting adrenal cortex [5, 7–9]. In addition to NGF, other agents apparently can promote neurite outgrowth from postnatal chromaffin cells, including acidic and basic fibroblast growth factors (FGF) [10, 11], ciliary neurotrophic factor (CNTF) [12], and tachykinin (NK1) receptor agonists [13]. Similar neural differentiation can be induced in chromaffin cells cocultured with Schwann cells, peripheral nerve, C6 glioma cells and their conditioned media, presumably by providing a source of neurite promoting factors [14–17]. The composition of the extracellular matrix may also be important in process outgrowth from chromaffin cells since fibroblasts producing fibronectin, laminin, and cell-adhesion molecules (N-CAM) can provide a supportive substrate and promote neurite outgrowth [18].

It is clear from this discussion that environmental factors have a great deal of influence on the morphologic phenotype of chromaffin cells in culture. Thus, it may be possible to alter neurite outgrowth and promote host-graft integration when these cells are transplanted to the CNS. This approach has been attempted using exogenous trophic factor infusion as well as co-grafts with trophic factor producing cells or tissues. Although the main focus of most of these studies is towards the enhancement of chromaffin cell survival in the host CNS (see Sect. below), there is some evidence of phenotypic differentiation using these approaches. For example, NGF treatment enhances survival and produces fiber outgrowth when adrenal medullary tissue is implanted into either the anterior chamber of the eye [19, 20] or the dopamine-denervated striatum [21]. Recent studies using recombinant human NGF demonstrate that the extent of neurite outgrowth is dose-dependent on NGF concentrations [22]. Other approaches for the provision of trophic support to chromaffin cells implanted in the striatum, an NGF poor CNS site, include co-grafting with peripheral nerve [14], C6 glioma cells [23], astrocytes genetically altered to secrete NGF [24, 25], and genetically modified NGF-producing fibroblasts [26]. Using these approaches, in addition to improved chromaffin cell survival, most reported some enhanced neuronal differentiation and process outgrowth from the implanted chromaffin cells. However, the extent of this phenotypic differentiation is variable and generally appears less robust than that observed *in vitro*. It is likely that neuritic outgrowth from CNS grafts depends on several factors, including locally achieved concentrations of NGF, age of donor cells, and the provision of extracellular matrix components. For example, Cunningham *et al.* [24, 25] reported extensive neurite outgrowth from chromaffin cells co-grafted with high NGF-producing transgenic astrocytes [25], but this was only observed in co-grafts with young postnatal, not adult, chromaffin cells [24]. Differences in the type of neuritic profiles obtained, and extent of differentiation to a sympathetic neuronal phenotype have been reported using co-grafts with NGF-producing fibroblasts [26]. The neurites observed were more characteristic of mature, differentiated neurons containing microtubule arrays, neurofilaments, small clear vesicles, and long, fine processes, compared to the short, thick neuritic processes using NGF injections or peripheral nerve co-grafts. One possible explanation is the provision of extracellular matrix material such as collagen, in addition to the NGF, by the co-grafted fibroblasts [26]. In this regard, even the limited neuritic outgrowth observed in other studies may be dependent on the presence of appropriate extracellular matrix components, as the majority of these studies utilize either solid tissue pieces or suspensions of adrenal medullary tissue, which are not homogenous chromaffin cells, and contain numerous other cell types including fibroblasts. In support for this, preliminary studies in our laboratory have indicated that neither NGF infusion nor co-grafts with peripheral nerve segments enhance neuritic outgrowth in grafts of bovine chromaffin cells substantially isolated from other adrenal medullary cells (Ortega *et al.*, unpublished observations). While direct comparisons cannot be made with other studies due to additional variables including species, donor age, and immunogenicity, it is interesting to note that phenotypic differentiation can be induced in bovine chromaffin cells *in vitro* with the addition of either peripheral nerve explants or NGF and bFGF (Ortega *et al.*, unpublished observations).

Neurochemical plasticity

Catecholamines

In addition to morphologic plasticity, chromaffin cells also exhibit remarkable neurochemical plasticity, another feature that may allow for the tailoring of cells in transplantation applications. A classically studied phenomenon is that of the influence of adrenal corticosteroids on catecholamine biosynthesis. When adrenal medullary cells are isolated from adrenal cortex, the production of catecholamines is shifted over time from high to low epinephrine:norepinephrine ratios (EPI:NE), presumably owing to reduced levels of the conversion enzyme phenylethanolamine-N-methyltransferase (PNMT) [5, 6, 9, 27, 28]. Thus, in isolated chromaffin cell cultures, there is a gradual decline in EPI:NE, both in terms of cellular content [8, 9] and catecholamine release [29] from approximately 6:1 to 2:1 without the addition of exogenous trophic factors or glucocorticoids to the media (see Fig. 1). However, addition of dexamethasone to chromaffin cell cultures can markedly increase epinephrine production and enhance PNMT activity [5, 8, 30]. In addition to shifts in the EPI:NE ratios, removal of the glucocorticoid influence of the adrenal cortex may result in increased levels of dopamine production, and has been suggested as a rationale for using chromaffin cells as a graft source for dopamine in Parkinsonian disorders [1, 2, 31]. Although findings in our laboratory have not revealed substantial increases in dopamine release from chromaffin cells either *in vitro* or in CNS grafts [29, 32–34], increases in relative cellular dopamine content *in vitro* [8, 35] and dopamine and dopamine metabolite levels in graft regions and dialysates [31, 36–38] have been reported. In other studies, transplants of bovine chromaffin cells in the rat striatum do not appear to contain or release pharmacologically relevant dopamine or metabolite levels [39, 40], although such implants do contain elevated NE and EPI [40]. Thus, to some extent, species or age differences may be important in relative catecholamine production levels; e.g. comparisons between chromaffin cell cultures from neonatal rats and adult cattle indicated that relative dopamine content could be increased from less than 1% to over 25% in 1–2 week old cultures of neonatal rat chromaffin cells compared to only 6% in adult bovine chromaffin cell cultures [35].

These studies suggest that it may be possible to tailor chromaffin cells neurochemically to provide the appropriate complement of catecholamines for different neural transplantation applications. In a recent study in our laboratory, the influence of *in vitro* maintenance prior to transplantation on the neurochemical outcome of chromaffin cell grafts was determined [29]. Catecholamine secretion patterns, particularly EPI:NE ratios, were compared over time *in vitro* and in transplants using a brain slice preparation. Results of this study are shown in Fig. 1. As shown in previous work, a steady decline in EPI:NE release was obtained from bovine chromaffin cells maintained *in vitro*, reaching stable ratios of approximately 2:1 by 21 days for up to 44 days in culture. In contrast, release ratios from grafted bovine chromaffin cells (4 weeks post-grafting in the midbrain periaqueductal gray) was dependent on prior maintenance in culture. When cells were grafted within 2 days of harvesting, they retained their initial high EPI:NE release ratios, significantly higher than chromaffin cells maintained *in vitro* for similar 4 week periods. However, when cells were maintained in culture for 21 days prior to grafting to allow *in vitro* EPI:NE levels to decline, this reduced ratio was also obtained in grafted chromaffin cells. These results suggest that the CNS may not provide sufficient environmental factors to influence the spectrum of catecholamine secretion once it is altered by prior *in vitro* conditions. In addition, these findings suggest the possibility that the preparation and maintenance of chromaffin cells can be manipulated in culture before transplantation to optimize the desired pattern of catecholamine secretion. Since studies in other laboratories have suggested that the EPI and NE cells are two separable populations in the adrenal medulla [41, 42], it is also feasible that enriched preparations of fractions can be utilized for transplantation.

Acetylcholine

Chromaffin cells have also been utilized in neural transplantation studies for the provision of acetylcholine [43–46]. The differentiation to cholinergic phenotypes is apparently also owing to the plasticity of chromaffin cells in response to environmental cues. In the presence of NGF and heart cell conditioned medium, adrenal chromaffin cells from neonatal or adult rats can be converted into cholinergic sympathetic neurons *in vitro* [5]. Similarly, when transplanted into NGF-rich CNS sites such as the hippocampus, chromaffin cells can develop neuron-like cholinergic phenotypes [44]. With the injection of cholinergic differentiation factor/leukemia inhibitory factor (CDF/LIF), a protein secreted by cultured heart cells, the conversion of chromaffin cells with cholinergic phenotypes in hippocampal grafts can be increased [43].

Neuropeptides

In addition to classical neurotransmitters, chromaffin cells also produce numerous neuroactive peptides as well as trophic factors, which can be utilized for neural transplantation. For example, the synthesis and release of opioid peptides is notable, and has been utilized in neural transplantation studies for the alleviation of pain (see Sect. below). The adrenal gland synthesizes a range of large, intermediate, and small opioid peptides derived from proenkephalin [47–49]. While the level of proenkephalin gene expression and opioid peptide synthesis in chromaffin cells is species dependent, with the rat adrenal gland apparently among the lowest [50–53], this is also subject to environmental factors. In particular, denervation of the adrenal gland produces marked increases in both preproenkephalin mRNA and enkephalin containing peptides in the rat adrenal medulla [54, 55]. This can be mimicked in explant cultures of rat adrenal medullary tissue [56, 57]. Both preproenkephalin mRNA and enkephalin-containing peptides are markedly increased after 2–4 days in organ culture. In our laboratory, similar increases in preproenkephalin mRNA have been found in adrenal medullary transplants in the rat spinal cord

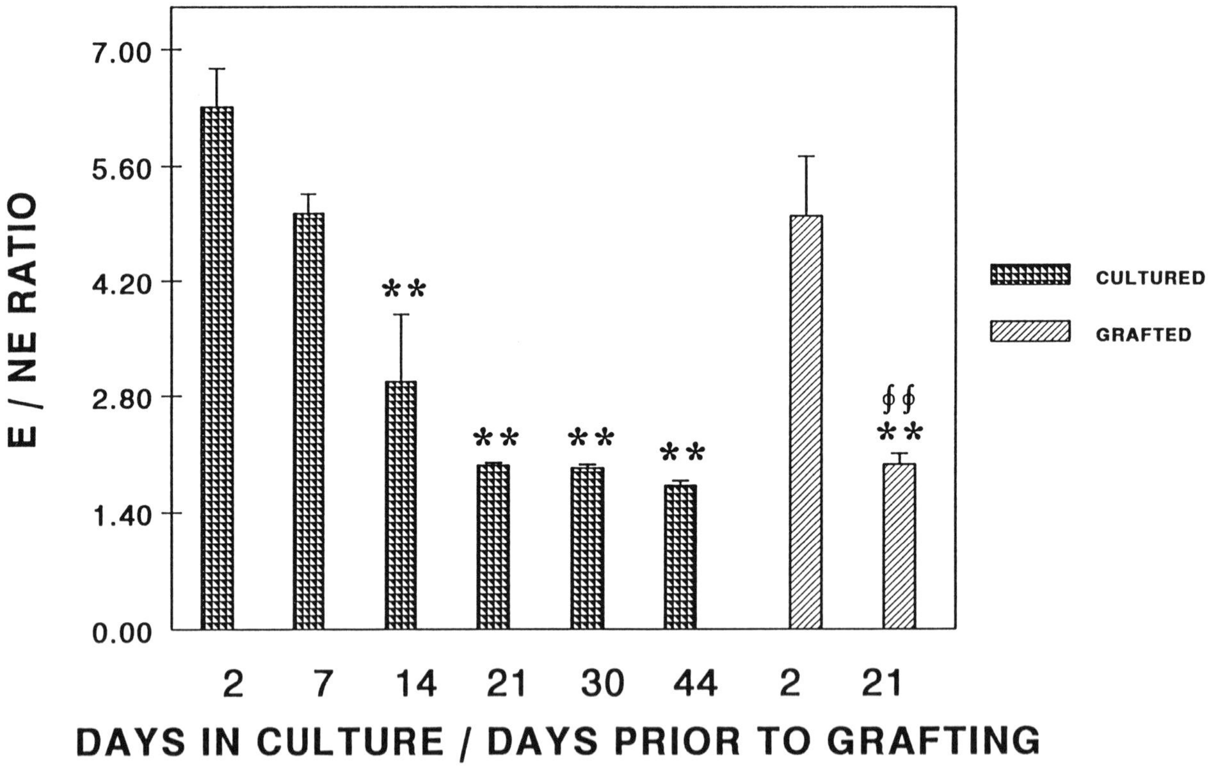

Fig. 1. Changes in epinephrine/norepinephrine (E/NE) release ratios in cultured bovine chromaffin cells (patterned bars) and grafted bovine chromaffin cells (diagonal bars) at various times following maintenance *in vitro*. ** $p < 0.01$ compared to ratios from 2 day cultures; §§ $p < 0.01$ compared to release from cells grafted after 2 days in culture. Overall F (df = 5, 19) = 28.77. Total catecholamine release from cultured cells (ng/5 min/5 $\times$ 10^5 cells) = 386.06 $\pm$ 70.24 (2 days); 347.76 $\pm$ 76.42 (7 days); 242.55 $\pm$ 80.47 (14 days); 299.32 $\pm$ 79.97 (21 days); 284.42 $\pm$ 75.49 (30 days); and 263.51 $\pm$ 73.24 (44 days). Total catecholamine release from brain slices (ng/5 min/1 $\times$ 10^5 initially implanted) = 14.30 $\pm$ 3.81 (from 2 day cultures) and 14.82 $\pm$ 3.62 (from 21 day cultures). Adapted from Sagen and Ortega, J Neurochem 63: 1159–1162, 1994.

[58]. Additional increases in proenkephalin mRNA and enkephalin biosynthesis in rat or bovine chromaffin cells can be induced by glucocorticoids [59, 60], vasoactive intestinal peptides [61], or bFGF [62]. Thus, it is likely that alterations in neuropeptide production as well as catecholamine production can be manipulated in chromaffin cells for neural transplantation applications.

Trophic factors

Finally, adrenal medullary chromaffin cells produce a variety of neurotrophic factors, neuropeptides, and cytokines with neurotrophic activity, a "trophic cocktail" [63], which may be useful in aiding repair in CNS injury and neurodegenerative diseases. Indeed, the provision of repair factors by adrenal medullary grafts has been suggested as a possible mechanism for the functional improvements observed in Parkinson's disease models and some human patients, even with poor chromaffin cell survival [3, 64–66]. Thus, increased host striatal tyrosine hydroxylase fiber staining can be observed independently of chromaffin cell survival, suggesting the sprouting of spared dopaminergic fibers in response to trophic factor support from the grafts and/or surgical procedures. Several growth factors and cytokines are found in chromaffin cells, including bFGF, transforming growth factor-β (TGF-β), interleukins, and a CNTF-like agent [63]. In particular, bFGF is found in chromaffin granules and can be increased by stimulation of cell surface receptors [67, 68]. The provision of this trophic factor by neural transplants may be useful in the protection and restoration of injured neurons, e.g. dopaminergic neurons in Parkinson's disease [63]. However, thus far release of bFGF *in vitro* from chromaffin cells has not been obtained, even following stimulation with high K^+ or cholinergic agonists [63, 67]. Nevertheless, chromaffin cells can apparently release neurotrophic factors since media conditioned by purified bovine chromaffin cells can promote the survival of several neuronal cultures of CNS and PNS origin [69].

Chromaffin cell grafting in behavioral and clinical models

Parkinson's disease

Trophic factor and co-grafting approaches to improve graft viability

The above review suggests that adrenal medullary chromaffin cells may be useful as a graft source for biogenic amines, neuropeptides, or trophic factors in the treatment of neurological disorders. Thus, chromaffin cells have been extensively utilized for transplantation into the CNS. The majority of this work has been focused on the use of adrenal medullary transplants in the striatum or lateral ventricles for the restoration of the nigrostriatal dopaminergic system in Parkinson's disease. Owing to the initial promising results in animal models, several hundred Parkinsonian patients received adrenal medullary grafts, primarily autografts (for review, see [1–3]). Unfortunately, this procedure produced only moderate improvement at best in the majority of closely followed cases. This may be in part due to poor graft survival, as most autopsies performed thus far have revealed few or no surviving chromaffin cells [66, 70–74]. Thus, this approach has largely waned in recent years in favor of more promising findings using fetal nigral grafts. Nevertheless, techniques to improve chromaffin cell survival in CNS grafts may improve the feasibility of this approach, and have been the primary focus of efforts in this field. These include both improved preparation techniques and the provision of trophic factor support, either by infusion or co-transplantation with trophic factor-producing cells or tissues. In particular, in animal models, the delivery of NGF appears to markedly enhance chromaffin cell survival in CNS grafts, as well as improve behavioral outcomes. Direct NGF administration by chronic infusion resulted in a significantly improved and sustained reduction in rotational behavior in rats with unilateral lesions of the nigrostriatal dopaminergic system [21]. This infusion also increased the number of surviving chromaffin cells in the grafts. In initial clinical trials, intraputamenal NGF infusion for one month following adrenal medullary autografting appeared to produce some sustained beneficial effects in at least one patient with advanced Parkinson's disease [75, 76].

Peripheral nerve is an alternative source of trophic factors, and has the advantage of providing a sustained cellular source of these agents in the local environment when co-grafted with adrenal medullary tissue. The feasibility of this approach has been demonstrated in both rodent and primate models [14, 77–84]. In MPTP (1-methyl-4-phenyl-1,2,3,6-tetrahydropyridine)-treated mice, co-grafted minced sciatic nerve tissue significantly enhanced chromaffin cell survival in adrenal medullary grafts [77]. When the donor sciatic nerve was retransected 24 hours prior to harvesting to increase NGF content, numerous viable chromaffin cells could be found in co-grafts for at least 12 months post-grafting [78, 79]. In addition, host dopaminergic fiber density and striatal dopamine concentrations were increased in these co-grafted animals. Using a similar approach in MPTP-treated hemiparkinsonian non-human primates, intrastriatal co-grafting of autologous adrenal medullary tissue fragments with minced sural nerve increased chromaffin cell survival 4–8-fold [14]. In addition, improved motor performance as assessed by a learned motor task could be demonstrated in cografted monkeys [84].

Cografting approaches have recently been applied in limited clinical studies with promising findings [85–88]. Using autologous intercostal nerve segments, adrenal medullary-peripheral nerve co-grafts in the caudate/putamen produced significant improvement in motor performance for up to 24 months following implantation in 3–4 of 5 patients (depending on the measurement employed) with advanced Parkinson's disease [87]. In another study, modest improvement was also reported in a one-year follow-up using similar cograft tissues and a transcallosal implantation approach [85]. This group also recently reported modest recovery in a patient using cografts with pretransected sural nerve tissue [86]. Autopsy material from this patient revealed numerous surviving tyrosine-hydroxylase staining chromaffin cells and a dense network of tyrosine hydroxylase immunoreactive fibers most likely of host origin at one year post-transplantation. Using autologous sural nerve and human fetal adrenal tissue, another group reported early improvement with some long-term recurrences in a group of 10 patients with severe Parkinson's disease [88].

Other potential cografting sources to improve chromaffin cell survival include C6 glioma cells [23], genetically altered NGF-producing astrocytes [24, 25], and genetically engineered NGF-producing fibroblasts [26]. In rats with unilateral 6-hydroxydopamine lesions of the nigrostriatal pathway, increased numbers of tyrosine hydroxylase positive adrenal chromaffin cells in the striatum were found in animals receiving cografts of amitotic C6 glioma cells producing neurotrophic factors [23]. In addition, there was a more marked reduction in ipsilateral rotations after amphetamine in these animals than in non-cografted groups. Similarly, cografts with genetically altered astrocytes enhanced chromaffin cell survival 3–12 fold in the dopamine-denervated striatum, and was paralleled by a 60% reduction in apomorphine-induced rotational behavior [24, 25]. Chromaffin cell survival was also markedly enhanced by cografts of primary fibroblasts genetically engineered to produce high levels of NGF [26].

Improvements in graft preparations

A second major thrust towards enhanced chromaffin cell survival in the CNS are improved methods for preparation and handling prior to transplantation. This includes the perfusion of adrenal medulla to remove blood cells which could potentially promote graft rejection [89], isolation of chromaffin cells from other more immunogenic cell types in the adrenal medulla [90–96], and the preparation of elongated ribbons of adrenal medullary tissue to increase host-graft contact [97, 98]. Using the latter ribbon technique, significant behavioral improvements and graft survival was obtained in primates. Using perfusion and isolation techniques, long-term survival of chromaffin cells from xenogeneic sources may be possible. In our laboratory, we have used this approach to transplant bovine chromaffin cells into the rat CNS. Xenogeneic donor sources are particularly promising in light of the potential morbidity associated with adrenalectomy for autografts and the limited availability of human allogeneic donors. However, in spite of the relative immune privilege enjoyed by the CNS, the long-term survival of xenografts in the absence of immunosuppression is difficult to achieve. Using a chromaffin cell isolation procedure which results in greater than 95% pure population of chromaffin cells from the bovine adrenal medulla, our laboratory has demonstrated long-term survival of bovine chromaffin cells in the CNS with only short-term immunosuppression [92, 93, 96]. Dense clusters of tyrosine hydroxylase positive chromaffin cells can be found following transplantation into several CNS sites, including the NGF-poor striatum, without the application of exogenous trophic factors [93, 95, 96]. An example of a bovine chromaffin cell graft in the caudate nucleus is shown in Fig. 2. This was obtained 12 months post-transplantation from a rat that was immunosuppressed daily with cyclosporine A (10 mg/kg, i.p.) for one month following surgery. These findings indicate that xenogeneic chromaffin cells can survive for prolonged periods in the CNS, and suggest that the removal of nonchromaffin supporting cells in the adrenal medulla may facilitate this survival. In order to explore this possibility, survival of bovine chromaffin cells was compared in animals receiving either perfused adrenal medullary cell suspensions (containing all cell types in the adrenal medulla), isolated chromaffin cells, or isolated chromaffin cells recombined with the adherent nonchromaffin adrenal medullary supporting cells. Results indicated poor or no chromaffin cell survival, as well as macrophage infiltration, and cellular debris in both groups containing nonchromaffin supporting cells, and suggested that the presence of these cells may be detrimental to the long-term survival of xenogeneic chromaffin cells [95, 96]. In particular, endothelial cells are notoriously antigenic, and may stimulate an immune rejection even in the CNS. Using an in vitro lymphocyte proliferation assay, recent studies in our laboratory have indicated that bovine endothelial cells isolated from either the adrenal medulla or aorta can potently stimulate proliferation of lymphocytes from both rat spleen and human blood [90, 91]. In contrast, this was not seen with exposure to isolated bovine chromaffin cells. Together these studies suggest that the purification of chromaffin cells from antigenic cell types in the adrenal medulla can improve survival of these cells in CNS grafts, even when obtained from xenogeneic sources.

Another approach to improve survival of xenogeneic chromaffin cells in the CNS is the use of immunoisolation techniques (for review, see chapters in Flanagan *et al.* [99]). Permselective polymer membranes can be used, which have selective diffusion properties that provide a protective barrier for the encapsulated cells from circulating lymphocytes, complement proteins and antibodies but simultaneously allow free passage of nutrients and released secretory products. Using this approach, bovine chromaffin cells were transplanted into the striatum of animals unilaterally lesioned with 6-hydroxydopamine [100]. Either cell-loaded or empty microcapsules were used, and rotational behavior was assessed using apomorphine. Results are shown in Fig. 3. Rotations were significantly reduced by 60%

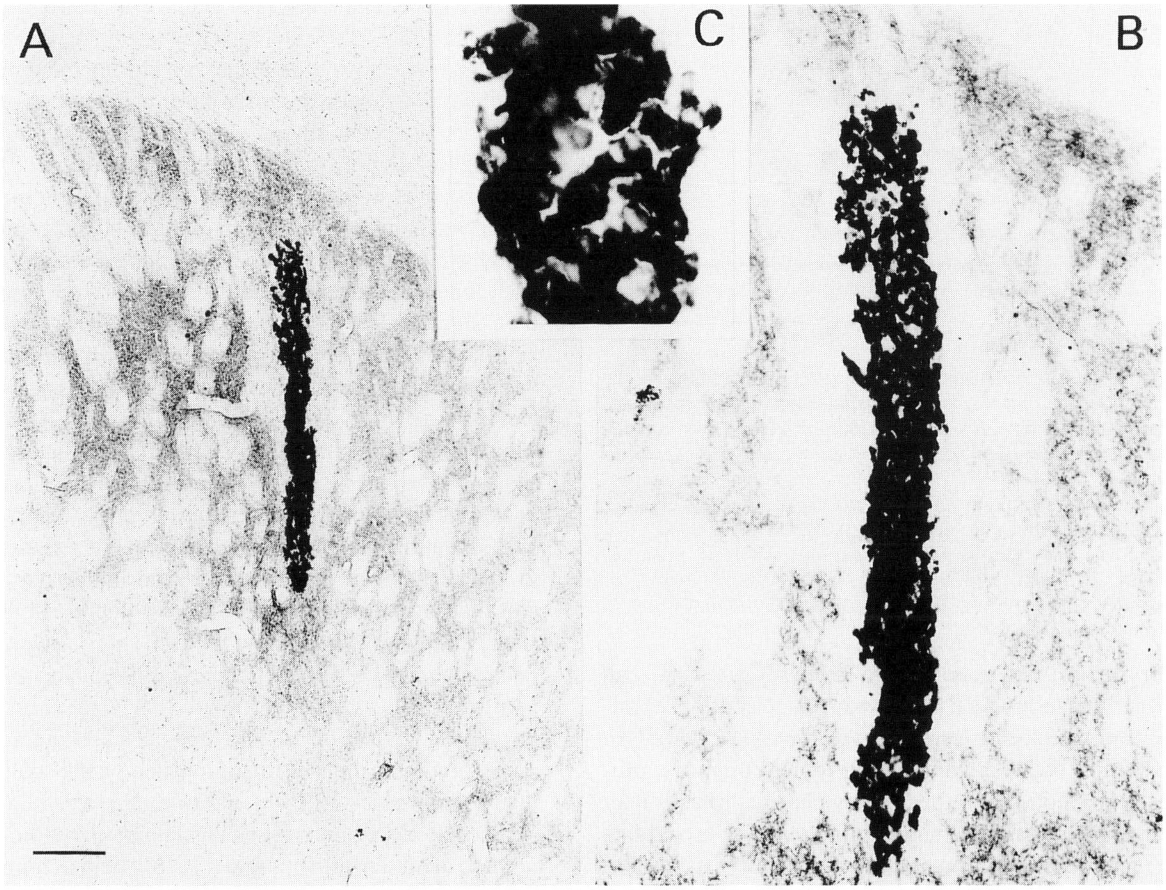

Fig. 2. (A) Low, (B) medium and (C) high power photomicrographs of TH-stained sections through transplants in rats receiving isolated bovine adrenal medullary chromaffin cells alone. Note the numerous healthy, rounded morphology of the chromaffin cells at 12 months post-transplantation, despite immunosuppression for only 1 month post-implantation. Scale bar in (A) represents the following magnifications: (A) = 250 μm; (B) = 110 μm; (C) = 16.7 μm. Taken from Schueler *et al.*, Cell Transpl 4: 55–64, 1995.

in animals with chromaffin cell-loaded capsules for at least 4 weeks post-transplantation. At this point, numerous viable tyrosine hydroxylase positive chromaffin cells could be found in the implanted microcapsules. No changes in rotational behavior were found in animals with empty capsule implants.

As another possible approach, chromaffin cells preadhered to microcarrier beads appears to improve cell survival in transplants to the striatum of unilaterally lesioned rats [101]. In addition, reduced apomorphine-induced rotations were obtained in animals receiving chromaffin cell-microcarrier beads compared to microcarriers alone.

Pain

Preclinical studies in chronic pain models

A second application for adrenal medullary or chromaffin cell transplantation is in the treatment of pain. The original rationale for this approach was that, as described above, chromaffin cells synthesize and release opioid peptides in addition to catecholamines. Both of these classes of agents are known to produce antinociception when injected into the spinal subarachnoid space, presumably via interaction with host spinal α-adrenergic (α_2) receptors and opioid (μ/δ) receptors. In addition, the co-activation of these recep-

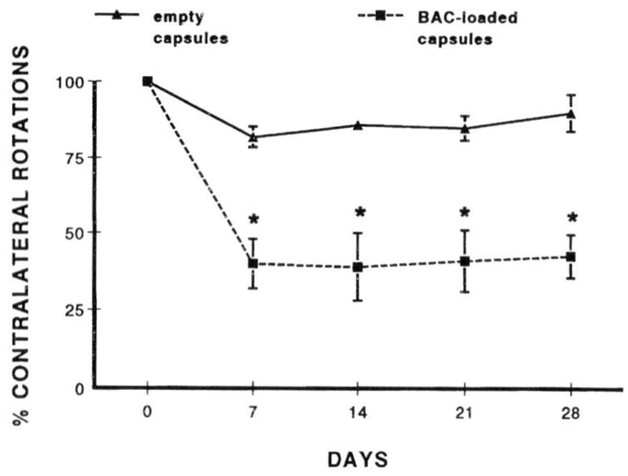

Fig. 3. Apomorphine-induced rotational behavior following the transplantation of empty and chromaffin-cell loaded microcapsules as a function of time. A significant reduction was observed between the implanted chromaffin cell-loaded capsules and the empty microcapsules ($p < 0.001$). The postimplantation turning rate was significantly different from baseline in animals that received encapsulated bovine chromaffin cells (* $p < 0.01$), but not in animals that received empty capsules. Adapted from Aebischer *et al.*, Brain Res 560: 43–49, 1991.

tors, even with subeffective levels of agonists, can synergize to produce profound antinociception, possibly with reduced tolerance development [102–105]. In addition to catecholamines and opioid peptides, other neuropeptides synthesized by chromaffin cells may have antinociceptive activity, including neuropeptide Y, neurotensin, and somatostatin. Finally, the potential role of neurotrophic factors produced by chromaffin cells must be considered. For example, recent reports have suggested that neurotrophic factors can have direct antinociceptive actions [106].

In our laboratory, we have utilized chromaffin cell transplants for the reduction of pain in several animal models, and in preliminary clinical trials [32, 34, 107–116]. The earliest studies in our laboratory utilized standard analgesiometric assays to assess whether released neuroactive substances from adrenal medullary allografts could exert behavioral changes [32, 109, 110]. These included the paw-pinch, hotplate, and tail-flick tests, and results indicated that the level of acute pain tolerance in animals with adrenal medullary transplants was much greater than in control animals following injection of a low dose of nicotine to stimulate release from chromaffin cells. Furthermore, this antinociceptive response was most likely due, at least in part to the co-release of opioid peptides and catecholamines from transplanted cells, since it could be attenuated by opiate antagonist naloxone or α-adrenergic antagonist phentolamine, and antinociceptive potencies correlated with CSF catecholamine and Met-enkephalin levels in spinal cord superfusates [32, 109, 113].

More recently, we have utilized chronic pain models that more closely resemble chronic pain syndromes found in the clinical situation. In our laboratory, we use a model for chronic joint inflammation and a peripheral neuropathy model. The adjuvant arthritic rat model was chosen for chronic inflammation since it shares many similarities with human rheumatic disease [117–119], and it has been utilized for the study of chronic pain since these rats present a variety of symptoms similar to those seen in human chronic pain conditions, including appetite depression and loss of body weight, irritability, and hyperventilation [120–124]. Thus, in the adjuvant-induced arthritic rat, manifestations of pain can be quantified by measuring changes in body weight and hyperventilation. It has been suggested that the weight loss and subsequent retardation of weight gain in these rats may be indicative of a chronic state of pain resulting from a depressed drive state, and can be attenuated by sectioning ascending pain pathways [121, 122]. Hyperventilation responses in arthritic rats have been shown to correlate accurately with the course of inflammatory disease and accurately predict responses to appropriate pharmacologic intervention [125–127]. In addition, hyperventilation is associated with numerous chronic pain syndromes in man, including arthritis and possibly pain in humans [128]. Adjuvant arthritis is induced in normal rats by injection of *Mycobacterium butyricum* intradermally into base of the tail. This leads to inflammation of the synovial tibio-tarsal joint with a detectable onset (measured as increased joint circumference) within 1 week, peak severity at 3–5 weeks, and decline following 8–10 weeks post-inoculation. Fig. 4A shows the time course and progression of the inflammatory process in animals with control striated muscle or adrenal medullary allografts [111, 115]. Note that the adrenal medullary transplants do not alter the course of the arthritic process. The effect of adrenal medullary or control transplants on weight loss in arthritic rats is shown in Fig. 4B. Similar to that reported for adjuvant arthritic rats in other studies [120, 122, 125], arthritic rats with control transplants exhibited retardation in weight gain during the early phases of the disease for the first 3 weeks post-inoculation, indicating no significant weight gain during this period. In contrast, animals with adrenal medullary transplants gained significant weight during this period compared to pre-inoculation. Differences in weight gain were significant between adrenal medullary and control transplant groups, particularly apparent during the acute inflammatory phases, since body weights of control transplanted animals recovered as the inflammation began to subside. Changes in ventilatory responses during adjuvant arthritis phases are illustrated in Fig. 4C. Respiratory parameters were determined using whole body plesthomography. In animals with control transplants, hyperventilation as assessed by tidal volume approximately paralleled the time course and severity of the inflammatory process. Increased tidal volume was apparent by one week post-inoculation, reached a peak at 2 weeks, and remained elevated at high levels until 7 weeks post-inoculation. In contrast, no significant increases in tidal volume were observed in adrenal medullary transplanted animals throughout the duration of the study, and remained at the low levels found in normal, non-arthritic controls.

These studies indicate that adrenal medullary transplants in the spinal subarachnoid space can reduce basal pain levels resulting from chronic inflammation. An example of an adrenal medullary transplant in the spinal subarachnoid space of an arthritic rat is shown in Fig. 5. The chromaffin cells, stained with a tyrosine hydroxylase antibody, appear to retain their *in situ* cuboidal morphology, and cluster in the CSF surrounding the host spinal cord.

The second chronic pain model used in our laboratory is the chronic constriction nerve injury model described by Bennett and Xie [129]. Chronic constriction nerve injury involves a peripheral nerve injury induced by loosely ligating the sciatic nerve unilaterally with chromic gut ligatures (for details see [107, 129]). This results in abnormal pain behaviors closely paralleling clinical neuropathic pain syndromes such as causalgia, including allodynia (perception of innocuous stimulus as painful) and hyperalgesia (increased sensitivity to a painful stimulus). These abnormal pain states are temporary, recovering by 7–9 weeks when normal responses to sensory stimuli are re-established. Using adrenal medullary allografts, prolonged alleviation of both allodynia and hyperalgesia was observed in animals with peripheral neuropathy, compared to control transplanted animals [107]. In addition, xenogeneic chromaffin cells can also effectively reduce these pain symptoms [108]. Fig. 6 shows some results from these studies. Allodynia was determined by placing animals on a normally innocuous cold copper surface, and comparing the duration of hindpaw lifting and guarding on the nerve-injured side with the intact side. As shown in the top panel of Fig. 6, unilateral nerve injury results in increased lifting on the injured side compared to the intact side, at 1 and 2

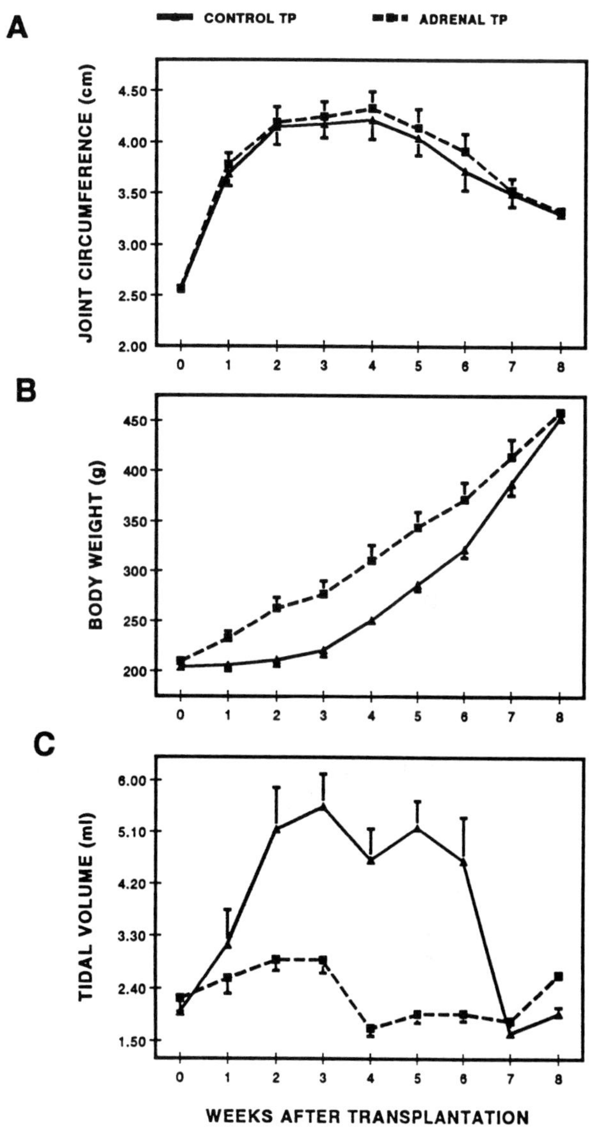

Fig. 4. Time course (in weeks) of alterations in (A) tibiotarsal joint circumference, (B) body weight (mean ± S.E.M.), and (C) tidal volume following induction of polyarthritis with complete Freund's adjuvant. 0 indicates pre-inoculation values. Animals received either adrenal medullary transplants (squares, $n = 14$) or control striated muscle transplants (triangles, $n = 10$) in the lumbar spinal subarachnoid space at the time of adjuvant inoculation. Adapted from Wang and Sagen, Pain 63: 313–320.

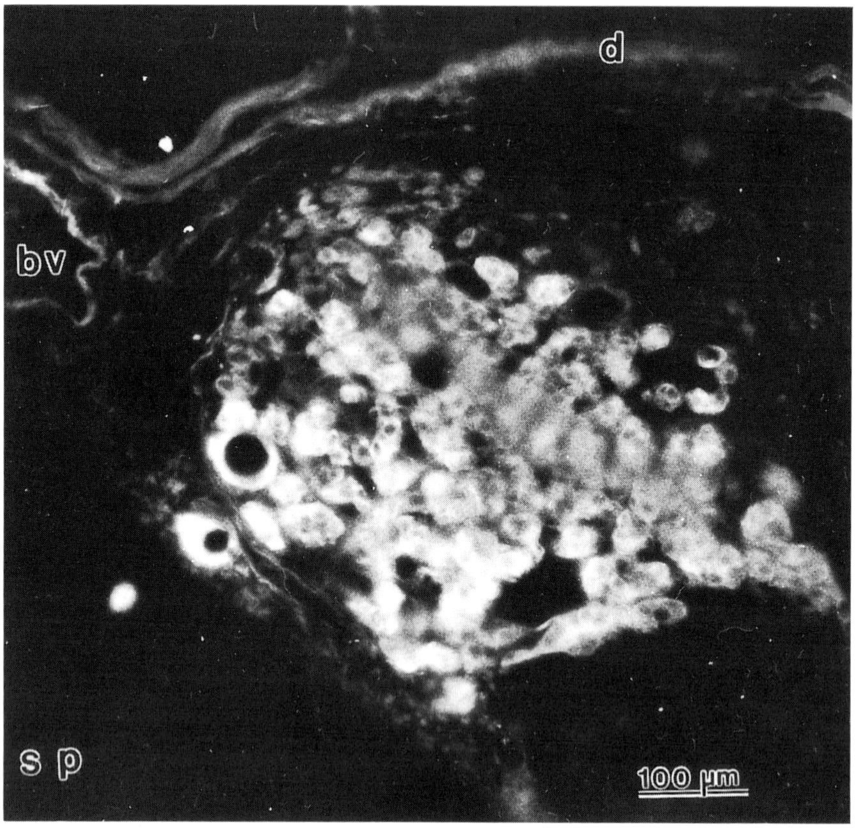

Fig. 5. Appearance of an adrenal medullary transplant 10 weeks after grafting in the spinal subarachnoid space of an arthritic rat. Chromaffin cells are immunocytochemically stained with a tyrosine hydroxylase antibody and a rhodamine-linked secondary antibody. sp = spinal cord; d = dura; bv = blood vessel. Taken from Wang and Sagen, Pain 63: 313–320, 1995.

weeks post-injury. Two weeks following induction of peripheral nerve injury, rats were implanted with either bovine chromaffin cells or control bovine fibroblasts. In rats with control transplants, allodynia persisted for the duration of the study. In contrast, this allodynia was reversed to pre-injury control levels in animals with chromaffin cell transplants, and remained low for the rest of the study. Hyperalgesia was determined by comparing the latency of both hindpaws to withdrawal from a noxious thermal stimulus. In this case, hyperalgesia is indicated by negative difference scores, as the nerve-injured side responds more rapidly. Prior to nerve injury, difference scores are near zero as expected. However, at 1 and 2 weeks post-ligation, negative difference scores, or hyperalgesia, is apparent. Within 1 week following chromaffin cell implantation, (3 weeks post-nerve injury), this hyperalgesia was significantly attenuated. In contrast, hyperalgesia persisted in animals with control transplants. These results indicated that abnormal exaggerated responses to both innocuous and noxious stimuli following peripheral nerve damage can be reduced by chromaffin cell transplants, and suggests that these transplants may be a viable approach in the treatment of traditionally difficult to manage pain syndromes such as those resulting from injury to the nervous system.

In summary, the results in chronic pain models in our laboratory have indicated beneficial effects of adrenal medullary transplants in alleviating abnormal pain behaviors. Studies in other laboratories have generally supported this concept using several models [130–133]. Using the tail flick test, a recent report suggests that acute antinociceptive responses in adrenal medullary transplanted animals are mediated by the release of neuroactive substances such as monoamines from the transplants, since antinociceptive effects are could be enhanced by amitriptyline treatment in adrenal medullary, but not sham-

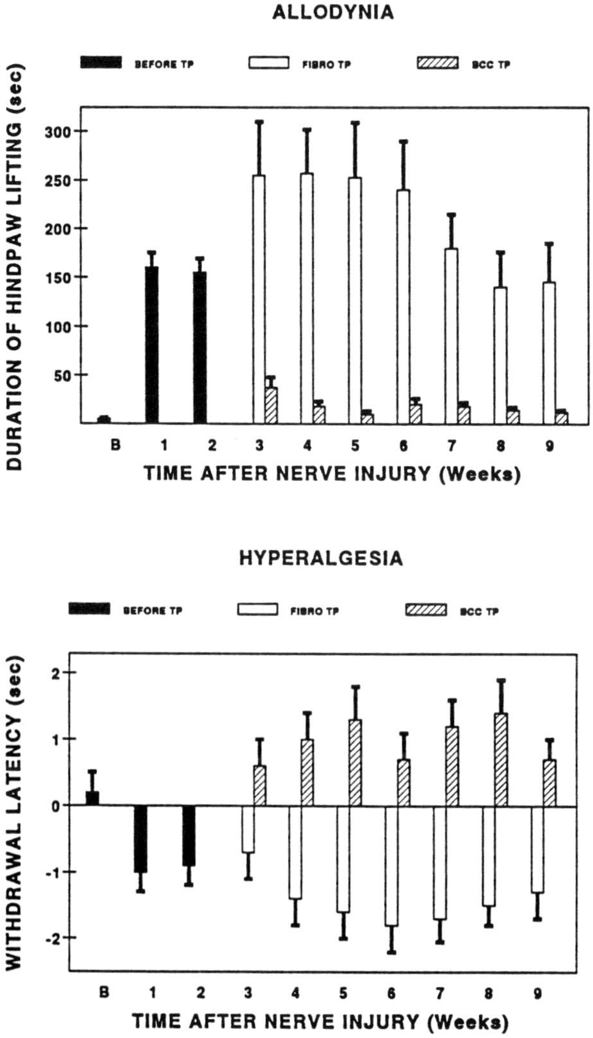

Fig. 6. Changes in degree of cold allodynia (top panel) and thermal hyperalgesia (bottom panel) over time following unilateral peripheral nerve injury. The abscissa is time (weeks) after sciatic nerve ligation of the right hind paw. Control bovine fibroblast ($n = 12$) or chromaffin cell ($n = 20$) implants were given two weeks after nerve ligation. Difference scores were calculated by subtracting left control hind paw scores from right ligated hind paw scores. No differences between right and left paw withdrawals before ligation were seen. Data are represented as mean ± S.E.M. 'B' = baseline. Cold allodynia was assessed by the total duration of hind paw lifting on a cold (5 °C) metal surface. Thermal hyperalgesia was assessed by measurement of the animal's response to a radiant heat source aimed at the hind paw plantar skin. Adapted from Hama and Sagen, Brain Res 651: 183–193, 1994.

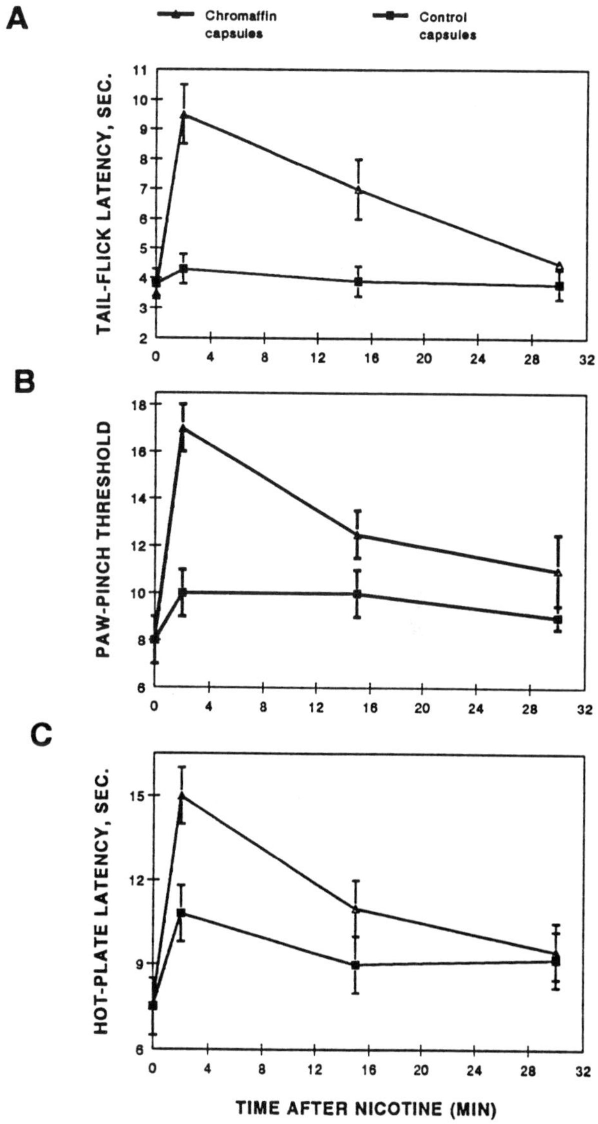

Fig. 7. The effect of encapsulated bovine chromaffin cells implants on pain sensitivity. The ordinate is the threshold for response to noxious stimuli as measured by the tail flick test (A), paw pinch test (B), and hot plate test (C) 8 weeks following transplantation into the subarachnoid space of the spinal cord. Each point represents the mean ± S.E.M. for chromaffin cell loaded capsules (triangles, $n = 29$) or empty capsules (circles, $n = 16$). The abscissa is the time course of responses to noxious stimuli before (at 0) and following nicotine stimulation. Adapted from Sagen *et al.*, J Neurosci 13: 2415–2423, 1993.

implanted animals [131]. Using a 'tonic' pain model, the formalin test, Vaquero *et al.* [133] reported a reduction in basal pain responses in animals that had received adrenal medullary allografts in the spinal subarachnoid space 2 months earlier, compared to control rats. In addition, this improvement could be reversed by opiate antagonist naloxone, suggesting the grafts spontaneously secrete opioid peptides, and that analgesic effects of adrenal medullary implants are partially mediated by activity at host opioid receptors. Another group, Ginzburg and Seltzer [130], using a peripheral

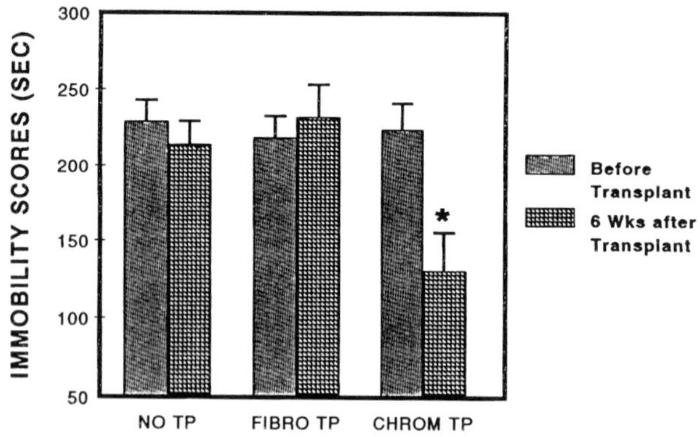

Fig. 8. The effect of chromaffin cell transplants on immobility scores in the forced swimming test (FST). Immobility scores were recorded during a 5-minute interval on the second day of the FST. Groups of animals that subsequently underwent in vivo microdialysis included the nontransplanted (no TP, $n = 5$) isolated bovine adrenal medullary fibroblast transplanted (Fibro TP, $n = 5$), and isolated bovine chromaffin cell transplanted (Chromaffin TP, $n = 5$). Values represent the mean ± S.E.M. for each group before (stippled bars) and 6 weeks following transplantation (geometric bars). Significant differences in post-transplantation FST scores between groups are denoted by asterisks (* $p < 0.05$). Adapted from Sortwell *et al.*, Exp Neurol 130: 1–8, 1994.

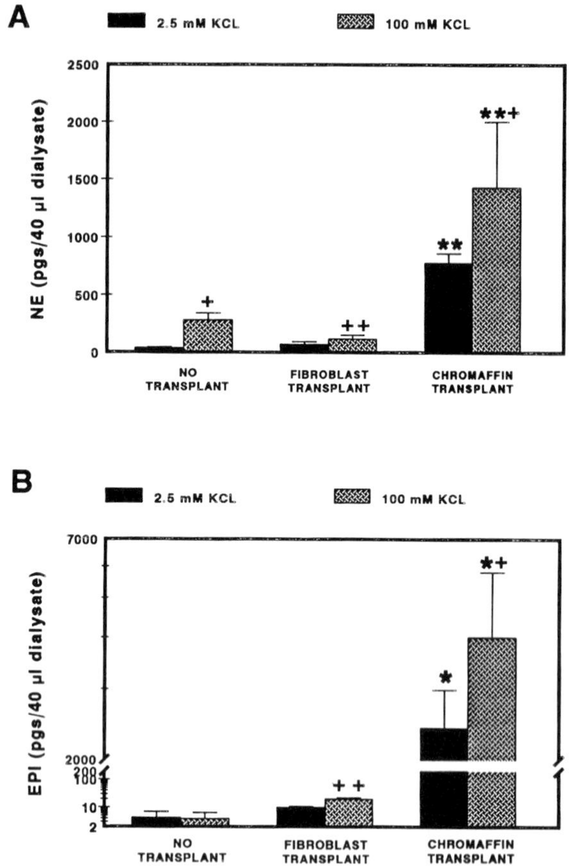

Fig. 9. The effect of bovine chromaffin cell transplants on dialysate levels of catecholamines in the frontal cortex. (A) Norepinephrine levels (pg/40 μl dialysate) and (B) epinephrine levels for three different treatment groups. Groups of animals included nontransplanted ($n = 5$), bovine adrenal medullary fibroblast transplanted ($n = 5$), and bovine chromaffin cell transplanted ($n = 5$). Values represent mean ± S.E.M. for each group at 2.5 mM KCl concentrations in the dialysate (solid bars) and at 100 mM KCl concentrations (hatched bars). Significant differences between transplant groups are denoted by asterisks (* $p < 0.05$; ** $p < 0.01$). Significant differences following potassium stimulation are denoted by plus signs (+ $p < 0.05$; ++ $p < 0.01$). Adapted from Sortwell *et al.*, Exp Neurol 130: 1–8, 1994.

nerve transection model, reported suppression of neuropathic pain behaviors in animals with transplants in the subarachnoid space. This group observed a marked reduction in autotomy throughout the 8 week follow-up period post-nerve injury in adrenal medullary implanted animals compared to sham-implanted and non-implanted animals. An interesting observation was an apparent increase in the density of catecholamine-positive varicosities in the host dorsal horn at termination sites of bulbospinal noradrenergic pathways. This may indicate sprouting of endogenous host processes induced possibly by trophic factors from the grafts.

Interestingly, recent studies in our laboratory have also suggested the possibility that trophic factors or other agents released by the transplanted chromaffin cells may contribute to their beneficial effects via neuroplastic remodeling. For example, adrenal medullary transplants can reduce the apparent activation of immediate-early gene c-*fos* in spinal cord neurons of arthritic rats [134]. It is thought that the activation of immediate-early genes may be indicative of abnormal activation of spinal neurons in response to noxious stimuli [135–137]. In addition, using the peripheral neuropathy model, increased levels of putative nitric oxide (NO) synthase, NADPH-diaphorase, are induced following nerve injury, which may be indicative of increased NO synthesis. This, in turn, could lead to further pathological changes in the spinal cord, and prolonged abnormal hyperexcitability in spinal neurons characteristic of persistent pain states [138]. In animals with adrenal medullary, but not control transplants, the induction of NADPH-diaphorase was reduced to levels similar to intact, non-injured animals [139]. Together, these results suggest that adrenal medullary transplants may intervene in the cascade of neuropathological events consequent to chronic peripheral injury or inflammation, and promote CNS recovery from these insults.

Another approach for transplanting xenogeneic cells into the spinal subarachnoid space is the use of immunoisolatory polymer capsules. This approach has been used for the transplantation of bovine chromaffin cells in the spinal subarachnoid space [140, 141]. Using acute pain models (Fig. 7), encapsulated bovine chromaffin cells produced antinociception following the injection of low doses of nicotine, in contrast to control (empty) capsules. These findings are similar to those described earlier using adrenal medullary allografts [109]. Encapsulated bovine chromaffin cells have also been shown to reduce chronic neuropathic pain syndromes when transplanted to the spinal subarachnoid space [141].

Clinical studies

The promising findings from our rodent studies led to the initiation of limited clinical trials in terminal cancer patients with intractable pain at the University of Illinois hospital [112, 116]. Consenting patients were chosen on the basis of their need for escalating doses of opioids which were providing insufficient pain relief. Of 5 patients in the study, 4 experienced progressive pain reduction (visual analog scale), 3 remained pain-free until death (4.5, 11, and 12 months post-transplantation). The results of these preliminary studies, although subject to interpretation due to the lack of controls, potential placebo effects, etc., were overall encouraging, and suggest that neural transplantation may provide an alternative or adjunct therapy in the management of intractable pain. Other groups have recently reported similar findings using modifications of this approach. Lazorthes *et al.* [142, 143] have reported a clinical trial involving 8 terminal cancer patients who received adrenal medullary allografts in the spinal subarachnoid space via similar procedures. A multidisciplinary pain evaluation (up to 1 year follow-up in 2 of the patients) revealed a progressive decrease in pain scores, concomitant with decrease or stabilization in analgesic drug intake, and increased levels of Met-enkephalin in the CSF in the majority of patients. However, activated lymphocytes (CD4+ and CD8+ subset of T cells) were also detected in some CSF samples, suggestive of a host immune response. It should be noted that there was earlier report [144] describing a disappointing outcome, however it is difficult to evaluate, as viability of the pretransplant tissue was not determined and the patient succumbed to his disease within a very short period. In another group, Aebischer *et al.* [145, 146] describe the implantation of encapsulated bovine chromaffin cells into 10 cancer patients, 8 in the spinal subarachnoid space and two to the lateral ventricles. Of these, 7 (all with implants in spinal subarachnoid space) showed decreased intake of pain medication and improved pain scores. Interestingly, among the most successful were patients with pain of neurogenic origin (personal communication). It is also worth noting that, in all three of the long-term clinical reports, there was some delay in observation of full analgesic benefits of the transplants, in spite of rapid increases in catecholamine and/or opioid peptide levels (patients experienced a moderate pain reduction in early post-transplantation phases, and progressively greater pain reduction in the following weeks). Although not directly comparable to rodent models, these findings again suggest the possible contribution of neuroplastic remodeling in the beneficial effects of the transplants.

Depression

Adrenal medullary and chromaffin cell transplants have also been utilized to provide a source of catecholamines for antidepressant effects in animal models [33, 147–151]. For these studies, both the learned helplessness model [152] and the forced swimming test [153] were employed. Using the learned helplessness model, it was shown that transplants of adrenal medullary allografts to the frontal neocortex of rats could prevent escape deficits induced in non-transplanted and control transplanted rats [149]. Similarly, adrenal medullary, but not control transplants, in the frontal neocortex could reduce immobility in the forced swimming test [150]. This was most likely due to catecholamine release from the transplants, since it could be reversed by pretreatment with specific adrenergic antagonists [147]. The transplantation of xenogeneic bovine chromaffin cells produced similar effects when assessed by the forced swimming test [33, 151]. Fig. 8 shows immobility scores before and six weeks following transplantation of either bovine chromaffin cells or bovine adrenal fibroblasts into the frontal neocortex of immunosuppressed rats. For comparison, a group of animals receiving no transplant was included and tested at both time points. To determine immobility scores, an animal is first conditioned in a cylinder containing water for 15 min 24 hours prior to testing. For testing, the animal is again placed in the water cylinder, this time for 5 min, during which the amount of time spent in a characteristic immobile posture is recorded. It has been shown that clinical-

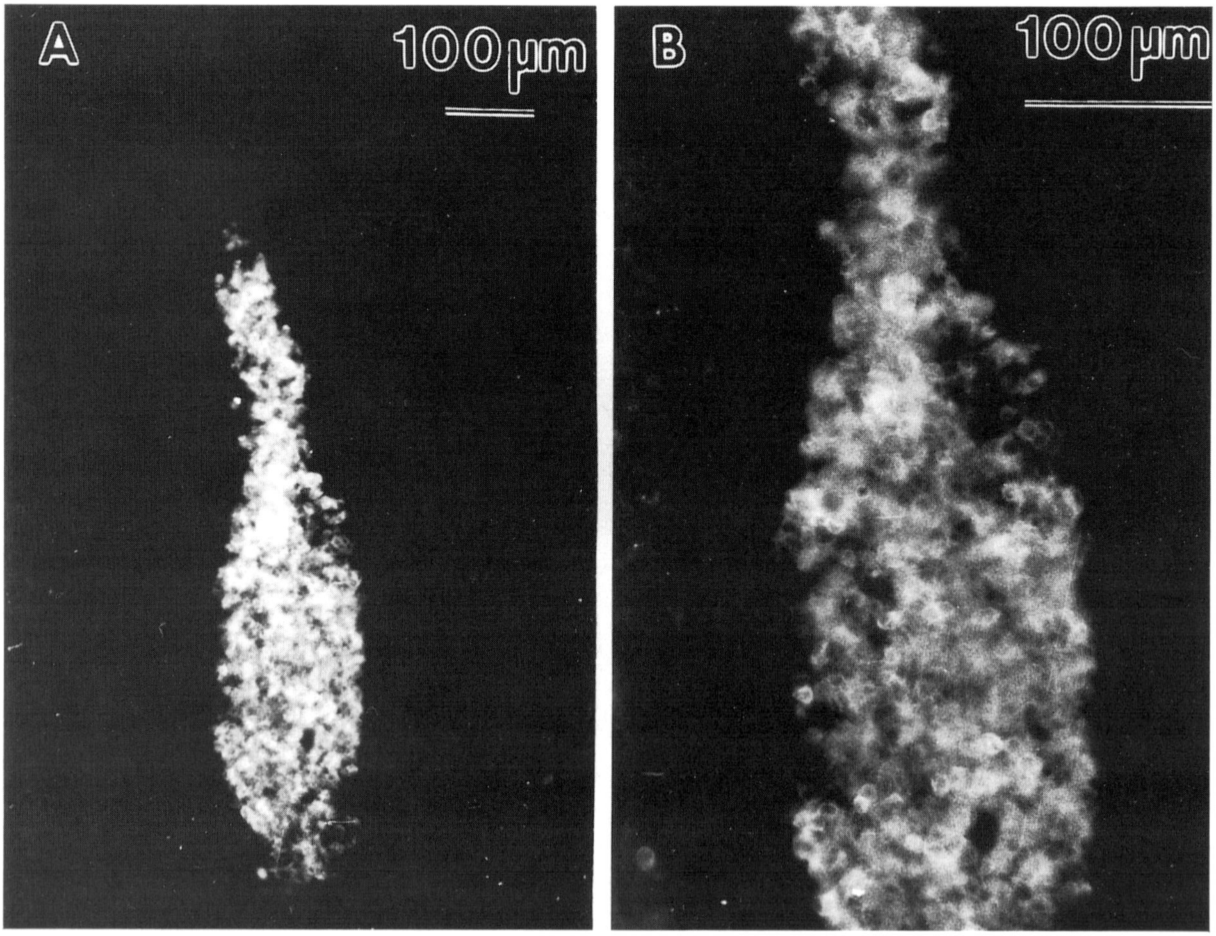

Fig. 10. Isolated bovine chromaffin cell implant in the rat frontal cortex 8 weeks following implantation. The graft contains dense clustering of cells robustly stained with a tyrosine hydroxylase antibody and a rhodamine linked secondary antibody. B is a higher magnification of the graft in A showing individual tyrosine hydroxylase positive bovine chromaffin cells within the graft. Taken from Sortwell *et al.*, Exp Brain Res 103: 59–69, 1995.

ly effective antidepressant therapies reliably reduce immobility scores, and thus a reduction in immobility is thought to mirror an increase in antidepressant activity [154]. Results revealed that bovine chromaffin cell grafts, but not control grafts, could significantly reduce immobility scores in animals six weeks post-transplantation (Fig. 8). To determine whether these chromaffin cell transplants could provide local release of catecholamines in the frontal neocortex, extracellular catecholamine levels were determined using *in vivo* microdialysis [33]. Results shown in Fig. 9 demonstrate that both NE and EPI levels are markedly increased in dialysates collected in the region of the chromaffin cell grafts, compared to both animals with fibroblast transplants or no transplants. In all cases, stimulation with high K^+ further increased catecholamine release, suggesting release from neural structures rather than blood. An example of a bovine chromaffin cell transplant in the rat frontal neocortex is shown in Fig. 10. The chromaffin cells are reacted with a tyrosine hydroxylase antibody, and rhodamine-linked secondary antibody. Together, the results of these studies suggest that chromaffin cell transplants in the frontal neocortex could have antidepressant effects by providing a local source of catecholamines.

Other applications

Chromaffin cell and adrenal medullary transplants have also been attempted for Huntington's disease

[155–157] and for restoration of memory deficits [43, 45, 158]. Encapsulated bovine chromaffin cells transplanted into the striatum in rats could reduce neuronal loss following quinolinic acid lesions of the striatum, possibly via neuroprotection [157]. However, thus far, intrastriatal adrenal medullary transplants do not appear promising in the treatment of Huntington's disease patients. The rationale for utilizing adrenal medullary grafts in the restoration of memory dysfunction is based in part on the ability of these cells to differentiate into a cholinergic phenotype. When implanted into the lesioned hippocampus, performances in a radial maze test were markedly improved compared to sham-operated controls [45]. This beneficial effect could be further improved by treating the grafts with cholinergic differentiation factor/leukemia inhibitory factor, which increases the conversion of adrenal medullary cells to cholinergic phenotypes [43].

In summary, adrenal medullary chromaffin cells have been attempted as donor sources for several applications in CNS injury and disease due to their adaptability and production of a wealth of potentially therapeutic neuroactive substances. Advances in our understanding of their pluripotential nature and neurochemical plasticity in response to specific environmental cues should lead to better utilization for neural transplantation therapies.

References

1. Barker R and Dunnett S (1993) The biology and behaviour of intracerebral adrenal transplants in animals and man. Rev Neurosci 4: 113–146.
2. Freed WJ, Poltorak M and Becker JB (1990) Adrenal medulla grafts: A review. Exp Neurol 110: 139–166.
3. Hansen JT and Gash DM (1994) Adrenal medulla in neural grafting and neural plasticity. Micros Res Tech 29: 155–160.
4. Anderson DJ and Axel R (1985) Molecular probes for the development and plasticity of neural crest derivatives. Cell 42: 649–662.
5. Doupe AJ, Landis SC and Patterson PH (1985) Environmental influences in the development of neural crest derivatives; glucocoritcoids, growth factors, and chromaffin cell plasticity. J Neurosci 5: 2119–2142.
6. Stachowiak MK, Hong JS and Viveros OH (1990) Coordinate and differential regulation of phenylethanolamine N-methyltransferase, tyrosin hydroxylase and proenkephalin mRNAs by neural and hormonal mechanisms in cultured bovine adrenal medullary cells. Brain Res 510: 27–288.
7. Unsicker K, Krisch B, Otten U *et al.* (1978) Nerve growth factor-induced fiber outgrowth from isolated rat adrenal chromaffin cells: impairment by glucocorticoids. Proc Natl Acad Sci USA 75: 1445–1460.
8. Tischler AS, Perlman, RL, Nunnemacher G *et al.* (1982) Long-term effects of dexamethasone and nerve growth factor on adrenal medullary cells cultured from young rats. Cell Tiss Res 225: 525–542.
9. Unsicker K, Griesser GH, Lindmar R *et al.* (1980) Establishment, characterization and fibre outgrowth of isolated bovine adrenal medullary cells in long-term cultures. Neuroscience 5: 1445–1460.
10. Claude P, Parada IM, Gordon KA *et al.* (1988) Acidic fibroblast growth factor stimulates adrenal chromaffin cells to proliferate and extend neurites but is not a long term survival factor. Neuron 1: 783–790.
11. Stemple DL, Mahanthappa HK and Anderson DJ (1988) Basic FGF induces neuronal differentiation, cell division and NGF dependence in chromaffin cells: A sequence of events in sympathetic development. Neuron 1: 517–525.
12. Unsicker K, Skaper SD and Varon S (1985) Neuronotrophic and neurite promoting factors: effects on early postnatal chromaffin cells from rat adrenal medulla. Dev Brain Res 17: 117–129.
13. Barker R, Dunnett S and Fawcett J (1993) A selective tachykinin receptor agonist promotes differentiation but not survival of rat chromaffin cells in vitro. Exp Brain Res 92: 467–472.
14. Kordower JH, Fiandaca MS, Notter MFD *et al.* (1990) NGF like trophic support from peripheral nerve for grafted rhesus adrenal chromaffin cells. J Neurosurg 73: 418–428.
15. Notter MFD, Hansen JT, Okawara S *et al.* (1989) Rodent and primate adrenal medullary cells in vitro; phenotypic plasticity in response to coculture with C6 glioma cells or NGF. Exp Brain Res 76: 38–46.
16. Paramore CG, Turner, DA and Madison, RD (1993) Induction of neuronal morphology in adrenal chromaffin cells cocultured with denervated Schwann cells. Exp Neurol 121: 288–294.
17. Unsicker K, Vey, J, Hofmann H-D *et al.* (1984) C6 glioma cell conditioned medium induces neurite outgrowth and survival or rat chromaffin cells in vitro; comparison with effects of nerve growth factor. Proc Natl Acad Sci USA 81: 2242–2246.
18. Poltorak M, Shimoda K and Freed WJ (1990) Cell adhesion molecules (CAMs) in adrenal medulla *in situ* and *in vitro* enhancement of chromaffin cell L1/Ng-CAM expression by NGF. Exp Neurol 110: 52–72.
19. Strömberg I, Ebendal T, Seiger Å *et al.* (1985) Nerve fiber production by intraocular adrenal medullary grafts: Stimulation by nerve growth factor or sympathetic denervation of the host iris. Cell Tiss Res 241: 241–249.
20. Strömberg I, Ebendal T, Olson L *et al.* (1990) Chromaffin grafts: survival and nerve fiber formation as a function of donor age, nerve growth factor and host sympathetic denervation. In: Dunnett SB, Richards S-J (eds) Progress in Brain Research, Vol 82, pp 87–94. Amsterdam: Elsevier.
21. Strömberg I, Herrera-Marschitz M, Ungerstedt U *et al.* (1985) Chronic implants of chromaffin tissue into the dopamine-denervated striatum. Effects of NGF on graft survival, fiber outgrowth and rotational behavior. Exp Brain Res 60: 335–349.
22. Förander P, Björklund L and Strömberg, I (1994) Dose-dependent effects of recombinant human NGF on grafted adult adrenal medullary tissue. Exp Neurol 126: 168–177.
23. Bing G, Notter MFD, Hansen JT *et al.* (1990) Cografts of adrenal medulla with C6 glioma cells in rats with 6-hydroxydopamine-induced lesions. Neuroscience 34: 687–697.

24. Cunningham LA, Hansen JT, Short MP *et al.* (1991) The use of genetically altered astrocytes to provide nerve growth factor to adrenal chromaffin cells grafted into the striatum. Brain Res 561: 192–202.
25. Cunningham LA, Short MP, Breakefield XO *et al.* (1994) Nerve growth factor released by transgenic astrocytes enhances the function of adrenal chromaffin cell grafts in a rat model of Parkinson's disease. Brain Res 658: 219–231.
26. Niijima K, Chalmers GR, Peterson DA *et al.* (1995) Enhanced survival and neuronal differentiation of adrenal chromaffin cells cografted into the striatum with NGF-producing fibroblasts. J Neurosci 15: 1180–1194.
27. Pohorecky LA, Piezzi RS and Wurtman RJ (1970) Steroid induction of phenylethanolamine-N-methyl transferase in adrenomedullary explants: independence of adrenal innervation. Endocrinology 86: 1466–1468.
28. Wong DL, Hayashi RJ and Ciaranello RD (1985) Regulation of biogenic amine methyltransferases by glucocorticoids via S-adenosylmethionine and its metabolizing enzymes, methionine adenosyltransferase and S-adenosylhomocysteine hydrolase. Brain Res 330: 209–216.
29. Sagen J and Ortega JD (1994) Influence of the CNS environment on chromaffin cell survival and catecholamine secretion patterns. J Neurochem 63: 1159–1162.
30. Kelner KL and Pollard HB (1985) Glucocorticoid receptors and regulation of phenylethanolamine-N-methyltransferase activity in cultured chromaffin cells. J Neurosci 5: 2161–2168.
31. Freed WJ, Karoum F, Spoor HE *et al.* (1983) Catecholamine content of intracerebral adrenal medulla grafts. Brain Res 269: 184–189.
32. Sagen J, Kemmler JE and Wang H (1991) Adrenal medullary transplants increase spinal cord CSF catecholamine levels and reduce pain sensitivity. J Neurochem 56: 623–627.
33. Sortwell CE, Petty F, Kramer G *et al.* (1994) In vivo release of catecholamines from xenogeneic chromaffin cell grafts with antidepressive activity. Exp Neurol 130: 1–8.
34. Wang H and Sagen J (1994) Optimization of adrenal medullary allograft conditions for pain alleviation. J Neural Transpl Plastic 5: 49–64.
35. Müller, TH and Unsicker K (1981) High-performance liquid chromatography with electrochemical detection as a highly efficient tool for studying catecholaminergic systems. I. Quantification of noradrenaline, adrenaline, and dopamine in cultured adrenal medullary cells. J Neurosci Meth 4: 39–52.
36. Becker JB and Freed WJ (1988) Adrenal medulla grafts enhance functional activity of the striatal dopamine system following substantia nigra lesions. Brain Res 462: 401–406.
37. Curran EJ, Albin RL and Becker JB (1993) Adrenal medulla grafts in the hemiparkinsonism rat: Profile of behavioral recovery predicts restoration of the symmetry between the two striata in measures of pre-and postsynaptic dopamine function. J Neurosci 13: 3864–3877.
38. Nishino H, Ono T, Shibata R *et al.* (1988) Adrenal medullary cells transmute into dopaminergic neurons in dopamine-depleted rat caudate and ameliorate motor disturbances. Brain Res 445: 325–327.
39. Decombe R, Rivot JP, Aunis D *et al.* (1990) Importance of catecholamine elease for the functional action of intrastriatal implants of adrenal medullary cells: pharmacological analysis and *in vivo* electrochemistry. Exp Neurol 107: 143–153.
40. Fitzgerald LR, Glick SD and Schneider AS (1989) Effect of striatal implatation of bovine adrenal chromaffin cells on turning behavior in a rat model of Parkinson's disease. Brain Res 481: 373–377.
41. Moro MA, López MG, Gandía L *et al.* (1990) Separation and culture of living adrenaline- and noradrenaline-containing cells from bovine adrenal medullae. Anal Biochem 185: 243–248.
42. Moro MA, Garcia AG and Langley OK (1991) Separation of two chromaffin cell populations isolated from bovine adrenal medulla. J Neurochem 57: 363–369.
43. Collery M, Delacour J and Jousselin-Hosaja M (1994) Cholinergic differentiation factor/leukemia inhibitory factor enhances functional effects of adrenal medulla grafts after hippocampal lesions in rats. Neuroscience 63: 667–677.
44. Jousselin-Hosaja M, Mailly P and Tsuji S (1993) Mouse adrenal chromaffin cells can transform to neuron-like cholinergic phenotypes after being grafted into the brain. Cell Tiss Res 274: 199–205.
45. Jousselin-Hosaja M, Collery M and Delacour J (1994) Effects of adrenal medulla grafts on memory capacities of rats after hippocampal lesions. Neuroscience 59: 275–284.
46. Manhanthappa N, Gage FH and Patterson PH (1990) Adrenal chromaffin cells as multipotential neurons for autografts. In: Dunnett SB, Richards S-J (eds) Progress in Brain Research, Vol 82, pp 33–37. Amsterdam: Elsevier.
47. Birch NP and Christie DL (1986) Characterization of the molecular forms of proenkephalin in bovine adrenal medulla and rat adreal, brain, and spinal cord with a site-directed antiserum. J Biol Chem 261: 12213–12221.
48. Liston D, Patey G, Rossier J *et al.* (1984) Processing of proenkephalin is tissue specific. Science 225: 734–737.
49. Stern AS, Lewis RV, Kimura S *et al.* (1980) Opioid hexapeptides and heptapeptides in adrenal medulla and brain: possible implications on the biosynthesis of enkephalins. Arch Biochem Biophys 205: 606–613.
50. Franklin SO, Yoburn BC, Zhu Y-S *et al.* (1991) Preproenkephalin mRNA and enkephalin in normal and denervated adrenals in the Syrian Hamster: comparison with central nervous system tissues. Mol Brain Res 10: 241–250.
51. Hexum TD, Yang H-YT and Costa E (1980) Biochemical characterization of enkephalin-like immunoreactive peptides of adrenal glands. Life Sci 27: 1211–1216.
52. Viveros OH, Diliberto EJ, Hazum E *et al.* (1979) Opiate-like materials in the adrenal medulla: evidence for storage and secretion with catecholamines. Mol Pharmacol 16: 1101–1108.
53. Yang H-YT, Hexum T and Costa E (1980) Opioid peptides in adrenal gland. Life Sci: 1119–1125.
54. Kilpatrick DL, Howells RD, Fleminger G *et al.* (1984) Denervation of rat adrenal glands markedly increases preproenkephalin mRNA. Proc Natl Acad Sci USA 81: 7221–7223.
55. Lewis RV, Stern AS, Kilpatrick DL *et al.* (1984) Marked increases in large enkephalin-containing polypeptides in the rat adrenal gland following denervation. J Neurosci 1: 80–82.
56. Wang X-T, Unnerstall J, Pappas GD *et al.* (1994) Comparisons between tyrosine hydroxylase and preproenkephalin mRNA in rat adrenal medullary explants in vitro and in CNS transplants. Soc Neurosci Abstr 20: 1706.
57. Zhu Y-S, Branch AD, Robertson HD *et al.* (1992) Time course of enkephalin mRNA and peptides in cultured rat adrenal medulla. Mol Brain Res 12: 173–180.
58. Wang, X-T, Unnerstall JR, Pappas GD *et al.* (1995) Expression of tyrosine hydroxylase and preproenkephalin mRNA

in rat adrenal medullary transplants in spinal subarachnoid space. Amer Soc Neural Transpl Abstr 2: 38.

59. Inturrisi CE, Branch AD, Robertson HD *et al.* (1988) Glucocorticoid regulation of enkephalins in cultured rat adrenal medulla. Mol Endocrinol 2: 633–639.
60. Naranjo JR, Moccetti I, Schwartz JP *et al.* (1986) Permissive effect of dexamethasone on the increase of proenkephalin mRNA induced by depolarization of chromaffin cells. Proc Natl Acad Sci USA 83: 1513–1517.
61. Wan DC-C and Livett BG (1989) Vasoactive intestinal peptide stimulates proenkephalin A mRNA expression in bovine adrenal chromaffin cells. Neurosci Lett 101: 218–222.
62. Puchacz, E, Stachowiak EK, Florkiewicz RZ *et al.* (1993) Basic fibroblast growth factor (bFGF) regulates tyrosine hydroxylase and proenkephalin mRNA levels in adrenal chromaffin cells. Brain Res 610: 39–52.
63. Unsicker K (1993) The trophic cocktail made by adrenal chromaffin cells. Exp Neurol 123: 167–173.
64. Bohn MC, Cupit L, Marciano F *et al.* (1987) Adrenal medulla grafts enhance recovery of striatal dopaminergic fibers. Science 237: 213–216.
65. Fiandaca MS, Kordower JH, Hansen JT *et al.* (1988) Adrenal medullary autografts into the basal ganglia cebus monkey: injury-induced regeneration. Exp Neurol 102: 76–91.
66. Kordower JH, Cochran E, Penn RD *et al.* (1991) Putative chromaffin cell survival and enhanced host-derived TH fiber innervation following a functional adrenal medullary autograft for Parkinson's disease. Ann Neurol 29: 405–412.
67. Stachowiak MK, Moffett J, Joy A *et al.* (1994) Regulation of bFGF gene expression and subcellular distribution of bFGF protein in adrenal medullary cells. J Cell Biol 127: 203–223.
68. Westermann R, Johannsen M, Unsicker K *et al.* (1990) Basic fibroblast growth factor (bFGF) immunoreactivity is present in chromaffin granules. J Neurochem 55: 285–292.
69. Lachmund A, Gehrke D, Krieglstein K *et al.* (1994) Trophic factors from chromaffin granules promote survival of peripheral and central nervous system neurons. Neuroscience 62: 361–370.
70. Hirsch FC, Duyeckerts C, Javoy-Agid F *et al.* (1990) Does adrenal graft enhance recovery of dopaminergic neurons in Parkinson's disease? Ann Neurol 27: 676–682.
71. Hurtig H, Joyce J, Sladek JR *et al.* (1989) Postmortem analysis of adrenal medulla to caudate autograft in a patient with Parkinson's disease. Ann Neurol 25: 607–614.
72. Jankovic J, Grossman R, Goodman C *et al.* (1989) Clinical, biochemical, and neuropathologic findings following transplantation of adrenal medulla to the caudate nucleus for the treatment of Parkinson's disease. Neurology 39: 1227–1234.
73. Peterson DI, Price ML and Small CS (1989) Autopsy findings in a patient who had an adrenal to brain transplant for Parkinson's disease. Neurology 39: 235–238.
74. Waters C, Itabashi HH, Apuzzo MLJ *et al.* (1990) Adrenal to caudate transplantation – postmortem study. Movement Dis 5: 248–250.
75. Olson L, Backlund E-O, Ebendal T *et al.* (1991) Intraputaminal infusion of nerve growth factor to support adrenal medullary autografts in Parkinson's disease. One year follow-up of first clinical trial. Arch Neurol 48: 373–381.
76. Olson L, Hoffer BJ, Backlund E-O *et al.* (1992) Intraputamenal infusion of nerve growth factor to support adrenal medullary autografts in Parkinson's disease. Restor Neurol Neurosci 4: 194.
77. Date I, Felten SY and Felten DL (1990) Cografts of adrenal medulla with peripheral nerve enhance the survivability of transplanted adrenal chromaffin cells and recovery of the host nigrostriatal dopaminergic system in MPTP-treated young adult mice. Brain Res 537: 33–39.
78. Date I, Yoshimoto Y, Gohda Y *et al.* (1993) Long-term effects of cografts of pretransected peripheral nerve with adrenal medulla in animal models of Parkinson's disease. Neurosurgery 33: 685–690.
79. Date I, Miyoshi, Y, Imaoka T *et al.* (1995) Chromaffin cell survival prolonged by nerve growth factor from pretransected sciatic nerve. Cell Transpl 4: S19–S21.
80. Doering LC (1992) Peripheral nerve segments promote consistent long-term survival of adrenal medulla transplants in the brain. Exp Neurol 118: 253–260.
81. Ellis JE, Byrd L and Bakay RAE (1992) A method for quantitating motor deficits in a nonhuman primate following MPTP-induced hemiparkinsonism and co-grafting. Exp Neurol 115: 376–387.
82. Lopez-Lozano JJ, Bravo G and Abascal J (1992) The CPH Neural Transplant Group: First clinical trial of co-grafting of autologous adrenal medulla and peripheral nerve in Parkinson's disease. Restor Neurol Neurosci 4: 207.
83. Watts RL, Bakay RAE, Herring CJ *et al.* (1990) Preliminary report on adrenal medullary grafting and cografting with sural nerve in the treatment of hemiparkinson monkeys. Prog Brain Res 82: 581–591.
84. Watts RL, Mandir AS and Bakay RAE (1995) Intrastriatal cografts of autologous adreanl medulla and sural nerve in MPTP-induced Parkinsonian macaques: behavrioal and anatomical assessment. Cell Transpl 4: 27–38.
85. Date I, Yoshimoto Y, Imaoka T *et al.* (1994) Cografts of adrenal medulla with peripheral nerve for Parkinson's disease. Cell Transpl 3: S47–S49.
86. Date I, Imaoka T, Miyoshi Y *et al.* (1995) Chromaffin cell survival and host dopaminergic fiber recovery in a Parkinsonian patient treated by cografts of adrenal medulla and pretransected peripheral nerve: an autopsy case. Amer Soc Neural Transpl 2: 16.
87. Subramanian T, Watts, RL, Bakay RAE *et al.* (1995) Stereotactic intrastriatal cografts of autologous adrenal medulla (AM) and peripheral nerve (PN) improves motor performance in Parkinson's disease (PD). Amer Soc Neural Transpl 2: 15.
88. Wa-cheng Z, Yu-ji D, Jia-kang C *et al.* (1994) Intracerebral co-grafting of Schwann's cells and fetal adrenal medulla in the treatment of Parkinson's disease. Chin Med J 107: 583–588.
89. Lopez-Lozano JJ, Bravo G, Abascal J *et al.* (1992) The CPH Neural Transplantation Group: Comparison of long-term outcome of neural transplants in Parkinson's disease using two different donor tissues. Restor Neurol Neurosci 4: 207.
90. Czech K, Pappas GD, Michalewicz P *et al.* (1994) Assessing xenograft rejection in bovine chromaffin cell neural transplants using one-way mixed lymphocyte reaction. Soc Neurosci Abstr 20: 260.
91. Czech K, Pappas GD and Sagen J (1995) The xenogeneic response to bovine adrenal chromaffin cells using an in vitro lymphocyte proliferation assay. Amer Soc Neural Transpl 2: 30.
92. Ortega JD, Sagen J and Pappas GD (1992) Short-term immunosuppression enhances long-term survival of bovine chromaffin cell xenografts in rat CNS. Cell Transpl 1: 33–41.
93. Ortega JD, Sagen J and Pappas GD (1992) Survival and integration of bovine chromaffin cells transplanted into rat CNS regions responsive to chromaffin cell secretory products. J Comp Neurol 322: 1–12.

94. Sagen J, Pappas GD and Ortega JD (1990) Host-graft relationships of isolated bovine chromaffin cells in rat periaqueductal gray. J Neurocytol 19: 697–707.
95. Schueler SB, Ortega JD, Sagen J *et al.* (1993) Robust survival of isolated bovine adrenal chromaffin cells following intrastriatal transplantation: A novel hypothesis of adrenal graft viability. J Neurosci 13: 4496–4510.
96. Schueler SB, Sagen J, Pappas GD *et al.* (1995) Long-term viability of isolated bovine adrenal medullary chromaffin cells following intrastriatal transplantation. Cell Transpl 4: 55–64.
97. Dubach M and German DC (1990) Extensive survival of chromaffin cells in adreal medulla "ribbon" grafts in monkey neostriatum. Exp Neurol 110: 167–180.
98. Dubach M (1992) Behavioral effects of multiple adrenal medullary grafts in longtailed Macaques. J Neural Transpl Plastic 3: 249–250.
99. Flanagan TR, Emerich DF, Winn SR (eds) (1994) Providing Pharmacological Access to the Brain: Alternative Approaches. Methods in Neurosciences. Vol 21, 508 pp. San Diego: Academic Press.
100. Aebischer P, Tresco PA, Sagen J *et al.* (1991) Transplantation of microencapsulated bovine chromaffin cells reduces lesion-induced rotational asymmetry in rats. Brain Res 560: 43–49.
101. Cherksey BD (1994) Microcarrier pre-adhesion enhances long term survival of adult cells implanted into the mammalian brain. Amer Soc Neural Transpl 1: 20.
102. Drasner K and Fields HF (1988) Synergy between the antinociceptive effects of intrathecal clonidine and systemic morphine in the rat. Pain 32: 309–312.
103. Sherman SE, Loomis CW, Milne B *et al.* (1988) Intrathecal oxymetazoline produces analgesia via spinal α-adrenoreceptors and potentiates spinal morphine. Eur J Pharmacol 148: 271–380.
104. Wilcox GL, Carlsson K-H, Jochim A *et al.* (1987) Mutual potentiation of antinociceptive effects of morphine and clonidine on motor and sensory responses in rat spinal cord. Brain Res 405: 84–93.
105. Yaksh T and Reddy SVR (1981) Studies in the primate on the analgetic effects associated with intrathecal actions of opiates, α-adrenergic agonists and baclofen. Anesthesiology 54: 451–467.
106. Siuciak JA, Altar CA, Wiegand SJ *et al.* (1994) Antinociceptive effect of brain-derived neurotrophic factor and neurotrophin-3. Brain Res 663: 326–330.
107. Hama AT and Sagen J (1993) Reduced pain-related behavior by adrenal medullary transplants in rats with experimental painful peripheral neuropathy. Pain 52: 223–231.
108. Hama A and Sagen J (1994) Alleviation of neuropathic pain symptoms by xenogeneic chromaffin cell grafts in the spinal subarachnoid space. Brain Res 651: 183–193.
109. Sagen J, Pappas GD and Perlow MJ (1986) Adrenal medullary tissue transplants in rat spinal cord reduce pain sensitivity. Brain Res 384: 189-194.
110. Sagen J, Pappas GD and Pollard HB (1986) Analgesia induced by isolated bovine chromaffin cells implanted in the rat spinal cord. Proc Natl Acad Sci 83: 7522-7526.
111. Sagen J, Wang H and Pappas GD (1990) Adrenal medullary implants in rat spinal cord reduce nociception in a chronic pain model. Pain 42: 69–79.
112. Sagen J, Pappas GD and Winnie AP (1993) Alleviation of pain in cancer patients by adrenal medullary transplants in the spinal subarachnoid space. Cell Transpl 2: 259–266.
113. Sagen J and Kemmler JE (1989) Increased levels of met-enkephalin-like immunoreactivity in the spinal cord CSF of rats with adrenal medullary transplants. Brain Res 502: 1–10.
114. Wang H and Sagen J (1994) Absence of appreciable tolerance and morphine cross-tolerance in rats with adrenal medullary transplants in the spinal cord. Neuropharmacology 33: 681–692.
115. Wang H and Sagen J (1995) Attenuation of pain-related hyperventilation in adjuvant arthritic rats with adrenal medullary transplants in the spinal subarachnoid space. Pain 63: 313–320.
116. Winnie AP, Pappas GD and Das Gupta TK *et al.* (1993) Alleviation of cancer pain by adrenal medullary transplants in the spinal subarachnoid space: A preliminary report. Anesthesiology 79: 644–653.
117. Jones RS and Ward JR (1963) Studies on adjuvant-induced polyarthritis in rats. II. Histogenesis of joint and visceral lesions. Arth Rheum 6: 23–35.
118. Pearson CM (1963) Experimental joint disease, observations on adjuvant-induced arthritis. J chron Dis 16: 863–874.
119. Rosenthale ME and Capetola RJ (1982) Adjuvant arthritis: immunopathological and hyperalgesic features. Fed Proc 41: 2577–2582.
120. Calvino B, Crepon-Bernard M-O and Le Bars D (1987) Parallel clinical and behavioural studies of adjuvant-induced arthritis in the rat: possible relationship with 'chronic pain'. Behav Brain Res 24: 11–29.
121. Colpaert FC (1987) Evidence that adjuvant arthritis in the rat is associated with chronic pain. Pain 28: 201–222.
122. Dardick SJ, Basbaum AI and Levine JD (1986) The contribution of pain to disability in experimentally induced arthritis. Arth Rheum 29: 1017–1022.
123. De Castro Costa M, De Sutter P, Gybels J *et al.* (1981) Adjuvant-induced arthritis in rats: a possible animal model of chronic pain. Pain 10: 173–185.
124. Pircio AW, Fedele CT and Bierwagen ME (1975) A new method for the evaluation of analgesic activity using adjuvant-induced arthritis in the rat. Eur J Pharmacol 31: 207–215.
125. Colpaert FC and Van den Hoogen RHWM (1983) Ventilatory response to adjuvant arthritis in the rat. Life Sci 32: 957–963.
126. Colpaert FC and Van den Hoogen RHWM (1983) Time course of the ventilatory response to adjuvant arthritis in the rat. Life Sci 33: 1065–1073.
127. Colpaert FC, Bervoets KJW and Van den Hoogen RHWM (1987) Pharmacological analysis of hyperventilation in arthritic rats. Pain 30: 243–258.
128. Glynn CJ, Lloyd JW and Folkhard S (1981) Ventilatory response to intractable pain. Pain 11: 201–211.
129. Bennett GJ and Xie Y-K (1988) A peripheral mononeuropathy in rats that produces disorders of pain sensation like those seen in human. Pain 33: 87–107.
130. Ginzburg R and Seltzer Z (1990) Subarachnoid spinal cord transplantation of adrenal medulla suppresses chronic neuropathic pain behavior in rats. Brain Res 523: 147–150.
131. Ortega-Alvaro A, Gibert-Rahola J, Chover AJ *et al.* (1994) Effect of amitriptyline on the analgesia induced by adrenal medullary tissue transplanted in the rat spinal subarachnoid space. Exp Neurol 130: 9–14.
132. Ruz-Franzi JI and Gonzalez-Darder JM (1991) Study of the analgesic effect of the implant of adrenal medulla into the subarachnoid space in rats. Acta Neurochirurigica 52: 39–41.

133. Vaquero J, Arias A, Oya S *et al.* (1991) Chromaffin cell allografts into the arachnoid of spinal cord reduce basal pain responses in rats. Neuroreport 2: 149–151.
134. Sagen J and Wang H (1995) Adrenal medullary grafts suppress c-*fos* induction in spinal neurons of arthritic rats. Neurosci Lett 192: 1–4.
135. Basbaum AI (1994) Immediate-early genes and pain: What's all the "Fos" about? APS J 3: 49–52.
136. Bullitt E (1990) Expression of c-*fos*-like protein as a marker for neuronal activity following noxious stimulation in the rat. J Comp Neurol 296: 517–530.
137. Menètrey D, Bannon A, Levine JD *et al.* (1989) Expression of c-fos protein in interneurons and projection neurons of the rat spinal cord in response to noxious somatic, articular, and visceral stimulation. J Comp Neurol 285: 177–195.
138. Meller ST and Gebhart GF (1993) Nitric oxide (NO) and nociceptive processing in the spinal cord. Pain 52: 127–136.
139. Hama AT and Sagen J (1994) Induction of spinal NADPH-diaphorase by nerve injury is attenuated by adrenal medullary transplants. Brain Res 640: 345–351.
140. Sagen J, Wang H, Tresco P *et al.* (1993) Transplants of immunologically isolated xenogeneic chromaffin cells provide a long-term source of pain-reducing neuroactive substances. J Neurosci 13: 2415–2423.
141. Sagen J, Hama AT, Winn SR *et al.* (1993) Pain reduction by spinal implantation of xenogeneic chromaffin cells immunologically-isolated in polymer capsules. Soc Neurosci Abstr 19: 234.
142. Lazorthes Y, Bès JC, Tafani M *et al.* (1994) Transplantation of human chromaffin cells for intractable cancer pain control. Amer Pain Soc Abstr 13: A43.
143. Lazorthes Y, Bès JC, Sagen J *et al.* (1995) Transplantation of human chromaffin cells for intractable cancer pain control. Acta Neurochir: in press.
144. Vaquero J, Martinez R, Oya S *et al.* (1988) Transplantation of adrenal medulla into spinal cord for pain relief: disappointing outcome. Lancet 2: 1315.
145. Aebischer P, Buchser E, Joseph JM *et al.* (1994) Transplantation in humans of encapsulated xenogeneic cells without immunosuppression. Transplantation 58: 1275–1277.
146. Aebischer P, Buchser E, Joseph JM *et al.* (1994) Xenotransplantation of encapsulated bovine chromaffin cells for the treatment of intractable pain: A phase I clincal trial. Soc Neurosci Abstr 20: 1707.
147. Dougherty DD, Sortwell CE and Sagen J (1995) Pharmacologic specificity of antidepressive activity by monoaminergic neural transplants. Psychopharmacology: in press.
148. Sagen J, Dougherty DD, Sortwell CE *et al.* (1994) Dissociation between antidepressant effects and monoaminergic receptor changes in rats with monoaminergic transplants in the frontal neocortex. 3rd Int Behav Neurosci Abstr 3: 22.
149. Sagen J, Sortwell CE and Pappas GD (1990) Monoaminergic neural transplants prevent learned helplessness in a rat depression model. Biol Psychiat 28: 1037–1048.
150. Sortwell CE and Sagen J (1993) Induction of antidepressive activity by monoaminergic transplants in rat neocortex. Pharmacol Biochem Behav 46: 225–230.
151. Sortwell CE, Pappas GD and Sagen J (1995) Chromaffin cell xenografts in the rat neocortex can produce antidepressive activity in the forced swimming test. Exp Brain Res 103: 59–69.
152. Seligman MEP and Beagley G (1975) Learned helplessness in the rat. J Comp Physiol Psychol 88: 534–541.
153. Porsolt RD, Lepinchon M and Jalfre M (1977) Depression: a new model sensitive to antidepressant treatments. Nature 266: 730–732.
154. Borsini F and Meli A (1988) Is the forced swimming test a suitable model for revealing antidepressant activity? Psychopharmacology 94: 147–160.
155. Allen (1988) Science News 133: 268.
156. Sanberg PR and Norman AB (1988) Adrenal transplants for Huntington's disease? Nature 335: 122.
157. Sanberg PR, Emerich DF, McDermott PE *et al.* (1991) Transplantation of polymer encapsulated bovine adrenal cells prevent quinolinic acid induced lesions of the striatum. Soc Neurosci Abstr 17: 903.
158. Welner SA, Koty ZC and Boksa P (1990) Chromaffin cell grafts to cerebral cortex reverse lesion-induced memory deficits. Brain Res 527: 163–166.

Section IV:

Corneal Transplantation

R. P. Lanza and W. L. Chick (eds.), Yearbook of Cell and Tissue Transplantation 1996/1997, 93–103.

Chapter 9

Corneal transplantation

Keryn A. Williams
Department of Ophthalmology, Flinders University of South Australia, 5042 Australia

The cornea, the clear window at the front of the eye (Fig. 1), may become opaque as a result of accidental damage or a disease process. To restore vision, corneal transplantation (keratoplasty) may be performed either as a full-thickness (penetrating) or partial thickness (lamellar) procedure. The indications and outcomes differ markedly, although both procedures require human corneas as donor material. This brief review of developments in clinical and experimental corneal transplantation since 1990 will concentrate primarily on penetrating keratoplasty, as far more penetrating than lamellar procedures are performed. First, aspects of donor corneal procurement and eye banking will be considered. Second, the results of available graft and visual outcome studies will be summarised. Third, information relating to tissue matching and immunosuppression for corneal transplantation will be examined, and advances in our understanding of the immunological basis of corneal graft rejection and their implications for clinical practice reviewed. Finally, various extensions of the basic procedure, including limbal transplantation, xenotransplantation and the prospects for an artificial cornea will be described.

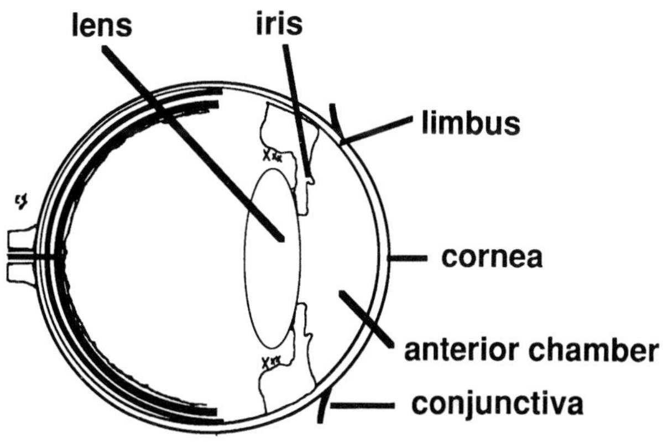

Fig. 1. Diagram of the human eye in cross-section. The cornea, the clear window at the front of the eye, is contiguous with the sclera. It is bathed on its anterior surface by the tear film and on its posterior surface by the aqueous humour, which fills the anterior chamber. In a penetrating keratoplasty procedure, the central cornea is replaced.

Corneal procurement

There is still no reliable artificial cornea, so that corneal transplantation depends on an adequate supply of donated human corneas. As with solid organs for transplantation, donor corneas are in short supply in some parts of the world. Cultural mores that do not favour donation remain a substantial obstacle to increasing the donor pool [1, 2], but even in societies where organ and tissue donation are reasonably well accepted, waiting lists for corneal transplantation may still exist. The authors of a 1992 study of the then-current laws governing organ and tissue donation throughout the world distinguished those countries where family consent for donation was paramount as the least successful in obtaining tissue for transplantation [1]. The results of a North American investigation into the effects of legislative changes in a community which had provided over 20,000 corneal donors in almost 30 years [3] suggested that legislation allowing limited presumed consent (permission for the collection of corneas from deceased people when no family

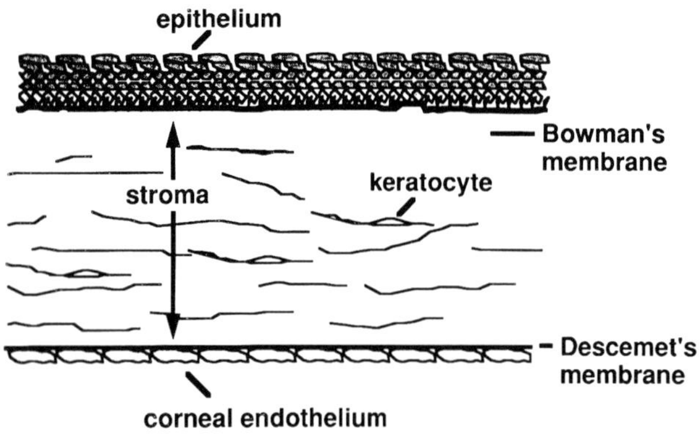

Fig. 2. Diagram of the structure of the human cornea.

objection was known) was associated with a sixfold increase in the number of corneas available for transplantation. That good publicity for transplantation can be a powerful method of increasing corneal donor numbers is shown by the British experience [4]; equally, the dangers of poor publicity have been graphically described by Patel [5]. Proclamation of appropriate legislation, sensitivity to local community attitudes and careful attention to public relations all appear to be essential components of successful corneal procurement programs.

Eye banking

Eye banks are generally responsible for the ethical, safe and equitable collection, assessment, preservation and distribution of human corneas for transplantation. The normal cornea is bounded on its anterior surface by the corneal epithelium and on its posterior surface by a monolayer of corneal endothelium (Fig. 2). The bulk of the cornea, the stroma, is composed of an ordered array of collagen fibrils interspersed with a few fibroblast-like cells, the keratocytes. The corneal endothelium in particular is essential for corneal survival and function because it contains an active metabolic pump responsible for maintaining the corneal stroma in a state of relative dehydration. Failure of the endothelial pump caused by endothelial cell loss or damage results in waterlogging of the stroma. An oedematous stroma is opaque. As the human corneal endothelium does not divide *in vivo*, especial care not to damage this essential monolayer must be taken during corneal collection and preservation.

One decision that must be made in any eye bank is whether to remove corneas or whole globes from the donor. The former may be less disfiguring and may be preferred by the donor's family. However, a small study of corneal procurement techniques suggests that enucleation of the globe followed by removal of the corneoscleral rim in the laboratory results in lesser degrees of microbial contamination and corneal endothelial cell trauma than does *in situ* excision of the cornea in the mortuary [6].

Primary non-functioning corneal grafts are clearly of great concern to any eye bank because of the possibility that poor handling in the bank may have caused irreparable tissue damage. A recent study [7] of primary non-functioning grafts initiated after an apparent clustering of such cases revealed an incidence of 2.7% (21/778) over a two-year period, with a 13% incidence in the donor pairs. Perhaps unexpectedly, identity of individual surgeons rather than specific tissue handling was strongly associated with graft failure from primary non-function. Of possible relevance here is the finding that the "center effect" is an important predictor of corneal graft outcome [8].

Many ophthalmic surgeons express a strong preference for corneal tissue that has been collected from young donors, especially if the potential recipient is young. The recent finding that the cause of donor death (but not donor age) influences the metabolic potential of the cornea is thus of some interest [9]. The effect of donor age is still somewhat controversial, with at least one large prospective study finding no influence on graft survival [8].

Corneal preservation

Corneas for transplantation may be stored for up to 24 hours at 4 °C as whole globes in a moist pot [10], an antiquated method seldom used unless a lamellar graft procedure is contemplated. Most corneas are stored as corneoscleral disks suspended in one of several corneal preservation media. McCarey-Kaufman (M-K) medium [11], DexSol [12] and Optisol [13] amongst others are designed for hypothermic storage at 4 °C, whereas other media are designed for corneal organ culture at 31–34 °C [14, 15]. These various media, all based on recipes originally formulated for tissue culture, differ primarily in the nature of their additives and especially in the type and concentration of colloidal osmotic agents. M-K medium, which allows four-day storage, is a time-honoured but inexpensive option still widely used in many countries. DexSol and Optisol allow preservation times to be extended to one to two weeks. Organ culture, which originated in Minneapolis, is used extensively in Britain and Western Europe to preserve corneas for up to four weeks. Some experimental work has been performed with an intracellular solution-type medium [16], but such solutions are not used clinically. Corneal cryopreservation [17], a somewhat troublesome and technically demanding technique, is seldom employed.

A number of workers have sought to establish whether any particular medium is preferable to another in terms of the quality of the cornea following storage. In a comparison of M-K medium and organ culture, no statistically significant difference in visual acuity, corneal thickness or corneal endothelial cell density between corneas stored in either system was found over a one to two year period after transplantation [18]. Modifications of DexSol have conferred no apparent benefit compared with the original solution [19]. Optisol-stored corneas may be slightly thinner than corneas stored in other hypothermic media [13, 20] and may show less lysosomal enzyme activity but do not appear to perform differently after transplantation [20]. There is probably no single system of corneal preservation that can currently be considered to be undeniably better than any other, and the decision about which to employ is best made locally.

Infection control in eye banks

In most countries, it is obligatory to test potential corneal donors for seropositivity for the human immunodeficiency virus (HIV) and hepatitis B and C, and to take a history designed to preclude individuals with known risk factors for these or other potentially transmissible diseases including Creutzfeldt-Jakob disease (CJD). HIV has been unequivocally identified in ocular tissues, including the cornea [21]. There has been no recorded instance of recipient HIV seroconversion as a direct result of corneal transplantation [22] despite inadvertent transplantation of corneas from infected donors [23, 24]. It appears (perhaps fortuitously) that existing infection control measures coupled with low viral infectivity or viral load in the corneas of infected individuals have thus far prevented transmission. The incidence of HIV infection (as determined by positive serology) has been estimated at less than 1% in several studies of corneal donors in Western societies [25, 26], but will vary temporally and geographically. The situation with respect to transmission of hepatitis B and C by corneal transplantation is not entirely clear, but transmission of the former at least has probably occurred [27]. CJD is also believed to have been transmitted by corneal transplantation [28]. A simple screening test for CJD in corneal donors is not available, but the recent finding that homozygosity at codon 129 of the amyloid gene is strongly correlated with susceptibility to iatrogenic disease suggests at least that genetic testing and counselling of possibly infected recipients may be feasible [29].

Sporadic reports of herpes simplex virus (HSV) infection following transplantation of an HSV-infected cornea have surfaced previously [30, 31] but have been difficult to verify. The detection by polymerase chain reaction of HSV DNA in corneas from individuals without a known history of ocular infection, coupled with good evidence for development of disease in a graft recipient as a result of corneal transplantation, suggest transmission of this virus does occur [32]. However, it is not currently clear how testing for HSV can be achieved without actually destroying the donor cornea.

Graft and visual outcome of penetrating keratoplasty

The large corneal graft registries operating in Australia [8], The Netherlands [33], Britain [34], Canada [35] and elsewhere provide important base-line information about patterns of corneal graft practice and graft and visual outcome. One of the most important aspects to be considered in any investigation of graft outcome is length of follow up: inspection of virtually any set of Kaplan-Meier plots of corneal graft survival will indicate that outcome continues to change over at least the first five postoperative years, so that citation of one-year survival rates may be a quite inadequate measure of overall outcome.

Graft outcome related to specific indication for graft

The most common indications for penetrating keratoplasty in most countries are keratoconus, pseudophakic bullous keratopathy and failed previous graft. Keratoconus is one of the most common indications for corneal transplantation in virtually every study reported, and graft survival in the group of patients with keratoconus uncomplicated by hydrops is exceptionally good, often better than 95% at five years [8]. However, Lass and his colleagues have emphasised the importance of appropriate non-surgical treatment in the first instance for these patients [36]. Graft survival for patients with bullous keratopathy is of the order of 70% at five years in patients with a posterior chamber lens *in situ* after uncomplicated extracapsular cataract extraction [37] and somewhat less when an unselected cohort is considered. Outcome is poorer with increasing ipsilateral graft number [8, 33, 34, 35]. Representative survival curves for these common indications for corneal transplantation are shown in Fig. 3. Recent studies have focused on outcome for some of the rarer indications for keratoplasty including granular dystrophy [38], aniridia [39], trauma [40], perforation [41], ulcerative keratitis [42] and microbial keratitis [43].

Risk factors for corneal graft failure

In the Australian database [8], one-year and five-year Kaplan-Meier graft survival rates are 91% and 72% respectively for 2248 followed penetrating corneal grafts. The most common reasons for graft failure are rejection (33%), glaucoma (11%), nonviral infections (10%), endothelial cell failure (8%) and herpetic infection (7%), although in 19% of cases the reason for graft failure is unclear. The variables that best predict penetrating corneal graft failure in Cox proportional hazards regression analysis are aphakia or the presence of an anterior chamber or iris-clip intraocular lens, very small or very large grafts, a history of previous ipsilateral graft, an indication for graft that is not keratoconus or any of the corneal dystrophies, inflammation at the time of graft, and a postoperative rise in intraocular pressure. In the British database [34], 214 grafts have failed amongst 2385 followed, and graft survival is 89% at 1 year. Decreased risk of failure is associated with surgeons reporting most grafts and increased risk is associated with regrafts, young patients (less than 10 years), non visual reasons for grafting, endothelial failure as the indication for graft and deep vascularization. The Quebec data [35] confirm the negative influence of a history of glaucoma on corneal graft survival. Other important risk factors are regrafts, large graft size, HLA-A and -B incompatibility, prior uveitis, and vitrectomy at the time of graft.

Visual outcome after penetrating keratoplasty

The Australian experience [8] is that in a large unselected cohort at least 50% of recipients will require spectacle or contact lens correction for best-corrected visual acuity after penetrating keratoplasty. Approximately 20% of the cohort can expect to achieve 6/6 (20/20) or better (best-corrected) in the grafted eye, about 60-65% will achieve 6/18 (20/60) or better and about 25% will achieve 6/60 (20/200) or worse. The overwhelming majority of grafts are performed to improve vision, and in about 80% of the cohort this goal is actually achieved, albeit sometimes to a limited extent [44].

The British register [34] shows that visual outcome is worse in older patients and is associated with cosmesis as the reason for transplantation, superficial vascularization preoperatively and corneal endothelial failure as a result of the presence of an intraocular lens. Visual acuity is better when the other eye has been grafted previously or when the indication for graft is keratoconus or a stromal dystrophy.

Astigmatism continues to be an important factor with a negative influence on visual rehabilitation after penetrating keratoplasty. Although many different surgical techniques designed to minimise astigmatism postgraft have been examined, the possible influence of pre-existing astigmatism in the donor eye is generally ignored. An interesting recent study has demonstrated the feasibility of using a portable keratometer to measure the curvature of donor eyes, thereby avoiding transplantation of highly astigmatic donor corneas [45]. Accurate and consistent readings were obtained even when intraocular pressures were unable to be recorded on the donor eyes and these readings matched postoperative curvatures of the grafted corneas to within +/− 0.5 dioptre. Larger degrees of graft oversizing (0.5 mm) are associated with significantly greater mean refractive errors postoperatively than are smaller degrees of oversizing (0.25 mm) [46] and there is some evidence that patients who undergo double continuous suture closure of the graft enjoy faster visual rehabilitation and suffer less astigmatism than do patients who undergo a combined continuous and interrupted suture closure technique [47].

Complications of penetrating keratoplasty

Many investigators have emphasised the importance of controlling inflammation in the postoperative period because of the strong correlation between inflammatory episodes and graft rejection [48]. A recent study has concluded that anterior chamber dysgenesis syndromes, combined cataract and lens implant surgery, anterior vitrectomy, anterior segment revision, anterior chamber lens implant removal and peripheral anterior synechiae formation are all risk factors for glaucoma after corneal transplantation [49].

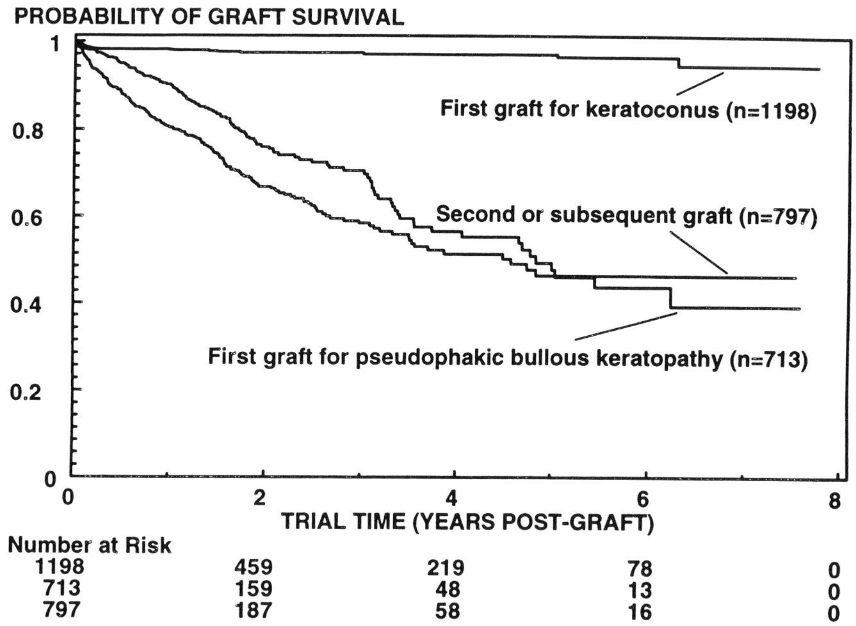

Fig. 3. Kaplan-Meier survival plots for penetrating corneal grafts performed in patients with keratoconus, pseudophakic bullous keratopathy and previous ipsilateral graft failure. Data are taken from the Australian Corneal Graft Registry database, with permission.

Animal models of corneal transplantation

Experimental work in animal models has provided many insights into the mechanisms of corneal graft rejection, the role of tissue matching and the efficacy or otherwise of new regimens of immunosuppression. Orthotopic corneal transplantation can be performed in the mouse [50], rat [51], rabbit [52], cat [53] and monkey [54]. We have also recently developed a model in the sheep. Additionally, some researchers use heterotopic models in which corneas are transplanted to subcutaneous pockets [55] or under the kidney capsule [56]. The relative merits of some of these models have been discussed elsewhere; in general, orthotopic models are to be preferred [57, 58].

The role of tissue matching in corneal transplantation

Unlike solid organ transplantation where organs are often collected from brain-dead donors, most corneas are obtained from donors post-mortem, for up to 12 hours after death. The collection of sufficient blood for essential serological testing and for tissue matching is sometimes difficult under these circumstances. Classic serological typing for HLA class I and II antigens can now be performed on cultured retinal pigment epithelial cells collected from the back of the donor eye, obviating the need for typing on post-mortem donor blood [59, 60]. Another approach involves application of a polymerase chain reaction-based procedure to type cadaveric donor material for HLA-DQ antigens [61]. However, although new methodology has revolutionised the tissue typing process so that it can be performed with ever-increasing precision, the benefits of matching for corneal transplantation remain uncertain.

Arguably one of the most important recent papers on corneal transplantation was the 1992 article describing the outcome of a multi-center trial designed to evaluate the effect of donor-recipient HLA matching and serological cross-matching on the survival of corneal grafts in high-risk recipients [62]. All patients received topical steroid therapy according to a standard protocol. Matching for HLA-A, -B and -DR antigens had no effect on overall graft survival, the incidence of irreversible rejection or the incidence of rejection episodes. The cross-match positive group had fewer graft failures from any cause and fewer rejection episodes than the negative group, but the differences were not statistically significant. An unexpected finding was that ABO blood group matching appeared to reduce the risk of corneal graft failure. The finding from this study that HLA matching has no influence on corneal graft survival conflicts with many other reports of a significant beneficial influence, particularly for class I antigens [33, 63–5]. Interestingly, however, Bradley and his colleagues have recently reported a negative effect of matching for HLA class II antigens in corneal transplantation [66].

Animal models, usually the best way in which to sort out conflicting results from the clinic, also return contradictory findings. In the rabbit, tissue matching (defined by the degree of mixed lymphocyte reactivity) does play a role in corneal graft outcome [67] but in the mouse [68] and rat, [69, 70] minor histocompatibility antigen mismatches appear far more important in determining graft outcome than do major mismatches, although the latter are not irrelevant. In the rat, corneal allografts do not induce sensitisation readily but once sensitisation has occurred, rejection will occur across class I antigen mismatches [71]. Pure major histocompatibility complex (MHC) class II disparities do not appear to induce rejection in this species [72] unless the host has been pre-immunized [73]. Interestingly, a recent report has demonstrated that MHC matched corneal grafts in the rat are relatively resistant to chemical immunosuppression: the reasons are still unclear [74].

Immunosuppression for corneal transplantation

There is no standard regimen of immunosuppression for prevention of corneal graft rejection, nor any single treatment for on-going rejection. In a recent poll of members of the Castroviejo Society, individual responses to a questionnaire seeking information about current practices of immunosuppression showed surprisingly wide variations in treatment preferences [75]. Topical corticosteroids remain the most widely used agents for prophylactic treatment of clinical corneal grafts [75]; despite early enthusiasm [76], topical cyclosporin A has proved somewhat disappointing. A single intravenous pulse of 500 mg of methylprednisolone has been shown to be more effective than a course of oral prednisone in reversing rejection in patients who present early with severe corneal endothelial rejection [77], and an extension of this work has now shown no benefit of multiple intravenous boluses of corticosteroid [78]. A number of units use variations of multiple systemic prophylactic immunosuppression for selected recipients in whom the risk of irreversible graft rejection is considered otherwise to be so high as to be inevitable and in whom the benefits of restoration of vision are believed to outweigh the risks of morbidity of the treatment [79, 80].

A variety of the newer chemical immunosuppressants including FK506 [81], rapamycin [82], deoxyspergualin [83], the platelet-activating factor antagonist BN52021 [84], liposome-encapsulated dichloromethylene diphosphonate [85] and the isoxal derivative leflunomide [86, 87], together with alternative methods of delivery of older agents such as cyclosporin A [88], have been examined in experimental models of corneal transplantation with varying results (Table 1). The use of monoclonal antibodies to prevent or reverse corneal graft rejection is also an area of increasing interest [89]. Monoclonal antibodies have been shown to prolong experimental corneal graft survival [90, 91] and to reverse ongoing rejection episodes [92] in experimental animals. Alternative strategies aimed at inducing tolerance rather than immunosuppression have included active enhancement with donor-specific blood transfusion [93] and UV irradiation of the cornea to inactivate the passenger cells believed to be partially responsible for sensitisation [94, 95, 96]. One of the photoreceptors for UV radiation is believed to be cis-urocanic acid and in fact cis-urocanic acid is immunosuppressive for experimental corneal grafts [97, 98].

New insights into corneal graft rejection

Correlates of corneal allograft rejection have recently been extensively reviewed elsewhere [58, 99]. There is no doubt that corneal graft rejection is a T cell-mediated phenomenon involving both CD4 and CD8 subsets, with concomitant upregulation of MHC antigens and various adhesion molecules in the cornea. Studies on the kinetics of cells infiltrating rejecting rat corneal allografts have shown that the CD4-positive response predominates early in the rejection process, followed later by CD8-positive cells [100]. Precisely which cytokines or other molecular mediators are actually responsible for the death of corneal endothelial cells remains unknown. There is some evidence that hydrolases released from infiltrating cells contribute to morphological disruption observed during corneal graft rejection in the rat [101].

The route by which foreign donor MHC or minor histocompatibility antigens are recognised following corneal transplantation remains uncertain but is gener-

Table 1. Effects of various new immunosuppressive agents on corneal graft survival in experimental models

Drug	Dose/ administration	Outcome/comment
FK506 [81]	3 mg/kg/day systemic (IP)	Prolonged rat allograft survival compared with controls, but all grafts rejected after cessation of treatment
Rapamycin [82]	2 mg/kg/day systemic (IM) for 25 days	Significant prolongation of rat allograft survival compared with controls; all grafts rejected while rats still on treatment
15-deoxyspergualin [83]	3–4 mg/kg/day systemic (IM)	Significant prolongation of rat graft survival; allografts rejected when therapy stopped
BN52021 [84]		All rabbit allografts rejected, but corneal neovascularization significantly reduced
CL2MDP [85]	0.1 ml liposome-encapsulated; 5 × subconjunctival	Indefinite survival of rat allografts compared with controls which rejected at a mean of 12 days post-graft
Leflunomide [86, 87]	2.5–10 mg/kg, PO 10 mg/kg/day, PO for 32 days	Prolonged survival of rat corneal allografts; significantly prolonged rat allograft survival, some grafts rejected after therapy ceased
Cyclosporin A [88]	2 mg/ml liposome-encapsulated, 5 × day, 10 days, topical	Significantly prolonged rat allograft survival, all grafts rejected after cessation of therapy

ating much interest [58]. There is some indirect evidence in both man [102, 66] and experimental animals [103, 104] that the indirect pathway of antigen processing, in which donor antigens are processed and presented by recipient antigen presenting cells, may play a role in sensitisation to the graft, especially to minor antigens. The operation of such a mechanism in the host response to a corneal graft may help to explain the correlation between an increasing number of host passenger cells in the bed of the graft and subsequent human graft failure [102], the increased likelihood of rejection observed in human corneal grafts that are well-matched for HLA-DR antigens, compared with poorly-matched grafts [66], and the importance of minor histocompatibility antigens in corneal graft rejection [68, 69, 70, 104].

Extending the options

Conjunctival and limbal grafts

One of the most fascinating advances in corneal biology has come with the realisation that the corneal epithelium regenerates from a sub-population of stem cells located in the basal layer of the limbal epithelium [105, 106]. It is now clear why central corneal epithelial cells cannot be propagated *in vitro*, undergoing senescence at the first or second passage, whereas limbal epithelial cells exhibit a substantial proliferative capacity [107]. Some patients in whom corneal disease is essentially confined to the ocular surface are believed to suffer from limbal stem cell dysplasia [106]. In such cases, transplantation of limbal epithelial tissue containing stem cells is a therapeutic option which can allow regeneration of a normal epithelium in the affected eye [108].

Limbal autografts using tissue taken from a normal contralateral eye can induce regeneration of a normal ocular surface in both experimental animals [109] and man [108, 110], and the technique is now being extended to allografts [111, 112, 48]. The indications for limbal transplantation are still being expanded, and such techniques are likely to replace the use of non-ocular tissue such as buccal mucosa for repair of the ocular surface [113]. Major unanswered questions relating to limbal stem cell transplantation include whether such grafts survive in the longterm and whether limbal allografts undergo immunological rejection. Allograft failure possibly caused by rejection has been described [114].

Xenografts

One method for increasing the number of corneas available for transplantation might be to explore the possibility of using corneas from species other than man. Corneal xenografts from guinea pigs to rats do not undergo immediate hyperacute rejection [115] but certainly fail in an accelerated fashion as do chicken to rat corneal grafts, probably as a result of damage mediated by preformed antibody and complement [116]. A subsequent cell-mediated response is also observed [117]. Human to rhesus monkey corneal xenografts exhibit variable survival depending to some extent on graft size, with smaller grafts faring better than larger grafts. Survival times measured in months for some grafts suggest that hyperacute rejection does not occur in this species combination [54]. Grafts of xenogeneic corneal endothelial cells only have also been reported. Rat corneas, seeded with cultured human endothelial cells prior to orthotopic transplantation into syngeneic recipients, remained clear for five days before failing, suggesting an accelerated immune reaction [118]. Overall, the results of corneal xenotransplantation thus far do not seem to hold exceptional promise for the future unless developments in the construction of genetically modified animal donors bear fruit.

Artificial corneas

Development of an artificial cornea has the potential to solve the problems of shortage of donor corneas, limited corneal storage times, and corneal graft failure from rejection and other causes. Considering the success of the artificial intraocular lens and the contact lens, both completely synthetic, it is perhaps surprising that the manufacture of a successful artificial cornea has thus far proved so difficult. The major difficulty has been the construction of a satisfactory interface between the natural and artificial surfaces. A number of approaches are being taken, with most researchers opting for a semi-synthetic cornea in which, for example, synthetic collagen gels serve as support structures for corneal epithelial cells [119, 120, 121].

Gene transfer

Gene therapy may in the future be able replace corneal transplantation as a successful treatment for some corneal disorders. Corneal storage diseases and some of the corneal dystrophies are likely to prove amenable to this form of therapy. Somatic cell transfer has been shown to reverse signs, including corneal pathology, of murine mucopolysaccharidosis type VII [122].

Conclusions

Corneal transplantation is a widely used, frequently successful treatment for many blinding diseases of the cornea, although it must be emphasised that graft failures do occur, especially in recipients with a history of inflammatory eye disease. Many of the problems associated with tissue procurement, matching, immunosuppression, immunological rejection and the objective measurement of graft outcome that are encountered in corneal transplantation are very similar to those that face surgeons and biologists involved in the transplantation of other organs.

Acknowledgements

Diagrams were drawn by Dr. E. Johnstone; Mrs. S. Muehlberg generated the Kaplan-Meier survival plot of the ACGR data. Critical appraisals of the manuscript provided by Dr. Justine Smith and Professor D.J. Coster, and the editorial assistance of Miss Wendy Laffer, are gratefully acknowledged. This work was supported by the NH&MRC.

References

1. Lee PP, Yang JC, McDonnell PJ *et al.* (1992) Worldwide legal requirements for obtaining corneas: 1990. Cornea 11: 102–7.
2. Coster DJ and Williams KA (1992) Donor cornea procurement: some special problems in Asia. Asia Pacific J Ophthalmol 4: 7–12.
3. Farge EJ, Silverman LM, Khan MM *et al.* (1994) The impact of state legislation on eye banking. Arch Ophthalmol 112: 180–5.

4. Armitage WJ, Rogers CA, Riggulsford MJ *et al.* (1991) Cornea donation boosted by positive publicity for transplantation. Lancet 338 (2): 1220.
5. Patel T (1993) France's troubled transplant trade. New Scientist 139: 12–13.
6. Lane SS, Mizener MW, Dubbel PA *et al.* (1994) Whole globe enucleation versus in situ corneal excision: a study of tissue trauma and contamination. Cornea 13: 305–9.
7. Mead MD, Hyman L, Grimson R *et al.* (1994) Primary graft failure: a case control investigation of a purported cluster. Cornea 13: 310–16.
8. Williams KA, Muehlberg SM, Wing SJ *et al.* (1993) The Australian Corneal Graft Registry, 1990 to 1992 report. Aust NZ J Ophthalmol 21 (suppl): 1–48.
9. Redbrake C, Becker J, Salla S *et al.* (1994) The influence of the cause of death and age on human corneal metabolism. Invest Ophthalmol Vis Sci 35: 3553–6.
10. Filatov VP (1937) Transplantation of the cornea from preserved cadavers' eyes. Lancet 1: 1395–7.
11. Aquavella JV, Van Horn DL and Haggerty CJ (1975) Corneal preservation using M-K medium. Am J Ophthalmol 80: 791–9.
12. Skelnik DL, Pearlstein CS, Mindrup EA *et al.* (1988) Corneal preservation at 4 ° C with chondroitin sulfate containing medium supplemented with dextran and epidermal growth factor (EGF). Invest Ophthalmol Vis Sci 29 (suppl): 111.
13. Lindstrom RL, Kaufman HE, Skelnik DL *et al.* (1992) Optisol corneal storage medium. Am J Ophthalmol 114: 345–56.
14. Lindstrom RL, Doughman DJ, Skelnik DL *et al.* (1986) Minnesota system corneal preservation. Br J Ophthalmol 70: 47–54.
15. Pels E, Schuchard Y (1984/5) The effects of high molecular weight dextran on the preservation of human corneas. Cornea 3: 219–27.
16. Taylor MJ, Hunt CJ and Madden PW (1989) Hypothermic preservation of corneas in a hyperkalaemic solution (CPTES): II. Extended storage in the presence of chondroitin sulphate. Br J Ophthalmol 73: 792–802.
17. Taylor MJ (1986) Clinical cryobiology of tissues: preservation of corneas. Cryobiology 23: 323–53.
18. Rijneveld WJ, Beekhuis WH, van Rij G *et al.* (1992) Clinical comparison of grafts stored in McCarey Kaufman medium at 4 degrees C and in corneal organ culture at 31 degrees C. Arch Ophthalmol 110: 203–5.
19. Lass JH, Musch DC, Gordon JF *et al.* (1994) Epidermal growth factor and insulin use in corneal preservation. Results of a multi center trial. The Corneal Preservation Study Group. Ophthalmology 101: 352–9.
20. Lass JH, Bourne WM, Musch DC *et al.* (1992) A randomized, prospective, double masked clinical trial of Optisol vs DexSol corneal storage media. Arch Ophthalmol 110: 1404–8.
21. Salahuddin SZ, Palestine AG, Heck E *et al.* (1986) Isolation of the human T-cell leukemia/lymphotropic virus type III from the cornea. Am J Ophthalmol 101: 149–52.
22. Caron MJ and Wilson R (1994) Review of the risk of HIV infection through corneal transplantation in the United States. J Am Optom Assoc 65: 173–8.
23. Schwarz A, Hoffman F, L'Age-Stehr *et al.* (1987) Human immunodeficiency virus transmission by organ donation. Outcome in cornea and kidney recipients. Transplantation 44: 21–4.
24. Simonds RJ, Holmberg SD, Hurwitz RL *et al.* (1992) Transmission of human immunodeficiency virus type I from a seronegative organ and tissue donor. New Engl J Med 326: 726–32.
25. Danneffel MB and Sugar A (1990) Incidence of HIV antibody positive eye/cornea donors in hospital versus medical examiner cases. Cornea 9: 271–2.
26. Williams KA, White MA, Badenoch PR *et al.* (1990) Donor cornea procurement: six year review of the role of the eye bank in South Australia. Aust NZ J Ophthalmol 18: 77–89.
27. Badenoch PR (1995) Corneal transplantation and infectious hepatitis. Br J Ophthalmol 79: 2.
28. Duffy P, Wolf J, Collins G *et al.* (1974) Possible person-to-person transmission of Creutzfeldt-Jakob disease. New Engl J Med 290: 692–3.
29. Brown P, Cervenakova L, Goldfarb LG *et al.* (1994) Iatrogenic Creutzfeldt Jakob disease:an example of the interplay between ancient genes and modern medicine. Neurology 44: 291–3.
30. Tullo AB, Marcyniuk B, Bonshek R *et al.* (1990) Herpes virus in a corneal donor. Eye 4: 766–7.
31. Cantin EM, Chen J, McNeill J *et al.* (1991) Detection of herpes simplex virus DNA sequences in corneal transplant recipients by polymerase chain reaction assays. Curr Eye Res 10: 15–21.
32. Cleator GM, Klapper PE, Dennett C *et al.* (1994) Corneal donor infection by herpes simplex virus: herpes simplex virus DNA in donor corneas. Cornea 13: 294–304.
33. Volker-Dieben HJ, D'Amaro J and Kok-Van Alphen CC (1987) Hierarchy of prognostic factors for corneal allograft survival. Aust NZ J Ophthalmol 15: 11–18.
34. Vail A, Gore SM, Bradley BA *et al.* (1994) Corneal graft survival and visual outcome. A multicenter Study. Corneal Transplant Follow up Study Collaborators. Ophthalmology 101: 120–7.
35. Boisjoly HM, Tourigny R, Bazin R *et al.* (1993) Risk factors of corneal graft failure. Ophthalmology 100: 1728–35.
36. Lass JH, Lembach RG, Park SB *et al.* (1990) Clinical management of keratoconus. A multicenter analysis. Ophthalmology 97: 433–45.
37. KA Williams, SM Muehlberg, RF Lewis *et al.* (1994) Influence of lens status on graft and visual outcome within a corneal graft register. Transplant Proc 27: 1389–91.
38. Lyons CJ, McCartney AC, Kirkness CM *et al.* (1994) Granular corneal dystrophy. Visual results and pattern of recurrence after lamellar or penetrating keratoplasty. Ophthalmology 101: 1812–17.
39. Kremer I, Rajpal RK, Rapuano CJ *et al.* (1993) Results of penetrating keratoplasty in aniridia. Am J Ophthalmol 115: 317–20.
40. Doren GS, Cohen EJ, Brady SE *et al.* (1990) Penetrating keratoplasty after ocular trauma. Am J Ophthalmol 110: 408–11.
41. Nobe JR, Moura BT, Robin JB *et al.* (1990) Results of penetrating keratoplasty for the treatment of corneal perforations. Arch Ophthalmol 108: 939–41.
42. Raizman MB, Sainz de la Maza M and Foster CS (1991) Tectonic keratoplasty for peripheral ulcerative keratitis. Cornea 10: 312–16.
43. Kirkness CM, Ficker LA, Steele AD *et al.* (1991) The role of penetrating keratoplasty in the management of microbial keratitis. Eye 5: 425–31.
44. Williams KA, Muehlberg SM, Lewis RF *et al.* (1994) How successful is corneal transplantation? A report from the Australian Corneal Graft Register. Eye 9: 219–27.

45. Dave AS and McCulley JP (1994) Demonstration of feasibility of application of a portable keratometer to cadaveric donor corneas. Cornea 13: 379–82.
46. Javadi MA, Mohammadi MJ, Mirdehghan SA *et al.* (1993) A comparison between donor recipient corneal size and its effect on the ultimate refractive error induced in keratoconus. Cornea 12: 401–5.
47. Assil KK, Zarnegar SR and Schanzlin DJ (1992) Visual outcome after penetrating keratoplasty with double continuous or combined interrupted and continuous suture wound closure. Am J Ophthalmol 114: 63–71.
48. Coster DJ (1994) Doyne lecture. Influences on the development of corneal transplantation. Eye 8: 1–11.
49. Kirkness CM and Ficker LA (1992) Risk factors for the development of postkeratoplasty glaucoma. Cornea 11: 427–32.
50. She SC, Steahly LP and Moticka EJ (1990) A method for performing full thickness, orthotopic, penetrating keratoplasty in the mouse. Ophthalmic Surg 21: 781–5.
51. Williams KA and Coster DJ (1985) Penetrating corneal transplantation in the inbred rat: a new model. Invest Ophthalmol Vis Sci 26: 23–30.
52. Khodadoust AA (1968) Penetrating keratoplasty in the rabbit. Am J Ophthalmol 66: 899–905.
53. Bahn CF, Meyer RF, MacCallum DK *et al.* (1982) Penetrating keratoplasty in the cat. A clinically applicable model. Ophthalmology 89: 687–99.
54. Li C, Xu JT, Kong FS *et al.* (1992) Experimental studies on penetrating heterokeratoplasty with human corneal grafts in monkey eyes. Cornea 11: 66–72.
55. Chandler JW, Ray-Keil L and Gillette TE (1982/3) Experimental corneal allograft rejection: description of a murine model and a new hypothesis of immunopathogenesis. Curr Eye Res 2: 387–97.
56. Duguid IGM, Cuthbertson RA, Guymer RH *et al.* (1990) A model of the corneal allograft reaction. Transplant Proc 22: 2105–6.
57. Williams KA, Johnstone EW, Guymer RH *et al.* (1995) Corneal transplantation in small animals. In: Green MK and Mandel TE (eds) Experimental Transplantation Models in Small Animals, pp 107–32. Chur: Harwood Academic Publishers.
58. Williams KA and Coster DJ (1993) Clinical and experimental aspects of corneal transplantation. Transplantation Reviews 7: 44–64.
59. Zavazava N, Westphal E, Duncker G *et al.* (1992) Post mortem HLA tissue typing of retinal pigment epithelial cells. Scand J Immunol (Suppl) 11: 192–4.
60. Baumgartner I, Asenbauer TT, Kaminski SL *et al.* (1992) Retinal pigment epithelial cells in post mortem HLA typing of corneal donors. Invest Ophthalmol Vis Sci 33: 1940–5.
61. Rakoczy P, Garlepp M and Constable I (1992) HLA DQA tissue typing of cadaveric eye bank donor material with polymerase chain reaction. Curr Eye Res 11: 445–52.
62. The Collaborative Corneal Transplantation Studies Research Group (1992) The collaborative corneal transplantation studies (CCTS). Effectiveness of histocompatibility matching in high risk corneal transplantation. Arch Ophthalmol 110: 1392–403.
63. Sanfilippo F, MacQueen JM, Vaughn WK *et al.* (1986) Reduced graft rejection with good HLA-A and B matching in high-risk corneal transplantation. New Engl J Med 315: 29–35.
64. Boisjoly HM, Roy R, Dubé I *et al.* (1986) HLA-A, B and DR matching in corneal transplantation. Ophthalmology 93: 1290–7.
65. Beekhuis WH, van Rij G, Renardel-de-Lavalette JG *et al.* (1991) Corneal graft survival in HLA-A and HLA-B-matched transplantations in high risk cases with retrospective review of HLA-DR compatibility. Cornea 10: 9– 12.
66. Bradley BA, Vail A, Gore SM *et al.* (1995) Negative effect of HLA-DR matching on corneal transplant rejection. Transplant Proc 27: 1392–4.
67. Maske R, Hill JC and Horak S (1994) Mixed lymphocyte culture responses in rabbits undergoing corneal grafting and topical cyclosporine treatment. Cornea 13: 324–30.
68. Sonoda Y and Streilein JW (1992) Orthotopic corneal transplantation in mice – evidence that the immunogenetic rules of rejection do not apply. Transplantation 54: 694–704.
69. Katami M (1991) Corneal transplantation - immunologically privileged status. Eye 5: 528–48.
70. Nicholls SM, Bradley BB and Easty DL (1991) Effect of mismatches for major histocompatibility complex and minor antigens on corneal graft rejection. Invest Ophthalmol Vis Sci 32: 2729–34.
71. Ross J, He YG and Niederkorn JY (1991) Class I disparate corneal grafts enjoy afferent but not efferent blockade of the immune response. Curr Eye Res 10: 889–92.
72. Katami M, Lim SM, Kamada N *et al.* (1990) A pure class II MHC disparity does not induce rejection of cornea or heart grafts in the rat. Transplant Proc 22: 2200–1.
73. Ross J, Callanan D, Kunz H *et al.* (1991) Evidence that the fate of class II disparate corneal grafts is determined by the timing of class II expression. Transplantation 51: 532–6.
74. Nicholls SM, Bradley BA and Easty DL (1995) Apparent resistance to immunosuppression of MHC-matched corneal transplants. Transplantation 59: 325–8.
75. Rinne JR and Stulting RD (1992) Current practices in the prevention and treatment of corneal graft rejection. Cornea 11: 326–8.
76. Belin MW, Bouchard CS and Phillips TM (1990) Update on topical cyclosporin A. Background, immunology, and pharmacology. Cornea 9: 184–95.
77. Hill JC, Maske R and Watson PG (1991) The use of a single pulse of intravenous methylprednisolone in the treatment of corneal graft rejection. A preliminary report. Eye 5: 420–4.
78. Hill JC and Ivey A (1994) Corticosteroids in corneal graft rejection: double versus single pulse therapy. Cornea 13: 383–8.
79. Hill JC (1989) The use of cyclosporine in high-risk keratoplasty. Am J Ophthalmol 107: 506–10.
80. Coster DJ and Williams KA (1991) Impediments to improved corneal transplantation. In: Khoo CY, Ang BC, Cheah WM, Chew PTK and Lim ASM (eds) New Frontiers in Ophthalmology, pp 90–3. Amsterdam: Elsevier Science Publishers BV.
81. Nishi M, Herbort CP, Matsubara M *et al.* (1993) Effects of the immunosuppressant FK506 on a penetrating keratoplasty rejection model in the rat. Invest Ophthalmol Vis Sci 34: 2477–86.
82. Olsen TW, Benegas NM, Joplin AC *et al.* (1994) Rapamycin inhibits corneal allograft rejection and neovascularization. Arch Ophthalmol 112: 1471–5.
83. Holland EJ, Olsen TW, Sterrer J *et al.* (1994) Suppression of graft rejection using 15 deoxyspergualin in the allogeneic rat penetrating keratoplasty model. Cornea 13: 28–32.

84. Cohen RA, Gebhardt BM and Bazan NG (1994) A platelet activating factor antagonist reduces corneal allograft inflammation and neovascularization. Curr Eye Res 13: 139–44.

85. Van der Veen G, Broersma L, Dijkstra CD *et al.* (1994) Prevention of corneal allograft rejection in rats treated with subconjunctival injections of liposomes containing dichloromethylene diphosphonate. Invest Ophthalmol Vis Sci 35: 3505–15.

86. Coupland SE, Klebe S, Karow AC *et al.* (1994) Leflunomide therapy following penetrating keratoplasty in the rat. Graefes Arch Clin Exp Ophthalmol 232: 622–7.

87. Niederkorn JY, Lang LS, Ross J *et al.* (1994) Promotion of corneal allograft survival with leflunomide. Invest Ophthalmol Vis Sci 35: 3783–5.

88. Milani JK, Pleyer U, Dukes A *et al.* (1993) Prolongation of corneal allograft survival with liposome encapsulated cyclosporine in the rat eye. Ophthalmology 100: 890–6.

89. Williams KA and Coster DJ (1994) The use of monoclonal antibodies in corneal transplantation. Clinical Immunotherapeutics 2: 32–41.

90. Hoffmann F, Kruse HA, Meinhold H *et al.* (1994) Interleukin 2 receptor targeted therapy by monoclonal antibodies in the rat corneal graft. Cornea 13: 440–6.

91. Guymer RH and Mandel TE (1991) Monoclonal antibody to ICAM 1 prolongs murine heterotopic corneal allograft survival. Aust NZ J Ophthalmol 19: 141–4.

92. Williams KA, Standfield SD, Wing SJ *et al.* (1992) Patterns of corneal graft rejection in the rabbit and reversal of rejection with monoclonal antibodies. Transplantation 54: 38–43.

93. Ayliffe W, McLeod D and Hutchinson IV (1992) The effect of blood transfusions on rat corneal graft survival. Invest Ophthalmol Vis Sci 33: 1974–8.

94. Niederkorn JY, Callanan D and Ross JR (1990) Prevention of the induction of allospecific cytotoxic T lymphocyte and delayed type hypersensitivity responses by ultraviolet irradiation of corneal allografts. Transplantation 50: 281–6.

95. Hill JC, Sarvan J, Maske R *et al.* (1994) Evidence that UV B irradiation decreases corneal Langerhans cells and improves corneal graft survival in the rabbit. Transplantation 57: 1281–4.

96. Williams KA, Ash JK, Mann TS *et al.* (1987) Cells infiltrating inflamed and vascularized corneas. Transplant Proc 19: 2889–91.

97. Williams KA, Lubeck D, Noonan FP *et al.* (1990) Prolongation of rabbit corneal graft survival following systemic administration of urocanic acid. In: Usui M, Ohno S and Aoki K (eds) Ocular Immunology Today, pp 103–6. Amsterdam: Elsevier Science Publishers BV.

98. Guymer RH and Mandel TE (1993) Urocanic acid as an immunosuppressant in allotransplantation in mice. Transplantation 55: 36–43.

99. Larkin DF (1994) Corneal allograft rejection. Br J Ophthalmol 78: 649–52.

100. Holland EJ, Olsen TW, Chan CC *et al.* (1994) Kinetics of corneal transplant rejection in the rat penetrating keratoplasty model. Cornea 13: 317–23.

101. Coupland S, Billson F and Hoffmann F (1994) Hydrolase participation in allograft rejection in rat penetrating keratoplasty. Graefes Arch Clin Exp Ophthalmol 232: 614–21.

102. Williams KA, White MA, Ash JK and Coster DJ (1989) Leucocytes in the graft bed are associated with corneal graft failure: analysis by immunohistology and actuarial graft survival. Ophthalmology 96: 38–44.

103. Williams KA, Ash JK, Mann TS *et al.* (1987) Antigen-presenting capabilities of cells infiltrating inflamed corneas. Transplant Proc 19: 255.

104. Sonoda Y, Sano Y, Ksander *et al.* (1995) Chararcterization of cell-mediated responses elicited by orthotopic corneal allografts in mice. Invest Ophthalmol Vis Sci 36: 427–34.

105. Cotsarelis G, Cheng SZ, Dong G *et al.* (1989) Existence of slow-cycling limbal epithelial basal cells that can be preferentially stimulated to proliferate: Implications on epithelial stem cells. Cell 57: 201–9.

106. Tseng SCG (1989) Concept and application of limbal stem cells. Eye 3: 141–57.

107. Lindberg K, Brown ME, Chaves HV *et al.* (1993) In vitro propagation of human ocular surface epithelial cells for transplantation. Invest Ophthalmol Vis Sci 34: 2672–9.

108. Kenyon KR and Tseng SCG (1989) Limbal autograft transplantation for ocular surface disorders. Ophthalmology 96: 709–23.

109. Tsai RJ, Sun TT and Tseng SC (1990) Comparison of limbal and conjunctival autograft transplantation in corneal surface reconstruction in rabbits. Ophthalmology 97: 446–55.

110. Williams KA, Davis GJ and Coster DJ (1993) Storage, surgery, outcome, and complications of corneal and conjunctival grafts. Current Opinion in Ophthalmol 4: 75–83.

111. Turgeon PW, Nauheim RC, Roat MI *et al.* (1990) Indications for keratoepithelioplasty. Arch Ophthalmol 108: 233–6.

112. Tsai RJ and Tseng SC (1994) Human allograft limbal transplantation for corneal surface reconstruction. Cornea 13: 389–400.

113. Shore JW, Foster CS, Westfall CT *et al.* (1992) Results of buccal mucosal grafting for patients with medically controlled ocular cicatricial pemphigoid. Ophthalmology 99: 383–95.

114. Thoft RA and Sugar J (1993) Graft failure in keratoepithelioplasty. Cornea 12: 362–5.

115. Ross JR, Howell DN and Sanfilippo FP (1993) Characteristics of corneal xenograft rejection in a discordant species combination. Invest Ophthalmol Vis Sci 34: 2469–76.

116. Larkin DF, Takano T, Standfield SD *et al.* Experimental orthotopic corneal xenotransplantation in the rat: mechanisms of graft rejection. Transplantation 60: 491-7.

117. Takano T and Williams KA (1995) Mechanism of corneal endothelial destruction in rejecting rat corneal allografts and xenografts: a role for CD4-positive cells. Transplant Proc 27: 260–1.

118. Insler MS and Lopez JG (1991) Extended incubation times improve corneal endothelial cell transplantation success. Invest Ophthalmol Vis Sci 32: 1828–36.

119. Mohay J, Lange TM, Soltau JB, *et al.* (1994) Transplantation of corneal endothelial cells using a cell carrier device. Cornea 13: 173–82.

120. Ohji M, SundarRaj N, Hassell JR *et al.* (1994) Basement membrane synthesis by human corneal epithelial cells in vitro. Invest Ophthalmol Vis Sci 35: 479–85.

121. Thompson KP, Hanna KD, Gipson IK *et al.* (1993) Synthetic epikeratoplasty in rhesus monkeys with human type IV collagen. Cornea 12: 35–45.

122. Wolfe JH, Sands MS, Barker JE *et al.* (1992) Reversal of pathology in murine mucopolysaccharidosis type VII by somatic cell gene transfer. Nature 360: 749–53.

Section V:

Fetal Tissue

R. P. Lanza and W. L. Chick (eds.), Yearbook of Cell and Tissue Transplantation 1996/1997, 107–116.

Chapter 10

Fetal tissue transplantation

Thomas E. Mandel
Transplantation Unit, The Walter and Eliza Hall Institute of Medical Research, Parkville, 3050, Australia

The replacement of damaged organs, tissues and cells by transplantation is now well established and has become routine treatment for a wide and increasing range of hitherto often fatal diseases. Although many transplant procedures are still regarded as experimental, more are becoming established as treatments of choice. This trend seems destined to continue and will be limited as much or more by a dearth of suitable donors as by any other consideration. Transplantation began over a century ago with corneal grafts [1], and the range of procedures has increased to include the replacement of most solid organs, sometimes even including whole organ complexes such as blocks of abdominal viscera, as well as individual tissues such as the islets of Langerhans and specific cells such as blood stem cells. Indeed, since the early 1960s when kidney transplantation began in earnest, the progress of transplantation has been rapid and seemingly inexorable. The major problems now are the need to develop safe and effective immunosuppression that will ideally result in immune tolerance, and the dearth of suitable organ and tissue donors that is increasingly becoming more desperate as the indications for transplantation widen. This ever increasing donor shortage has led many investigators to explore the feasibility of using "unconventional" sources of cells, tissues and organs including animal donors (xenografts) and particularly to explore the use of fetal tissue of either human or animal origin.

Fetal tissue transplants raise some unique issues including the very contentious ethical problems that must be faced when using human tissue that almost always has to be obtained from medically aborted fetuses [2, 3]. Strong views are expressed by both the proponents and opponents of the use of fetal tissue and need to be discussed and considered rationally so that the issue of fetal tissue use is not hijacked by protagonists of either extreme viewpoint. I believe that a good case can be made for the use of fetal tissue in specific circumstances but equally firmly believe that this does not give *carte blanche* to the indiscriminate use of fetal tissue, and particularly not to the proposition that fetal tissue can be obtained by performing abortions specifically to obtain it. However, if an abortion is being performed for independent and legitimate reasons and this decision is not based on or influenced by a perceived need for tissue, if tissue becomes available as a result it seems to me to be utterly wasteful not use it for a potentially life saving procedure. It is true, however, that so far the clinical application of fetal tissue transplantation has been quite limited but extensive experimental studies in animal models have clearly indicated its potential value and there are some indications that this may also be true in humans. Fetal tissue transplantation in humans is still largely experimental but it does appear likely that in the near future it will move to broader clinical applicability. Indeed, there are increasing instances where fetal tissue grafts have been used for the treatment of Parkinson's disease with apparent benefit, and fetal pancreas has be used for the treatment of insulin-dependent diabetes mellitus but so far with little well documented evidence of success. Many of the issues relating to fetal tissue research, including transplantation, have been recently reviewed by Fine [4] and ethical issues with the use of fetal tissues, particularly as these relate to CNS grafts, have been discussed by Hoffer and Olson [5].

In this chapter I will focus predominantly on recent studies (mainly since 1990) in both human and animal fetal tissue transplantation and attempt to indicate the present state of these procedures, their current problems, and the likely advances and prospects for the near future. However, apart from noting that major ethical problems remain to be resolved I will not attempt to make a case for or against the ethical use of fetal tissue transplantation. These issues have been reviewed recently [6] and previous studies with an excellent historical perspective and a broad-ranging overview of the uses of fetal tissue are discussed at length by Fine [4] and older studies (pre-1960) have been comprehensively reviewed by Woodruff [7]. Thus, the concept of the use of fetal tissue is not new but, in stark contrast to the use of adult organs, there has been relatively little well documented success, at least in clinical studies. Despite this, over the past 4 years well over 1000 citations on fetal tissue transplantation have appeared in Medline and research with such tissue is increasing rapidly. Based on the data from animal studies it seems only a matter of time before similar good results will be attained in humans.

The reasons for use of fetal tissue

Despite this apparent lack of success, at least in clinical studies, there are two predominant reasons why fetal tissue may have a valid place in transplantation. The most cogent is that some fetal cells and tissues have properties that are lost as they mature. In particular, certain cells have a finite capacity for proliferation that is lost either at or relatively soon after birth so that expansion of these cells or even their survival does not occur when mature tissue is used. Most notably this is a property of grafts of central nervous system (CNS) tissues but other grafts such as the islets of Langerhans may also have a greatly diminished capacity for proliferation as they mature. Whether this can be reversed is still contentious but there is recent evidence that growth or trophic factors exist that can rescue cells from death and perhaps enable even mature cells to begin a renewed cycle of growth [8-10].

A second useful property of fetal tissue is its apparent relative resistance to ischemic damage. This means that tissue of good viability can more readily be obtained from aborted fetuses that are non-heart beating (this is a legal requirement for the dissection of fetuses in many legislations) than is the case with adult tissues. In addition, since fetal tissue transplants will be performed as free grafts of tissue fragments they will need to survive an avascular period before there is an adequate ingrowth of host-derived vessels. Thus the capacity to survive relative ischemia is not only useful in obtaining viable tissue but is also beneficial while vascularization of the graft develops.

Other potential benefits that are either unique to or particularly well developed in fetal tissue include "plasticity" i.e. the capacity of the tissue to undergo variable differentiation, and anatomical features such as the poor development of cell processes that enable the tissue to be dispersed into small fragments or even single cells without undue damage. Both of these considerations are of particular benefit in grafts of neural tissues where the capacity to establish appropriate functional connections with host tissues is needed and where avoidance of severe and possibly irreparable damage to the graft during its preparation is also required. In addition, it has been suggested that fetal tissue may be of lower immunogenicity than adult tissue. There is little evidence for this, particularly when relatively well differentiated tissues are used, as was shown in a comparison of fetal and adult murine islets [11], although some evidence for reduced immunogenicity was reported in fetal kidney allografts and attributed to a reduced content of "passenger leucocytes" [12].

Problems with fetal tissue grafts

Despite these potentially valuable attributes, fetal tissues also present problems. These fall into two groups; firstly difficulty in obtaining suitable tissue because of the way in which many abortions are performed, and secondly problems due to the lack of functional differentiation due to immaturity of the grafts.

Obtaining suitable tissue from the majority of terminations that are performed in the first trimester i.e. under about 12 weeks gestation may be difficult because most such abortions are performed by suction curettage and the tissue is obtained in small fragments. Although these are sterile it is usually difficult and may be impossible to identify the organs needed, particularly when small organs such as the pancreas or specific regions of the brain are sought. HogenEsch *et al.* [13] found that 50% of curette specimens from suction abortuses were too fragmented to be useful. Fetal liver fragments are easier to locate and may yield suitable tissue (e.g. hemopoietic stem cells) particularly when the termination procedure is modified to reduce tissue damage [14]. With the more mature second trimester abortions where the fetus is delivered intact the problem often lies with the method used to induce the abortion. Usually this is with prostaglandin induction of labor that often results in fetal death in utero with prolonged intrauterine warm ischemia and consequent cell death and pancreas from such fetuses was rarely viable [15, 16].

Alternative sources of fetal tissue that have been suggested include spontaneous terminations, stillbirths and ectopic pregnancies but there is little likelihood of obtaining suitable tissue from these [2, 3]. Most spontaneously aborted fetuses die in utero and the tissue when obtained has suffered a prolonged period of warm ischemia and is usually dead. In ectopic pregnancy specimens there is also often great difficulty in identifying specific fetal parts in addition to problems with genetic abnormalities [2].

Thus, while fetal tissue is a potentially important source of material for transplantation and in many countries very many abortions are performed, in practice only a small proportion are likely to yield tissue suitable for transplantation. Nevertheless, even a small yield may be valuable and should not be neglected. However, for potentially common applications e.g. islet transplantation, it is most unlikely that human fetal tissue will be available in anywhere near the amounts required and the use of alternative sources such as fetal tissue from animal sources needs to be explored. The use of animal tissues including fetal tissue, i.e. xenotransplantation, is indeed becoming more extensively investigated and for applications such as islet transplantation in diabetes may be quite suitable. There is some evidence that suggests that in the absence of immediate "hyperacute" rejection that occurs with vascularized organs between discordant donor-recipient pairs, rejection of neovascularized xenogeneic tissue may be more readily controlled that the rejection of allogeneic tissues since xenograft rejection appears to be more dependent on CD4+ve T cells [17, 18].

Transplantation of neural tissues

At present it is in CNS transplantation that the most useful information is available on the application of fetal tissue transplants in humans and has been recently extensively reviewed [19]. In most cases this has been for the treatment of Parkinson's disease (PD) and followed early studies that showed that fetal grafts were beneficial in a drug-induced model of PD in rats [20–22].

PD is due to the selective loss of dopamine secreting cells that are localized to a small area of the brain and because of its focal nature and the defined transmitter defect produced their replacement should be able to cure the disease. Since the function sought from the graft is quite specific and the location of the disease is focal, PD should be an ideal disease in which to test the efficacy of cell replacement therapy. Theoretically the ideal replacement for the destroyed cells would be their precursors since they could proliferate and differentiate in situ. Many experimental studies were indeed performed with some evidence of success [reviewed in 23], and clinical trials began in the late 1980s.

In the initial clinical trials autografted fragments of adrenal medulla were used as a source of dopamine producing cells and some benefit was reported [24] but this was not confirmed by other groups. Fetal tissue grafts appeared to give better results and soon many groups started clinical trials. At present the value of fetal transplants in the treatment of PD is still controversial [25] but there is increasing evidence that they may be functional and the time is right for a thorough study of the methodology in animal models and in humans [26].

There is growing consensus that there is some benefit to be had in both spontaneous and drug-induced (1-methyl-4phenyl-1, 2, 3, 6-tetrahydropyridine [MPTP]) PD with some evidence of medium term benefit though not a cure and many studies have now been published on small and generally uncontrolled clinical trials (27–34). A number of studies performed in primates with induced PD produced by MPTP have shown apparent reversal of symptoms [35]. However it should be noted that in all human cases of spontaneous PD that were treated by fetal mesencephalic tissue transplantation the disease was very advanced and perhaps it is asking too much to expect total relief of symptoms in these patients. In addition it is conceivable that the disease could recur in the grafts and indeed there is some evidence to suggest that the results in drug-induced PD where the disease is not progressive are better.

Many problems of fetal tissue transplants in PD remain and include the probable need for tissue from multiple donors although many of the patients treated had grafts from only a single donor. There may also be a need to transplant bilaterally though this would not be a problem if adequate amounts of tissue were available. The best gestational age also needs to be defined but perhaps tissue from younger (6–8 weeks gestation) rather than older fetuses (11–18 weeks gestation) may be preferable though these too have been used with claims of some success [36]. The use of first trimester donors would certainly simplify acquisition of tissue since most abortions are performed during this stage of pregnancy but tissue acquisition remains a problem. Thus, although at present the evidence for beneficial effects of fetal transplants in PD is mounting, there is also some evidence that other treatments such as posteroventral pallidotomy may be as effective [37]. Clearly, the final word on fetal transplants in arguably the best documented model of this treatment is still not in.

Transplantation in other CNS diseases

Transplantation of fetal tissue has also been attempted in a range of other CNS diseases in humans and in animal models of the relevant conditions including Huntington's disease (HD) [38, 39], multiple sclerosis [40–42], Alzheimer's disease, amyotropic lateral sclerosis, hereditary ataxia, other destructive lesions of the CNS e.g. stroke and even schizophrenia have been suggested as being suitable for this sort of treatment but extensive studies in experimental animals where ever possible are needed before clinical trials are attempted [43]. Whether in most of these usually more diffusely distributed conditions replacement of neural tissue by transplantation has a place remains to be seen.

Since a number of neurologic diseases, including PD, are caused by the lack of secreted factors it may be possible to transfect autologous cells with the appropriate genes so that these genetically engineered cells could be transplanted without the risk of rejection. While this type of treatment is appealing and some studies have been reported [44, 45], with some promising results in animals [46] it is still some way from clinical application. A conceptually similar technique would be to transfect brain cells in situ with the appropriate genetic material but the technical problems involved are still to be mastered.

Rejection of non-autologous tissue remains a problem despite the allegedly immunologically privileged state of the brain but may potentially be overcome by encapsulating foreign transplanted cells within cell and antibody-impermeable barriers [47, 48]. Whatever the source of cells that are used in immunoisolation devices, fetal tissue may well have a place since it may be possible to use low loading densities with potentially proliferating cells and as vascularization around the device develops and oxygenation within the device improves, growth of the cells may increase their functional mass. Thus, many approaches are being tested and it is likely that one or more will succeed, perhaps with the aid of the newly identified and still often poorly characterized growth factors.

Fetal islet transplantation

Undoubtedly the greatest potential use of fetal transplants may be in insulin-dependent diabetes mellitus (IDDM). Vast numbers of patients with IDDM are currently being treated with insulin injections; a treatment that is far from ideal since constant maintenance of normal blood glucose levels is rarely possible and the devastating complications of IDDM are linked to inadequate glycemic control [49]. Islet replacement using adult allogeneic pancreas is successful and is being increasingly used but there is a severe shortage of donors and this will always limit its application, at least while human cadaveric donors are required [50]. Indeed, whole or segmental pancreas transplantation may best be reserved for patients that already need a kidney allograft since a successful graft greatly improves their quality of life but has little effect on the diabetic complications already present in these individuals. Human fetal pancreas is unlikely to be available in adequate amounts even if it were found to be suitable. There are a number of studies in which fetal pancreas has been used in clinical trials but the data are generally not well documented despite some claims that success has been achieved [51, 52]. Nevertheless, there is a great deal of evidence from experiments in animals that clearly shows that fetal pancreas is effective and that human fetal pancreas can reverse diabetes in rodent models [53]. These reports therefore show that the fetal islets can function adequately and even across wide species barriers but much work needs to be done before these results can be applied to humans. Recent studies with fetal islet or proislet grafts in animals have been comprehensively reviewed [54, 55].

Over the past 17 years we have been using fetal pancreas grafts in murine models. We isolate the islets and, perhaps more importantly their ductal precursors, in organ culture using a method that selectively preserves these cells while eliminating the developing acinar tissue as well as reducing the number of other irrelevant cells (e.g. endothelium, fibroblasts etc) [56]. In mice the grafts are generally placed under the renal capsule although this is not a physiological site and may not be optimal but has the advantages that it is simple and the graft can be removed by a unilateral nephrectomy to prove graft function [57]. We showed initially that isografts of fetal pancreas in mice made diabetic with streptozotocin were able to be reversed to euglycemia with a renal subcapsular graft of a small piece of fetal pancreas equivalent to one half or less of the fetal organ. The reversal to euglycemia took some time, generally 2–4 weeks, presumably while the fetal tissue proliferated and developed functional maturity. We subsequently used allogeneic grafts and showed that with pretransplant immunomodification of the graft by exposure of the fetal pancreas to high concentrations of O_2 in vitro, allograft survival and reversal of diabetes without a need for immunosuppression was possible in some strains of mice [58]. This result was similar to data with adult islets that were either hand picked or that had been immunomodified by various pretransplant culture conditions including exposure to high concentrations of O_2, exposure to low temperature, irradiation with UV light etc. [59]. Thus, fetal pancreas behaved in a manner similar to that seen with adult islets with the exception that reversal to euglycemia, which is rapid when adequate numbers of adult islets are used, took longer when fetal tissue was used.

The development of functional responsiveness in fetal islet grafts may take a long time and even in mice the reversal to euglycemia may take weeks [60] or even months [61] depending on the gestational age, amount of tissue transplanted and the way the graft is prepared. Similar prolonged time to adequate function is seen with fetal islet xenografts. This is in marked contrast to the usually very rapid return to euglycemia that occurs when adequate numbers of adult islets are used where reversal of diabetes generally occurs within 1–2 days. It is not known how long it may take for a fetal islet transplant to function in a human recipient in the absence of rejection and/or recurrent disease but it seems likely that the time would be long, perhaps months or even years may elapse before adequate function develops. This may be a major problem with fetal islet grafts, allogeneic or xenogeneic, in patients that are in late stages of IDDM with multiple complications already present, i.e. in patients that now are accepted for a pancreas transplant, and may be one reason why there is so little evidence of fetal islet graft function in transplanted patients. Prolonged time to develop function would not be such a problem in early diabetes as these patients would be well enough to tolerate slow development of function. It is not known what determines the rate at which fetal grafts develop functional responsiveness but gestational age and tissue mass are likely to be important factors. Another may be the normal physiologic development of the fetal organ. It is known that growth and development are to some extent "pre-programmed" in the donor tissue and it is conceivable that the maturation of the graft will be determined by the normal rates of growth of the donor species. Thus, if this is true the development of a fetal human pancreas may be quite slow whereas the development of a graft from a donor species in which growth is rapid (e.g. pig) may also be maintained in the recipient. Thus, a xenograft from such a species may develop function sooner than an allograft in a human recipient. We have some data from xenografts of late gestation fetal pig pancreas that shows that reversal of diabetes in spontaneously diabetic NOD mice can occur within a few weeks (Mandel and Koulmanda, unpublished observations).

Fetal islet xenografts

As noted above, fetal human pancreas is not going to be available in adequate quantity, at least in most Western countries. Xenogeneic donors are a likely alternative and for human use it is probable that pigs will be the ideal species [62]. Pig fetal islets have indeed already been tested in humans with some evidence of graft survival by histology and C-peptide assays but with no functional success [63]. In these studies tissue culture of fetal pig pancreas was used to generate islet cell clusters and these were transplanted either intraportally or under the renal capsule in recipients of a renal allograft. Urinary C-peptide was detected as was histologic evidence of the graft in a few patients but no patient showed any clinical benefit. It is conceivable that the amount of tissue used was insufficient and it is also likely that insulin resistance due both to a chronic diabetic state and to the side effects of the immunosuppressants used could have contributed to the failure of a clinical response.

In animal experiments, however, fetal xenografts were demonstrated to be effective when athymic (nude) mice or rats were used in which rejection of the graft was not a problem [64–68]. These experiments demonstrated quite clearly that cross-species grafts of fetal islets were functional even when very diverse species combinations were used e.g. human or pig to mouse. Thus, the physiologic aspects of such grafts showed that there was adequate regulation of blood glucose levels and indeed that in most experiments the recipient's blood glucose was in the range appropriate for the donor, e.g. human BG levels in mice. This suggests that some autonomy exists in the regulation of blood glucose that is built into the donor tissue and this may determine the sorts of donor species that may be used.

The use of fetal xenografts in spontaneously diabetic recipients as models of human autoimmune IDDM were also tested. These were generally rodent models; either murine (NOD) or rat (BB) models. In both of these models there is spontaneous immune-mediated destruction of the recipient's β cells and the animals are believed to represent reasonable models of the human disease [69]. Therefore it is possible to test not only the effect of immunosuppression on graft rejection but also determine whether there is any evidence of recurrent autoimmune disease in the transplant.

We have shown previously that in prediabetic NOD mice transplanted with isogeneic fetal NOD pancreas and simultaneously with fetal MHC-mismatched allogeneic tissue and with fetal pig pancreas there is a hierarchy of responses in animals that had received peritransplant anti-CD4 treatment. In these prediabetic NOD/Wehi male mice there was good preservation of the xenograft to 28 days post-transplant at which time most of the MHC-mismatched allografts had been destroyed while fetal isografts showed perigraft accumulations of lymphoid cells similar in extent to that present in the host's pancreas [17]. This shows that in the absence of antibody-mediated complement-dependent hyperacute rejection there was better preservation of the xenograft than the allograft and implies that the xenograft was also resistant to recurrent disease. If this is true in other models it may be that transplantation of xenogeneic islets is the most likely way of solving the problem of donor shortage with the possible added benefit that the graft may be resistant to recurrent disease. It is conceivable that tolerance to the xenogeneic tissue may be produced as has been shown in discordant grafts of islets in a number of rodent models [70–75]. Whether these data are also true for large animal outbred recipients such as humans remains to be seen but it is likely that obtaining adequate graft survival will be much harder in such studies.

Thus, for islet transplantation, probably the most widely needed situation where replacement of a defective cell mass may have the greatest benefit, the use of fetal tissue needs much more study. It seems likely that eventually the promising results obtained in rodent models with fetal pancreas grafts, particularly of xenogeneic origin, will be reproduced in large animal models but much more effort is required before this becomes a reality and such data are needed before clinical trials are contemplated.

Hepatocyte and hemopoeitic stem cell grafts

There are a number of scenarios where fetal hepatocytes could be used [76, 77]. One is the use of the fetal liver (FL) as a source of hemopoietic stem cells since it contains many blood cell precursors but few if any lymphoid cells, thus reducing the risk of graft versus host disease in the recipient [78]. FL has, for example, been used in experimental thallasemia [79]. While this was a potent potential use of FL, the recent advances in harvesting of peripheral blood hemopoeitic stem cells [80–82] has diminished the attraction of fetal liver. FL also a potential source of hepatocytes and has been used in, for example, congenital hypoalbuminemia [83], and fetal liver fragments have even been used recently in patients in fulminant hepatic failure [84] but with only modest success. Many patients have been treated for a range of inborn errors of metabolism but the transplants have had to be repeated to maintain an effect in most individuals [85]. The massive capacity of adult hepatocytes to proliferate perhaps takes the edge off the use of fetal tissue but it is possible that FL may be a useful source of cells particularly as it is easy to harvest even from first trimester terminations.

Fetal hemopoeitic cell replacement into fetuses affected with potentially lethal conditions has been tested in a number of studies in both animals and even in humans with some evidence of success [76, 86–89]. The immunologically immature state of the fetus as a recipient may make it less able to reject a stem cell transplant and even xenogeneic stem cell transplants across wide species barriers (human to sheep) have been reported where the FL-derived donor cells established chimerism in the recipient but interestingly only appeared in the circulation after birth and appeared initially to home to the recipient's bone marrow [90]. Allogeneic FL transplants in patients have also been reported with claims of success for severe immunodeficiency and thallassemia [91].

Fetal thymus transplantation

The early pivotal studies by Miller and Good that independently showed that the thymus was critical in allowing the development of T cells suggested that thymus transplantation could be used to treat T cell immunodeficiencies. This was quickly shown in mice where athymic (nude) mice could be reconstituted with thymus transplants, often using either fetal or neonatal tissue. However, the clinical application of this proved to be more difficult [92]. Nevertheless, the use of fetal thymus has been attempted on many occasions over the past 15 or so years for the treatment of severe T cell immunodeficiency with some recent reports of success and with evidence of chimerism by chromosome analysis [93, 94].

Other fetal tissue transplants

Many other examples of fetal tissue transplantation have been reported over the years and in the past couple of years these have included; syngeneic adrenal glands in rats [95], allogeneic fetal intestine, to reduce the load of lymphoid tissue that is characteristically present in adult intestine, with no evidence of graft-versus-host disease [96], olfactory bulb grafts in rats with evidence of function [97], orthotopic grafts of fetal cardiomyocytes that showed integration with the host cells [98], and retinal pigment cells that could be useful if successful in the treatment of retinitis pigmentosa, a common cause of blindness [99].

Many of these experiments are of predominantly academic interest but the application of cardiomyocyte or retinal cell grafts could have widespread application that will be limited by shortage of allogeneic donors as much as by problems with graft acceptance. The information gained from these studies, even if not likely to be directly useful in clinical situations may well be of great value in defining the biologic rules of fetal tissue transplantation.

Conclusions – The state of the art towards the end of the 20th century

It is clearly still too early to say where exactly fetal transplants are headed but there is gathering evidence that fetal tissue grafts will have a significant place in modern medical treatment. Obviously there are major problems with the use of fetal tissues, particularly if these are to come from aborted human fetuses, but these considerations are societal rather than scientific. In any case it is conceivable that xenogeneic fetal tissues many be useful in at least some circumstances and this would largely negate the ethical problems with the use of human fetal tissue. The unique attributes of fetal tissues, particularly their plasticity and potential capacity for growth, makes the use of such tissues appealing and continued research is essential.

References

1. Von Hippel A (1888) Eine neue Methode der Hornhauttransplantation. Albrecht v Graefes Arch Ophthalmol 34: 108–130.
2. Garry DJ, Caplan AL, Vawter DE and Kearney W (1992) Are there really alternatives to the use of fetal tissue from elective abortions in transplantation research? N Engl J Med 327: 1592–1595.
3. Branch DW, Ducat L, Fantel A, Low WC, Zhou FC, Dayton DH and Gill TJ 3rd. (1995) Suitability of fetal tissues from spontaneous abortions and from ectopic pregnancies for transplantation. Human Fetal Tissue Working Group. JAMA 273: 66–68.
4. Fine A (1994) Human fetal tissue research: practice, prospects, and policy. Cell Transplantation 3: 113–145.
5. Hofer BJ and Olson L (1991) Ethical issues in brain-cell transplantation. Trends Neurol Sci 14: 384–388.
6. Turner DA and Kearney W (1993) Scientific and ethical concerns in neural fetal tissue transplantation. Neurosurg 33: 1031–1037.
7. Woodruff MA (1960) The transplantation of tissues and organs. Springfield, IL: CC Thomas.
8. Oppenheim RW, Houenou LJ, Johnson JE, Lin LFH, Li LX and Lo AC, Newsome AL, Prevette DM and Wang SW (1995) Developing motor neurons rescued from programmed and axotomy-induced cell death by GDNF. Nature 373: 344–346.
9. Henderson CE, Phillips HS, Pollock RA, Davies AM, Lemeulle C, Armanini M, Simpson LC, Moffet B, Vandlen RA, Koliatsos VE *et al.* (1994) GDNF: a potent survival factor for motoneurons present in peripheral nerve and muscle. Science 266: 1062–1064.
10. Oppenheim RW, Yin QW, Prevette D and Yan Q (1992) Brain-derived neurotrophic factor rescues developing avian motoneurons from cell death. Nature 360: 755–757.
11. Simeonovic CJ and Lafferty KJ (1988) Immunogenicity of mouse fetal pancreas and proislets. A comparison. Transplantation 45: 824–827.
12. Velasco AL and Hegre OD (1989) Decreased immunogenicity of fetal kidneys: the role of passenger leukocytes. J Pediatr Surg 24: 59–63.
13. HogenEsch RI, Staal MJ, Kema IP, Buys CH and Go KG (1993) Utility of fragmented human fetal tissue as a potential dopaminergic brain graft in Parkinson's disease. Stereotact Funct Neurosurg 61: 1–11.
14. Westgren M, Ek S, Bui TH, Hagenfeldt L, Markling L, Pschera H, Seiger A, Sundstrom E and Ringden O (1994) Establishment of a tissue bank for fetal stem cell transplantation. Acta Obstet Gynecol Scand 73: 385–388.
15. Mandel TE and Georgiou HM (1983) Insulin secretion by fetal human pancreatic islets of Langerhans in prolonged organ culture. Diabetes 32: 915–920.
16. Lim SM, Heng KK, Lim NK, Seah ML, Wee A , Li SQ, Soh P, Rauff A and Vengadasalam D (1991) An in vitro assessment of human fetal pancreatic islets of Langerhans in culture. Ann Acad Med Singapore 20: 465–471.
17. Mandel TE and Koulmanda M (1992) Survival of xeno-, allo-, and isografts in NOD, and xenografts in other mouse strains

following immunosuppression with anti-CD4 monoclonal antibody. Diab Nutrition Metab 5 (suppl 1): 91–96.
18. Pierson RN, Winn HJ, Russell PS and Auchincloss H Jr. (1989) Xenogeneic skin graft rejection is especially dependent on CD4+ T cells. J Exp Med 170: 991–996.
19. Koutouzis TK, Emerich DF, Borlongan CV, Freeman TB, Cahill DW and Sanberg PR (1994) Cell transplantation for central nervous system disorders. Crit Rev Neurobiol 8: 125–162.
20. Bjorklund A and Stenevi U (1979) Reconstruction of the nigrostriatal dopamine pathway by intracerebral nigral transplants. Brain Res 177: 555–560.
21. Bjorklund A, Stenevi U, Dunnett SB and Iversen SD (1981) Functional reactivation of the deafferented neostriatum by nigral transplants. Nature 289: 497–499.
22. Perlow MJ, Freed WJ, Hofer BJ, Seiger A, Olson L and Wyatt RJ (1979) Brain grafts reduce motor abnormalities produced by destruction of nigrostriatal dopamine system. Science 204: 643–647.
23. Redmond DE Jr, Roth RH, Spencer DD, Naftolin F, Leranth C, Robbins RJ, Marek KL, Elsworth JD, Taylor JR, Sass KJ *et al.* (1993) Neural transplantation for neurodegenerative diseases: past, present, and future. Ann NY Acad Sci 695: 258–266.
24. Madrazo I, Drucker-Colin R, Diaz V, Martinez-Mata J, Torres C and Becerril JJ (1987) Open microsurgical autograft of adrenal medulla to the right caudate nucleus in two patients with intractable Parkinson's disease. N Engl J Med 316: 831–834.
25. Goetz CG, De Long MR, Penn RD and Bakay RA (1993) Neurosurgical horizons in Parkinson's disease. Neurology 43: 1–7.
26. Fiandaca MS and Gash DM (1992) New insights and technologies in brain grafting. Clin Neurosurg 39: 482–508.
27. Hitchcock ER, Kenny BG, Clough CG, Hughes RC, Henderson BT and Detta A (1990) Stereotactic implantation of fetal mesencephalon. Stereotact Funct Neurosurg 54–55: 282–289.
28. Fiandaca MS (1991) Brain grafting for Parkinson's disease. Experimental, clinical, and immunological considerations. Transplantation 51: 549–556.
29. Lindvall O, Rehncrona S, Brundin P, Gustavii B, Astedt B, Widmer H, Lindholm T, Bjorklund A, Leenders KL, Rothwell JC *et al.* (1990) Neural transplantation in Parkinson's disease; the Swedish experience. Prog Brain Res 82: 729–734.
30. Widner H, Tetrud J, Rehncrona S, Snow B, Brundin P, Gustavii B, Bjorklund A, Lindvall O and Langston JW (1992) Bilateral fetal mesencephalic grafting in two patients with parkinsonism induced by 1-methyl-4-phenyl-1,2,3,6-tetrahydropyridine (MPTP) N Engl J Medi 327: 1556–1563.
31. Freed CR, Breeze RE, Rosenberg NL, Schneck SA, Kriek E, Qi JX, Lone T, Zhang YB, Snyder JA, Wells TH *et al.* (1992) Survival of implanted fetal dopamine cells and neurologic improvement 12 to 46 months after transplantation for Parkinson's disease. N Engl J Med 327: 1549–1555.
32. Spencer DD, Robbins RJ, Naftolin F, Marek KL, Vollmer T, Leranth C, Roth RH, Price LH, Gjedde A, Bunney BS *et al.* (1992) Unilateral transplantation of human fetal mesencephalic tissue into the caudate nucleus of patients with Parkinson's disease. N Engl J Med 327: 1541–1548.
33. Iacoco RP, Tang ZS, Mazziotta JC, Grafton S and Hoehn M (1992) Bilateral fetal grafts for Parkinson's disease: 22 months results. Stereotact Funct Neurosurg 58: 84–87.
34. Hitchcock ER, Kenny BG, Henderson BT, Clough CG, Hughes RC and Detta A (1991) A series of experimental surgery for asdvanced Parkinson's disease by foetal mesencephalic transplantation. Acta Neurochirg 52 (suppl): 54–57.
35. Taylor JR, Elsworth JD, Roth RH, Sladek JR Jr, Collier TJ and Redmond DE Jr (1991) Grafting of fetal substantia nigra to striatum reverses behavioral defecits induced by MPTP in primates: a comparison with other types of grafts as controls. Exp Brain Res 85: 335–348.
36. Henderson BT, Clough CG, Hughes RC, Hitchcock ER and Kenny BG (1991) Implantation of human fetal ventral mesencephalon to the right caudate nucleus in advanced Parkinson's disease. Arch Neurol 48: 822–827.
37. Iacono RP, Lonser RR, Mandybur G, Morenski JD, Yamada S and Shima F (1994) Stereotactic pallidotomy results for Parkinson's exceed those of fetal graft. Am Surg 60: 777–82.
38. Sanberg PR, Koutouzis TK, Freeman TB, Cahill DW and Norman AB (1993) Behavioral effects of fetal neural transplants:relevance to Huntington's disease. Brain Res Bull 32: 493–496.
39. Madrazo I, Franco-Bourland RE, Cuevas C *et al.* (1991) Fetal neural grafting for the treatment of Huntington's disease (HD) – Report of the first case. Soc Neurosci 17: 902 (abstr).
40. Gumpel M, Baumann N, Raoul M and Jacque C (1983) Survival and differentiation of oligo-dendrocytes from neural tissue transplanted into newborn mouse brain. Nerosci Lett 37: 307–311.
41. Gumpel M, Lachapelle F, Gansmuller A, Baulac M, Baron-Van, Evercooren A and Baumann N (1987) Transplantation of human embryonic oligodendrocytes into shiverer brain. Ann NY Acad Sci 495: 71–85.
42. Rosenbluth J, Hasegawa M, Shirasaki N, Rosen CL and Liu Z (1990) Myelin formation following transplantation of normal fetal glia into myelin deficient rat spinal cord. J Neurocytol 19: 718–730.
43. Lindvall O (1991) Prospects of transplantation in human neurodegenerative diseases. Trends Nerosci 14: 376–384.
44. Chen LS, Ray J, Fisher LJ, Kawaja MD, Schinstine M, Kang UJ and Gage FH (1991) Cellular replacement therapy for neurologic disorders: potential of genetically engineered cells. J Cell Biochem 45: 252–257.
45. Bankiewicz K, Mandel RJ and Sofroniew MV (1993) Trophism, transplantation, and animal models of Parkinson's disease. Exp Neurol 124: 140–9.
46. Frim DM, Simpson J, Uhler TA, Short MP, Bossi SR, Breakefield XO and Isacson O (1993) Striatal degeneration induced by mitochondrial blockade is prevented by biologically delivered NGF. J Neurosci Res 35: 452–8.
47. Emerich DF, Winn SR, Christenson L, Palmetier MA, Gentile FT and Sanberg PR (1992) A novel approach to neural transplantation in Parkinson's disease: use of polymer-encapsulated cell therapy. Neuroscience Biobehav Rev 16: 437–447.
48. Aebischer P, Winn SR, Tresco PA, Jaeger CB and Greene LA (1991) Transplantation of polymer encapsulated neurotransmitter secreting cells: effect of the encapsulation technique. J Biomech Engin 113: 178–183.
49. Anonymous (1993) The effect of intensive treatment of diabetes on the development and progression of long-term complications in insulin-dependent diabetes mellitus. The Diabetes Control and Complications Trial Research Group. New Engl J Med 329: 977–986.
50. Robertson RP (1991) Pancreas transplantation in humans with diabetes mellitus. Diabetes 40: 1085–1089.
51. Farkas G, Szabo M and Voros P (1992) Ten years' clinical experience of fetal islet transplantation. Transplant Proc 24: 2954–2955.

52. Hu YF, Gu ZF, Zhang HD and Ye RS (1989) Fetal islet cell transplantation in China. Transplant Proc 21: 2605–2607.
53. Peterson CM, Jovanovic-Peterson L and Formby B (eds) (1988) Fetal islet transplantation. Implications for diabetes. Springer-Verlag: New York.
54. Simeonovic CJ, Brown DJ, Teittinen KUS and Wilson JD (1992) Preparation and transplantation of fetal proislets. In: Ricordi C (ed) Pancreatic islet cell transplantation, pp 238–248. Austin, TX: RG Landes Co.
55. Tuch BE and Simpson AM (1992) Experimental fetal islet transplantation. In: Ricordi C (ed) Pancreatic islet cell transplantation, pp 279–290. Austin, TX: RG Landes Co.
56. Mandel TE and Koulmanda M (1994) Islet isolation from the fetal pancreas. In: Lanza RP and Chick WL (eds) Pancreatic islet transplantation, vol 1., pp 107–124. Austin, TX: RG Landes Co.
57. Mandel TE (1994) Renal subcapsular transplantation. In: Green MJ and Mandel TE (eds) Experimental transplantation models in small animals, pp 147–167. Melbourne: Harwood Academic Publishers.
58. Mandel TE (1988) Fetal islet transplantation in diabetic mice: a model for human islet transplants. In: Peterson CM, Jovanovic-Peterson L and Formby B (eds) Fetal islet transplantation. Implications for diabetes, pp 165–184. New York: Springer-Verlag.
59. Lacy PE (1993) Status of islet cell transplantation. Diabetes Rev 1: 76–92.
60. Mandel TE (1983) Growth of organ cultured foetal mouse pancreas isografts in diabetic and non-diabetic recipients. Aust J Exp Biol Med Sci 61: 497–508.
61. Simeonovic CJ, Mackie DJ and Wilson JD (1990) Functional properties of proislet isografts in diabetic mice. Diabetes Res Clin Pract 8: 275–281.
62. Cooper DKC, Ye Y, Rolf LL Jr and Zuhdi N (1991) The pig as potential organ donor for man. In: Cooper DKC, Kemp E, Reemtsma K and White DJG (eds) Xenotransplantation, pp 481–500. Berlin: Springer-Verlag.
63. Groth CG, Korsgren O, Tibell A, Tollemar J, Moller E, Bolinder J, Ostman J, Reinholt FP, Hellerstrom C and Andersson A (1994) Transplantation of porcine fetal pancreas to diabetic patients. Lancet 344: 1402–1404.
64. Thompson SC and Mandel TE (1990) Fetal pig pancreas. Preparation and assessment of tissue for transplantation, and its in vivo development and function in athymic (nude) mice. Transplantation 49: 571–581.
65. Sandler S, Andersson A, Schnell A, Mellgren A, Tollemar J, Borg H, Petersson B, Groth CG and Hellerstrom C (1985) Tissue culture of human fetal pancreas. Development and function of B-cells in vitro and transplantation of explants to nude mice. Diabetes 34: 1113–1119.
66. Tuch BE, Osgerby KJ and Turtle JR (1988) Normalization of blood glucose levels in nondiabetic nude mice by human fetal pancreas after induction of diabetes. Transplantation 46: 608–611.
67. Tuch BE and Monk RS (1991) Regulation of blood glucose to human levels by human fetal pancreatic xenografts. Transplantation 51: 1156–1160.
68. Tuch BE, Monk RS and Beretov J (1991) Reversal of diabetes in athymic rats by transplantation of human fetal pancreas. Transplantation 52: 172–175.
69. Bernard CCA, Mandel TE and Mackay IR (1992) Experimental models of human autoimmune disease: overview and prototypes. In: Rose NR and Mackay IR (eds) The Autoimmune diseases II, pp 47–106. San Diego: Academic Press.
70. Lacy PE, Ricordi C and Finke EH (1989) Effect of transplantation site and anti-L3T4 treatment on survival of rat, hamster and rabbit islet xenografts in mice. Transplantation 47: 761–766.
71. Ricordi C, Lacy PE, Sterbenz K and Davie JM (1987) Low-temperature culture of human islets or in vivo treatment with L3T4 antibody produces a marked prolongation of islet human-to-mouse xenograft survival. Proc Natl Acad Sci USA 84: 8080–8084.
72. Falqui L, Finke EH, Carel JC, Scharp DW and Lacy PE (1991) Marked prolongation of human islet xenograft survival (human-to-mouse) by low-temperature culture and temporary immuno-suppression with human and mouse anti-lymphocyte sera. Transplantation 51: 1322–1324.
73. Simeonovic CJ, Ceredig R and Wilson JD (1990) Effect of GK1.5 monoclonal antibody dosage on survival of pig proislet xenografts in CD4+ T cell-depleted mice. Transplantation 49: 849–856.
74. Faustman D and Coe C (1991) Prevention of xenograft rejection by masking donor HLA class I antigens. Science 252: 1700–1702.
75. Linsley PS, Wallace PM, Johnson J *et al.* (1992) Immunosuppression in vivo by a soluble form of the CTLA-4 T cell activation molecule. Science 257: 792–795.
76. Rojansky N and Schenker JG (1993) The use of fetal tissue for therapeutic applications. Int J Gynaecol Obstet 41: 233–240.
77. Touraine JL, Roncarolo MG, Bacchetta R, Raudrant D, Rebaud A, Laplace S, Cesbron P, Gebuhrer L, Zabot MT, Touraine F *et al.* (1993) Fetal liver transplantation: biology and clinical results. Bone Marrow Transplant 11 (Suppl 1): 119–122.
78. Hillyer CD and Wells SJ (1993) Alternative sources of hematopoietic stem cells for bone marrow transplantation and rescue. J Hematother 2: 491–499.
79. Bethel CA, Muragesh D, Harrison MR, Mohandas N and Rubin ER (1993) Fetal hematopoietic stem cell transplantation into beta-thalassemic mice. J Pediatr Surg 28: 1232–1237.
80. Williams SF, Zimmerman T, Grad G and Mick R (1993) Source of stem cell rescue: bone marrow versus peripheral blood progenitors. J Hematother 2: 521–523.
81. Hillyer CD and Wells SJ (1993) Alternative sources of hematopoietic stem cells for bone marrow transplantation and rescue. J Hematother 2: 491–499.
82. Gale RP, Henon P and Juttner C (1992) Blood stem cell transplants come of age. Bone Marrow Transplant 9: 151–155.
83. Otsu I, Ikebukuro H, Hoshi T, Ogawa K, Mito M and Nozawa M (1992) Long-term effects of fetal liver transplantation in congenitally albumin-deficient rats. Transplant Proc 24: 2962–2963.
84. Habibullah CM, Syed IH, Qamar A and Taher-Uz Z (1994) Human fetal hepatocyte transplantation in patients with fulminant hepatic failure. Transplantation 58: 951–952.
85. Touraine JL, Laplace S, Rezzoug F, Sanhadji K, Veyron P, Royo C, Maire I, Zabot MT, Vanier MT, Rolland MO *et al.* (1991) The place of fetal liver transplantation in the treatment of inborn errors of metabolism. J Inherit Metab Dis 14: 619–626.
86. Flake AW and Zanjani ED (1993) In utero transplantation of hematopoietic stem cells. Crit Rev Oncol Hematol 15: 35–48.
87. Touraine JL (1992) Rationale and results of in utero transplants of stem cells in humans. Bone Marrow Transplant 10 (Suppl 1): 121–126.
88. Zanjani ED, Ascensao JL, Flake AW, Harrison MR and Tavassoli M (1992) The fetus as an optimal donor and recipient of

hemopoietic stem cells. Bone Marrow Transplant 10 (Suppl 1): 107–114.

89. Raudrant D, Touraine JL and Rebaud A (1992) In utero transplantation of stem cells in humans: technical aspects and clinical experience during pregnancy. Bone Marrow Transplant 9 (Suppl 1): 98–100.

90. Zanjani ED, Pallavacini MG, Ascensao JL, Flake AW, Harrison MR and Tavassoli M (1992) Human-ovine xenogenic transplantation of stem cells in utero. Bone Marrow Transplant 9 (Suppl 1): 86–89.

91. Touraine JL, Raudrant D, Rebaud A, Roncarolo MG, Laplace S, Gebuhrer L, Betuel H, Frappaz D, Freycon F, Zabot MT *et al.* (1992) In utero transplantation of stem cells in humans: immunological aspects and clinical follow-up of patients. Bone Marrow Transplant 9 (Suppl 1): 121–126.

92. Hong R (1986) Reconstitution of T-cell deficiency by thymic hormone or thymus transplantation therapy. Clin Immunol Immunopathol 40: 136–141.

93. Higuchi S, Yanabe Y, Tsuchiya H, Akahoshi I, Udaka K, Migita M and Matsuda I (1993) Irradiated fetal thymus transplantation in a patient with combined immunodeficiency with predominant T cell defect. Acta Paediatr Jpn 35 (1): 39–44.

94. Roncarolo MG, Bacchetta R, Bigler M, Touraine JL, de Vries JE and Spits H (1991) A SCID patient reconstituted with HLA-incompatible fetal stem cells as a model for studying transplantation tolerance. Blood Cells 17 (2): 391–402.

95. Trammer A, Kellnar S, Welsch U and Schwarz HP (1994) Transplantation of fetal adrenal glands in syngeneic rat strains. Eur J Pediatr Surg 4: 249–252.

96. Yang R, Liu Q, Rescorla FJ and Grosfeld JL (1994) Lack of graft-versus-host disease after fetal intestine transplantation. J Pediatr Surg 29: 1157–1160.

97. Doucette R, Kott J and Westrum L (1994) The development of glial fibrillary acidic protein-positive cells and the appearance of laminin-like immunoreactivity in fetal olfactory bulb transplants. Brain Res 649: 334–338.

98. Soonpaa MH, Koh GY, Klug MG and Field LJ (1994) Formation of nascent intercalated disks between grafted fetal cardiomyocytes and host myocardium. Science 264: 98–101.

99. Bhatt NS, Newsome DA, Fenech T, Hessburg TP, Diamond JG, Miceli MV, Kratz KE and Oliver PD (1994) Experimental transplantation of human retinal pigment epithelial cells on collagen substrates. Am J Ophthalmol 117: 214–221.

Section VI:

Germ Cells

R. P. Lanza and W. L. Chick (eds.), Yearbook of Cell and Tissue Transplantation 1996/1997, 119–124.

Chapter 11

Transplantation of spermatogonial stem cells

Neelakanta Ravindranath, Ghenima Dirami, Meng-Chun Jia and Martin Dym
Department of Cell Biology, Georgetown University School of Medicine, 3900 Reservoir Road, NW, Washington, DC 20007, U.S.A.

Over the centuries, scientists have attempted to replace damaged or diseased tissues and organs with grafts from other individuals or from lower animal species. The early history of organ transplantation dates back to the second century before Christ [1]. Transplantation experiments involving the testis have been of two types, those in which the testis was the tissue to be transplanted and those in which the testis was the recipient site. The first successful transplantation experiments recorded are those by John Hunter, who transplanted the testis of a cock into the belly of a hen [2]. The next important series of transplantation experiments were those carried out by Berthold (see Setchell, 1984 for a review) [3]. However, very little attention was paid to these studies at the time. It was only at the end of the nineteenth century that a number of transplant studies were again reported [4–6]. In these early studies, transplantation of the testes of mammals appeared to be quite difficult and it was only with the work of Cevelotto [7], Steinach [8], and Sand [9] that successful results were reported. Even in the successful grafts, in many instances the seminiferous tubules eventually atrophied. Subsequently, Turner [10] and Dameron [11, 12] demonstrated that complete spermatogenesis could also be achieved in testicular grafts placed in the anterior chamber of the eye. Williams [13] found that some spermatogenic activity persisted in tubular fragments placed in transparent chambers in rabbit's ears. The first person to use the testis as a site of transplantation appears to have been Sand [9, 14] who found that an ovary transplanted into the substance of the testis developed follicles. Thereafter, skin [15], isolated islets of Langerhans [16], and parathyroid glands [17, 18] were transplanted into the rat testis. This suggested that the testis may be an immunologically privileged organ. This characteristic appeared not to be altered by prenatal irradiation; parathyroid grafts survived in testes from which the germ cells had been eliminated by prenatal irradiation as well as those in normal testes [19]. In the ram, thyroid autografts survived whereas the allografts did not [20]. Experiments involving transplanting testicular tissue to the testis was initiated by Voronoff [21]. Fragments of testes of young rams were transplanted to the tunica albuginea of the testes of old animals and this restored the virility and general well-being of the recipients, as well as increasing their wool production. Voronoff soon extended his studies to humans, using tissue from monkeys. When monkey testes were transplanted into the scrota of hypogonadal males, their appetite returned and their general physical condition improved [21]. Subsequent studies were carried out by Turner and Asakawa [22] and Aron and Luxembourger [23] who showed that testes from new-born rats attained complete spermatogenesis when transplanted into the testes of adults. Although histological and functional proof of graft survival was lacking in these early attempts at testis transplants, they paved the way for transplanting a testis into an anorchic man [24].

Need for testis or spermatogonial stem cell transplantation

The transplantation of a testis or of germ cells into an individual may be necessary for one of the following reasons: 1. *Anorchia*: A significant number of individuals are born without testes (anorchia). These individuals require testosterone supplementation for the normal functioning of their body [24]. 2. *Infertility*: In men with primary testicular failure because of germinal epithelial damage or complete germ cell arrest, medical or surgical treatment of infertility is often not possible. Similarly, men who have developed androgen resistance do not respond to high doses of testosterone. Thus, they are rendered infertile. In the absence of any possible medical treatment, transplantation of testis or germ cells is the only alternative approach available. 3. *Reproductive senescence in men*: During and after mid-life, men generally undergo impairment in reproductive function with loss of fertility. To restore fertility and vitality, testicular transplantation may be a feasible option. 4. *Gene therapy*: The development of the spermatogonial stem-cell transplantation technique [25, 26] and the possibility of manipulating the genetic makeup of the spermatogonia will have many potential applications in biology, medicine, and agriculture [27].

Types of testis transplantation

Organ transplantation

Whole testis transplantation was performed by Attaran and colleagues [28] in dogs, however, the transplants were mostly rejected. Successful testicular transplantation in rats have been reported by a number of investigators [29–31]. Heterologous transplantation of monkey testis into old men [21] with successful graft survival for as long as two and half years was cited in an editorial in 'Lancet, 1991'. In another report, in one of the two genetically identical twins born without testes, transplantation of a testis from the other twin resulted in normal serum testosterone levels within two hours of surgery [24]. Furthermore, normal sperm counts were observed in the recipient's semen by 230 days after surgery.

Stem cell transplantation

While whole organ replacement therapy is necessary in such conditions as anorchia, it is essential to discover the cause of infertility and target it directly. The etiology of infertility can be classified under the following headings: pretesticular (including endocrine causes), testicular causes affecting spermatogenesis, and post-testicular (ductal factors and sexual dysfunction). Testicular causes affecting spermatogenesis is the most frequent reason for male infertility [32].

Spermatogenesis is a complex process of cellular renewal and differentiation that begins with the divisions of the type A spermatogonial stem cells and ends as late spermatids are released into the seminiferous tubule lumen as spermatozoa [33]. This process consists of: (1) the proliferation of type A spermatogonial stem cells to renew themselves (stem cell renewal), to undergo an apoptotic form of cell death, or to differentiate into intermediate spermatogonia leading to the formation of type B spermatogonia and preleptotene spermatocytes [34]; (2) further differentiation of the preleptotene spermatocytes through meiosis to the formation of newly-formed round spermatids; and (3) the morphogenesis of the round spermatids into mature spermatozoa [35]. The transplantation of spermatogonial stem cells from a normal donor will obviate the need for organ replacement in the treatment of infertility caused by a defect in spermatogenesis. In addition to the treatment of infertility, spermatogonial stem cell transplantation will eventually be a useful tool in preventing the transmission of genetic illnesses to the offspring [27]. This cannot be achieved unless a culture system that allows expansion of spermatogonial stem cells and selection of specifically modified clones is developed.

Source of spermatogonial stem cells

The type A spermatogonial stem cells could be obtained from a donor testis. Although the procedure for isolation of spermatogonial cells from the human testis has not yet been standardized, methods are available to isolate spermatogonia from immature mouse and rat testis [36–39] and we have recently succeeded in isolating type A spermatogonia from the pig testes (unpublished observations). We have modified the procedure of Bellvé and colleagues [39] and purified rat type A spermatogonia to $> 95\%$ homogeneity [40]. Briefly, the decapsulated testes were suspended in DMEM/F12 containing collagenase (1.5 mg/ml) and DNAse (1 μg/ml), and incubated at 34 °C for 15 min in a shaking water bath operated at 100 cycles/min. After two washes in DMEM/F12 medium, seminiferous cord fragments, mostly devoid of interstitial cells, were incubated in DMEM/F12 medium containing collagenase (1.5 mg/ml), hyaluronidase (1.5 mg/ml), trypsin (0.5 mg/ml), and DNAse (1 μg/ml) for 20–30 min using the conditions described above. The dispersed cells were washed twice with medium and filtered through

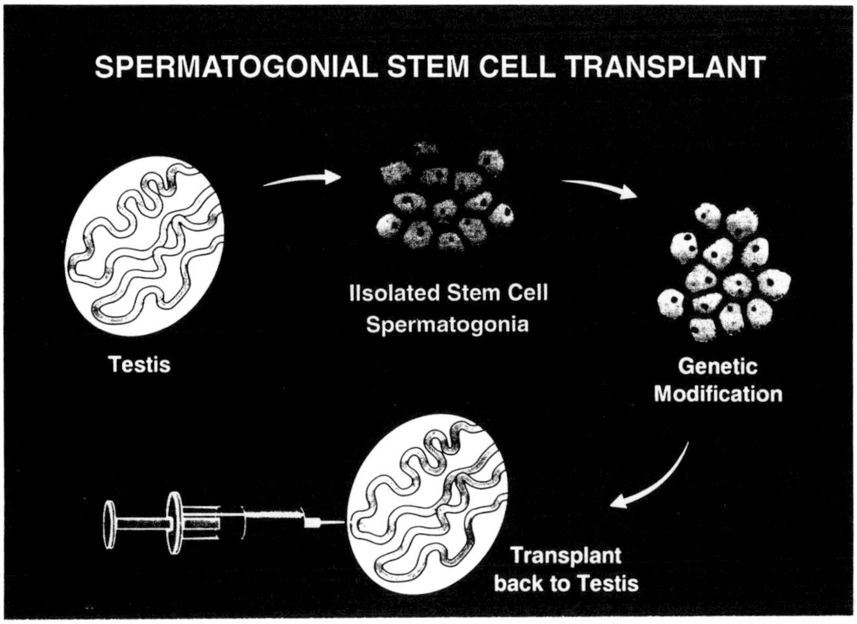

Fig. 1. Injection of 'microfil' compound into the rete testis of a rat resulted in a simultaneous retrograde filling of many seminiferous tubules [41]. After dehydration in ethanol and clearing in methyl salicylate, the undulating pattern of the rat's seminiferous tubules are clearly visible. The rete testis is the highly irregular region in the right central portion of the photograph. × 3.9.

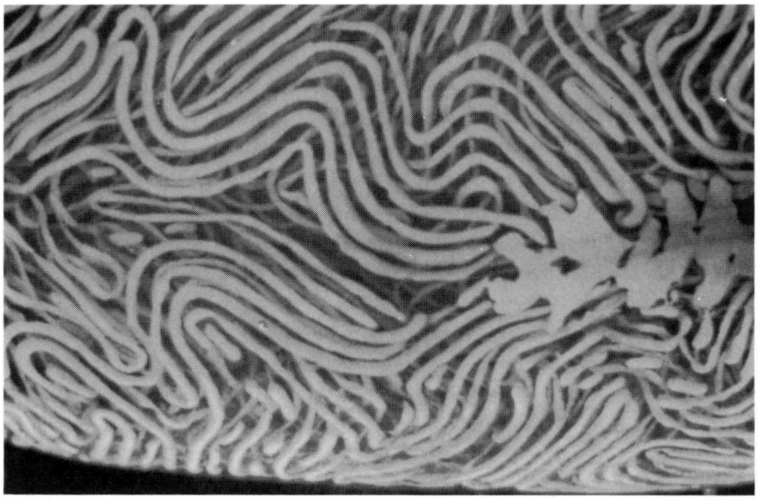

Fig. 2. A schematic drawing outlining the steps required for the transplantation of genetically-modified spermatogonial stem cells. Spermatogonial stem cells would first be removed from an individual using testicular biopsy. After purification, the stem cells will be expanded in culture, and modified for example via targeted homologous recombination of DNA sequences, and individual clones of cells can be selected and transplanted back to the same individual whose testes were made sterile by local irradiation.

80 μm and 40 μm nylon mesh, successively. The cells of the dissociated epithelium were then separated by sedimentation velocity at unit gravity at 4 °C, with the use of a 2–4% BSA gradient in DMEM/F12 medium. The cells were bottom-loaded into an SP-120 chamber in a volume of 30 ml, and a BSA gradient was generated using 275 ml of 2% and 4% BSA. The cells were allowed to sediment for a standard period of 2.5 hrs, and then 35 fractions of 15 ml volume were collected at 90 sec intervals. The cells in each fraction were examined under a phase contrast microscope, and fractions containing cells of similar size and morphology were pooled and spun down by low-speed centrifugation, and then resuspended in DMEM/F12 medium. Type A spermatogonial cells were identified by immunostaining for the c-*kit* receptor, a marker for the type A spermatogonia in the immature rat [40].

Procedure for the transplantation of spermatogonial stem cells

Recent studies by Brinster and Zimmerman [26] and Brinster and Avarbock [25] have shown that the injection of spermatogonial stem cells into the testes of sterile recipient mice gives rise to functional sperm in the recipients. They prepared a mixture of testis cells containing the spermatogonial stem cells, Sertoli cells, and myoid cells from 4–12 day old mice by the enzymatic procedure of Bellvé *et al.* [39] with minor modifications. The cells were injected directly into individual seminiferous tubules. The recipient mouse was anesthetized and the testis was exteriorized through a midline abdominal incision. Using an appropriate pipette for injection, the testis cells were injected into a number of individual seminiferous tubules using a dissecting microscope. Trypan blue was used to follow the progress of seminiferous tubule filling. Following injection of donor testis cells, recipient mice were maintained for 48 to 230 days. The testes were examined histologically for the presence of spermatogenesis. One-third of the testes injected showed colonization by donor cells.

An alternative approach for injection of spermatogonial stem cells would be injection directly into the rete testis and the subsequent retrograde filling of many seminiferous tubules simultaneously (Fig. 1) [41]. Each seminiferous tubule is a single coiled loop that opens at both ends into a series of collecting channels known as the rete testis. The rete testis is a tubular system which connects the straight tubular ends of seminiferous tubules to the efferent ducts. The efferent ducts lead into the epididymis. The veins draining the testis of the rat converge in an exceedingly complex pattern at the testicular apical pole just prior to piercing the tunica albuginea and entering the pampiniform plexus. In the rat a clear spherical area, 1 mm in diameter, is delineated by these venous channels and the cranial portion of the rete testes is invariably found immediately deep to the tunica albuginea in this region. A 30-gauge needle could be introduced into the rete testis at this region. Injection into the rete testis in this manner would fill many seminiferous tubules uniformly in comparison to injection into a number of individual tubules.

Possible implications of spermatogonial stem cell transplantation / germ cell gene therapy

There are many ethical and legal implications associated with germ cell gene therapy using spermatogonial stem cells, and appropriate regulating agencies will have to be empowered to monitor these developments. Eventually, germ line modifications in the progeny could be brought about by manipulating spermatogonial stem cells, for example, via targeted homologous recombination of DNA sequences. The drawing (Fig. 2) outlines schematically the steps in the procedure that could be developed. Testis spermatogonial stem cells could be removed from an individual using a biopsy procedure and allowed to proliferate in culture prior to genetic modification. The specifically modified clones would then be transplanted back to the same individual whose testes were made sterile by local irradiation. If the procedures function properly, the transplanted stem cells would insinuate themselves at the base of the seminiferous tubules (now sterile) and initiate new waves of spermatogenesis. The sperm produced would have the corrected gene that could then be transferred to the progeny. However, at the moment, very little is known about the factors regulating the spermatogonial stem cells. Before the spermatogonia could be used for gene therapy studies, much more information is required regarding the physiology of this omnipotent stem cell.

Another possible use of spermatogonial stem cell transplant would be as a treatment for certain types of male infertility. For example, in individuals diagnosed with Hodgkin's disease which require radiation treatment resulting in sterility due to germ cell depletion, a biopsy of the testis could be removed and the spermatogonial stem cells could be allowed to proliferate in culture. After successful radiation treatment of the disease, the spermatogonia could then be transplanted back to the patient to reinitiate spermatogenesis and in this manner fertility could be restored.

Acknowledgement

This work was supported in part by NIH grant PO1 HD24633 to M.D.

References

1. Flye MW (1989) History of transplantation. In: Flye MW (ed) Principles of Organ Transplantation, pp 1–17. Philadelphia, PA: W.B. Saunders Company.
2. Qvist G (1981) John Hunter 1728–1793. London: William Heinemann.
3. Setchell BP (1984) Male Reproduction. In: Langley LL (ed) Benchmark Papers in Human Physiology, pp 1–401. New York: Van Nostrand Reinhold Company.
4. Lode A (1895) Zur Transplantation der Hoden bie Hahnen. Wien Klin Wochenschr 8: 345.
5. Hanau A (1897). Versuche uber den Einfuss der Geschlechtsdrusen auf die secundaren Sexualcharactere. Arch Ges Physiol 65: 516–617.
6. Foges A (1898) Zur Hodentransplantation bei Hahnen. Centrlbl Physiol 12: 898–901.
7. Cevelotto G (1909) Uber Verpfanzungen und Gefuerungen der Hoden, Frank. Zeitschr Path 3: 331–337.
8. Steinach E (1910) Geschelchtstrieb und echtsekundare Geschelchtsmerkmale als folgender innersekretorischen Funktion der Keimdrusen. Centrlbl Physiol 24: 551–566.
9. Sand K (1919) Experiments on the internal secretion of the sexual glands, especially on experimental hermaphroditism. J Physiol 53: 257–263.
10. Turner CD (1938) Intra-ocular homotransplantation of prepuberal testes in the rat. Am J Anat 63: 101–159.
11. Dameron JT (1950) Homologous transplantation of fetal endocrine tissue to adult non-related host. Proc Soc Exp Biol Med 73: 343–345.
12. Dameron JT (1951) The anterior chamber of the eye for investigative purposes. A site for transplantation of fetal endocrine tissues and cancer, and for the study of tissue reaction. Surgery 30: 787–799.
13. Williams RG (1950) Studies of living interstitial cells and seminiferous tubules in autogenous grafts of testis. Am J Anat 86: 343–367.
14. Sand K (1923) Experiments on the endocrinology of the sexual glands. Endocrinol 7: 273–301.
15. Nicholas RJ and Scothorne R (1969). Studies on the testis as an "immunologically privileged" site. J Anat 104: 194 (abstract).
16. Moskalewski S (1969) Studies on the culture and transplantation of isolated islets of Langerhans of the guinea pig. Proc Rou Nederl Akad Wct 72: 157–171.
17. Dib-Kuri A, Revilla A and Chavex-Peon F (1975) Successful rat parathyroid allografts and xenografts to the testis without immunosuppression. In: Schlesinger M and Billingham RE (eds) Transplantation Today, pp 753–756. New York: Grune and Stratton.
18. Whitmore WF and Gittes RF (1979) Intratesticular grafts: the testis as an exceptional immunologically privileged site. Trans Am Assoc Genito-Urinary Surg 70: 76–80.
19. Whitmore WF, Karsh L and Gittes RF (1985) The role of germinal epithelium and spermatogenesis in the privileged survival of intratesticular grafts. J Urol 134: 782–786.
20. Maddocks S and Setchell BP (1988) The rejection of thyroid allografts in the ovine testis. Immunol Cell Biol 66: 1–8.
21. Voronoff S (1925) Rejuvenation by Grafting. London: George Allen and Unwin.
22. Turner CD and Asakawa H (1963) Complete spermatogenesis in intratesticular homotransplants of fetal and neonatal testes in the rat. Proc Soc Exp Biol Med 112: 132–135.
23. Aron M and Luxembourger M-M (1968) Facteurs de l'evolution spermatogenetique et de l'evolution de la glande interstitielle avant la maturite sexuelle. Etude chez le rat par l'homotransplantation du testicule neo-natal dans le testicule de l'animal mur. Arch Anat Histol Embryol 51: 41–52.
24. Silber SJ (1978) Transplantation of a human testis for anorchia. Fertility and Sterility 30: 181–187.
25. Brinster RL and Avarbock MR (1994) Germ line transmission of donor haplotype following spermatogonial transplantation. Proc Natl Acad Sci USA 91: 11303–11307.
26. Brinster RL and Zimmerman JW (1994) Spermatogenesis following male germ cell transplantation. Proc Natl Acad Sci USA 91: 11298–11302.
27. Dym M (1994) Spermatogonial stem cells of the testis. Proc Natl Acad Sci USA 91: 11287–11289.
28. Attaran SE, Hodges DV, Crary LS Jr, Van Galder GC, Lawson RK and Ellis LR (1966) Homotransplants of the testis. J Urol 95: 387.
29. Lee S, Tung KSK and Orloff MJ (1971) Testicular transplantation in the rat. Transplant Proc 3: 586–590.
30. Altwein JE, Lee S and Gittes RF (1972) Testicular transplantation versus implantation: differentiation by gonadotrophin radioimmunoassay. Invest Urol 10: 91–95.
31. Gittes RF, Altwein JE, Yen SSC and Lee S (1972) Testicular transplantation in the rat: long-term gonadotrophin and testosterone radioimmunoassay. Surgery 72: 187–192.
32. Burger HG and Baker HWG (1987) The treatment of infertility. In: Greger WP, Coggins CH and Hancock EW (eds) Annual Review of Medicine, pp 29–46.
33. Dym M (1983) The Male Reproductive System. In: Weiss L (ed) Histology: Cell and Tissue Biology, pp 1000–1053. New York, NY: Elsevier Biomedical.
34. Dym M and Clermont Y (1970) Role of spermatogonia in the repair of the seminiferous epithelium following X-irradiation of the rat testis. Amer J Anat 128: 265–282.
35. Abe SI (1987) Differentiation of spermatogenic cells from vertebrates in vitro. Int Rev Cytol 109: 159–209.
36. Lam D, Furrer R and Bruce WR (1970) The separation, physical characterization, and differentiation kinetics of spermatogonial cells of the mouse. Proc Natl Acad Sci USA 65: 192–199.
37. Meistrich ML (1972) Separation of mouse spermatogenic cells by velocity sedimentation. J Cell Physiol 80: 299–312.
38. Romrell LJ, Bellve AR and Fawcett DW (1976) Separation of mouse spermatogenic cells by sedimentation velocity. A morphological characterization. Dev Biol 49: 119–131.
39. Bellve AR, Cavicchia JC, Millette CF, O'Brien DA, Bhatnagar YM and Dym M (1977) Spermatogenic cells of the prepuberal

mouse: isolation and morphological characterization. J Cell Biol 74: 68–85.

40. Dym M, Jia MC, Dirami G, Price JM, Rabin SJ, Mocchetti I and Ravindranath N (1995) Expression of *c-kit* receptor and its phosphorylation in immature rat type A spermatogonia. Biol Reprod 52: 8–19.

41. Dym M (1976) The mammalian rete testis - a morphological examination. Anat Rec 186: 493–524.

Section VII:

Hepatocytes

R. P. Lanza and W. L. Chick (eds.), Yearbook of Cell and Tissue Transplantation 1996/1997, 127–133.

Chapter 12

Hepatocyte transplantation

Michio Mito and Masayuki Sawa
The Second Department of Surgery, Asahikawa Medical College, 4-5, Nishikagura, Asahikawa, 078, Japan

Irreversible dysfunction of the liver leads to high mortality of patients, since the liver has complex functions, such as metabolism and synthesis of substances that are essential for the maintenance of life, and detoxification of toxic substances. Various kinds of artificial liver support systems have been developed to treat patients with hepatic failure, but the survival rate of patients has not improved as well as was expected [1].

Orthotopic liver transplantation has already become an established therapeutic method to treat patients with irreversible liver disease. However, the recent marked increase in the frequency of organ transplantation has caused a serious donor organ scarcity. Moreover, complete replacement of the whole liver is not always necessary for patients with acute hepatic failure or congenital hepatic enzyme deficiency.

The collagenase digestive method has enabled us to isolate hepatocytes from the liver with a high cell yield and viability, and thus hepatocyte transplantation (HTx) has been studied to develop an alternative therapeutic method for patients with liver disease. In the mid-1970s, Sutherland *et al.* first reported successful HTx for chemically induced acute hepatic failure (AHF) in rats and dogs [2]. Thereafter, many efforts were made to confirm the beneficial effect of HTx on the survival rate of animals with various kinds of experimentally induced hepatic failure or on congenitally enzyme deficient animals (Table 1). Figure 1 shows a photomicrograph of "the hepatized spleen" at 27 months after HTx. A large part of the spleen was occupied by grafted hepatocytes, but it took almost the life span of recipient rats for this proliferation to take place. The most important and difficult issues of HTx are how to induce rapid proliferation and long-term survival of transplanted hepatocytes, and how to prevent immunologic reactions. In this section, we will review the recent advance in the field of HTx and discuss the future aspects of HTx.

HTx for AHF

It is already well established that HTx is an effective treatment for animals with experimentally induced AHF; however the mechanism of its beneficial effect remains still unknown [2–6]. According to the report by Sutherland *et al.*, the survival rate of rats with dimethylnitrosamine-induced AHF markedly improved, from 6% to 63%, at 3 weeks after HTx, despite the fact that intraperitoneal transplanted hepatocytes could not be identified by histologic examination. Intraperitoneal administration of regenerating liver cytosol could improve the survival rate of rats with chemotherapeutically induced AHF [4]. Intrasplenic injection of hepatocyte culture supernatants also improved the survival rate of D-galactosamine (D-gal)-induced AHF rats from 5.4 to 42.9% and the authors suggested that this enhanced survival may be due to stimulation of the reticuloendothelial system (RES) or induction of unspecific metabolic or humoral changes [5].

The RES plays an important role in the immune and inflammatory reponses during liver injury. We investigated the correlation between RES activity and the improved survival of D-gal-induced AHF rats undergoing HTx [7]. Encapsulated rat hepatocytes within calcium alginate gel (4×10^6) were transplanted into the peritoneal cavity of syngeneic Wistar rats at 24 h after the induction of AHF with D-gal and the animals were observed for 2 weeks postoperatively. Intraperitoneal

Table 1. Historic review of hepatocyte transplantation

Report	Content
Ebate & Mito (1976)	Syngeneic i.s. HTx in rats
Matas (1976)	Congeneic i.p. HTx in Gunn rats
Groth (1977)	Allogeneic i.h. HTx in Gunn rats
Sutherland (1977)	Syngeneic i.p. HTx in rats with DMNA induced AHF
Makowka (1980)	Allo and xeno i.p. HTx in rats with anoxic induced AHF
Sommer (1980)	HTx in dogs with ischemic AHF
Miyazaki (1983)	Regenerating liver cytosol (i.p.) for rats with chemically induced AHF
Kasai & Mito (1987)	Autologous i.s. HTx in dogs and monkeys with portocaval shunt
Ricordi (1988)	Cotransplantation of clusters of hepatocytes and pancreatic islets
Dixit (1990)	Allogeneic i.p. microencapsulated HTx in Gunn rats
Demetriou (1991)	I.p. microcarrier-attached HTx in NARs
Chowdry (1991)	*Ex vivo* gene therapy in LDLR-deficient rabbits
Sandbichler (1992)	Intrapulmonary HTx in rats with chemically induced AHF

HTx: hepatocyte transplantation, DMNA: dimethylnitrosamine, AHF: acute hepatic failure, i.s.: intrasplenic, i.p.: intraperitoneal, i.h.: intrahepatic, allo: allogeneic, xeno: xenogeneic, NAR: Nagase analbuminemic rat, LDLR: low.

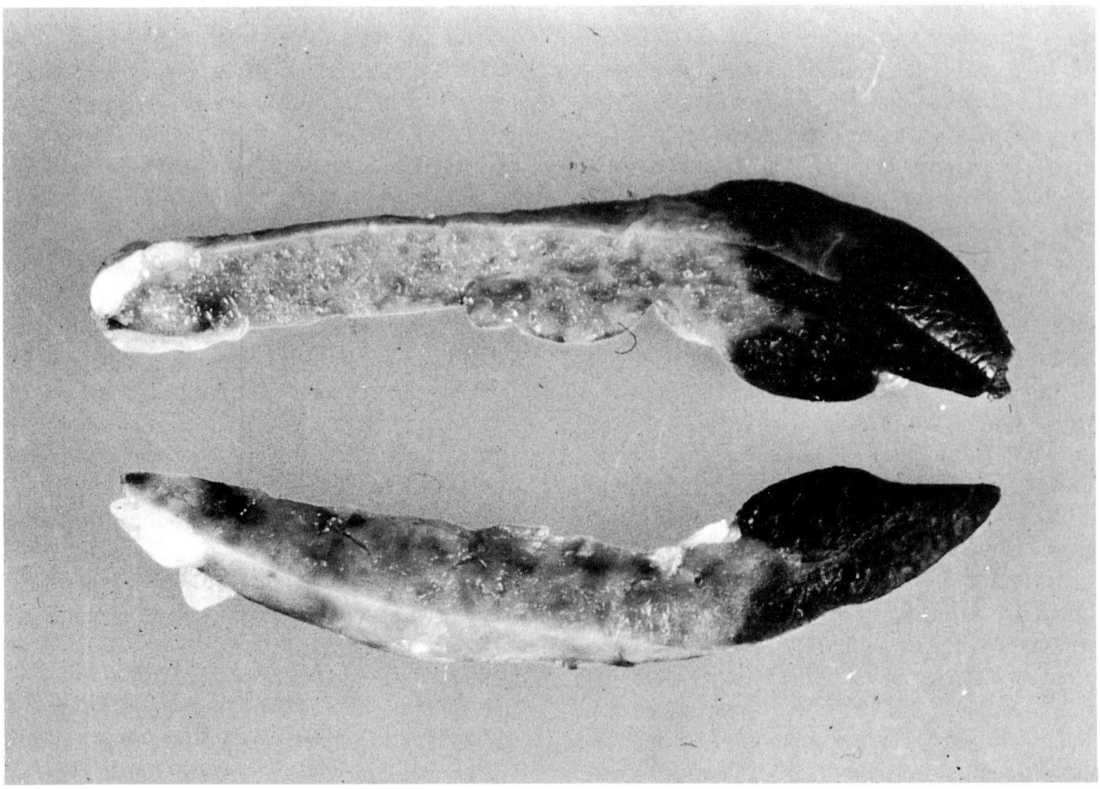

Fig. 1. A photomicrograph of "the hepatized spleen" at 27 months after intrasplenic hepatocyte transplantation.

transplantation of encapsulated hepatocytes activated the RES of rats with D-gal-induced AHF, which was evaluated by studying carbon ink clearance, and improved their survival rate from 5 to 75%. The increase of serum transaminase levels was reduced, and the AKBR of surviving rats recovered to above 1.0 at 12 h after HTx. Indomethacin, a well known prostaglandin inhibitor, treatment revealed significant low RES activity of rats with D-gal-induced AHF even after HTx, and their survival rate did not improve compared with nontransplanted rats (Figure 2). It is therefore suggested that the improvement in survival and liver function of rats with D-gal-induced AHF may be related to a cytoprotective effect of prostaglandin produced in response to transplantation of encapsulated hepatocytes.

Fetal liver has potentially high proliferative ability and lower immunogenicity compared with the adult liver, and it is transplanted as fragments, with the anticipation that the hepatocytes themselves can proliferate rapidly with the aid of nonparenchymal cells. Hagiwara *et al.* investigated the effect of isogeneic (iso) and xenogeneic (xeno) donor fetal liver fragments (FLF: approximately 1 mm^3) transplantation into the omentum of Wistar rats with D-gal-induced AHF [8]. Iso or xeno FLF-transplanted rats with AHF by a lethal dose of D-gal showed higher survival rates (70, 80% vs. 9.1%) than nontransplanted rats. Iso or xeno FLF caused marked improvement in the serum GOP, GPT, and total bilirubin levels of recipients with AHF by a sublethal dose of D-gal. Histological examination of the livers revealed severe damage in controls, while only slight damage was found in iso or xeno FLF-transplanted rats. The authors concluded that FLF transplantation into the omentum can be an alternative therapy for incurable liver insufficiency.

Based on the beneficial effect of HTx on experimental AHF, a clinical trial of fetal hepatocyte transplantation was attempted in patients with fulminant hepatic failure [9]. Candidates for HTx were patients with FHF of less than 2 weeks duration and having portosystemic encephalopathy (PSE) of grade III and IV. Those who did not give consent served as controls. Pooled single-celled blood group-matched human fetal hepatocyte suspension from of either sex and of gestational ages between 26 and 34 weeks with 95% viability were used. Fetal hepatocytes were administered intraperitoneally through a peritoneal dialysis catheter in a single dose of 60 million cells/kg of body weight in a concentration of 3 million cells/ml of Hank's medium. All of 3 patients with grade III coma survived after HTx; on the other hand, the survival rate of the nontransplanted patients was 50%. One patient with grade IVa coma survived with HTx, while those with grade IVb coma died despite HTx. A liver biopsy in surviving patients showed varying amounts of parenchymal necrosis, and follow-up of surviving patients for a period of one year did not show any clinical or biochemical morbidity. Fetal hepatocyte transplantation showed beneficial effects on the recovery from PSE in patients with FHF. It was suggested that the maximum advantage may be achieved if HTx is performed in PSE grade III.

Fetal liver has vigorous proliferative potential and is less immunogenetic; however, in many nations, donor fetal hepatocytes are not available due to ethical concerns. Rapid proliferation and long-term survival of transplanted hepatocytes is not required for the treatment of AHF by Htx, because intraperitoneal injection of each cytosol, culture supernatant, or bone marrow cells was also able to ameriolate experimentally induced AHF in animals [4,5]. Adult hepatocytes are able to relieve animals with experimentally induced AHF [2–4,6], and adult human hepatocytes were also successfully preserved for up to 2 months [10]. Thus, clinical application of HTx using adult human or xeno hepatocytes should also be attempted for the treatment of AHF patients.

HTx for CHF

Chronic hepatic failure (CHF) requires long-term and sufficient functional support by transplanted hepatocytes, and thus methods to transplant larger number of hepatocytes or to induce the rapid proliferation and long-term survival of transplanted hepatocytes have to be developed before clinical application of HTx for patients with CHF. Extracellular matrices, such as collagen, microcarriers, or Ca alginate hydrogel, have been studied as to their availability for long-term survival and immunoisolation of transplanted hepatocytes, because hepatocytes reveal their cellular functions well when they make contact with some extracellular matrices [11].

Uyama *et al.* investigated the usefulness of prevascularized polyvinyl alcohol (PVA) sponge (28.5 cm^2) for biometrics for HTx [12]. PVA sponges were preimplanted into subcutaneous tissue and between mesenteric leaves 1 week after portocaval shunt formation. After 5 days of prevascularization of the sponges, recipient rats underwent 70% partial hepatectomy and

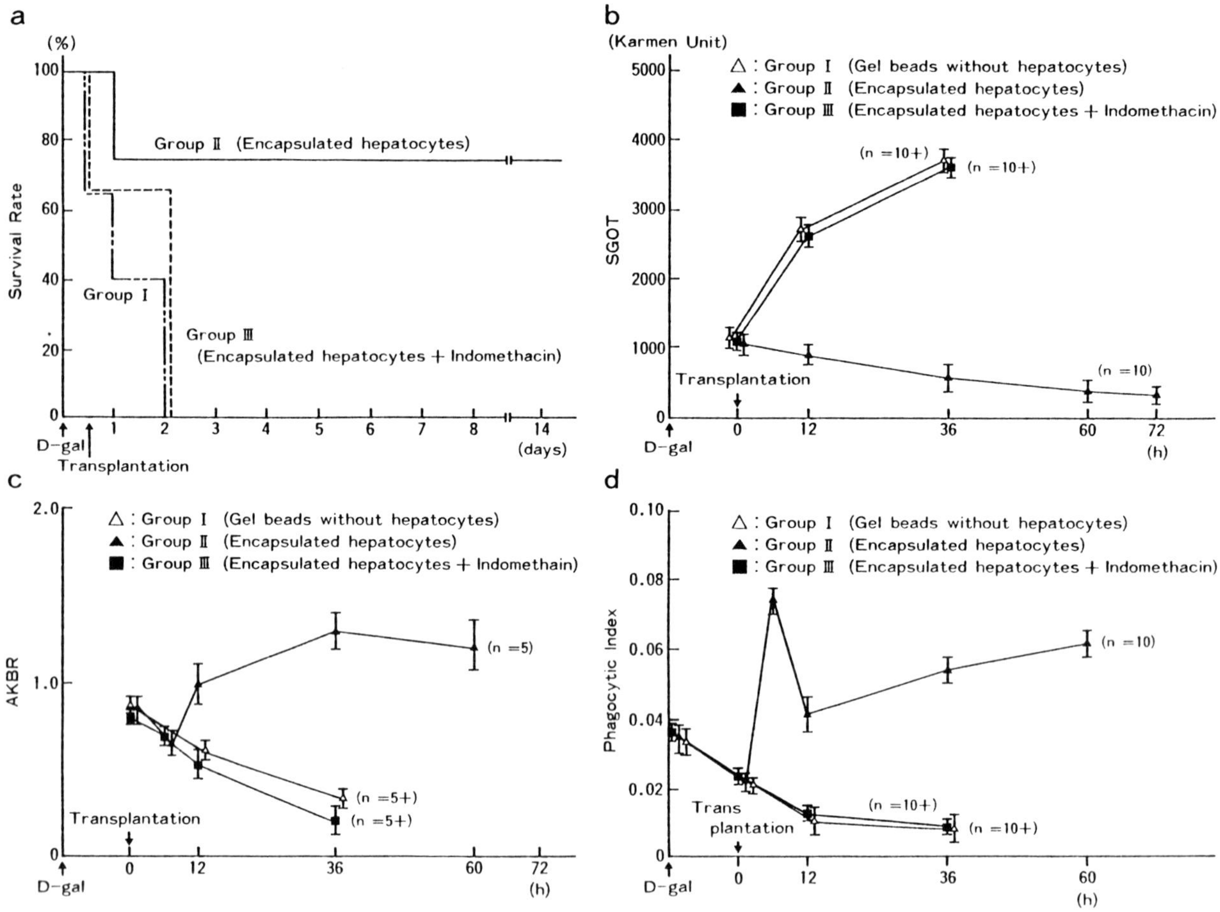

Fig. 2. (a) Survival of rats with D-gal/induced AHF ($p < 0.01$, group II vs. groups I and III). (b) Serum GOT(SGOT) levels of rats with D-gal/induced AHF ($p < 0.01$, group II vs. groups I and III). (c) Arterial ketone body ratio (AKBR) of rats with D-gal/induced AHF ($p < 0.05$, group II vs. groups I and III). (d) Phagocytic index of rats with D-gal/induced AHF ($p < 0.01$, group II vs. groups I and III).

received 5×10^8 hepatocytes (equivalent to an entire rat liver) in the prevascularized PVA sponges. Histologic examination revealed many hepatocytes layered in the vascularized tissue that had developed in the pores of the sponge, and tubular structures were also observed, suggesting early biliary tubular formation. The mesenteric sponge showed better engraftment of hepatocytes compared with the subcutaneous sponge. The calculated hepatocyte area on day 7 was larger than that on day 3, suggesting proliferation. In the Gunn rat experiment, serum bilirubin in rats that received hepatocytes from normal Wistar rats decreased to 74.7% of the pretransplantation level. A hepatocyte mass equivalent to a whole liver can be delivered into prevascularized polymer sponge devices. At day 7, between 10.8% and 18.0% of the immobilized hepatocytes were estimated to be grafted and functioning. Polytetrafluoroethylene (PTFE) was also investigated as to its availability as an alternative location for HTx [13]. They were implanted intraperitoneally several weeks prior to HTx, allowing time for the initial foreign body response to subside and for vascularization to occur. Hepatocytes isolated from the liver of inbred male Wistar rats were transplanted either intrasplenically or into PTFE supports already implanted intraperitoneally 4 weeks prior to HTx. Partial hepatectomy (70%) was performed at the time of HTx, or at 2, 4, or 6 weeks after HTx, and resulted in no increase in the labeling index of hepatocytes when recipients underwent partial hepatectomy at the time of HTx. On the other hand, when partial hepatectomy was done at 2, 4, or 6 weeks after HTx, the labeling index of hepatocytes in the speen and PTFE supports increased markedly.

These studies demonstrated a novel method to increase the quantity of hepatocytes for HTx by using prevascularized PVA sponges or PTFE. The first study showed that portocaval shunt and partial hepatectomy stimulated the proliferation of transplanted hepatocytes in the prevascularized sponges for 7 days, but at day 14, the calculated area decreased markedly. We also found a similar beneficial effect of hepatic stimulatory substance (HSS) on the proliferation of transplanted hepatocytes in the rat spleen [14]. HSS was given intravenously (2/week) to recipient rats after intrasplenic HTx. Image analysis of the hepatocyte-occupied area in the longitudinal cut surface of the spleen showed proliferation of transplanted hepatocytes at first 2 weeks after HTx; However no remarkable proliferation was found in the subsequent 2 weeks. The latter study demonstrated that timing of exposure of transplanted hepatocytes to a strong mitogenic stimulus is likely to be important for obtaining a maximal proliferative response by the engrafted cells. Further investigation is needed to develop an optimal method to stimulate the proliferation of ectopically transplanted hepatocytes continuously.

Allogeneic hepatocytes are rejected if transplanted without immunosuppression. Microencapsulation technique has also been applied for immunoisolation of transplanted hepatocytes as well as long-term maintenance of their function, and calcium alginate is a widely used material for microencapsulation of isolated hepatocyte [4,15,16]. Encapsulation of hepatocytes in a new semipermeable membrane, polyelectrolyte anionic hydrogel hollow fibers, was attempted for intraperitoneal allogeneic HTx to protect transplanted hepatocytes from rejection [17]. The viability of hepatocytes in hollow fibers assessed at days 15, 30, 60, and 90 remained greater than 80%. Morphological examination of hepatocytes in hollow fibers demonstrated well-maintained hepatocytes and their normal fine structures. Allogeneic hepatocytes in explanted hollow fibers cultured for 3 days secreted substantial levels of albumin, and there was no difference in albumin synthesis between allogeneic and syngeneic hepatocytes. For clinical HTx, xenogeneic as well as allogeneic hepatocytes have to be transplanted, and the results of this study showed the availibility of a new semipermeable membrane to protect transplanted allogeneic hepatocytes from rejection for up to 3 months. Encapsulation of hepatocytes in a hydrogel, or immobilization of hepatocytes to microcarriers, has been attempted to maintain cell viability and function for a long period of time, but the effect was still temporary [4,15,16,18].

Intrasplenic HTx is usually done by direct injection of hepatocyte suspension into the splenic parenchyma with clamping the splenic pedicles during inoculation of cells, to minimize losses via venous drainage from the spleen; however, lethal portal embolization sometimes occurs in larger animals, even with clamping of the splenic pedicles. Long-term morphological observation of the spleen revealed a small number of surviving hepatocytes only around the splenic vessels [19]. Briand *et al.* thus attempted intrasplenic HTx by retrograde injection of hepatocyte suspension into the splenic vein, to reduce cell leakage from the spleen, and assessed the possible beneficial effect of this support in hepatectomized dogs [20]. Autologous hepatocytes (2×10^9 cells) isolated from the resected liver after 65% partial hepatectomy were transplanted through a catheter, to allow cell reflux into the inferior terminal polar splenic vein of hepatectomized dogs. All dogs that received HTx by direct puncture of the splenic parenchyma died of portal embolization; on the other hand, 50% of dogs that received HTx by retrograde injection survived, and histological examination showed that the frequency of portal embolization reduced to 25%.

The liver may be an excellent site for grafting hepatocytes, because it provides the most suitable architecture for transplanted hepatocytes to survive and function. It has been reported that the presence of normal hepatocyte grafts in the spleens of enzyme-deficient animals resulted in biliary excretion of bilirubin glucuronides by the host's liver [21]. However, hepatocytes remaining at ectopic sites do not show biliary excretion. Hamaguchi *et al.* investigated whether hepatic transport can be established by transplanted hepatocytes in conjunction with the host's biliary system [22]. Hepatocytes (1×10^7) obtained from inbred Sprague-Dawlery rats (SDRs) were injected into the portal vein of Eizai hyperbilirubinemic rats (EHBRs) immediately after 68% partial hepatectomy. Blood samples were taken up to 120 days after transplantation, to determine the serum total bilirubin level. Biliary excretion in indocyanine green (ICG) and sulfobromophthalein (BSP) was examined 30 days after transplantation. Serum bilirubin concentrations decreased significantly after hepatectomy and intraportal hepatocyte injection. Excretion of biliary ICG and BSP significantly increased in EHBRs that underwent 68% partial hepatectomy and intraportal injection of SDR hepatocytes. Intraportal injection of SDR hepatocytes and partial hepatectomy showed approximately

20% of biliary ICG and BSP excretion in EHBRs compared with those in SDR. The authors concluded that hepatic transport of bile acid conjugates in EHBRs can be restored by HTx in conjunction with the recipient's excretory biliary system. EHBRs have an abnormality similar to the defect in Dubin-Johnson syndrome in humans. This study demonstrated that intraportal HTx is a possible way to restore hepatic transport of bile acid conjugates in EHBRs. However, without hepatectomy HTx showed only a slight increase in biliary ICG and BSP excretion, and partial hepatectomy is impractical in patients with liver insufficiencies.

Intraportal injection of hepatocyte suspension sometimes causes portal embolization and hepatic failure, and it is almost impossible for patients with CHF who already have severe portal hypertension to receive large quantity of hepatocytes through the portal vein. Therefore, a new method has to be established to increase the number of surviving hepatocytes in the liver and to avoid lethal portal embolization. Repeated intraportal HTx was developed, to avoid cell embolization in the portal vein and to increase the number of transplanted cells [23]. Hepatocytes (5×10^6) were infused through a vascular access port into the gastroduodenal vein of analbuminemic rats on days 1, 3, 5, 8, 10, and 12. The animals that received a single large dose of hepatocytes (2×10^7) showed liver tissue damage, whereas those subjected to repeated cell infusions revealed normal liver histology and a significantly enhanced increase in the plasma albumin level. This novel method of HTx enables increasing the amount of surviving hepatocytes in the liver and avoiding liver necrosis due to cell embolization.

Future aspect of HTx

The recent marked advance in gene engineering techniques has enabled us to *ex vivo* manipulation of cells isolated from genetically ill organs. In gene therapy protocols, in which therapeutic genes are delivered to the liver via genetically modified hepatocytes, expression of the introduced gene declines over time. Rhim *et al.* investigated the replicative capacity of adult liver cells and their medical usefulness as donor cells for transplantation, by transfer of adult mouse liver cells into transgenic mice that display an endogenous defect in hepatic growth and function [24]. The histological examination of the removed liver showed many more cells than the total number of transplanted cells, and these results suggested that extensive donor cell expansion occurred within recipient liver. The authors demonstrated the marked replicative potential of genetically modified hepatocytes when transplanted into the liver of diseased recipients, and the possibility of genetically engineered hepatocyte transplantation for correcting a metabolic defect, as long as a suitable microenvironment for growth of these cells can be provided.

Clinical trials of HTx have already been performed in limited cases of AHF and in cases with an inborn metabolic error; however, an effective method to induce rapid proliferation and long-term survival of transplanted hepatocytes has not yet been established. The establishment of exogenous as well as intrinsic treatment and gene engineering techniques that enable cells to proliferate rapidly is expected in the near future.

References

1. Mito M. Hepatic assist: Present and future. Artif Organs 1986; 10: 214–8.
2. Sutherland DER, Numata M, Najarian JS *et al.* Hepatocellular transplantation in acute liver failure. Surgery 1977; 82: 124–32.
3. Makowka L, Falk RE, Rotstein LE *et al.* Reversal of experimental acute hepatic failure in the rat. J. Surg Res 1980; 29: 479–87.
4. Miyazaki M, Makowka L, Venturi D *et al.* Reversal of lethal, chemotherapeutically induced acute hepatic necrosis in rats by regenerating liver cytosol. 1983; 94: 142–50.
5. Baumgartner D, O'Neil PML, Sutherland DER *et al.* Effects of intrasplenic injection of hepatocytes, hepatocyte fragments and hepatocyte culture supernatants on D-galactosamine-induced liver failure in rats. Eur Surg Res 1983; 15: 129–35.
6. Sandbichler P, Then P, Vogel W *et al.* Hepatocellular transplantation into the lung for temporary support of acute liver failure in the rat. Gastroenterol 1992; 102: 605-9.
7. Hirai S, Kasai S, Mito M. Encapsulated Hepatocyte Transplantation for the Treatment of D-Galactosamine-Induced Acute Hepatic Failure. Eur Surg Res 1993; 25: 193–202.
8. Hagiwara M, Shimura T, Tsuji K *et al.* Effects of iso and xeno fetal liver fragments transplantation on acute and chronic liver failure in rats. Cell Transplant 1894; 3: 283–90.
9. Habibullah CM, Syed IH, Qamar A *et al.* Human fetal hepatocyte transplantation in patients with fulminant hepatic failure. Transplantation 1994; 58: 951–2.
10. Ryan CM, Carter EA, Tompkins RG *et al.* Isolation and long-term culture of human hepatocytes. Surgery 1993; 113: 48–54.
11. Rojkind M, Gatmatian Z, Reid LM *et al.* Connective tissue biomatrix: its isolation and utilization for long-term cultures of normal rat hepatocytes. J. Cell Biol 1980; 87: 255.
12. Uyama S, Kaufman PM, Vacanti JP. Delivery of whole liver-equivalent hepatocyte mass using polymer devices and hepatotrophic stimulation. Transplantation 1993; 55: 932–5.
13. Borel Rinkes IM, Bijma A, Terpstra OT *et al.* Proliferative response of hepatocytes transplanted into spleen or solid sup-

port. J Surg Res 1994; 56: 417–231.

14. Jiang B, Kasai S, Mito M *et al.* Beneficial effect of hepatic stimulatory substances on the survival of intrasplenically transplanted hepatocytes Cell Transplant 1993; 2: 325–9.
15. Dixit R, Darvasi R, Gitnick G *et al.* Restoration of liver function in Gunn rats without immunosuppression using transplanted microencapuslated hepatocytes. Hepatology 1990; 12: 1342–9.
16. Dixit V, Darvasi R, Gitnick G *et al.* Cryopreserved microencapsulated hepatocytes – Transplantation studies in Gunn rats. *Transplantation* 1993; 55: 616–622.
17. Balladur P, Creme E, Nordlinger B *et al.* Transplantation of allogeneic hepatocytes without immunosuppression: Long-term survival. Surgery 1995; 117: 189–94.
18. Demetriou AA, Whiting J, Feldman D *et al.* Replacement of liver function in rat by transplantation of microcarrier-attached hepatocytes. Science 1986; 233: 1190–2.
19. Kasai S, Sawa M, Mito M *et al.* Intrasplenic hepatocyte transplantation in mammals. Transplant Proc 1981; 19: 992–4.
20. Briand D, Centeno NA, Astre C *et al.* Comparison of two methods of autologous intrasplenic hepatocellular transplantation in partially hepatectomized dogs. Eur Surg Res 1993; 25: 104–6.
21. Vroemen JP, Buurman WA, Kootsra G *et al.* Hepatocyte transplantation for enzyme deficiency disease in congenic rats. Transplantation 1986; 42: 130–5.
22. Hamaguchi H, Yamaguchi Y, Ogawa M *et al.* Hepatic biliary transport after hepatocyte transplantation in Eizai hyperbilirubinemic rats. Hepatology 1994; 20: 220–24.
23. Rozga J, Holzman M, Demetriou AA *et al.* Repeated intraportal hepatocyte transplantation in analbuminemic rats. Cell Transplant 1995; 4: 237–43.
24. Rhim JA, Sandgren EP, Degen JL *et al.* Replacement of diseased mouse liver by hepatic cell transplantation. Science 1994; 263: 1149–52.

Section VIII:

Islets of Langerhans

R. P. Lanza and W. L. Chick (eds.), Yearbook of Cell and Tissue Transplantation 1996/1997, 137–144.

Chapter 13

Clinical islet transplantation

Bernhard J. Hering[1], David C. Wahoff [2] and David E.R. Sutherland[2]

[1] *Department of Medicine, Justus Liebig University, Giessen, D-6300 Germany;* [2] *Department of Surgery, University of Minnesota, Minneapolis, MN 55455, U.S.A.*

The ultimate goal of beta cell transplantation is to improve the quality of life and to prevent the severe acute and the devastating chronic complications of diabetes, by utilizing isolated allogeneic human or xenogeneic porcine islets early in the course of type 1 diabetes. An additional aim is to minimize the use of long-term immunosuppression that is currently needed to prevent islet rejection and autoimmune recurrence of diabetes [1–3]. Several increasingly complex steps will have to be taken to reach this final objective and it is evident that the immunological obstacles, particularly those involved in pig-to-man islet xenotransplantation [4], are formidable and can only be overcome by continued research efforts directed towards the problems peculiar to isolated islets.

A number of remarkable achievements have been made in recent years. It is generally agreed that achieving and maintaining insulin independence for periods longer than one year in islet autograft [5] and in immunosuppressed islet allograft recipients [6–11] is considered a successful islet transplant. The definition of success should additionally include other indices of metabolic control. Islet transplants have resulted in C-peptide secretion and normalization of HbA1c levels with no episodes of hypoglycemia for periods of over 4 years [12]. This level of metabolic improvement, without the risk of hypoglycemia, is only achievable in less than five percent of patients with type 1 diabetes who are administered intensive insulin treatment with multiple daily insulin injections and frequent measurements of capillary glucose in combination with diet and exercise, as reported in the DCCT [13]. An islet transplant that does not allow complete withdrawal of exogenous insulin, but results in continued C-peptide secretion, may be beneficial. In a prospective study, a 3-month course of subcutaneous C-peptide administered to patients with type 1 diabetes ameliorated diabetic nephropathy and autonomic neuropathy [14]. These findings may become increasingly important when defining the indications for intensive insulin therapy, islet transplantation, and pancreas transplantation, respectively, in type 1 diabetic patients who have become incapacitated due to extremely labile diabetes (e.g. recurrent episodes of life-threatening severe hypoglycemia) [15]. Successful pancreas transplants have been shown to improve the quality of life in the previously diabetic recipients [16] and it is anticipated that a successful islet transplant would do the same. In addition, islet transplantation is associated with a lower incidence (nearly zero) of surgical morbidity. The challenge is to increase the success rate of islet transplants to that of pancreas transplants (currently 75% insulin independence rate at one year) [17].

In this chapter, we will first summarize the current status of clinical islet transplantation in different recipient categories before addressing future directives. A more detailed report on the effect of different variables on islet graft function by the Islet Transplant Registry is presented in a special chapter [18]. The status of fetal islet transplantation will not be detailed herein, because this issue is comprehensively reviewed in a separate chapter of this book [19].

Current status

Islet autografts

For patients with intractable pain from small duct chronic pancreatitis or failed drainage procedure, total or near-total pancreatectomy may be the only treatment option. Since 1977, when the first islet autotransplant was performed at the University of Minnesota, this approach has been applied in a number of centers in an attempt to prevent diabetes after total pancreatectomy. In the meantime, results on 149 islet autografts performed at 21 different institutions between 1977 and 1994 have been published and/or have been reported to the Islet Transplant Registry [20]. In the vast majority of cases, unpurified islet tissue was implanted into the liver via the portal vein. The University of Minnesota group has the largest experience, with a total of 48 islet autografts [5]. Over 80% of patients had partial or total pain relief after pancreatectomy. Of the 39 patients who underwent near-total (> 95% before 1986), or total pancreatectomy (since 1986), 20 (51%) were initially insulin-independent; between 2 and 10 years posttransplant the rate was 34%, with no grafts failing after 2 years. The main predictor of insulin independence was the number of islets transplanted. Of 14 patients who received > 300,000 islets, 74% were insulin-independent at > 2 years posttransplant. One pancreatectomized islet autograft recipient is still insulin-independent at > 10 years posttransplant.

Islet allografts in pancreatectomized recipients

Between 1990 and 1994, a total of 14 islet allografts have been performed in pancreatectomized recipients [20]. Islets have been implanted into a simultaneously transplanted liver, following upper abdominal exenteration in patients suffering from primary or secondary hepato-biliary malignancies. In about 50% of the cases, islets were derived from the liver donor only, and in the remaining cases islets isolated from two donor pancreases have been used. Nine of 14 recipients became insulin-independent, 7 at the University of Pittsburgh [6, 21], 1 at the University of Milan [22], and 1 at the Universities of Giessen and Wuerzburg [23]. Nearly all of these patients eventually expired with recurrent metastatic disease while still being insulin-independent. The longest duration of insulin independence was noted in the first Pittsburgh patient who died from recurrent malignancy while being insulin-independent almost five years following the combined liver-islet transplant [6]. At autopsy, insulin containing islets were readily detected in specimens of the liver [12].

Islet allografts in type 1 diabetic recipients

Between 1974 and 1994, a total of 215 islet allografts have been performed in type 1 diabetic recipients worldwide [20]. In 96 of these 215 recipients, pretransplant basal and/or stimulated C-peptide was either positive (i.e. > 0.2 ng/ml, e.g. as a result of residual C-peptide secretion or of a previous islet transplant) or unknown. Of the 119 pretransplant C-peptide negative type 1 diabetic islet allograft recipients, 14 became insulin independent after islet transplantation [20]. In these successful cases, 6,000 to 27,000 islet equivalents (150 μm islets), isolated from one to eight donor pancreata with a mean preservation time of < 8 hrs, have been grafted intraportally, and polyclonal anti-T cell antibodies have been used for induction immunosuppression. Variables such as recipient age, sex, duration of diabetes, number of donor pancreata, and islet purity did not influence one-year islet graft survival rates, as detailed in the Islet Transplant Registry report (see chapter 16). The time interval between the islet transplant and the complete withdrawal of exogenous insulin has varied from 10 to 400 days, and in one recipient, after a transient period of insulin independence from day 155 to 166, insulin independence has resumed at day 837 posttransplant [20]. The longest periods of insulin independence have been achieved with > 2 years (at the time of writing) at the University of Minnesota [10], with 2.3 years at the University of Alberta [7], and with 3.2 years at the University of Milan [8]. In an analysis by era, restricted to pretransplant C-peptide negative type 1 diabetic islet allograft recipients, the percentages of patients who showed basal C-peptide levels > 1 ng/ml at $\geq$ 1 month posttransplant and who became insulin independent for > 1 week in the 1974–84 era (n = 1 case) were 0% and 0%, in the 1985–89 era (n = 24 cases) were 21% and 8%, in the 1990–91 era (n = 41 cases) were 61% and 17%, in the 1992–93 era (n = 36 cases) were 58% and 11%, and in 1994 (n = 17 cases) were 35% and 6%, respectively. The data shows, that the trend toward improved islet graft function which occurred in the 1990–91 period did not continue afterwards.

A synopsis of the results achieved in different recipient categories

It is conceivable that a comparison between islet autotransplantation with its frequent successes and islet allotransplantation with its frequent failures can aid in finding the right track for imperative improvements in islet allotransplantation. From the additional consideration of the two main types of diabetes, i.e. pancreatectomy-induced diabetes versus type 1 diabetes, and the two solid organ grafts most commonly associated with an islet allograft, i.e. liver and kidney grafts (Table 1), even more valid directives may emerge.

The remarkable results of islet autotransplantation [5] are pertinent to the present situation in clinical islet allotransplantation, where long-lasting insulin independence remains only anectodal [6, 7, 9, 10, 11, 20, 21]. In addition, because techniques are lacking for adequate functional and immunological graft monitoring after islet allotransplantation, it is not possible to distinguish between technically or immunologically mediated graft failure. The outcome in recipients of > 300,000 unpurified autologous intraportally grafted islets [5] clearly indicates that: 1) technical factors can be overcome, 2) the liver as an implantation site does not preclude maintenance of islet graft function as the experience with intraportal islet autografts in pancreatectomized dogs suggested [24], 3) purification of islets does not constitute a prerequisite for successful islet engraftment, and 4) rejection of the islets and possibly also recurrence of autoimmunity are the barriers preventing consistent success in treating type 1 diabetics with islet allografts.

Although the data convincingly demonstrates that technical problems can be overcome the data does not exclude the possibility that technical (and metabolic) factors have caused failure of intraportal islet allografts. Besides the fact, that autografted islets have not had to face alloimmunity and autoimmunity, in cases of islet autotransplantation: 1) pancreata are not subjected to in situ flush, 2) pancreata are subjected to minimal cold ischemia only, 3) autografted islets are not exposed to density gradient media (at least in the majority of cases), 4) autograft recipients are not insulin resistant at the time of islet transplantation, 5) autografted islets are not implanted into a “diabetic” liver, 6) autograft recipients are suffering from various degrees of exocrine insufficiency, 7) autografted islets are not located downstream of the pancreas, 8) autograft recipients receive total parenteral nutrition during the first 7–10 days posttransplant, 9) autograft recipients receive i.v. insulin during the first 7–10 days posttransplant, and 10) autografted islets are not exposed to islet-toxic effects of immunosuppressive drugs. Although the potential relevance of these non-immunological differences between islet autografts and islet allografts has not been thoroughly elucidated, it is generally assumed that the barriers preventing long-lasting insulin independence in a great number of type 1 diabetic recipients are those of rejection and autoimmune recurrence.

In this context it should be emphasized that insulin independence following islet allotransplantation seems much more likely in pancreatectomized recipients compared with type 1 diabetic recipients (Table 1), although direct evidence that autoimmunity caused the difference in outcome is lacking. These clinical observations support the notion that more value must be attached to disease recurrence as a potential cause of islet allograft failure in immunosuppressed type 1 diabetic recipients.

Recent experimental work in the BB rat model by Bartlett *et al.* [25] investigated the loss of islet grafts due to autoimmunity. In initial studies islets isolated from diabetes-resistant BB rats were transplanted into diabetic-prone recipients. They found that these islet grafts were susceptible to autoimmunity even with cyclosporine therapy, whereas, whole pancreas transplants from diabetes-resistant rats were not susceptible. Further work in this model [26] showed that whole pancreas grafts abrogated autoimmunity by passive transfer to the host of an autoregulatory T cell subset (RT6+ T cells). Isolated islet cell grafts did not have the T cell subset. Using a BB-DR strain, which is immunodeficient for RT6+ T cells, as recipients, and the BB-Ac strain , which has RT6 +T cells, as tissue donors, they found that repletion of RT6+ T cells using lymph node cells prevented autoimmunity of transplanted islets as observed with whole pancreas grafts. So, isolated islets are no more susceptile than islets within a pancreas to autoimmune attack. The apparent difference was only due to the absence of RT6+ cells in isolated islet grafts. No such identifiable deficiency has been demonstrated in humans. However, we know that recurrent disease in human pancreatic islets is prevented by immunosuppression. Thus, autoimmunity is likely not the primary obstacle to successful islet transplants in immunosuppressed patients. We need to look for other reasons. If we can ever induce tolerance, or do anything with immunosuppression to prevent rejection, then recur-

Table 1. Insulin independence after intraportal transplantation of > 300,000 islets into totally pancreatectomized or pretransplant C-peptide negative recipients according to recipient category (1990–93 cases only)

Recipient category*	Ref.	No. (percentage) of insulin at ≥ 1 month posttransplant	Independent cases at ≥ 1 year posttransplant
Autograft-ITA-PanEx	5	86%	71%
Allograft-SIL-PanEx	20	8/10 (80%)	6/10 (60%)
Allograft-SIL-IDDM	20	0/4 (0%)	0/4 (0%)
Allograft-SIK/IAK-IDDM	20	8/51 (16%)	5/51 (10%)

* ITA: Islet Transplant Alone, SIL: Simultaneous Islet Liver, SIK: Simultaneous Islet Kidney, IAK: Islet After Kidney. PanEx: Diabetes induced by total pancreatectomy. IDDM: type 1 diabetes mellitus (without remaining C-peptide production pretransplant).

rence of disease will be a problem in pancreas and islets alike.

It is just as possible, based on the data summarized in Table 1, that, in type 1 diabetic recipients as compared to pancreatectomized islet allograft recipients islet engraftment is affected by long-standing microangiopathy [27] (Table 1). In addition, in type I diabetic recipients, [2] a progressive fall of a marginal transplanted beta-cell mass is caused by an underlying insulin resistance [28], lacking exocrine insufficiency, and transient hyperglucagonemia. A retrospective analysis of limited data is always fraught with difficulties. Therefore, only a carefully designed prospective study might finally reveal the real obstacles of successful islet transplantation in type 1 diabetic recipients.

Future directions

The principal feasibility of successful islet transplantation has clearly been proven by the demonstration of long-lasting insulin independence following islet autotransplantation and islet allotransplantation in pancreatectomized and in type 1 diabetic recipients. However, the number of transplants being performed worldwide has plateaued (approximately 30 per year) and the trend toward an increase in the percentage of type 1 diabetic islet allograft recipients with continued C-peptide secretion and insulin independence has not continued since 1990. Therefore, a revision of current concepts seems imperative if substantial progress is to be attained at in the not-too-distant future. Five issues deserve particular attention:

1. More sensitive measures for monitoring cellular grafts such as isolated islets

Decision making while taking care of an islet allograft recipient is severely hampered because deteriorations of graft function can likewise reflect glucose toxicity [28], unspecific cytokine mediated islet destruction [29], adrenergic islet reinnervation [30, 31], beta cell toxicity resulting from immunosuppressive drugs [32–34], recurrence of autoimmunity [25], and last but not least rejection, to name just a few. It is not only that islet graft failure is the inevitable ultimate consequence of these limitations in a number of cases, the unavailability of informative established measures such as graft biopsies prevents lessons from being learned from individual cases. It will be critically important, in particular for solitary islet transplants, to develop techniques for the early detection of islet rejection episodes, so that antirejection treatment can be initiated before irreversible graft failure has ensued.

2. Single-donor islet transplants

The lack of availability of effective lots of collagenase has prevented the consistent isolation of sufficient islets from a single pancreas that is believed to be required to reverse type 1 diabetes. As a consequence, multiple donor islet transplants have been performed in about 60% of the cases since 1990 [20]. Although the use of multiple donors has facilitated the achievement of insulin independence in a small number of patients, the majority of multiple donor islet allograft recipients remained insulin dependent [20]. This approach is inefficient compared to pancreas transplantation and, therefore, in the long-term incomprehensible in the light of organ shortage and at present this approach seriously interferes with a precise evaluation of the impact of potentially crucial donor factors such as ischemia time, HLA, mass, and functional integrity of transplanted islets etc. on graft outcome. The success of islet autotransplants in pancreatectomized recipients [5] and allotransplants in type 1 diabetic recipients [6, 8, 10, 11, 35] demonstrates that islet transplantation from a single donor is technically feasible. The islet mass that was transplanted in these cases should be obtainable from an increasing number of adequately selected and procured donor pancreata utilizing purified enzyme blends which have recently become commercially available [36] or in the future recombinant enzyme mixtures which are currently being evaluated [37] together with already established techniques for human islet isolation [38] and purification [39]. It is becoming more and more appreciated that the relevant quantity of islets required to reverse diabetes is the amount of islet tissue effectively engrafted rather than the number of islets transplanted [3]. Experimental data suggest that strict metabolic control can lessen the number of islets required to reverse diabetes [40, 41] and that maintaining a period of near-normoglycemia after islet transplantation enhances the performance of an islet graft that would otherwise be expected to fail [42]. The Giessen group has introduced Biostator treatment and total parenteral nutrition in order to achieve tight blood glucose control both in the peripheral and in the portal circulation [43]. Their first type I diabetic islet allograft recipient became insulin independent following intraportal transplantation of only 6,140 islet equivalents/kg [11].

3. Spleen as implantation site

A number of pitfalls are linked to the intrahepatic localization of islets: 1) A significant failure rate of initially successful intraportal compared to intrasplenic islet autografts has been noted in dogs and sub-human primates [22, 44–46]. 2) Intrahepatic islets are localized downstream from the pancreas which produces glucagon, somatostatin, and noradrenaline overflow, all having an inhibitory effect on beta-cell secretion [47]. 3) Intrahepatic islets are vascularized by a mixture of portal vein and hepatic artery blood containing an unknown but presumably low amount of oxygen along with intestinal metabolites and related byproducts such as postprandial glucose (portal venous blood levels are 20–80% higher than those in peripheral veins). In addition, islets are exposed to intestinal derived peptides, gastrointestinal derived endotoxins and orally administered drugs that are metabolized primarily by the liver [47–49]. 4) Intrahepatic islets are reinnervated by the hepatic nerve fibers which in the periportal areas seem to be mainly noradrenergic and which therefore have an inhibitory action on insulin secretion [30, 31]. 5) Insulin produced by intrahepatic islets is secreted in an anomalous environment and is handled and cleaved off by the liver in an unknown amount [47]. 6) Intrahepatic islets are exposed to Kupffer cells, the largest population of fixed tissue macrophages in the body [50]. 7) Intrahepatic autologous islets do not restore a glucagon response to insulin-induced hypoglycemia [51] in contrast to an increased glucagon resonse to hypoglycemia and increased rates of recovery from hypoglycemia after intrasplenic islet autotransplantation in totally pancreatectomized dogs [52]. 8) Intrahepatic islets in rats cause hepatocytic hyperproliferation (25-fold increased mitotic index) which is compensated for by apoptosis (8-fold increased apoptotic index). This phenomenon has so far only been observed in preneoplastic or neoplastic liver foci induced by carcinogens [53]. These considerations, most of which are derived from experimental studies, are partly weakened by the demonstration of long-lasting insulin independence following transplantation of an adequate islet mass in patients [5]. Nevertheless, these considerations would justify a study comparing the intrahepatic and the intrasplenic site in clinical islet allotransplantation.

4. Means to control nonspecific inflammatory damage

A problem peculiar to islet transplantation is that transplanted islets are particularly vulnerable to proinflammatory and inflammatory cytokines IL-1, IL-6, and TNF-a which are produced as a result of the non-specific wound reaction associated with the islet implantation procedure itself. The damaging effects of these cytokines seem to be mediated in large part through the elaboration of nitric oxide and free radicals [54, 55] by macrophages infiltrating islets early following transplantation. The lack of enzymes which protect other cell types from free radical injury [56, 57] explains the specific susceptibility of beta-cells to early inflammatory mediators leading to early-islet dysfunction and impaired islet engraftment. In rodents, this early islet graft destruction can be inhibited by the macrophage toxin silica [29] as well as the novel immunosuppressive agent 15-deoxyspergualin [58], which has anti-macrophage as well as anti-T-cell effects [59]. Additionally, administration of a specific nitric oxide inhibitor improves the early function of rodent isografts [60]. These data confirm the hypothesis that early islet dysfunction results from an early cell-mediated response, involving primarily macrophages or their by-products, that either injures islets or inhibits their function until classic T-cell mediated destruction ensues [29]. Pursuing this hypothesis might generate innovative new strategies to obviate the need for multiple donors.

5. Non-diabetogenic immunosuppressive regimens
The definite main attraction of islet transplantation is to potentially avoid chronic immunosuppression. The seminal contributions of Lafferty *et al.* [61] and Lacy *et al.* [62], of Lim and Sun [63], and of Posselt, Barker, and Naji *et al.* [64], show that this aim can be achieved in rodents by pretransplant immunomodulation of donor islets, by immunoisolation of donor islets, or by using immunopriviledged islet implantation sites. These approaches have in recent years entered the clinical stage, however the results give, at best, cause for careful optimism [9, 65–68]. Nevertheless, the excellent progress that has been made in the research laboratory and the diversity of approaches being utilized provide definitive hope that the goal can be achieved [1]. In the meantime, we must continue to seek better immunosuppressive regimens which: 1) protect islet allografts likewise from unspecific early and chronic inflammatory destruction, recurrence of autoimmunity and classic T-cell mediated-rejection, 2) do not impede insulin secretion and insulin action significantly, and 3) are less toxic to the patient than at present. The issue of immunosuppression for islet transplantation has recently been comprehensively reviewed [34]. The available data suggest that islet transplantation will only achieve significant levels of success if steroid-free maintenance immunosuppressive regimens can be developed and if combinations can be selected that either are lacking in diabetogenic tendencies, or because of therapeutic synergy, can be used at doses below the threshold for diabetogenicity. The most likely candidates, according to currrent understanding, are mycophenolic mofetil [69] in combination with low doses of cyclosporine and rapamycin [70].

Currently islet transplantation is fully justified in uremic diabetic recipients of kidney transplants, who will be obligated to immunosuppression anyway. There is no reason not to attempt to ameliorate diabetes as well as uremia. Pancreas transplantation is routinely applied to such patients at many institutions [17], but at the risk of surgical morbidity. For those who do not wish to undergo the surgical risk of a pancreas transplant, an islet transplant can be done with nearly zero risk. As the islet transplant success rate increases, it would replace pancreas transplantation and lead to liberalization of the criteria for total endocrine replacement therapy, including in non-uremic patients where pancreas transplantation is currently done only in those with extremely labile diabetes or hypoglycemia unawareness.

A controlled prospective study of the effects of islet transplantation versus intensified insulin treatment on quality-adjusted life years in non-uremic, non-kidney transplanted Type I diabetic patients would then also be justified and should be done. This will require the development of low risk, non-diabetogenic strategies to prevent: 1) non-specific inflammatory damage; 2) antigen-specific T-cell mediated rejection; and 3) autoimmune destruction of transplanted islets. Experimental work shows the feasibility of achieving these goals. The clinical success that has so far been achieved under less than ideal circumstances bodes well for the future of islet transplantation.

References

1. Lacy PE (1993) Status of islet cell transplantation. Diabetes Reviews 1: 76–92.
2. Warnock GL and Rajotte RV (1992) Human pancreatic islet transplantation. Transplantation Reviews 6: 195–208.
3. Sutherland DER, Gruessner RWG and Gores PF (1994) Pancreas and Islet Transplantation: An update. Transplantation Reviews 8: 185–206.
4. Hamelmann W, Gray DW, Cairns TD *et al.* (1994) Immediate destruction of xenogeneic islets in a primate model. Transplantation 58: 1109–1114.
5. Wahoff DC, Papalois BE, Najarian JS *et al.* (in press) Autologous islet transplantation to prevent diabetes after pancreatic resection. Ann Surg.
6. Ricordi C, Tzakis AG, Carroll PB *et al.* (1992) Human islet isolation and allotransplantation in 22 consecutive cases. Transplantation 53: 407–414.
7. Warnock GL, Kneteman NM, Ryan EA, Rabinovitch A and Rajotte RV (1992) Long-term follow-up after transplantation of insulin-producing pancreatic islets into patients with type 1 (insulin-dependent) diabetes mellitus. Diabetologia 35: 89–95.
8. Socci C, Falqui L, Davalli AM *et al.* (1991) Fresh human islet transplantation to replace pancreatic endocrine function in type 1 diabetic patients. Report of six cases. Acta Diabetol 28: 151–157.
9. Scharp DW, Lacy PE, Santiago JV *et al.* (1991) Results of our first nine intraportal islet allografts in type 1, insulin-dependent diabetic patients. Transplantation 51: 76–85.
10. Gores PF, Najarian JS, Stephanian E *et al.* (1993) Insulin independence in type I diabetes after transplantation of unpurified

islets from single donor with 15-deoxyspergualin. Lancet 341: 19–21.
11. Hering BJ. unpubl.
12. Kenyon NS, Ricordi C and Mintz DL (in press) Islet cell transplantation.
13. The DCCT Research Group (1993) The effect of intensive treatment of diabetes on the development and progression of long-term complications in insulin-dependent diabetes mellitus. N Engl J Med 329: 977–986.
14. Wahren J (1994) Experimental and clinical studies with C-peptide in type I diabetes. Presented at the Satellite Symposium to the 15th International Diabetes Federation Congress: Advances in Diabetes Therapy: From Insulin to Gluco-Active Peptides and Beyond. Kobe, Japan, Nov. 3–4.
15. Cryer PE (1994) Banting Lecture. Hypoglycemia: the limiting factor in the management of IDDM. Diabetes 43: 1378–1389.
16. Sutherland DER (1993) Pancreatic transplantation. Diabetes Reviews 152–165.
17. Gruessner A, Sutherland DER (1994) Pancreas transplant results in the United Network for Organ Sharing (UNOS) United States of America (USA) Registry compared with Non-USA data in the International Registry. In: Terasaki and Cecka (eds) Clinical Transplants 1994, chapter 4, pp 47–68. CA: UCLA Tissue Typing Laboratory.
18. Bretzel RG, Hering BJ, Federlin F International Islet Transplant Registry Report. *1996 Yearbook of Cell and Tissue Transplantation*, chapter 16. Kluwer Academic Publishers.
19. Mandel P Fetal Pancreatic islet tissue. *1996 Yearbook of Cell and Tissue Transplantation*, chapter 12. Kluwer Academic Publishers.
20. Hering BJ, Geier C, Schultz AO *et al.* (1995) Islet Transplant Registry Newsletter No. 6, 5: 1–28.
21. Tzakis AG, Ricordi C, Alejandro R *et al.* (1990) Pancreatic islet transplantation after upper abdominal exenteration and liver replacement [see comments]. Lancet 336: 402–405.
22. Socci C, Mazzaferro V, Regalia E *et al.* (1994) Insulin independence after islet-liver transplantation for metastatic neuroendocrine pancreatic tumor. Transplant Proc 26: 577–578.
23. Engemann R, Hering BJ, Timmermann W *et al.* (1995) Kombinierte Leber-Insel-Transplantation nach Oberbauchexenteration bei Papillencarcinom. Der Chirurg 66: 371–376.
24. Alejandro R, Cutfield RG, Shienvold FL *et al.* (1986) Natural history of intrahepatic canine islet cell autografts. J Clin Invest 78: 1339–1348.
25. Bartlett ST, Hadley GA, Dirden B *et al.* (1993) Composite kidney-islet transplantation prevents recurrent autoimmune beta-cell destruction. Surgery 114: 211–217.
26. Bartlett ST, Chim T, Dirden B *et al.* (in press) Inclusion of peripancreatic lymph node cells prevents recurrent autoimmune destruction of islet transplants: Evidence of donor chimerism. Surgery.
27. Cuthbertson RA, Koulmanda M and Mandel TE (1988) Detrimental effect of chronic diabetes on growth and function of fetal islet isografts in mice. Transplantation 46: 650–654.
28. Montana E, Bonner Weir S and Weir GC (1993) Beta cell mass and growth after syngeneic islet cell transplantation in normal and streptozocin diabetic C57BL/6 mice. J Clin Invest 91: 780–787.
29. Kaufman DB, Platt JL, Rabe FL, Dunn DL, Bach FH and Sutherland DE (1990) Differential roles of Mac−1+ cells, and CD4+ and CD8+ T lymphocytes in primary nonfunction and classic rejection of islet allografts. J Exp Med 172: 291–302.
30. Korsgren O, Jansson L, Andersson A and Sundler F (1992) Reinnervation of transplanted pancreatic islets: a comparison between islets implanted into the kidney, spleen, or liver. Transplant Proc 24: 1025–1026.
31. Gardemann A, Jungermann K, Grosse V *et al.* (1994) Intraportal transplantation of pancreatic islets into livers of diabetic rats. Reinnervation of islets and regulation of insulin secretion by the hepatic sympathetic nerves. Diabetes 43: 1345–1352.
32. Gray DW and Morris PJ (1987) Developments in isolated pancreatic islet transplantation. Transplantation 43: 321–331.
33. Robertson RP (1992) Seminars in medicine of the Beth Israel Hospital, Boston: Pancreatic and islet transplantation for diabetes–cures or curiosities? [see comments]. N Engl J Med 327: 1861–1868.
34. Gores PF and Sutherland DER (1994) Immunosuppression for islet transplantation. In: Lanza RP and Chick WL (eds) Pancreatic Islet Transplantation Volume II: Immunomodulation of Pancreatic Islets, pp 63–74. Austin: R.G. Landes Company.
35. Largiader F, Kolb E and Binswanger U (1980) A long-term functioning human pancreatic islet allotransplant. Transplantation 29: 76–77.
36. Fetterhoff TJ, Cavanagh J, Wile KJ *et al.* (1995) Human pancreatic dissociation using a purified enzyme blend. Presented at the Fifth International Congress on Pancreas and Islet Transplantation, Miami, June 18–22.
37. Hesse F, Burtscher H, Popp F *et al.* (1995) Recombinant enzymes for islet isolation: purification of a collagenase from clostridium histolyticum and cloning/expression of the gene. Presented at the Fifth International Congress on Pancreas and Islet Transplantation, Miami, June 18–22.
38. Ricordi C, Lacy PE, Finke EH, Olack BJ and Scharp DW (1988) Automated method for isolation of human pancreatic islets. Diabetes 37: 413–420.
39. London NJM, Robertson GSM, Chadwick DR (1993) Purification of human pancreatic islets by large scale continuous density gradient centrifugation. Horm metab Res 25: 61A.
40. Hayek A, Lopez AD and Beattie GM (1988) Decrease in the number of neonatal islets required for successful transplantation by strict metabolic control of diabetic rats. Transplantation 45: 940–942.
41. Rajab AA, Ahren B and Bengmark S (1989) Islet transplantation to the renal subcapsular space in streptozotocin-diabetic rats: short-term effects on glucose-stimulated insulin secretion. Diabetes Res Clin Pract 7: 197–204.
42. Juang JH, Bonner Weir S, Wu YJ and Weir GC (1994) Beneficial influence of glycemic control upon the growth and function of transplanted islets. Diabetes 43: 1334–1339.
43. Hering BJ, Bretzel RG, Hopt UT *et al.* (1994) New protocol toward prevention of early human islet allograft failure. Transplant Proc 26: 570–571.
44. Warnock GL, DeGroot T, Untch D, Ellis DK and Rajotte RV (1989) The natural history of pure canine islet autografts in hepatic or splenic sites. Transplant Proc 21: 2617–2618.
45. Kaufman DB, Morel P, Field MJ, Munn SR and Sutherland DE (1990) Purified canine islet autografts. Functional outcome as influenced by islet number and implantation site. Transplantation 50: 385–391.
46. Sutton R, Gray DW, McShane P, Peters M and Morris PJ (1987) The metabolic efficiency and long-term fate of intraportal islet grafts in the cynomolgus monkey. Transplant Proc 19: 3575–3576.
47. Luzi L, Secchi A, Pozza G (1992) Metabolic assessment of posttransplant islet function in humans. In: Ricordi C (ed) Pancreatic islet cell transplantation. 1892–1992. One century of transplantation for diabetes, pp 361–382. Austin: R.G. Landes.

48. London NJ, Robertson GS, Chadwick DR, Johnson PR, James RF, Bell PR (1994) Human pancreatic islet isolation and transplantation. Clin Transplant 8: 421–459.
49. White JJ and Dupre J (1968) Regulation of insulin secretion by the intestinal hormone, secretin: studies in man via transumbilical portal vein catheterization. Surgery 64: 204–213.
50. Clavien PA, Harvey PR and Strasberg SM (1992) Preservation and reperfusion injuries in liver allografts. An overview and synthesis of current studies. Transplantation 53: 957–978.
51. Pyzdrowski KL, Kendall DM, Halter JB, Nakhleh RE, Sutherland DE and Robertson RP (1992) Preserved insulin secretion and insulin independence in recipients of islet autografts [see comments]. N Engl J Med 327: 220–226.
52. Ansara MF, Saudek F, Newton M, Raynor AC, Cryer PE and Scharp DW (1994) Pancreatic islet transplantation prevents defective glucose counter-regulation in diabetic dogs. Transplant Proc 26: 664–665.
53. Dombrowski F, Lehringer Polzin M and Pfeifer U (1994) Hyperproliferative liver acini after intraportal islet transplantation in streptozotocin-induced diabetic rats. Lab Invest 71: 688–699.
54. Yamada K, Inada C, Otabe S, Takane N, Hayashi H and Nonaka K (1993) Effects of free radical scavengers on cytokine actions on islet cells. Acta Endocrinol Copenh 128: 379–384.
55. Corbett JA, Sweetland MA, Wang JL, Lancaster JR Jr and McDaniel ML (1993) Nitric oxide mediates cytokine-induced inhibition of insulin secretion by human islets of Langerhans. Proc Natl Acad Sci USA 90: 1731–1735.
56. Bonner Weir S, Baxter LA, Schuppin GT and Smith FE (1993) A second pathway for regeneration of adult exocrine and endocrine pancreas. A possible recapitulation of embryonic development. Diabetes 42: 1715–1720.
57. Malaisse WJ, Malaisse Lagae F, Sener A and Pipeleers DG (1982) Determinants of the selective toxicity of alloxan to the pancreatic B cell. Proc Natl Acad Sci USA 79: 927–930.
58. Kaufman DB, Field MJ, Gruber SA *et al.* (1992) Extended functional survival of murine islet allografts with 15-deoxyspergualin. Transplant Proc 24: 1045–1047.
59. Dickneite G, Schorlemmer HU, Weinmann E, Bartlett RR and Sedlacek HH (1987) Skin transplantation in rats and monkeys: evaluation of efficient treatment with 15-deoxyspergualin. Transplant Proc 19: 4244–4247.
60. Xenos ES, Stevens RB, Sutherland DE *et al.* (1994) The role of nitric oxide in IL-1 beta-mediated dysfunction of rodent islets of Langerhans. Implications for the function of intrahepatic islet grafts. Transplantation 57: 1208–1212.
61. Lafferty KJ, Cooley MA, Woolnough J and Walker KZ (1975) Thyroid allograft immunogenicity is reduced after a period in organ culture. Science 188: 259–261.
62. Lacy PE, Davie JM and Finke EH (1979) Prolongation of islet allograft survival following in vitro culture (24 degrees C) and a single injection of ALS. Science 204: 312–313.
63. Lim F and Sun AM (1980) Microencapsulated islets as bioartificial endocrine pancreas. Science 210: 908–910.
64. Posselt AM, Barker CF, Tomaszewski JE, Markmann JF, Choti MA and Naji A (1990) Induction of donor-specific unresponsiveness by intrathymic islet transplantation [see comments]. Science 249: 1293–1295.
65. Calafiore R, Basta G, Falorni A *et al.* (1991) Vascular graft of microencapsulated human pancreatic islets in non-immunosuppressed diabetic recipients. Preliminary results. Diab Nutr Metab 4: 45–48.
66. Soon Shiong P, Heintz RE, Merideth N *et al.* (1994) Insulin independence in a type 1 diabetic patient after encapsulated islet transplantation. Lancet 343: 950–951.
67. Scharp DW, Swanson CJ, Olack BJ *et al.* (1994) Protection of encapsulated human islets implanted without immunosuppression in patients with type I or type II diabetes and in nondiabetic control subjects. Diabetes 43: 1167–1170.
68. Arias Diaz J, Vara E, Balibrea JL *et al.* (1994) CT-guided fine needle approach for intrathymic islet transplantation in diabetic patients. Cell Transplantation 3: 213A.
69. Hao L, Calcinaro F, Gill RG, Eugui EM, Allison AC and Lafferty KJ (1992) Facilitation of specific tolerance induction in adult mice by RS-61443. Transplantation 53: 590–595.
70. Yakimets WJ, Lakey JR, Yatscoff RW *et al.* (1993) Prolongation of canine pancreatic islet allograft survival with combined rapamycin and cyclosporine therapy at low doses. Rapamycin efficacy is blood level related. Transplantation 56: 1293–1298.

R. P. Lanza and W. L. Chick (eds.), Yearbook of Cell and Tissue Transplantation 1996/1997, 145–152.

Chapter 14

Experimental islet transplantation

Gordon C. Weir and Susan Bonner-Weir
Joslin Diabetes Center, Brigham and Women's Hospital, Deaconess Hospital, Harvard Medical School, Boston, MA 02215, U.S.A.

Transplantation of the islets of Langerhans offers the possibility of curing diabetes, but the problem has turned out to be more complex than was envisioned in the early 1970s when the first successful experimental transplants were performed in rodents [1, 2]. Much energy has been devoted to transplantation of human islets obtained from cadaver donors into the liver, via the portal vein of immunosuppressed type I diabetics [3, 4]. These efforts have been disappointing because most have failed within a short period of time. The failures are interesting contrasts to the far more impressive results obtained with whole pancreas transplants and islet autotransplants [5, 6].

Even if the results with human islets were more successful, there would be a marked shortage of islet tissue because the number of available donor pancreases do not even come close to providing enough islets to supply all the patients who could benefit. Much attention is being paid to xenotransplantation, with porcine and bovine sources being especially attractive. Somehow the rejection of xenografts by the immune system will have to be controlled, with induction of tolerance, safer immunosuppressive medication, immunobarrier devices, or some combination of these. With all of these approaches there remain fundamental questions about what happens to transplanted islets even when they are not under immune attack. Normal islets are very complex microorgans with unique vasculature and innervation, that are able to exert precise control over glucose, lipid and fat metabolism. In order to be transplanted, islets must be removed from their normal position in the pancreas with collagenase and mechanical agitation, which means that they are stripped of their vasculature and nerve supply. They are then expected to engraft in a foreign tissue site and perform in the same manner as before. Remarkably, they often perform very well in experimental situations, but the disappointing results obtained with human islet allografts are a painful reminder that we must learn much more about what can go wrong during the transplantation process. This short review will focus upon important issues that are germane to any islet transplants, whether they be isografts, allografts or xenografts. These include questions about beta cell birth and death, adverse effects of hypoxia and hyperglycemia, and the concept of beta cell setpoint.

The challenge of hypoxia faced by islets immediately after transplantation

Assuming that immune attack on freshly transplanted islets is controlled, the most important threat is hypoxia. Islets are normally highly vascularized. Even though they occupy only about 1% of the pancreas, they receive about 10% of the blood supply [7]. Beta cells probably require more oxygen than most other cells because insulin secretion is dependent upon aerobic metabolism. An islet placed into a tissue bed is completely dependent upon oxygen provided by surrounding capillaries; it takes several days for significant angiogenesis to begin and a new vascular network that extends throughout the islet is in place only about 10 days after the transplant [8, 9]. Therefore, islets are particularly vulnerable in the first few days after transplantation. The fate of beta cells is dependent upon both the severity and duration of hypoxia, although little is known about the details of these events. Some

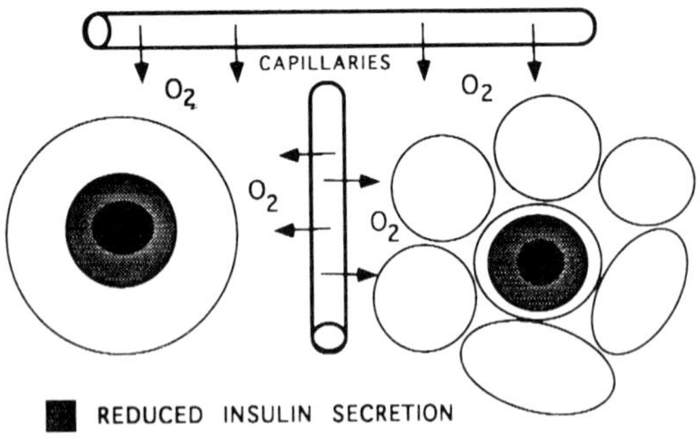

Fig. 1. Freshly transplanted islets are subject to hypoxia. One level of hypoxia can inhibit insulin secretion; a more severe level can cause necrosis. As depicted on the left, large islets of greater than 200 μm in diameter are subject to the adverse effects of hypoxia. As shown on the right, even smaller transplanted islets can suffer from anoxia if they are surrounded by avascular islets that compete for the available oxygen.

hypoxia will markedly curtail insulin secretion; more severe hypoxia will cause cell death [10–12].

The size of the islets is important because sufficient oxygen from surrounding capillaries can only be provided over a distance of about 80–100 μm, which means that oxygen can penetrate through a layer of cells only 8–10 cells thick [10–12]. Therefore, isolated islets greater than 160 μm in diameter will have some central hypoxia, perhaps only enough to cause mild reduction of insulin secretion, and islets over 200 μm will have enough hypoxia to have some central necrosis (Fig. 1). The larger the islet, the greater the loss of beta cells. Because isolated islets are typically transplanted as clumps, even some small islets can be subject to hypoxia if they are surrounded by other avascular islets. The effects of hypoxia may not be completely negative because it probably is a powerful stimulus to angiogenesis and the growth of host capillaries into the islet. The mechanisms responsible for this are unknown, but the synthesis of vascular endothelial growth factor (VEGF) is known to be enhanced by hypoxia in several cell types [13]. We suspect that VEGF, probably acting in concert with other factors, plays an important role in the angiogenesis of transplanted islets.

There is considerable variability in the efficacy of different transplant sites in accommodating islets [14–22]. The kidney subcapsular site and the liver accessed by the portal vein are particularly favorable locations. The spleen can also be a good site even though islets are sometimes carries downstream towards the liver. The peritoneal cavity and subcutaneous sites are less good and have been shown to require many more islets to lower glucose levels in diabetic rodents [16,22]. The reasons for these variable results are not known, but an obvious potential explanation is differences in blood flow. Perhaps strategies can be developed to improve blood flow in areas not normally considered for transplantation, which could make them more attractive sites.

Dynamic changes in islets immediately after transplantation

Recent studies performed by Davalli and others in our laboratory have documented the dynamic events that occur in the first few days of islet transplantation [23]. In these experiments, 400 syngeneic islets were transplanted under the kidney capsule of streptozocin (STZ) diabetic mice; this approach routinely cures diabetes within one or two days. At one and three days after transplantation there is impressive disruption of the architecture of these islets, with obvious necrosis being seen not only in the center of many islets, but even in other portions that presumably were ischemic. Then remodeling occurred as islet cells merged to form compact healthy-looking islet units that could be seen seven days after the transplantation. Beta cell mass fell by as much as 50% during this short time period.

Cell death can result from necrosis or apoptosis, and both occurred during these few days immediately following islet transplantation [23]. The differences between the two processes are not completely understood, but each has very distinctive characteristics [24]. Necrosis can be likened to an explosion of a cell with dilation of organelles and disruption of membranes. Apoptosis is more of an implosion, which is a more orderly process that is even associated with increased expression of certain genes, so-called "programmed cell death". In the early stages there is orderly enzymatic cleavage of DNA which results in a "laddering" pattern of DNA fragments on electrophoretic gels. Then, a fragmentation process occurs after condensation of nucleus and cytoplasm, with the fragments contained by intact membranes. The condensed and fragmented nuclei are called apoptotic bodies that can be identified by electron microscopy or by staining with propidium iodide. In these islet transplant experiments, it was not possible to quantify the relative contributions of necrosis and apoptosis to the overall death of cells. In the end, the pattern of cell death probably may not matter very much. Perhaps cells exposed to more severe hypoxia rapidly succumb to necrosis, whereas cells with slightly more oxygen progress down the apoptosis pathway. This raises questions about when the events of apoptosis are irreversible, and whether there is an early point when restoration of oxygen supply could reverse the process.

During this early transplant period there was also a marked reduction of insulin content and of insulin mRNA which were out of proportion to the loss of beta cells. This result probably indicates that insulin synthesis is very sensitive to hypoxia. Insulin secretion is known to be very sensitive to hypoxia, even before insulin stores are depleted [10–12], so there are multiple reasons for why insulin output could be compromised when transplantation is accompanied by islet ischemia.

The influence of hyperglycemia upon the early stages of islet engraftment

There is accumulating evidence that hyperglycemia has an adverse influence upon islets during the early stages of the engraftment process [25–33]. The results of the study of the events of early islet transplantation provided further support for this assumption [23]. In hyperglycemic recipients, that became normoglycemic within 1-2 days after transplantation of 400 syngeneic islets, it was found that the grafts contained more apoptotic bodies and had a more severe reduction of beta cell mass. This is not surprising. In the presence of hyperglycemia beta cells should have increased aerobic metabolism with an accompanying increase in oxygen consumption, that could create more severe local hypoxia and resultant cell death.

An adverse effect of hyperglycemia could also be demonstrated with another experimental approach, in which it was found that 200 islets, that are normally incapable of curing STZ diabetes in mice, could be successful if allowed to engraft in a more normoglycemic environment [33]. This was accomplished by placing grafts of 200 islets in both kidneys of a STZ diabetic mouse; this promptly cured the diabetes. After engraftment had taken place in both sites, one graft was removed and the remaining graft, which otherwise would have been inadequate, was able to maintain normoglycemia. The success of this remaining graft was presumably dependent upon the increase of beta cell replication and cell size that was found to take place when the first graft was removed. These results were similar to our earlier studies in rats in which increased beta cell replication and cell size occurred in transplanted islets exposed to hyperglycemia following partial pancreatectomy [34]. In summary, hyperglycemia has an important adverse influence upon the early engraftment process, probably due to increased aerobic metabolism in beta cells that allows hypoxia to take a greater toll. However once engraftment takes place, hyperglycemia can exert a positive effect by stimulating beta cell replication and increasing beta cell size.

Beta cell birth and death: replication, regeneration, senescence, necrosis and apoptosis

Beta cell mass is normally maintained by a complex balance between cell birth and death [35, 36]. It must be emphasized that beta cells have a finite life span and that there is constant turnover of beta cells throughout adult life. In rats the life span of beta cells has been estimated to be 1–3 months [37]; the life span in humans and pigs is longer but unknown. An appreciation of these dynamic events is important for understanding the fate of transplanted islets and the pathogenesis of all forms of diabetes. As depicted in Fig. 2 new beta cells arise from the replication of existing beta cells and from the differentiation of precursor cells in large pancreatic ducts. Beta cell mass is also determined by cell size and it is known that at least short-term hyperglycemia can cause beta cell hypertrophy [38]. Beta cell death results from either apoptosis or necrosis, processes described earlier. Some of the parameters can be measured accurately, others are difficult to estimate. Beta cell mass can be measured with standard morphometric approaches, as can beta cell size. The replication rate of existing beta cells can be accurately measured with 5-bromo-2′deoxyuridine (BrdU), with tritiated thymidine, or even with counts of mitotic figures after colchicine, as long as beta cells are identified with immunostaining [36, 39, 40]. Generation of new beta cells from pancreatic ducts can not yet be quantified and is awaiting the identification of a reliable marker. The death of beta cells is equally complex. We know that apoptosis can be quantified by counting propidium iodide stained apoptotic bodies in normal and pathological states. Unfortunately, it is not yet possible to convert this to a rate of cell death, because the longevity of these apoptotic bodies in islets is not known. Even though necrosis presumably occurs with hypoxia, immune attack, and chemical injury, we have little idea how much necrosis contributes to the regulation of beta cell mass in abnormal states. Evidence of necrosis is rarely seen in the islets, but because necrosis may occur very rapidly, the process could make a substantial contribution to overall death.

Senescence of cells is poorly understood but may be important for our understanding of beta cell failure in diabetes and transplantation. This represents a stage in a cell's life when replication is longer possible. It will be important to learn if differences exist in the function of young or senescent cells. It would be interesting if senescent cells were more prevalent in NIDDM and had less capacity for insulin secretion. Senescent cells presumably progress on to an apoptotic route of death but may have reduced defense against injury and therefore be especially vulnerable to necrosis. These comments are speculative but it would be very useful to have markers for senescent beta cells, because they could greatly enhance our understanding of the beta cell death that occurs during islet remodeling, during normal states, with aging, and with diabetes and islet transplantation.

Why beta cell regeneration from pancreatic ducts may be important for islet transplantation

There are important issues about how long transplanted islets can maintain their function, which raise practical questions about how often islets may need to be replaced. Syngeneic islet transplants placed under the kidney capsule in rodents can maintain normoglycemia for periods of a year or more. But a troublesome finding was obtained in dogs, in which autotransplantation of islets into the liver resulted in normoglycemia for only a limited period of time [41]. Islet allografts in humans have been largely unsuccessful, in contrast to the impressive results obtained with human autotransplants [6]. These findings are difficult to make complete sense of. Perhaps, allografted human islets are vulnerable to immune rejection and the toxic effects of immunosuppressive medication, but how can this be reconciled with the excellent long term results obtained with whole pancreas allografts? Perhaps the hepatic site makes the islets more susceptible to injury. Another potential factor is that the presence of pancreatic ducts in a whole pancreas may allow continual replenishment of beta cells for years. The islet autografts are often crude pancreas digests that may contain both islets and ducts, in contrast to the relatively pure islets typically used for islet allografts. This contaminating duct tissue may allow continual regeneration of beta cells. This hypothesis is speculative but testable.

The effects of hyperglycemia upon beta cell birth and death

We know very little about the effects of hyperglycemia upon the turnover rate of beta cells. This is of obvious importance for human NIDDM in which we assume there is a gradual fall in beta cell mass [42]. A similar situation must occur with rodent models of NIDDM such as db/db mice, ob/ob mice, and Zucker diabetic rats, although the changes have not be comprehensively studied. There are important questions about autoimmune diabetes, such as about how the hyperglycemia of the early stages of this disease might influence beta cell regenerative capacity and susceptibility to immune injury. Equally important questions face the field of pancreas and islet transplantation.

We know that short-term (days) exposure to hyperglycemia can enhance beta cell replication, and although this process must be driven by high glucose levels, a variety of permissive factors must also be involved. The effects of hyperglycemia upon the production of new beta cells from pancreatic ducts is less well understood. This pathway is known to be important in fetal and neonatal islet development and in the regeneration that occurs after partial pancreatectomy [36, 43]. The process must be orchestrated by a complex coordination of growth and differentiation factors. Perhaps the process is started by non-glucose stimuli with glucose playing an increasingly important role once precursor cells reach the early stages of beta cell differentiation. If beta cell growth is dependent upon glucose metabolism, it might be expected that beta cells would not be able to respond to glucose until GLUT2 and glucokinase are expressed. This is a very important issue for diabetes. Perhaps the regenerative potential of ducts is not utilized by hyperglycemia and therefore theoretically accessible to some combination of growth and differentiation factors; or perhaps it is used and completely expended once diabetes become chronic? Is the capacity for regeneration from ducts exhausted with age or the diabetic state, and if so, how did this happen?

There are equally interesting questions about the effects of chronic hyperglycemia upon the death rate of beta cells. In some circumstances death rate can be

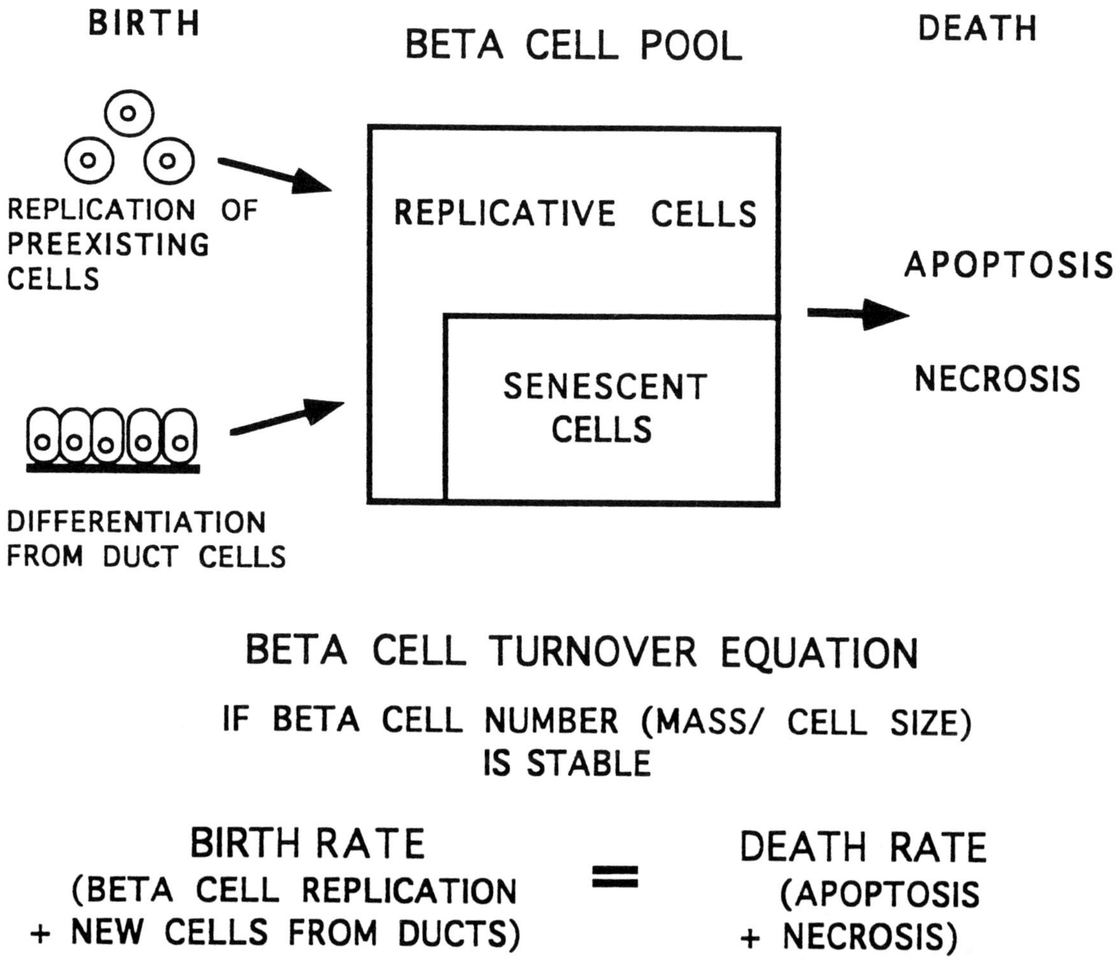

Fig. 2. Beta cells turn over continually, with beta cell mass being maintained by a balance between beta cell birth and death. The beta cell pool has replicative cells that have the capacity to divide and scenescent cells that do not. Beta cell size can vary, with hypertrophy being found after glucose stimulation and atrophy resulting from hypoglycemia. The beta cell turnover equation states that if beta cell number is constant, birth rate must equal death rate.

estimated. If beta cell birth rate is known and if beta cell mass and cell size are known to be stable, then the death rate must equal the birth rate [37, 44]. Unfortunately, the presence of pancreatic ducts introduces an unknown for beta cell birth rate and makes calculation of death rate impossible, even though it must be at least as high as the known replication rate of preexisting beta cells which can be measured with BrdU. With transplantation of pure islets without pancreatic ducts, as typically occurs with experimental transplants in rodents, an accurate birth rate can be determined, as can mass and cell size, so death rate can be estimated, even though the relative contributions of necrosis and apoptosis can not be determined. Using this approach for grafted islets in mice, we were able to study the effects of hyperglycemia and found that beta cell death was accelerated and led to a premature fall in beta cell mass [44]. This finding was for a circumscribed set of conditions and must now be further explored in a variety of other islet transplantation situations. For example, what happens with different species, for various transplant sites, and with immunobarrier devices?

Beta cell setpoint and islet transplantation

Somehow the beta cell has evolved to secrete precisely enough insulin to maintain glucose levels within a narrow range during such challenges as meals, fasting and exercise. The beta cell can be considered to have a setpoint, much like a thermostat, although there are some normal fluctuations in this setpoint that allow glucose levels to rise after meals and fall during fasting. The ability of beta cells to respond to glucose is dependent upon a tight linkage between glucose metabolism and insulin secretion, with the enzyme glucokinase playing a key role because it is the rate-limiting step for glycolysis [45]. Different mammalian species have different glucose levels, such that humans and pigs have glucose levels that are typically about 20–30 mg/dl lower than rats or mice. When pig or human islets are transplanted into athymic mice, the glucose levels of the recipient fall to those of the donor species [46, 47]. Thus, the setpoint of the donor beta cells appears to take control of host glucose metabolism.

Because of what is known about beta cell setpoint and the regulation of beta cell mass it would be expected that recipients of human or porcine beta cells in sufficient or even excess amounts would have perfectly normal metabolism. Indeed, some recipients of human allograft and autograft islets have maintained glucose levels in the normal range [3, 4, 6]. Nonetheless, there are some questions that deserve some attention.

How will recipients cope with excessive quantities of islets? We know that humans or rats with chronic hypoglycemia due to an insulin-producing tumor will develop beta cell atrophy [48, 49]. This adaptation fits perfectly with setpoint theory; the beta cell exposed to hypoglycemia should turn off many glucose-dependent genes and undergo atrophy, ensuring that insulin production will be lowered. One might expect that the same adaptation would occur if an excessive number of islets were transplanted, that too much insulin would produce mild hypoglycemia that would lead to reduction of insulin secretion and beta cell mass, and restoration of normoglycemia. It was surprising to find that this adaptation did not occur during an 8 wk time period after the transplantation of an excessive number of rat islets (2000) into STZ nude mice [46]. These recipients maintained glucose levels, even after oral and intraperitoneal glucose challenges, that were clearly lower than control mice and lower than what would be expected in normal rats. Therefore, either the beta cells mass was not down regulated as expected or the setpoint for glucose-induced insulin release was altered. Beta cell mass was measured in these experiments and the mass of the rat graft was about one mg, which was about 40% less than the beta cell mass in the pancreas of a normal nude mouse. It is impossible to say if there was an appropriate degree of downregulation of the beta cells in the graft.

How might beta cell setpoint be altered in an islet graft? A potential explanation is the normal relationship between the beta cells and non-beta cells of an islet graft is distorted. Normally within an islet, blood flows from the islet core that contains beta cells to the mantle which contains the cells that store and secrete glucagon, somatostatin and pancreatic polypeptide [50, 51]. Glucagon suppression by glucose has been hypothesized to depend upon intraislet secretion of insulin, which acts directly upon the alpha cell to inhibit glucagon secretion. The hypothesis further suggests that beta cells see very little intraislet glucagon because alpha cells are downstream from beta cells. There are important unanswered questions about the vascular relationship between the beta and non-beta cells of transplanted islets. If beta cells were downstream from alpha cells, the constant exposure of beta cells to glucagon could lead to excessive insulin secretion at any given glucose concentration. Glucagon is a potent insulin secretagogue that acts through cAMP. There is increasing evidence that generation of cAMP in beta cells can reduce the set point for glucose-induced insulin secretion [52]. In Miami in June, 1995, Stagner, Samols, and Mokshagumdam presented data suggesting that the normal relationship between beta and alpha cells are not reestablished in transplanted islets, and that alpha cells are not downstream form beta cells. The main question raised by this discussion is whether transplanted islets might hypersecrete insulin and cause problems with hypoglycemia. Hopefully, other counterregulatory mechanisms would provide sufficient protection against hypoglycemia.

An abnormal vascular arrangement between beta cells and alpha cells in transplanted islets may help explain the surprising finding that glucagon is not secreted in response to insulin-induced hypoglycemia from autotransplanted islets in humans [6]. This can not be explained by a lack of alpha cells; their presence has been demonstrated by unambiguous glucagon responses to arginine stimulation. If normal glucagon responses to hypoglycemia are somehow dependent upon a reduction of intraislet insulin secretion from the upstream beta cells, a rearranged vasculature could explain this failure to respond. The loss of a glucagon response to hypoglycemia may not be a problem for autograft recipients; no reports of hypoglycemia have emerged and epinephrine, a mainline defense against hypoglycemia, probably would provide adequate protection.

Summary

The possibility that islet transplantation will eventually be successful continues to offer hope for people with diabetes. Unfortunately, the task is very difficult. This short review has focused upon some of the problems faced by islets when they are transplanted. Questions about islet survival and function will continue to be a critical consideration for any of the many approaches to islet transplantation.

Acknowledgements

The research described in this review was supported by grants from the Juvenile Diabetes Foundation International, the National Institutes of Health (NIH DK-35449) and the Jesse B. Cox Foundation.

References

1. Ballinger WF and Lacy PE (1972) Transplantation of intact pancreatic islets in rats. Surg 72: 175–186.
2. Reckard CR and Barker CF (1973) Transplantation of isolated pancreatic islets across strong and weak histocompatibility barriers. Transplant Proc 5: 761–763.
3. Lacy PE (1995) Treating Diabetes with Transplanted Cells. Scientific American: 50–58.
4. Lacy PE (1993) Status of islet cell transplantation. Diabetes Rev 1: 76–92.
5. Sutherland DE (1992) Pancreatic Transplantation: State of the art. Transplant Proc 24: 762–766.
6. Pyzdroswki KL, Kendall DM, Halter JB *et al.* (1992) Preserved insulin secretion and insulin independence in recipients of islet autografts. N Engl J Med 327: 220–226.
7. Lifson N, Lassa CV and Dixit PK (1985) Relation between blood flow and morphology in islet organ of rat pancreas. Am J Physiol 249: E43–E48.
8. Menger MD, Jaeger S, Walter P *et al.* (1989) Angiogenesis and hemodynamics of microvasculature of transplanted islets of Langerhans. Diabetes 38 (Suppl. 1): 199–201.
9. Menger MD, Vajkoczy P, Beger C *et al.* (1994) Orientation of microvascular blood flow in pancreatic islet isografts. J Clin Invest 93: 2280–2285.
10. Dionne KE, Coulton CK and Yarmush ML (1993) Effect of hypoxia on insulin secretion by isolated rat and canine islets of Langerhans. Diabetes 42: 12–21.
11. Colton CK and Avgoustiniatos ES (1991) Bioengineering in development of the hybrid artificial pancreas. J Biomechanical Engineering 113: 152–170.
12. Dionne KE, Colton CK and Yarmush ML (1991) A Microperifusion system with environmental control for studying insulin secretion by pancreatic tissue. Biotechnol Prog 7: 359–368.
13. Mukhopadhyay D, Tslokas L, Zhou X-M *et al.* (1995) Hypoxic induction of human vascular endothelial growth factor expression through c-Src activation. Nature 375: 577–581.
14. Bretzel RG, Hering BJ, Stroedter D *et al.* (1992) Experimental islet transplantation in small animals. In: Ricordi C (ed) Pancreatic islet cell transplantation, pp 249–260. Austin, Texas: R.G. Landes Company.
15. Warnock GL, Ao Z, Cattral MS *et al.* (1992) Experimental islet transplantation in large animals. In: Ricordi C (ed) Pancreatic islet cell transplantation, pp 261–278. Austin, Texas: R.G. Landes Company.
16. Kemp CB, Knight MJ, Scharp DW *et al.* (1973) Effect of transplantation site on the results of pancreatic islet isografts in diabetic rats. Diabetol 9: 486.
17. Mellgren A, Schnell-Landstrom AH, Petersson B *et al.* (1986) The renal subcapsular site offers better growth conditions for transplanted mouse pancreatic islet cells than the liver or spleen. Diabetol 29: 670–672.
18. Coulombe MG, Warnock GL and Rajotte RV (1988) Reversal of diabetes by transplantation of cryopreserved rat islets of Langerhans to the renal subcapsular space. Diabetes Res 8: 9.
19. Woehrle M, Markman JF, Beyer K *et al.* (1990) The influence of the implantation site (kidney capsule vs. portal vein) on islet survival. Horm Metab Res (Suppl. 25): 163–165.
20. Hiller WF, Klempnauer J, Luck R *et al.* (1991) Progressive deterioration of endocrine function after intraportal but not kidney subcapsular rat islet transplantation. Diabetes 40: 134–140.
21. Lacy PE, Hegre OD, Gerasimidi-Vazeou A *et al.* (1991) Maintenance of normoglycemia in diabetic mice by subcutaneous xenografts of encapsulated islets. Science 254: 1782–1784.
22. Juang J-H, Bonner-Weir S, Vacanti JP *et al.* (1995) Outcome of subcutaneous islet transplantation improved by polymer device. Transplantation Proc: in press.
23. Davalli AM, Scaglia L, Zangen DH *et al.* (1995) Early changes in syngeneic islets grafts: effect of recipient's metabolic control on graft outcome. Transplantation Proc: in press.
24. Schwartzman RA and Cidlowski JA (1993) Apoptosis: The biochemistry and molecular biology of programmed cell death. Endocr Rev 14: 133–151.
25. Andersson A, Eriksson U, Petersson B *et al.* (1981) Failure of successful intrasplenic transplantation of islets from lean mice to cure obese-hyperglycemic mice, despite islet growth. Diabetol 20: 237–241.
26. Korsgren O, Jansson L and Andersson A (1989) Effects of hyperglycemia on function of isolated mouse pancreatic islets transplanted under kidney capsule. Diabetes 38: 510–515.
27. Gray DWR, Cranston D, McShane P *et al.* (1989) The effect of hyperglycemia on pancreatic islets transplanted into rats beneath the kidney capsule. Diabetol 32: 663–667.
28. Korsgren O, Jansson L, Sandler S *et al.* (1989) Hyperglycemia-induced B-cell toxicity. the fate of pancreatic islets transplanted into diabetic mice is dependent on their genetic background. J Clin Invest 86: 2161–2168.
29. Andersson A (1983) The influence of hyperglycemia, hyperinsulinemia and genetic background on the fate of intrasplenically implanted mouse islets. Diabetol 25: 269–272.
30. Hayek A, Lopez AD and Beattie GM (1988) Decrease in the number of neonatal islet required for successful transplantation by strict metabolic control of diabetic rats. Transplant 45: 940–942.
31. Ar'Rajab A and Ahren B (1992) Prevention of hyperglycemia improves the long term result of islet transplantation in streptozotocin-diabetic rats. Pancreas 7: 435–442.
32. Keymeulen B, Vetri M, Gorus F *et al.* (1993) The effect of insulin treatment on function of intraportally grafted islets in streptozotocin diabetic rats. Transplant 56: 60–64.

33. Juang J-H, Bonner-Weir S, Wu Y-J *et al.* (1994) Beneficial influence of glycemic control upon the growth and function of transplanted islets. Diabetes 43: 1334–1339.
34. Montana E, Bonner-Weir S and Weir GC (1994) Transplanted beta cell response to increased metabolic demand. J Clin Invest 93: 1577–1582.
35. Weir GC, Bonner-Weir S and Leahy JL (1990) Islet mass and function in diabetes and transplantation. Diabetes 39: 401–405.
36. Bonner-Weir S (1994) Regulation of pancreatic β-cell mass *in vivo*. Recent Prog Horm Res 49: 91–104.
37. Finegood DT, Scaglia L and Bonner-Weir S (1995) (Perspective) Dynamics of B-cell mass in the growing rat pancreas: estimation with a simple mathematical model. Diabetes 44: 249–256.
38. Bonner-Weir S, Deery D, Leahy JL *et al.* (1989) Compensatory growth of pancreatic B-cells in adult rats after short-term glucose infusion. Diabetes 38: 49–53.
39. Swenne I (1992) Pancreatic beta-cell growth and diabetes mellitus. Diabetol 35: 193–201.
40. Brockenbrough JS, Weir GC, Bonner-Weir S (1988) Discordance of exocrine and endocrine growth after 90% pancreatectomy in rats. Diabetes 37: 232–236.
41. Alejandro R, Cutfield RG, Shienvold FL *et al.* (1986) Natural history of intrahepatic canine islet cell autografts. J Clin Invest 78: 1339–1348.
42. Weir GC and Leahy JL (1994) Pathogenesis of non-insulin-dependent (type II) diabetes mellitus. In: Kahn CR and Weir GC (eds) Joslin's Diabetes Mellitus, 13th ed, pp 240–264. Philadelphia: Lea and Febiger.
43. Bonner-Weir S, Baxter LA, Schuppin GT *et al.* (1993) Two pathways for beta cell regeneration of the adult endocrine and exocrine pancreas: A possible recapitulation of embryonic development. Diabetes 42: 1715–1720.
44. Montana E, Bonner-Weir S and Weir GC (1993) Beta cell mass and growth after syngeneic islet cell transplantation in normal and streptozocin-diabetic C57BL/6 mice. J Clin Invest 91: 780–787.
45. Matschinsky F, Liang Y, Kesavan P *et al.* (1993) Glucokinase as pancreatic B cell glucose sensor and diabetes gene. J Clin Invest 92: 2092–2098.
46. Davalli AM, Ogawa Y, Scaglia L *et al.* (1995) Function, mass, and replication of porcine and rat islets transplanted into diabetic nude mice. Diabetes 44: 104–111.
47. Warnock GL, Dwayne DK, Cattral M *et al.* (1989) Viable purified islets of Langerhans from collagenase-perfused human pancreas. Diabetes 38: 136–139.
48. Miyaura C, Chen L, Appel M *et al.* (1994) Expression of reg/PSP, a pancreatic exocrine gene: relationship to changes in islet β-cell mass. Mol Endocrinol 5: 226–234.
49. Bedoya FJ, Matschinsky FM, Shimizu T *et al.* (1986) Differential regulation of glucokinase activity in pancreatic islets and liver of the rat. J Biol Chem 261: 10760–10764.
50. Samols E, Weir GC and Bonner-Weir S (1986) Intra-islet insulin-glucagon-somatostatin relationships. Clin Endocrinol Metab 15: 33–58.
51. Weir GC and Bonner-Weir S (1990) Islets of Langerhans: the puzzle of intraislet interactions and their relevance to diabetes. J Clin Invest 85: 983–987.
52. Prentki M and Matchinsky FM (1987) Ca 2+, cAMP, and phospolipid-derived messengers in coupling mechanisms of insulin secretion. Physiol Rev 67: 1185–248.

R. P. Lanza and W. L. Chick (eds.), Yearbook of Cell and Tissue Transplantation 1996/1997, 153–160.

Chapter 15

International islet transplant registry report

Reinhard G. Bretzel, Bernhard J. Hering, Andreas O. Schultz, Christoph Geier and Konrad Federlin
Third Medical Department, Justus-Liebig-University, Giessen, D-35385 Germany

Long-term studies strongly suggest that tight control of blood glucose levels achieved by conventional intensive insulin treatment, self-blood glucose monitoring and patient's education can significantly prevent the development and retard the progression of chronic complications of type 1 diabetes mellitus [1, 2]. The expense for this benefit was a threefold increase of the number of severe hypoglycemic episodes, a significant increase of the body weight, and dietary and other life-style restrictions affecting the quality of life.

By contrast, experimental studies in rodents have clearly demonstrated that replacement of the recipient's islets of Langerhans by transplantation of isolated pancreatic islets is followed by optimal metabolic control, arrest or even reversal of chronic diabetic complications, prolonged survival of the islet recipient and hypoglycemic episodes were not observed [3]. Over the past five years, significant advances have been made in the number and purity of islets that can be harvested from the human pancreas, thus encouraging several centers to resume clinical trials of islet cell transplantation in patients with type 1 diabetes mellitus [4]. The transplantation of isolated pancreatic islets in patients with type 1 diabetes mellitus holds significant potential advantages over whole-gland transplants; however, islet cell transplantation is still a clinical investigational procedure [5]. Nevertheless, insulin independence has been achieved in several type 1 diabetic patients, at different institutions, including our own [6–11].

Human islet allotransplants were reported to the International Pancreas and Islet Transplant Registry at the University of Minnesota, Minneapolis, USA, until 1989 [12], when responsibility for data collection and analysis was transferred to the newly established International Islet Transplant Registry at the Justus-Liebig-University Giessen, Germany [13]. The goals of the International Islet Transplant Registry (ITR) are to collect data on all islet transplants performed world-wide, to perform scientific analysis on collected islet transplant data, and to communicate information on islet transplantation to member institutions as well as the health care community in general. The data are regularly updated and an ITR Newsletter was issued at least once a year. This report is based on a recent analysis of the Islet Transplant Registry published in detail in the ITR Newsletter No. 6, 1995 [14]. It summarizes the status of clinical adult and fetal islet transplants performed world-wide by December 31, 1994, if not differently indicated.

Materials and methods

Adult islet transplantation

(1) Islet autografts in pancreatectomized patients
From 1977 through June 10, 1995, 140 adult islet autografts were performed at 20 different institutions in pancreatectomized patients suffering from small duct chronic pancreatitis (Table 1). In the vast majority of cases, unpurified islet tissue was implanted into the liver via the portal route.

(2) Islet allografts in pancreatectomized patients
Between 1990 and 1994, a total of 14 adult islet allografts have been performed in pancreatectomized recipients (cases included in Table 2). Islets were

Table 1. Summary of islet autografts according to institution and year through June 10, 1995. Data compiled from the literature and data communicated to the Registry. * Year published.

Center	Year of Tx	No. of Cases
Minneapolis	1977–1995	48
Los Angeles UCLA	1977–1992	5
Baltimore	1978–1981	8
New York	1978	1
Loma Linda	1979–1980	5
Genova	1979–1983	24
Berlin (Charité)	1979–1982	12
Detroit	1979–1984	8
Giessen	1980–1982	4
Boston	1981	5
Helsinki	1982–1983	2
Nagasaki	1982–1984	5
Paris	1984	5
Leningrad	1987	1
St. Louis	1988	2
Philadelphia	1989	2
Geneva	1991–1994	5
Pittsburgh	1991	2
Gent/Giessen	1993	1
Leicester	1994–1995	6
Total no. of cases:		140
Total no. of institutions:		20

implanted into the simultaneously transplanted liver following upper abdominal exenteration in patients suffering from primary or secondary hepato-biliary malignancies. In about 50% of the cases, islets were derived from the liver donor only, and in the remaining cases islets isolated from two donor pancreases have been used.

(3) Islet allografts in type 1 diabetic patients

From December 12, 1893, through December 31, 1994, 243 adult islet allografts and one adult islet xenograft (sheep to man) including historical cases have been performed in diabetic patients at 34 institutions world-wide (Table 2). By continent, 144 transplants were performed at 13 institutions in North America, 99 at 20 institutions in Europe, and one elsewhere. Unfortunately, some transplants were also performed in patients with non-insulin-dependent diabetes or in patients with insulin-dependent diabetes but significant residual C-peptide secretion. Therefore, the most relevant analysis was performed on 75 well-documented pretransplant C-peptide negative type 1 diabetic patients (basal and if available stimulated plasma C-peptide concentration $\leq$ 0.2 ng/ml) who received adult islet allografts between 1990 and 1993. It is assumed that nearly all islet allografts performed world-wide during this period were reported to the Registry.

Table 2. Summary of adult islet allografts (and one xenograft*) according to institution and year through December 31, 1994.

Institution	Year of Tx	No. of Cases
Bristol*	1893	1
Newcastle-upon-Tyne	1916	2
Padova	1927	2
New York	1935	1
Leiden	1944	1
Petah Tiqwah	1968	1
Minneapolis	1974–1994	41
Zurich	1977–1988	8
Genova	1978–1979	12
Hannover	1978	2
Detroit	1980–1985	7
Giessen	1980–1994	12
Berlin (Charité)	1982–1987	8
St. Louis	1985–1994	27
Miami	1985–1994	16
Paris	1988–1991	7
Perugia	1989–1994	6
Berlin (Moabit)	1989	1
Edmonton	1989–1994	6
Milan	1989–1994	19
St. Louis/London, Ontario	1990–1992	4
Pittsburgh	1990–1994	34
Leicester	1991–1994	3
Oxford	1991–1994	4
Charlestown	1991	2
Los Angeles (UCLA)	1992–1993	4
Madrid	1992–1994	4
Los Angeles (St. Vincent Med. Center)	1993–1994	2
Giessen/Wuerzburg	1993	1
Verona/Padua	1993	1
Homburg	1993	1
London, Ontario	1993	1
Brussels	1994	1
Omaha	1994	1
Total no. of cases:		244
Total no. of institutions:		34

Table 3. Summary of fetal islet allografts reported to the Registry according to institution and year through December 31, 1994.

Center	Year of Tx	No. of Cases
Genova	1979–1980	13
Shanghai	1981–1988	75
Stockholm	1982–1984	3
Szeged	1982–1992	25
Sydney	1983–1986	5
Melbourne	1984–1986	8
Albany	1985	1
Dallas	1985	6
Denver	1985–1987	16
Belgrade	1986–1994	10
Moscow	1988–1991	25
Total no. of cases:		187
Total no. of institutions:		11

Fetal islet transplantation

Probably more than 2,000 fetal islet allografts and xenografts have been accomplished world-wide of which the vast majority has been performed in Russia and China. Out of these, 187 fetal islet allografts have been reported to the Registry from 11 institutions (Table 3).

Results

Transplantation efficacy and the outcome of islet transplantation should be assessed in several ways: (1) patient life expectancy; (2) graft functional survival (i.e., insulin independence; significant C-peptide secretion); (3) normalcy of the patient's metabolic state; (4) impact on diabetic complications; and (5) patient's quality of life. Currently, the still limited number of adult and fetal islet allografts only allows to assess the efficacy of this appealing method for treating type 1 diabetes mellitus concerning its impact on patient survival, graft survival and metabolic control. No data are available to give an answer to the impact of islet grafts on diabetic complications and patient's quality of life.

Adult islet transplantation

(1) Islet autografts in pancreatectomized patients
The Registry just has begun to collect data on islet autografts. At present, the first analysis is not yet finished. But, taken from a very recently published study on 48 islet autografts performed at the University of Minnesota, Minneapolis, the long-term survival rate of autografts in terms of insulin independence was 34% with the longest graft function at present more than ten years posttransplant [15]. The main predictor of insulin independence was the number of islets transplanted (> 300,000 islets).

(2) Islet allografts in pancreatectomized patients
Nine of the 14 patients in whom islets were simultaneously allografted with the liver (SIL) became insulin-independent, 7 at the University of Pittsburgh, one at the University of Milan, and one patient transplanted in a cooperative study of our group at Giessen University with the surgical department at Wuerzburg University (Table 4). Nearly all of these patients eventually expired with recurrent metastatic disease while still being insulin-independent. The longest period of insulin independence was noted in the first Pittsburgh patient who died from recurrent malignancy while being insulin-independent almost five years following the simultaneous islet-liver transplant. At autopsy, insulin containing islets were readily detected in liver specimens.

(3) Islet allografts in type 1 diabetic patients
Nineteen type 1 diabetic patients are reported to be insulin independent for at least one week after adult islet allografts (Table 5). A one-year islet allograft survival in terms of insulin independence was observed in 12 cases. Insulin independence was achieved in islet-after-kidney (IAK) transplants, simultaneous islet-kidney (SIK) transplants, and simultaneous islet-liver (SIL) transplants (Table 6). The most effective site of implantation was the liver approached by the portal route.

A detailed analysis of the 75 pretransplant C-peptide negative type 1 diabetic patients allografted between 1990 and 1993 revealed a one-year patient survival rate, graft survival rate (basal plasma C-peptide concentration $\geq$ 1 ng/ml) and insulin independence of 95%, 28%, and 11%, respectively (Fig. 1). Fig. 2 illustrates the one-year course of serum creatinine following islet-after-kidney and simultaneous islet-kidney

Table 4. Insulin independence (> one week) after adult islet allotransplantation into patients with pancreatectomy-induced diabetes mellitus. Summary of cases through December 31, 1994.

Institution	Year of Tx	No. of Donors	IEQ/kg IEQ/kg	Recipient Category	Site of Tx	Induction Immunosuppression	Period of Insulin Independence (Months Post-Tx)		
Pittsburgh	1990	1	11,850	SIL	liver	FK 506 + P (single bolus)	2	–	58[a]
Pittsburgh	1990	1	6,568	SIL	liver	FK 506 + P (single bolus)	2	–	4[a]
Pittsburgh	1990	2	8,028	SIL	liver	FK 506 + P (single bolus)	2	–	15
Pittsburgh	1990	2	4,526	SIL	liver	FK 506 + P (single bolus)	5	–	15
Pittsburgh	1990	1	3,519	SIL	liver	FK 506 + P (single bolus)	3	–	10[a]
Pittsburgh	1990	2	15,447	SIL	liver	FK 506 + P (single bolus)	2	–	20[a]
Pittsburgh	1990	2	5,627	SIL	liver	FK 506 + P (single bolus)	1	–	15[a]
Milan	1992	2	11,526	SIL	liver	ALG (+P–C–A)	2	–	14
							21	–>	25[a]
Giessen/Wuerzburg	1993	2	9,571	SIL	liver	ALG + Anti–IL–2 (+P–C)	4	–	6

[a] died off insulin; [b] as of Dec. 31, 1994; IEQ = Islet Equivalents; SIL: Simultaneous Islet Liver; P: Prednisone; C: Cyclosporine A; A: Azathioprine.

Table 5. Insulin independence (> one week) after adult islet allotransplantation into type 1 diabetic patients (1). Summary of cases through December 31, 1994.

Institution	Year of Tx	No. of Donors Fresh	Cryo	IEQ/kg	Purity (%)	Period of Insulin Independence Days Post Tx					
Zurich[a]	1978	1	–	3,846[d]	1	290	–	590			
Paris[b]	1988	1	–	2,143[d]	80	214	–	1,495			
St. Louis	1989	2	–	12,661	95	10	–	25			
St. Louis	1990	1	+2	14,733	98	33	–	341			
St. Louis[c]	1993	1	+7	22,197	90	92	–>	500[e]			
St. Louis	1993	1	+2	26,494	87	274	–>	384[e]			
Edmonton	1990	1	+4	9,698	70	69	–	821			
Edmonton	1992	1	+5	9,867	55	155	–	166	+837	–>	984[e]
Milan	1990	1	–	10,767	95	120	–	330			
Milan	1990	2	–	8,607	75	60	–	1,178			
Milan	1991	1	+2	15,241	80	210	–	360	+480	–	635
Milan	1992	2	–	11,566	80	150	–>	835[e]			
Milan	1994	1	+2	15,945	50	42	–	90			
Miami	1990	3	–	18,700	55	42	–	78			
Miami	1990	3	–	18,891	50	87	–	125			
Mnpls	1992	1	–	7,882	1	326	–>	1,043[e]			
Mnpls[c]	1992	1	–	13,319	1	123	–	348			
Giessen	1992	1	–	6,158	92	400	–>	763[e]			
Verona/Padua[c]	1993	1	–	4,392	60	35	–>	532[e]			

[a] no C-peptide data; [b] cholangiocarcinoma, hemochromatosis, and type 1 diabetes; [c] pre-tx basal C-peptide > 0.2 ng/ml; [d] Islets/kg; [e] ongoing insulin independence as of Dec. 31, 1994.

Table 6. Insulin independence (> one week) after adult islet allotransplantation into type 1 diabetic patient (2). Summary of cases through December 31, 1994.

Institution	Year of Tx	Site of Tx	Recipient Category	# of Shared HLA–Ag AB	DR	Induction Immunosuppression	
Zurich	1978	spleen	SIK	0	0	ALG	(+P–A–Cyclophosph.)
Paris	1988	epipl. flap	SIL	1	1	ALG	(+P–C–A)
St. Louis	1989	liver	IAK	1/3	2/1	ALG	(+P)
St. Louis	1990	liver	IAK	1/2/2	1/1/0	ALG	(+P)
St. Louis	1993	liver	SIK	2 (fresh)	1 (fresh)	OKT3	(+P–C–A)
St. Louis	1994	liver	SIK	NA	NA	ALG	(+P–C–A)
Edmonton	1990	liver	SIK	3 (fresh)	0 (fresh)	ALG	(+P–C at day 10–A)
				1/0/2/0	0 (cryo)		
Edmonton	1992	liver	SIK	3	1 (fresh)	ALG	(+P–C at day 8–A)
				1/0/0/1/0	1/1/0/0/1 (cryo)		
Milan	1990	liver	IAK	1	0	ALG	(+P–C–A)
Milan	1990	liver	IAK	1/2	1/0	ALG	(+P–C–A)
Milan	1991	liver	SIK	NA	NA	ALG	(+P–C–A)
Milan	1992	liver	IAK	NA	NA	ALG	(+P–C–A)
Milan	1994	liver	IAK	NA	NA	ATG	(+P–C–A)
Miami	1990	liver	IAK	0/2/0	1/1/0	OKT3	(+P–C–A)
Miami	1990	liver	IAK	0/0/0	0/1/0	OKT3	(+P–C–A)
Mnpls	1992	liver	SIK	1	1	ALG	(+P–C–D)
Mnpls	1992	liver	SIK	2	0	ALG	(+P–C–A)
Giessen	1992	liver	IAK	2	1	ATG	(+P–C)
Verona-Padua	1993	liver	SIL	NA	NA		(P–C)

SIK: Simultaneous Islet Kidney; IAK: Islet after Kidney; P: Prednisone; C: Cyclosporine A; A: Azathioprine; D: Deoxyspergualin.

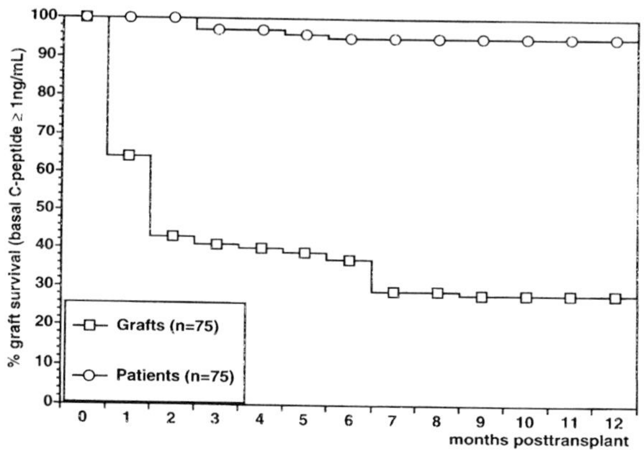

Fig. 1. One-year patient and islet allograft survival (basal plasma C-peptide ≥ 1 ng/ml) in 75 pretransplant C-peptide negative type 1 diabetic recipients (1990–93).

allografts in C-peptide negative type 1 diabetic recipients derived from data communicated to the Registry. Recipient age, sex, duration of diabetes, number of donor pancreases and islet purity did not significantly influence one-year graft survival rates.

However, insulin independence was achieved only if common characteristics identified in the individual protocols applied, were shared. These four common characteristicss are: (1) mean preservation time of the donor pancreas < 8 h; (2) islet mass ≥ 6,000 IEQ (islet equivalents) per kg body weight; (3) implantation site: liver via the portal route; (4) induction immunosuppression using ALG/ATG but not OKT3. Twenty-four of the 75 pretransplant C-peptide negative type 1 diabetic adult islet allograft recipients met all four characteristics. Twenty of these 24 patients (83%) demonstrated basal plasma C-peptide levels ≥ 1 ng/ml posttransplant (Table 7). At one-year follow-up, 11/24 (46%) had HbA1c levels < 7% and significant basal C-peptide secretion, respectively, and 7/24 (29%) were insulin-independent (Table 7). In this pre-selected group of patients, insulin independence (8/24) and insulin dependent recipients (16/24) did not significantly differ concerning age, body mass index, diabetes duration, pretransplant HbA1c, pretransplant daily insulin requirements, donor age, cold storage time, and IEQ/kg body weight; but, the former had significantly higher basal C-peptide levels at one

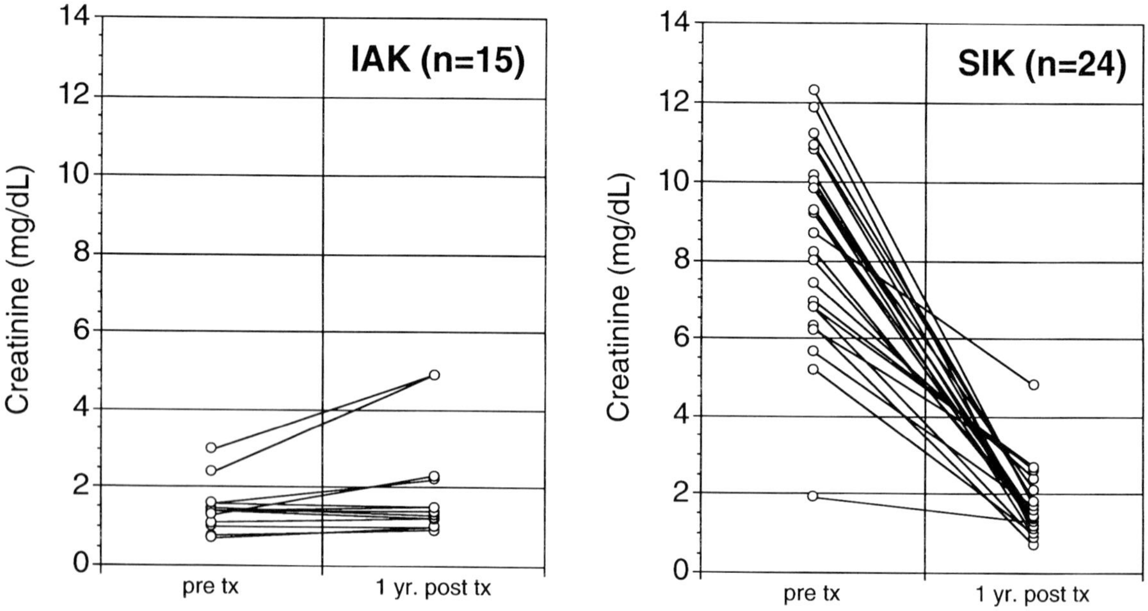

Fig. 2. One-year course of serum creatinine following IAK and SIK transplantation in C-peptide negative type 1 diabetic recipients. Δ serum creatinine levels both before and one year after islet allotransplantation were available in 15/25 IAK-recipients and in 24/43 SIK-recipients transplanted from 1990 to 1993.

Table 7. World-wide experience in islet allotransplantations 1990–93. Results in 24 C-peptide negative type 1-diabetic recipients fulfilling the four common characteristics of insulin independent recipients.

Success Rate	≥ 1 month %	post islet tx	≥ 1 year %
Basal C-peptide			
≥ 1 ng/ml	83		46
HbA1c < 7%			46
Insulin Independent	33		29

month posttransplant (2.7 ± 0.5 vs 1.4 ± 0.4 ng/ml; $p = 0.045$) and at one year posttransplant (2.4 ± 0.1 vs 0.5 ± 0.2 ng/ml; $p = 0.0001$), respectively.

By contrast, basal C-peptide levels ≥ 1 ng/ml and insulin independence were observed in only 20% and 1%, respectively, when at least one criterion was not fulfilled.

Current adult islet allotransplant activities are depicted in Table 8. Due to the prevailing activities in 1995, a doubling in the number of islet transplantations per year may be expected for the first time within five years.

Table 8. Islet transplant activities in 1994 and in 1995. Number of adult islet allografts according to institution and year from Jan 1, 1994, through June 10, 1995.

	94	01-05/95	Σ
Brussels	1	–	1
Edmonton	1	1	2
Giessen	5	8	13
London*	1	–	1
Los Angeles II	1	–	1
Madrid	1	–	1
Miami	3	6	9
Milano	4	2	6
Milano/Innsbruck	–	1	1
Milano/Nantes	–	1	1
Milano/Odense	–	5	5
Minneapolis	2	5	7
Omaha	1	–	1
Oxford	1	–	1
Perugia	1	–	1
Pittsburgh	4	–	4
St. Louis	2	–	2
$\Sigma n = 17$	28	31	59

* Ontario, Canada.

Fetal islet transplantation

Fetal islet transplant efficacy in terms of lowest daily insulin requirement and highest basal C-peptide *with-*

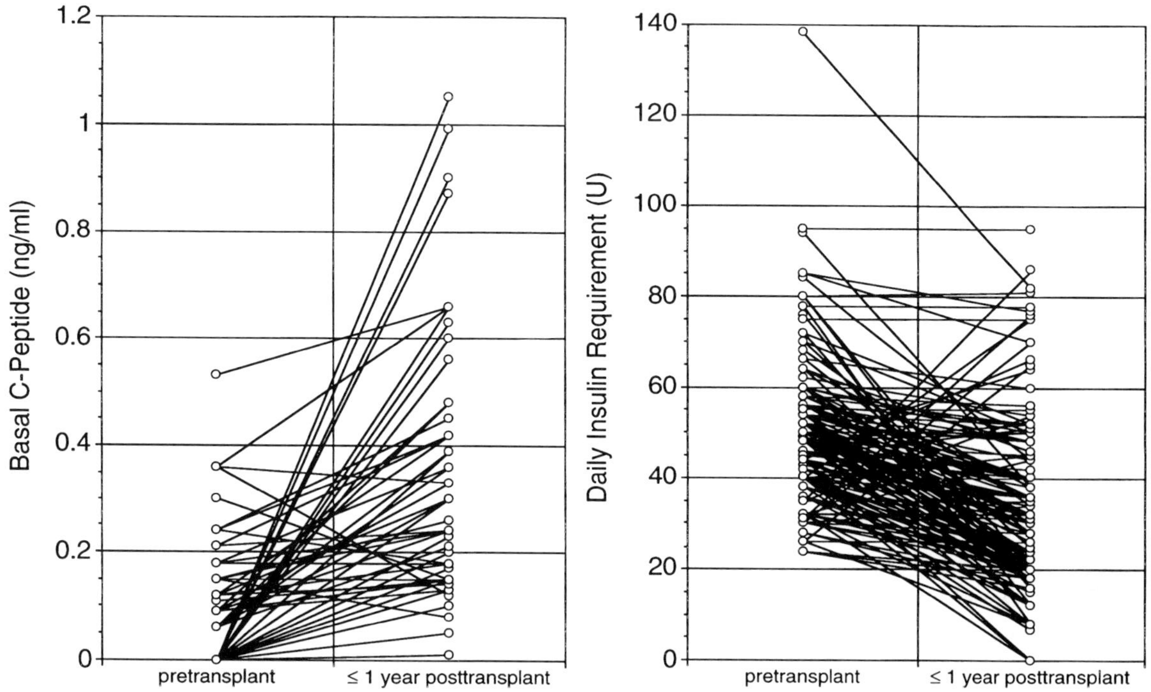

Fig. 3. Fetal islet allotransplant success in terms of highest basal plasma C-peptide and lowest daily insulin requirement and observed within one year posttransplant.

in one year posttransplant, available from 157 and 75 patients, respectively, are illustrated in Fig. 3. Pretransplant C-peptide levels in the three patients who became insulin independent after fetal islet allotransplantation (right panel of Fig. 3) were not determined. So far, insulin independence has not been achieved in a well-documented C-peptide negative type 1 diabetic recipient. However, recent data reported to the Registry demonstrate basal C-peptide levels exceeding 0.5 ng/ml in eight recipients without residual pretransplant C-peptide secretion (left panel of Fig. 3) lasting for periods of 2 to > 48 months.

Conclusions

Transplantation of isolated pancreatic islets appears to be the most direct and appealing effort for treating type 1 diabetes mellitus. It offers potential advantages over pancreatic organ transplantation. It is a minor rather than major procedure and proved of being safer than a pancreas transplantation. The principal feasibility of successful adult islet transplantation has clearly been demonstrated by long-lasting insulin independence following islet autografts and allografts in pancreatectomized and type 1 diabetic patients. But, insulin independence was achieved in only a few cases whereas significant (basal and stimulated) C-peptide secretion could be re-established in a reasonable number of cases. The hypothesis that an islet transplant providing continued C-peptide secretion thus allowing reduction but not complete withdrawal of exogenous insulin supply may exert remarkably beneficial effects was recently further substantiated. It could be demonstrated, both in animal experiments and in a prospective clinical study, that C-peptide administered subcutaneously to subjects with type 1 diabetes significantly ameliorated the course of or even reversed early diabetic secondary complications [16, 17].

Many questions are still to be addressed to augment the probability of achieving prolonged insulin independence after transplantation of adult islet allografts, and even more when fetal islet tissue is used. The strikingly different outcome in islet autograft recipients, in pancreatectomized islet allograft recipients, and in type 1 diabetic islet allograft recipients reveals once more the barriers preventing consistent success in treating type 1 diabetics with adult islet allografts: graft rejection, autoimmunity, and insulin resistance. Finally, clinical islet allotransplantation will only achieve significant levels of success in the not-too-distant future if (1) methods for the early detection of islet rejection episodes can be developed, and if (2) immunosuppressive regimens can be elaborated protecting islet allografts likewise from classic T-cell mediated-rejection/nonspecific inflammatory reaction and recurrence of autoimmunity, but which will not significantly impede insulin secretion and insulin action.

References

1. The Diabetes Control and Complications Trial Research Group (1993) The effect of intensive treatment of diabetes on the development and progression of long-term complications in insulin-dependent diabetes mellitus. N Engl J Med 329: 977–986.
2. Wang PH, Lau J and Chalmers TC (1993) Meta-analysis of effects of intensive blood glucose control on late complications of type I diabetes. Lancet 341: 1306–1309.
3. Bretzel RG, Hering BJ, Stroedter D, Zekorn T and Federlin KF (1992) Experimental islet transplantation in small animals. In: Ricordi C (ed) Pancreatic Islet Cell Transplantation, pp 249–260. Austin: R.G. Landes Company.
4. Ricordi C, Lacy PE, Finke EH, Olack BJ and Scharp DW (1988) Automated method for isolation of human pancreatic islets. Diabetes 37: 413–420.
5. American Diabetes Association (ADA) (1992) Pancreas transplantation for patients with diabetes mellitus. Diab Care 15: 1668–1672; ibid 1673.
6. Scharp DW, Lacy PE, Santiago JV, McCullough CS, Weide LG, Falqui L, Marchetti P, Gingerich RL, Jaffe AS, Cryer PE, Anderson CB, Flye MW (1990) Insulin-independence after islet transplantation into type I diabetic patient. Diabetes 39: 515–518.
7. Socci C, Falqui L, Davalli AM, Ricordi C, Braghi S, Bertuzzi F, Maffi P, Secchi A, Gavazzi F, Fresch M, Magistretti P, Socci S, Vignali A, di Carlo V and Pozza G (1991) Fresh human islet transplantation to replace pancreatic endocrine function in type I diabetic patients. Acta Diabetol 28: 151–157.
8. Warnock G, Kneteman NM, Ryan EA, Rabinovitch A and Rajotte RV (1992) Long-term follow-up after transplantation of insulin-producing pancreatic islets into patients with Type I (insulin-dependent) diabetes mellitus. Diabetologia 35: 89–95.
9. Alejandro R, Burke G, Shapiro ET, Strasser S, Nery J, Ricordi C, Esquenazi V, Miller J and Mintz DH (1992) Long-term survival of intraportal islet allografts in type I diabetes mellitus. In: Ricordi C (ed) Pancreatic Islet Cell Transplantation, pp 410–413. Austin: R.G. Landes Company.
10. Gores PF, Najarian JS, Stephanian E, Lloveras JJ, Kelly SL and Sutherland DE (1993) Insulin independence in type I diabetes after transplantation of unpurified islets from single donor with 15-deoxyspergualin. Lancet 341: 19–21.
11. Bretzel RG, Hering BJ, Brandhorst D, Brandhorst H, Bollen CC, Raptis G, Helf F, Grossmann R, Rau W and Federlin K (1994) Insulin independence in type 1 diabetes achieved by intraportal transplantation of purified pancreatic islets. Diabetologia 37 (suppl 1): A38.
12. Stegall MD, Sutherland DER and Hardy MA (1988) Registry report on clinical experience with islet transplantation. In: Schilfgaarde R van and Hardy MA (eds) Transplantation of the Endocrine Pancreas in Diabetes Mellitus, pp 224–233. Amsterdam: Elsevier.
13. Bretzel RG, Hering BJ and Federlin KF (1992) Islet transplantation registry report 1991. Diab Nutr Metab suppl 1: 177–181.
14. Hering BJ, Schultz AO, Geier C, Bretzel RG and Federlin K (1995) International Islet Transplant Registry, Newsletter No. 6, Vol. 5.
15. Wahoff DC *et al.* (1995) Autologous islet transplantation to prevent diabetes after pancreatic resection. Ann Surg: in press.
16. Williamson J (1994) C-peptide ameliorates vascular and neural dysfunction induced by diabetes. Satellite Symposium "Advances in Diabetes Therapy: From Insulin to Glucoactive Peptides and Beyond" to the 15th International Diabetes Federation Congress, Kyoto, Japan, Nov 3–4.
17. Wahren J (1994) Experimental and clinical studies with C-peptide in type I diabetes. Satellite Symposium "Advances in Diabetes Therapy: From Insulin to Glucoactive Peptides and Beyond" to the 15th International Diabetes Federation Congress, Kyoto, Japan, Nov 3–4.

Section IX:

Neural Transplantation

R. P. Lanza and W. L. Chick (eds.), Yearbook of Cell and Tissue Transplantation 1996/1997, 163–173.

Chapter 16

Cell transplantation: Brain

William J. Freed
Section on Preclinical Neuroscience, Neuropsychiatry Branch, NIMH Neuroscience Center at St. Elizabeths, Washington, DC 20032 U.S.A.

This chapter will consider briefly the use of transplantation as a means of repairing CNS defects arising through trauma or disease. The initial use of neural transplantation as an approach to the treatment of disease came about because of the development of an understanding of the pathophysiology of Parkinson's disease, and a realization of the possibility that the defect might be repaired via transplantation. Adaptations of the principles thus derived have been applied, more or less successfully, to animal models of several additional disorders.

Repairing the brain with cell grafts can be approached in essentially three ways, although the distinction between the three is not always sharply defined. First, cells may be required to form new connections with the host brain, or in some other manner form an integral, structural part of the host brain, and produce a functional restitution thereby. There have now been many applications of neural transplantation in this category, the first and leading example of which continues to be Parkinson's disease. Secondly, cells may be required to produce a soluble chemical substance, usually either a neurotransmitter precursor or a growth factor, which exerts an effect on brain cells. This second idea has been applied experimentally in a number of different ways; some examples will be discussed in the present chapter. The use of adrenal chromaffin cells for transplantation in Parkinson's disease and chronic pain, which will be discussed in chapter 8 (Sagen, this volume) is another version of this second approach.

A third form of neural transplantation involves the creation of an improved substrate for the regrowth of neuronal processes. This approach was derived, fundamentally, from the observation that regeneration of severed axons occurs with great avidity in peripheral nerve, but not in spinal cord. The general approach of using transplantation to restructure CNS tissue, for the purpose of improving the substrate for neuronal growth processes, is discussed in chapter 17 (Spinal cord). Although there are possible applications to this form of neural transplantation in the brain as well, these will not be covered in this chapter.

Neural tissue transplantation is a vast topic, being the subject of numerous books and review articles [1–3]. The present chapter can provide only a very brief introduction to the topic, including selected and very limited citations. As the prototype disorder, Parkinson's disease will be emphasized. Some details will be presented, so that the reader may gain an appreciation for the complex issues which have arisen in this very extensive literature, especially those issues associated with considerations of clinical use. Very brief discussions of a few additional examples of the potential applications of neural tissue transplantation are also included. The technology of developing cell lines for use in neural transplantation will also be considered briefly. The emphasis is on the more significant background developments, although highlights from the past year are also included.

Parkinson's disease

Background Parkinson's disease continues to be the

leading candidate for application of neural transplantation to disease. The methodology which now receives the most emphasis, by far, involves the use of fetal tissue from the ventral mesencephalon, which includes the substantia nigra (SN). An alternative technique, involving the use of catecholaminergic chromaffin cells from the adrenal medulla is currently being used to only a very limited degree, being apparently less effective. Nonetheless, it might conceivably be resurrected if improved methods are developed (chapter 8, Sagen). In many respects, Parkinson's disease seems to be a special case, in which transplanted neurons can function in relative isolation. Nonetheless, neural transplantation in Parkinson's disease has many features in common with a number of other applications of neural transplantation.

The clinical features of Parkinson's disease are related to a loss of dopaminergic synapses in the striatum. Thus, the general idea has been to implant cells into the brain which can produce and deliver dopamine to these synapses. This can be accomplished, perhaps, either by implanting cells which form synapses with host neurons and deliver dopamine synaptically, or by implanting cells which simply produce and release dopamine into the extracellular space. In the latter case, dopamine would have to reach synaptic sites by diffusion. Although it would probably be "better" to deliver dopamine through actual regulated synaptic connections, dopamine is probably at least partially effective when delivered diffusely. Diffusion may supplement synaptic delivery, and may even occur to some degree in the normal brain. Additional recent ideas for influencing Parkinson's disease via transplantation include delivery of trophic factors, such as glial-derived neurotrophic factor, to save dopaminergic neurons from degenerating [4], or the use of Sertoli cells from the testes, which seem to produce a remarkable recovery of function after transplantation in an animal model of Parkinson's disease, possibly by promoting sprouting of host dopaminergic fibers [5].

Preclinical studies

Interest in the use of neural transplantation for Parkinson's disease began in 1979 and 1980, when several reports first suggested that neural transplants could be used to alleviate abnormalities related to destruction of the nigrostriatal dopamine system in rodent models [6–10]. The tissue used for transplantation was the ventral mesencephalon of fetal animals, which contains the developing substantia nigra (which will be referred to here as SN). In 1980, long-term studies showed that SN grafts could become integrated into the host brain on a more-or-less permanent basis, and that brain tissue grafts thus could become permanent, functional constituents of the brain of another animal [10].

Many studies of SN grafts have since expanded on these initial findings. Dopaminergic neurons can be transplanted into the striatum either as solid fragments or as suspensions of dissociated cells [8, 11, 12]. These transplanted dopaminergic neurons send neurites into the host brain, are spontaneously active, synthesize and release dopamine, and form new synapses with host neurons [13–17].

Initial studies of SN grafts employed rotational behavior tests, which measure the asymmetry induced by destruction of the nigrostriatal system on one side of the brain [18]. Animals receive unilateral lesions of the nigrostriatal dopamine system, and are then given drugs such as amphetamine or apomorphine to stimulate asymmetric movements. Grafts of fetal dopaminergic neurons can alleviate these rotational behavior abnormalities quite effectively. When other behavioral measures are used, some improvements are also observed, although grafts appear to be considerably less effective in improving other measures of functional deficits [11, 19–21]. In particular, bilateral SN lesions in rodents produce a syndrome of extreme debilitation involving inactivity, adipsia, aphagia, and loss of grooming and other behaviors. This behavioral bilateral-lesion syndrome can hardly be alleviated at all by SN grafts in adult animals. When SN grafts are produced in neonatal animals, however, they quite effectively protect against later induction of the bilateral lesion syndrome [22, 23].

Rat studies have proven to be generally accurate predictors of the results of clinical trials in humans. There are, however, several specific issues such as the optimal donor age for human fetal donor tissue for which rodent experiments are not adequate. Further development of procedures for human use has been approached in two ways. First, a number of studies have employed human fetal tissue for transplantation into rat hosts, with various forms of immunosuppression to promote xenograft survival [24–27]. These studies have defined several important factors: Most notably, it has been clearly demonstrated that human fetal tissue must be younger than 10 weeks, and preferably about eight weeks of fetal age for optimal survival of transplanted dopaminergic neurons [24, 25]. These studies have also indicated a number of additional important factors: For example, it appears that development of human fetal donor tissue occurs over a relatively protracted time course, as compared to rodent donor tissue. The degree of reafferentation of the host brain also appears to increase with the size of the donor species. In addition, these experiments have been used to resolve minor technical issues related to the transplantation of human fetal tissue [28].

Sub-human primates have also been used; however, the reasons for these experiments are slightly different, since primate models are particularly suited to development of procedures for the host surgery, as compared to properties of the graft tissue per se. Primate studies have also accomplished the important purpose of affirming the validity of the principle that fetal tissue grafts can be effective in alleviating the manifestations of SN lesions in higher animals [29–32]. Thus, because primate studies have been relatively successful, it has been possible to proceed to studies in humans with the knowledge that the principles learned from studies in rats; e.g., that SN grafts can be effective, are generally applicable in higher animals. Moreover, studies in primates can be used to lessen the requirements for controlled studies, and hence for performance of "sham" surgery in human subjects.

There are two or three significant potential limitations to fetal tissue transplantation for Parkinson's disease which must be considered. The first, and simplest, is that limited numbers of dopaminergic neurons survive transplantation. If one fetal SN were transplanted into one host, and all of the dopaminergic neurons survived, one would expect to need one fetal SN for transplantation into each host. It turns out that only about 10% of the transplanted dopaminergic neurons generally survive; because of this limited cell survival most clinical studies have used multiple fetal specimens for transplantation into each human graft recipient. A second limitation is that the ability of transplanted dopaminergic neurons to extend neurites into the host brain appears to be limited. Normally, even when good survival of transplanted dopaminergic neurons is obtained, the neurites produced by these cells extend only a part of the way into the host target tissue. This limitation has been thought to be related to limitations of the mature striatum as a target for neurite extension; indeed, immature striatal targets are more completely reafferented than the adult host [33, 34]. A third limitation is that the connections of transplanted dopaminergic neurons are incomplete. In the normal brain, dopaminergic neurons not only send outputs to the striatum, but also receive inputs from striatal neurons and other brain regions. Ectopic placement of grafted neurons inevitably limits integration of the transplanted neurons into the host brain, although some inputs from striatal neurons may develop [15].

Finally, it should be noted that despite the fact that clinical trials have already been initiated, animal experiments may suggest possibilities for improving the efficacy of SN grafts. For example, antioxidants have been found to improve survival of grafted neurons [35]. Another possibility that could potentially be used in human patients is co-transplantation of fetal SN with segments of peripheral nerve: Several experiments in rats suggest that this procedure can improve graft efficacy [36, 37]. Recently, Collier and coworkers [38] have reported that co-transplantation of fetal SN and peripheral nerve also improves graft efficacy in a primate model.

Clinical studies

There have now been a considerable number of clinical studies of transplantation of fetal SN, or ventral mesencephalon, in Parkinson's disease. Well over 200 patients world-wide have received grafts of fetal tissue. Details of all of the clinical trials cannot be described here; only a few of the major trials will be briefly described.

There are major variations in technique: One of the principal factors which probably influences the relative success is the age of the donor tissue. Since studies involving transplantation of human fetal tissue into rodent host animals strongly suggest that donors of fetal tissue for transplantation in Parkinson's disease models should be no older than nine weeks gestational age [24, 25], clinical studies in which more mature tissue has been used have probably obtained limited clinical improvement because of this factor.

In 1992, Widner, Lindvall, and coworkers [39] at Lund University reported on the effects of fetal tissue grafts in two patients with parkinsonism induced by the illicit drug 1-methyl-4-phenyl-1,2,3,6-tetrahydropyridine (MPTP). MPTP-induced parkinsonism has two advantages as a test for the clinical use of neural transplantation: One is that the disease is relatively stable, rather than progressive as in idiopathic Parkinson's disease, and a second is that the degeneration is much less widespread, being localized almost entirely to the dopaminergic neurons of the SN. Substantial improvements were seen in these patients. Of particular interest was the relatively slow development of improvements, especially in one of the patients, with much of the clinical improvement occurring subsequent to one year. Lindvall and coworkers [40, 41] also used similar methods for study the efficacy of fetal tissue transplants in idiopathic Parkinson's disease. Although little improvement was seen in the first few patients examined [40], later patients showed considerable improvements, some of which also developed gradually beyond one year after transplantation [41]. Improvements prior to one year were observed as well, but there is a possibility that these were due to non-specific changes, since improvements during the first six months have been seen in virtually all studies [42].

Another series of patients received fetal tissue transplants at Yale University [43]. This study was especially noteworthy for several reasons; first, because the procedures used were tested directly in primates before being used in humans, and also because of the experimental design, which included a comparison between a group of patients that received transplants and a "comparison" group that did not receive surgery. The comparison group was tested in parallel, and measures were used such as videotaping the comparison patients with their heads bandaged to allow for unbiased evaluation of the clinical changes. In this study, solid tissue fragments were used for transplantation. Interestingly, there were improvements in the comparison patients as well as in the patients that received transplants. For some measures of outcome, especially rating scales, the transplant patients improved only slightly more than the comparison group. Quantitiative tests of motor function, however, demonstrated distinctly larger improvements in the transplanted patients. The age of the donor tissue was variable, but in some cases the donors were near to the probable upper age limit for dopaminergic neuron survival. There was a single autopsy case, from the patient that received the oldest donor tissue, and only a single tyrosine hydroxylase (TH)-positive neuron was found [44].

The patients in this study were also systematically evaluated for the development of psychiatric disturbances following transplantation; few psychiatric problems were found during the immediate postoperative period, although three of the nine patients developed psychiatric disturbances (primarily depression) a number of months after surgery. This contrasts with adrenal medulla grafts, which tend to produce psychiatric and psychological disturbances shortly following surgery [45].

C. Freed and coworkers [46, 47] at the University of Colorado in Denver have transplanted fetal mesencephalic tissue in a large series of patients. The first patient received grafts of dissociated cells, while later patients received grafts which consisted of "noodles" of tissue extruded from a syringe. Significant improvements were also observed in some of these patients. Notably, the first patient also showed a slowly-developing long term improvement in walking, which began to appear about one year after transplantation. A controlled trial has been initiated using this methodology. For the controlled trial, the control group patients will receive skull drilling only, and will have the option of receiving active grafts after one year. The entire study will be blinded, so that neither the patients nor the physicians involved directly with the patients will know which patients received grafts. Results will not be available for some time.

One of the most interesting ongoing studies of fetal tissue transplantation will undoubtedly turn out to be the study which is being conducted at the University of South Florida, by Freeman and coworkers [48]. The technique employed involves transplantation of dissociated cells, from fetuses of eight weeks of gestation or earlier. Several fetuses were used for each recipient, with grafts distributed throughout the putamen. Tissue was kept in cold storage for up to 48 hours. Patients were immunosuppressed for six months after transplantation. As of the spring of 1994, five patients had received grafts. Preliminary clinical results suggest that substantial improvements were seen, especially in clinical status during "off" periods. Some of the improvements occurred rapidly within a few weeks after transplantation.

The reason for the current importance of this study is a chance occurrence: One patient in the South Florida study died of unrelated causes 18 months after transplantation, in such a manner that the brain was obtained for histological study very rapidly [49]. This patient had been evaluated clinically a few months before death, and had shown a considerable degree of improvement. Histochemical studies of the graft post-mortem revealed excellent graft survival; approximately 250,000 dopaminergic neurons were found. Reafferentation of the host brain was extensive. The graft was very well integrated into the host brain structure, and there was no evidence of graft deterioration or scar formation around the graft. The properties of this graft were similar to the best examples of grafts in both primates and lower animals, and suggest that the general approach being employed is reasonable, and that at least some of the ongoing clinical studies will be valid tests of the clincal efficacy of SN transplantation.

Alternative disease applications

Huntington's disease

The "next" disease for application of neural transplantation is often suggested to be Huntington's disease. The genetic basis for the disorder is now known, consisting of a trinucleotide repeat at the end of chromosome 4 [50]. Availability of this information has not as yet led to a treatment. In Huntington's disease, the pathological cause of the disorder is primarily a progressive loss of neurons in the striatum. Perhaps, replacement of these neurons by transplantation could improve or retard progression of the disorder. Moreover, there is little to be lost by trials of neural transplantation, as the disease is inexorably progressive with no effective treatment.

Grafts of striatal tissue have been shown to produce behavioral improvements in animal models of Huntington's disease [51, 52]. There are several general obstacles to clinical application. The major one is that treatment of Huntington's disease with neural transplants requires the transplant to do much more than for Parkinson's disease. Whereas in Parkinson's disease transplants perhaps need only produce dopamine, and perhaps deliver it at synapses to produce a therapeutic benefit, in Huntington's disease, intrinsic striatal neurons are lost. The medium spiny neurons of the striatum, which is the major cell type which degenerates, form important connections with other neurons in several different places in the brain. Since these connections form initially in developing animals, the conditions for their reformation do not exist in adult animals. The good news is that it may not be necessary to competely reconstruct these connections to get some degree of functional restitution. And, grafts of striatum into the striatum of animals with striatal lesions do form some connections with appropriate parts of the brain [53] and produce some degree of behavioral restitution [54].

A possibility with great potential is the use of cells which secrete trophic factors to prevent degeneration in Huntington's disease, rather than replacing cells which have been lost [55–57]. A study by Frim *et al.* [56] showed that the neurotoxic effects of quinolinic acid could be attentuated by the transplantation of fibroblasts which had been modified to produce NGF. The NGF-producing fibroblasts were implanted, and seven days later the animals received lesions of the striatum with quinolinic acid. When the animals were examined after 18 days, the lesions were much smaller in animals that had received NGF-producing fibroblasts as compared to controls. In the Frim *et al.* [56] study, the fibroblasts that were used were, however, capable of producing NGF for only short periods after transplantation. Another study by Emerich *et al.* [55] obtained a similar attenuation of striatal lesions, by transplanting encapsulated cells from a hamster kidney fibroblast cell line into which a human NGF gene had been introduced by calcium phosphate transfection. The methods employed in the latter study can result in continued NGF production over an extended time period after transplantation [58]. Since the degeneration which occurs in Huntington's disease occurs slowly, over a long period of time, a means of obtaining growth factor production over a period of years would be needed for clinical application. Also, the degree to which quinolinic acid lesions mimic the process of degeneration which occurs in Huntington's disease is not known, so that it is somewhat uncertain whether this approach would be effective in human patients.

Oligodendrocyte replacement

There has been some interest in treating demyelinating diseases by transplantation of undifferentiated oligodendrocytes or oligodendrocyte precursors into the site of demyelination. This approach might, conceivably be applicable to disorders in which localized demyelination leads to loss of function as, for example, in multiple sclerosis. Lachapelle *et al.* [59] showed that transplanted oligodendrocytes were capable of remyelinating CNS in shiverer mutant mice. Remyelination of spinal cord by transplanted oligodenrocytes can also result in enhanced conduction velocity in the myelin-deficient rat [60]. It has recently been shown that purified glial cells, oligodendrocyte precursors, and oligodendrocyte precursor cell lines can remyelinate the CNS in animals with demyelinating lesions or with genetic myelin deficiencies [61–63].

Cerebellar degeneration

The possibility of repairing cerebellar circuitry in cerebellar degeneration, although thought to be a remote possibility with regard to actual clinical use, has recently advanced markedly by studies in an animal model of Purkinje cell degeneration [64]. Transplants of Purkinje cells into the cerebellar cortex have been studied previously [65], but are only partially successful. Purkinje cells transplanted into the cerebellar cortex receive inputs from granule neurons, but are incapable of forming axonal output connections with the cerebellar deep nuclei. Transplants made in this manner are thus incapable of being functionally effective.

Triarhou [64] has recently reported that a more complete reconstruction of cerebellar circuitry can be obtained by transplanting fetal Purkinje cells into the cerebellar deep nuclei. When this is done, some of the transplanted Purkinje cells can produce axonal connections with the deep nuclei, so that the ouputs of the Purkinje cells are partially reconstituted. Moreover, some of the transplanted Purkinje cells also migrate superiorly into or towards the cerebellar cortex, so that they can also receive inputs from the granule cell axons. Thus, it may be possible to restore the inhibitory outputs from Purkinje cells to cerebellar deep nuclei by this method. It may be possible, therefore to obtain a degree of functional restitution following cerebellar injury with this technique [64].

Cell lines

Encapsulated cells

The PC12 cell line is a tumor cell line derived from a rat pheochromocytoma. Although these cells produce large amounts of dopamine, they are tumor cells, and would not be a candidate for transplantation. Moreover, since they are rat cells, they would be rejected in humans and could not be used in humans for that reason as well.

An interesting means of avoiding both of these problems has been described in several studies by Aebischer and coworkers [66]. These investigators have encapsulated PC12 cells in semi-permeable membranes, which permit nutrients and other substances to diffuse and to reach the cells, but accomplish the dual purposes of containing the cells, and thus preventing unlimited growth, and also preventing graft rejection. The interesting aspect of this technology is that if the capsules fail, in a human patient, the grafts would be rejected, since they are xenografts from rats. If a human tumor cell line were similarly used, the entire approach would of course be untenable, since escape of a single cell from the capsule would be capable of forming a lethal tumor. Some efficacy of encapsulated PC12 cells in a primate model has also recently been reported [67]. Considerable additional work will be required to demonstrate that this approach is safe, effective, and consistent over the long-term before it can be used in human patients. This approach has been used in other systems as well, especially for the delivery of growth factors into the brain, a situation where physical contact between transplanted cells and the host brain is not required [55, 58].

Non-immortalized cell lines

For many reasons, it would be desirable to be able to produce cells for use in neural transplantation in culture. This would require either propagation of primary cells by manipulation in tissue culture, or the use of immortalized cell lines. Recently, it has become possible to propagate cells from the CNS in tissue culture by manipulating the tissue culture conditions, without immortalization by viral oncogenes [68]. These techniques can be applied to fetal or immature CNS tissue, and also to tissue from adult animals, although this technique is more limited with the use of tissue from adults. In adults, the technique is generally restricted to the use of peri-ventricular tissue, which contains a population of cells which can replicate in culture. These cells can give rise to cells with glial and neuronal properties [69, 70]. It may be possible to propagate cells in tissue culture with useful properties through the use of these and similar techniques.

Immortalized cell lines

Immortalization of neurons essentially requires the used retroviral vectors [71]. Thus, the most common technique which has been used to immortalize neural cell lines for transplantation is retroviral transfer of SV40 large T antigen or the v-*myc* oncogene [72–76]. On the other hand it is possible to immortalize glial cells, which divide more readily in tissue culture, using a number of techniques including repeated passage in tissue culture [77] and introduction of oncogenes with plasmid vectors [78]. Immortalization with retroviral vectors requires a single round of replication for incorporation of the viral construct into the host cell DNA. Immortalization with plasmid vectors or other techniques usually requires multiple rounds of cell division in culture. Generally, therefore, cells are immortalized by dissociation and growth in tissue culture, under conditions which promote at least some division of the target cells.

Two contrasting approaches to development of immortalized cell lines for neural transplantation have emerged. One idea is that undifferentiated cell lines could be used, with the idea that these cells will differentiate into appropriate phenotypes *in vivo* following transplantation. To some degree, this does occur, and has been observed in several experiments [73, 74, 76, 79]. Multipotent immortalized neural precursor cells have recently been transplanted into the brain of an animal model of a lysosomal storage disorder and produced a widespread correction of the enzyme deficiency [80].

The second approach involves an emphasis on the replacement of particular types of cells, with particular and consistent phenotypes. This requires the development of cells which consituitively express the desired phenotype, so that the properties of interest are expressed irrespective of the prevailing conditions, both in vitro and in vivo. Immortaliztion of cells with specific neuronal phenotypes requires immortalization during the final round of cell division, which may be difficult to obtain in tissue culture. Generally, neurons are not committed to a particular phenotype until near the final cell division, so that neurons immortalized from early stages of fetal development will be multipotential. There are several examples of immortalized neurons and glial cells with specific phenotypes, for example, a GABAergic cell line described by Giordano *et al.* [72]. An interesting "intermediate" example is the RN33b cell line, developed by Whittemore and colleagues [74], which has a glutamatergic phenotype but also retains a considerable degree of plasticity. When these cells are transplanted into the brain, they can assume a number of different morphological phenotypes, depending on the exact area of the brain into which they are transplanted.

An approach which has received much attention is the use of a temperature-sensitive form of the SV40 large T antigen. It is thought that this can be used to produce cell lines which will divide in tissue culture at 33 °C, but will revert to a "mortal" state, and also cease dividing at the non-permissive 39.5 °C temperature. In practice, however, this does not always occur, and cell lines immortalized with the temperature-sensitive form of SV40 large T antigen may continue to express the immortal phenotype at 39.5 °C [81]. Certain cell lines immortalized with this oncogene do appear to differentiate at 39.5 °C [74].

Delivery of foreign gene products

One of the major areas in which neural cell transplantation is likely to be used in the future is as a means to deliver soluble chemical mediators to the brain on a continuous basis. Several studies have employed genetically altered cells as a means of delivering trophic factors into the brain, such as the studies of NGF delivery in models of Huntington's disease, discussed above, and in models of cholinergic degeneration [58, 82, 83]. One of the difficulties in delivery of a foreign gene product via transplanted cells is that the cells tend to gradually inactivate introduced genes following transplantation, probably due in part to gradual inactivation of the mouse maloney leukemia virus LTR promoter which is often used [84, 85].

A number of studies have also attempted to deliver neurotransmitters or the neurotransmitter precursor L-DOPA, through transplanted cells [86, 87]. In most cases, the production of L-DOPA after transplantation has been seen only for short time periods. Anton *et al.* [88] has used an immortalized cell line from the ventral mesencephahlon for subsequent modification to enable to cells to produce L-DOPA following transplantation for use in Parkinson's disease models. Initially, in this cell line expression of TH was quite minimal. TH production was increased by subsequent introduction of the TH cDNA with a retroviral vector. These cells were capable of expressing TH for at least two months after transplantation into the brain, and produced behavioral improvement in animal models on a relatively sustained basis [88].

One of the most effective methods described so far for delivering a foreign gene product into the brain involves the use of muscle cells. These cells appear to be relatively permissive to the introduction of foreign DNA. A TH cDNA in a plasmid vehicle was transfected into muscle cells using lipofection. The CMV promoter was used to drive TH expression. In this system, TH was stably expressed by the transplanted muscle cells for long periods: Muscle cells into which the TH cDNA was introduced were effective in a rat behavioral model for at least 10 weeks [89], which is longer than has been seen in most other studies.

Advances in basic science of cellular differentiation

A basic requirement of neural transplantation is the obtaining of cells with the capacity to differentiate in the appropriate direction. Crude dissection of appropriate cells from the fetal brain, although certainly effective, is probably not ultimately the most efficient means of obtaining usable cells. In addition, when cells are immortalized, they continue to cycle indefinitely. Differentiation requires cell cycle arrest in G0/G1 phase. In addition to cell cycle arrest, however, additional signals are usually required for differentiation and expression of appropriate phenotypes. It may ultimately be necessary to understand and to be able to apply these signals to cell lines in order to obtain populations of cells with phenotypes appropriate for neural transplantation.

One approach to the generation of cells is to exploit the factors which control cell differentiation during development, and to apply these factors in such a manner that usable cells can be generated in larger numbers from fetal brain tissue. Another possibility is that these methods could be used to direct the phenotypic development of immortalized cells.

For example, floor plate cells, present in the base of the brain during development, are capable for guiding the development of dopaminergic neurons [90]. It was shown that dopaminergic neurons develop in the vicinity of floor plate cells. Contact with floor plate cells was shown to be required for developing neurons to develop a dopaminergic phenotype. Therefore, a non-diffusable, contact-mediated factor which is present in floor plate cells embryonic age nine days or so has the ability to induce differentiation of cells into the dopaminergic neuronal phenotype [90].

Several transcription factors which exert an influence over whether developing cells with express neuronal phenotypes have also been identified. One of the most interesting is the neuron-restrictive silencer factor (NRSF), which binds to a DNA sequence which suppresses the transcription of a large number of neuron-specific genes [91]. This factor appears to play a major role in repressing the development of neuronal characteristics in non-neuronal cells. Ultimately, it may very well be possible to employ these or similar factors to guide the development of neurons which could then be used for transplantation.

Conclusions

It has not been possible to consider all of the potential applications of neural transplantation in this chapter. Numerous additional diseases are potential candidates for the use of neural transplantation: Examples include epilepsy, Alzheimer's disease, stroke, spinal cord injury, and neuroendocrine disorders. Neural transplantation has also been performed experimentally in many models for which no counterpart exists in human disease, and as such has been very useful as a technique to help understand the neural systems which are involved in certain behaviors [1–3, 92].

Especially insofar as the issue of applications to disease is concerned, neural transplantation initially seemed likely to be useful only in very restricted circumstances, such as Parkinson's disease and neuroendocrine disorders, in which secretion of biochemical mediators from transplanted cells was thought to be the principal requirement for transplant function. As more has been learned about technical factors and cellular mechanisms, additional strategies for employing neural transplants have developed. In the future, improvements in the understanding of cellular mechanisms of CNS dysfunction are likely to suggest further possibilities for employing neural transplantation as a treatment for CNS disorders.

References

1. Freed WJ (1993) Neural transplantation: Prospects for clinical use. Cell Transplantation 2: 13–31.
2. Koutouzis TK, Emerich DF, Borlongan CV *et al.* (1994) Cell transplantation for central nervous system disorders. Crit Rev Neurobiol 8: 125–62.
3. Dunnett SB, Bjorklund A (eds) (1994) Functional Neural Transplantation. New York: Raven Press.
4. Stromberg I, Bjorklund L, Johansson M *et al.* (1993) Glial cell line-derived neurotrophic factor is expressed in the developing but not adult striatum and stimulates developing dopamine neurons in vivo. Exp Neurol 124: 401–12.
5. Sanberg PR, Borlongan CV, Saporta S *et al.* (1995) Sertoli cells: An alternative cell source for neural transplantation in Parkinson's disease. Paper presented at American Society for Neural Transplantation, Vol. 2, p 30. Exp Neurol 135: 169.
6. Perlow MJ, Freed WJ, Hoffer BJ *et al.* (1979) Brain grafts reduce motor abnormalities produced by destruction of nigrostriatal dopamine system. Science 204: 643–7.
7. Bjorklund A and Stenevi U (1979) Reconstruction of the nigrostriatal dopamine pathway by intracerebral nigral transplants. Brain Res 177: 555–60.
8. Bjorklund A, Schmidt RH and Stenevi U (1980) Functional reinnervation of the neostriatum in the adult rat by use of intraparenchymal grafting of dissociated cell suspensions from the substantia nigra. Cell Tissue Res 212: 9–45.
9. Freed WJ (1983) Functional brain tissue transplantation: Reversal of lesion-induced rotation by intraventricular substantia nigra and adrenal medulla grafts, with a note on intracranial retinal grafts. Biol Psychiat 18: 1205–67.
10. Freed WJ, Perlow MJ, Karoum F *et al.* (1980) Restoration of dopaminergic function by grafting of fetal rat substantia nigra to the caudate nucleus: Long term behavioral, biochemical, and histochemical studies. Ann Neurol 8: 510–19.

11. Dunnett SB, Bjorklund A, Stenevi U and Iversen SD (1981) Behavioural recovery following transplantation of substantia nigra in rats subjected to 6-OHDA lesions of the nigrostriatal pathway. I. Unilateral lesions. Brain Res 215: 147–61.
12. Heim RC, Willingham G and Freed WJ (1993) A comparison of solid intraventricular and dissociated intraparenchymal fetal substantia nigra grafts in a rat model of Parkinson's disease: Impaired graft survival is associated with high baseline rotational behavior. Exp Neurol 122: 5–15.
13. Freed WJ (1991) Substantia nigra grafts and Parkinson's disease: From animal experiments to human therapeutic trials. Rest Neurol Neurosci 3: 109–34.
14. Freund TF, Bolam JP, Bjorklund A *et al.* (1985) Efferent synaptic connections of grafted dopaminergic neurons reinnervating the host neostriatum: A tyrosine hydroxylase immunocytochemical study. J Neurosci 5: 603–16.
15. Mahalik TJ, Finger TE, Stromberg I *et al.* (1985) Substantia nigra transplants into denervated striatum of the rat: Ultrastructure of graft and host connections. J Comp Neurol 240: 60–70.
16. Wuerthele SM, Freed WJ, Olson L *et al.* (1981) Effect of dopamine agonists and antagonists on the electrical activity of substantia nigra neurons transplanted into the lateral ventricle of the rat. Exp Brain Res 44: 1–10.
17. Zetterstrom T, Brundin P, Gage FH *et al.* (1986) In vivo measurement of spontaneous release and metabolism of dopamine from intrastriatal nigral grafts using intracerebral dialysis. Brain Res 362: 344–9.
18. Ungerstedt U and Arbuthnott GW (1970) Quantitative recording of rotational behavior in rats after 6-hydroxy-dopamine lesions of the nigrostriatal dopamine system. Brain Res 24: 485–93.
19. Dunnett SB, Bjorklund A, Stenevi U *et al.* (1981) Behavioural recovery following transplantation of substantia nigra in rats subjected to 6-OHDA lesions of the nigrostriatal pathway. II. Bilateral lesions. Brain Res 229: 457–70.
20. Dunnett SB, Whishaw IQ, Rogers DC *et al.* (1987) Dopamine-rich grafts ameliorate whole body motor asymmetry and sensory neglect but not independent limb use in rats with 6-hydroxydopamine lesions. Brain Res 415: 63–78.
21. Mandel RJ, Brundin P and Bjorklund A (1990) The importance of graft placement and task complexity for transplant-induced recovery of simple and complex sensorimotor deficits in dopamine denervated rats. Eur J Neurosci 2: 888–94.
22. Schwarz S and Freed WJ (1987) Brain tissue transplantation in neonatal rats prevents a lesion-induced syndrome of adipsia, aphagia, and akinesia. Exp Brain Res 65: 449–54.
23. Rogers DC, Martel FL and Dunnett SB (1990) Nigral grafts in neonatal rats protect from aphagia induced by subsequent adult 6-OHDA lesions: The importance of striatal location. Exp Brain Res 80: 172–6.
24. Brundin P, Barbin G, Strecker RE *et al.* (1988) Survival and function of dissociated rat dopamine neurons grafted at different developmental stages or after being cultured in vitro. Dev Brain Res 39: 233–43.
25. Freeman TB, Sanberg PR, Nauert GM *et al.* (1995) The influence of donor age on the survival of solid and suspension intraparenchymal human embryonic nigral grafts. Cell Transplantation 4: 141–54.
26. Stromberg I, van Horne C, Bygdeman M *et al.* (1991) Function of intraventricular human mesencephalic xenografts in immunosuppressed rats: An electrophysiological and neurochemical analysis. Exp Neurol 112: 140–52.
27. Van Horne C, Mahalik T, Hoffer B *et al.* (1990) Behavioral and electrophysiological correlates of human mesencephalic dopaminergic xenograft function in the rat striatum. Brain Res Bull 25: 325–34.
28. Brundin P and Sauer H (199?) Grafting human fetal brain tissue: A practical guide. In: Lindvall O, Bjorklund A and Widner H (eds) Intracerebral Transplantation in Movement Disorders, pp 171–82. Amsterdam: Elsevier.
29. Taylor JR, Elsworth JD, Roth RH *et al.* (1991) Grafting of fetal substantia nigra to striatum reverses behavioral deficits induced by MPTP in primates: A comparison with other types of grafts as controls. Exp Brain Res 85: 335–48.
30. Bakay RAE, Fiandaca MS, Barrow DL *et al.* (1985) Preliminary report on the use of fetal neural tissue transplantation and correct MPTP induced primate model of Parkinsonism. Appl Neurophysiol 48: 358–1.
31. Sladek JR, Collier TJ, Haber SN *et al.* (1986) Survival and growth of fetal catecholamine neurons transplanted into primate brain. Brain Res Bull 17: 809–18.
32. Annett LE, Martel FL, Rogers DC *et al.* (1994) Behavioral assessment of the effects of embryonic nigral grafts in marmosets with unilateral 6-OHDA lesions of the nigrostriatal pathway. Exp Neurol 125: 228–46.
33. De Beaurepaire R and Freed WJ (1987) Embryonic substantia nigra grafts innervate embryonic striatal co-grafts in preference to mature host striatum. Exp Neurol 95: 448–54.
34. Yurek DM, Collier TJ and Sladek JR Jr (1990) Embryonic mesencephalic and striatal co-grafts: Development of grafted dopamine neurons and functional recovery. Exp Neurol 109: 191–9.
35. Nakao N, Frodl EM, Duan W-M *et al.* (1994) Lazaroids improve the survival of grafted rat embryonic dopamine neurons. Proc Natl Acad Sci USA 91: 12408–12.
36. Collier TJ and Springer JE (1991) Co-grafts of embryonic dopamine neurons and adult rat sciatic nerve into the denervated striatum enhance behavioral and morphological recovery in rats. Exp Neurol 102: 76–91.
37. Van Horne, CG, Stromberg I, Young D *et al.* (1991) Functional enhancement of intrastriatal dopamine-containing grafts by the co-transplantation of sciatic nerve tissue in 6-hydroxydopamine-lesioned rats. Exp Neurol 113: 143–54.
38. Collier TJ, Elsworth JD, Taylor JR *et al.* (1994) Peripheral nerve-dopamine neuron co-grafts in MPTP-treated monkeys: Augmentation of tyrosine hydroxylase-positive fiber staining and dopamine content in host systems. Neuroscience 61: 875–89.
39. Widner H, Tetrud J, Rehncrona S *et al.* (1992) Bilateral fetal mesencephalic grafting in two patients with Parkinsonsim induced by 1-methyl-4-phenyl-1,2,3,6-tetrahydropyridine (MPTP). N Eng J Med 327: 1556–63.
40. Lindvall O, Widner H, Rehncrona S *et al.* (1992) Transplantation of fetal dopaminergic neurons in Parkinson's disease: One-year clinical and neurophysiological observations in two patients with putaminal implants. Ann Neurol 31: 155–65.
41. Lindvall O, Sawle G, Widner H *et al.* (1994) Evidence for long-term survival and function of dopaminergic grafts in progressive Parkinson's disease. Ann Neurol 35: 172–80.
42. Freed WJ (1990) Fetal brain grafts and Parkinson's disease. Science 250: 1434.
43. Spencer DD, Robbins RJ, Naftolin F *et al.* (1992) Unilateral transplantation of human fetal mesencephalic tissue into the caudate nucleus of patients with Parkinson's disease. New Engl J Med 327: 1541–48.

44. Redmond DE Jr, Leranth C, Spencer DD, Robbins R, Vollmer T, Kim JH, Roth R, Dwork AJ and Naftolin F (1990) Fetal neural graft survival. Lancet 336: 820–2.
45. Price LH, Spencer DD, Marek KL *et al.* (1995) Psychiatric status after human fetal mesencephalic tissue transplantation in Parkinson's disease. Biol Psychiatry: in press.
46. Freed CR, Breeze RE, Rosenberg NL *et al.* (1990) Transplantation of human fetal dopamine cells for Parkinson's disease: Results at 1 year. Arch Neurol 47: 505–12.
47. Freed CR, Breeze RE, Rosenberg NL *et al.* (1992) Survival of implanted fetal dopamine cells and neurologic improvement 12 to 46 months after transplantation for Parkinson's disease. New Engl J Med 327: 1549–55.
48. Freeman TB, Sanberg PR, Snow BJ *et al.* (1994) The USF protocol for fetal nigral transplantation in Parkinson's disease. Exp Neurol 129: 6.
49. Kordower JH, Freeman TB, Snow BJ *et al.* (1995) Neuropathological evidence of graft survival and striatal reinnervation after the transplantation of fetal mesencephalic tissue in a patient with Parkinson's disease. N Eng J Med 332: 1118–24.
50. The Huntington's Disease Collaborative Research Group (1993) A novel gene containing a trinucleotide repeat that is expanded and unstable on Huntington's disease chromosomes. Cell 72: 971–83.
51. Giordano M, Houser SJ and Sanberg PR (1988) Intraparancymal fetal striatal transplants and recovery in kainic acid-lesioned rats. Brain Res 446: 183.
52. Isacson O, Dunnett SB and Bjorklund A (1986) Graft induced behavioral recovery in an animal model of Huntington's disease. Proc Natl Acad Sci USA 83: 2728.
53. Wictorin K (1992) Anatomy and connectivity of intrastriatal striatal transplants Prog Neurobiol 38: 611.
54. Bjorklund A, Campbell K, Sirinathsinghji DJ, Fricker RA and Dunnett SB (1994) Functional capacity of striatal transplants in the rat Huntington model. In: Dunnett SB and Bjorklund A (eds) Functional Neural Transplantation, pp 157–95. New York: Raven Press.
55. Emerich DF, Hammang JP, Baetge EE *et al.* (1994) Implantation of polymer-encapsulated human nerve growth factor-secreting fibroblasts attenutates the behavioral and neuropathological consequences of quinolinic acid injections into rodent striatum. Exp Neurol 130: 141–50.
56. Frim DM, Short MP, Rosenberg WS *et al.* (1993) Local protective effects of nerve growth factor-secreting fibroblasts against excitotoxic lesions in the rat striatum. J Neurosurg 78: 267–73.
57. Schumacher JM, Short MP, Hyman BT *et al.* (1991) Intracerebral implantation of nerve growth factor-producing fibroblasts protects striatum against neurotoxic levels of excitatory amino acids. Neuroscience 45: 561–70.
58. Winn SR, Hammang JP, Emerich DF *et al.* (1994) Polymer-encapsulated cells genetically modified to secrete human nerve growth factor promote the survival of axotomized septal cholinergic neurons. Proc Natl Acad Sci USA 91: 23–8.
59. Lachapelle F, Gumpel M, Baulac C *et al.* (1983) Transplantation of CNS fragments into the brain of shiverer mutant mice: Extensive myelination by implanted oligodendrocytes I. Dev Neurosci 6: 325–34.
60. Utzschneider DA, Archer DR, Kocsis JD *et al.* (1994) Transplantation of glial cells enhances action potential conduction of amyelinated spinal cord axons in the myelin-deficient rat. Proc Natl Acad Sci USA 91: 53–7.
61. Archer DR, Leven S and Duncan ID (1994) Myelination by cryopreserved xenografts and allografts in the myelin-deficient rat. Exp Neurol 125: 268–77.
62. Groves AK, Barnett SC, Franklin JM *et al.* (1993) Repair of demyelinating lesions by transplantation of purified O-2A progenitor cells. Nature 362: 453–5.
63. Tontsch U, Archer DR, Dubois-Dalcq M *et al.* (1994) Transplantation of an oligodendrocyte cell line leading to extensive myelination. Proc Natl Acad Sci USA 91: 11616–20.
64. Triarhou LC (1995) Cerebellar transplantation in hereditary ataxia and the recovery of function: Why do the deep cerebellar nuclei represent a better graft site than the cerbellar cortex. Paper presented at the American Society for Neural Transplantation, Vol. 2, No. S-16. Exp Neurol 135: 171.
65. Sotelo C and Alvarado-Mallart R-M (1988) Integration of grafted Purkinje cell into the host cerebellar circuitry in Purkinje cell degeneration mutant mouse. Prog Brain Res 78: 141–54.
66. Aebischer P, Tresco PA, Winn SR *et al.* (1991) Long-term cross-species brain transplantation of a polymer-encapsulated dopamine-secreting cell line. Exp Neurol 111: 269–75.
67. Aebischer P, Goddard M, Signore AP *et al.* (1994) Functional recovery in hemiparkinsonian primates transplanted with polymer-encapsulated PC12 cells. Exp Neurol 126: 151-8.
68. Reynolds BA and Weiss S (1992) Generation of neurons and astrocytes from isolated cells of the adult central nervous system. Science 255: 1707–10.
69. Reynolds BA, Tetzlaff W and Weiss S (1992) A multipotent EGF-responsive striatal embryonic progenitor cell produces neurons and astrocytes. J Neurosci 12: 4565–74.
70. Von Visger JR, Yeon DS, Oh TH *et al.* (1994) Differentiation and maturation of astrocytes derived from neuroepithelial progenitor cells in culture. Exp Neurol 128: 34–40.
71. Cepko CL (1989) Immortalization of neural cells via retrovirus-mediated oncogene transduction. Ann Rev Neurosci 12: 47–65.
72. Giordano M, Takashima H, Herranz A *et al.* (1993) Immortalized GABAergic cell lines derived from rat striatum using a temperature-sensitive allele of the SV40 large T antigen. Exp Neurol 124: 395–400.
73. Snyder EY, Deitcher DL, Walsh C *et al.* (1992) Multipotent neural cell lines can engraft and participate in development of mouse cerebellum. Cell 68: 33–51.
74. Whittemore SR and White LA (1993) Target regulation of neuronal differentiation in a temperature-sensitive cell lines derived from medullary raphe. Brain Res 615: 27–40.
75. Whittemore SR, Neary JT, Kleitman N *et al.* (1994) Isolation and characterization of conditionally immortalized astrocyte cell lines from adult human spinal cord. Glia 10: 211–26.
76. Renfranz PJ, Cunningham MG and McKay RDG (1991) Region-specific differentiation of the hippocampal stem cell line HiB5 upon implantation into the developing mammalian brain. Cell 66: 713–29.
77. Louis JC, Magal E, Muir D *et al.* (1992) CG-4, a new bipotential glial cell line from rat brain, is capable of differentiating in vitro into either mature oligodendrocytes of type-2 astrocytes. J Neurosci Res 31: 193–204.
78. Major EO, Miller AE, Mourrain P *et al.* (1985) Establishment of a line of human fetal glial cells that supports JC virus multiplication. Proc Natl Acad Sci USA 82: 1257–61.
79. Onifer SM, Whittemore SR and Holets VR (1993) Variable morphological differentiation of a Raphe-derived neuronal cell line following transplantation into the adult rat CNS. Exp Neurol 122: 130–42.
80. Snyder EY, Taylor RM and Wolfe JH (1995) Neural progenitor cell engraftment corrects lysosomal storage throughout the MPS VII mouse brain. Nature 374: 367–70.

81. Freed WJ, Giordano M, Takashima H *et al.* (19??) Conditional immortalization strategies: Cell lines derived using SV40 large T tsA58. Abstracts, American Society for Neural Transplantation, vol. 2, p 18. Exp Neurol 135: 164.
82. Ernfors P, Ebendal T, Olson L *et al.* (1989) A cell line producing recombinant nerve growth factor evokes growth responses in intrinsic and grafted central cholinergic neurons. Proc Natl Acad Sci USA 86: 4756–60.
83. Rosenberg MB, Friedman T, Robertson RC *et al.* (1988) Grafting genetically modified cells to the damaged brain: Restorative effects of NGF expression. Science 242: 1575–8.
84. Palmer TD, Rosman GJ, Osborne WR *et al.* (1991) Genetically modified skin fibroblasts persist long after transplantation but gradually inactivate introduced genes. Proc Natl Acad Sci USA 88: 1330–4.
85. Schnistine M and Gage FH (1993) Factors affecting proviral expression in primary cells grafted into the CNS. Res Publ Assoc Nerv Ment Dis 71: 311–23.
86. Fisher LJ, Jinnah HA, Kale C *et al.* (1991) Survival and function of intrastriatal grafts of primary fibroblasts genetically modified to produce L-DOPA. Neuron 6: 371–80.
87. Horrellou P, Brundin P, Kalen P *et al.* (1990) In vivo release of DOPA and dopamine from genetically engineered cells grafted to the denervated rat striatum. Neuron 5: 393–402.
88. Anton R, Kordower JH, Maidment NT *et al.* (1994) Neural-targeted gene therapy for rodent and primate hemiparkinsonism. Exp Neurol 127: 207–18.
89. Jiao S, Gurevich V and Wolff JA (1993) Long-term correction of rat model of Parkinson's disease by gene therapy. Nature 362: 450–53.
90. Hynes M, Poulsen K, Tessier-Lavigne M *et al.* (1995) Control of neuronal diversity by the floor plate: Contact-mediated induction of midbrain dopaminergic neurons. Cell 80: 95-101.
91. Schoenherr CJ and Anderson DJ (1995) The neuron-restrictive silencer factor (NRSF): A coordinate repressor of multiple neuron-specific genes. Science 267: 1360–3.
92. Dunnett SB and Richards S-J (eds) (1990) Neural Transplantation: From Molecular Basis to Clinical Applications. Amsterdam: Elsevier.

R. P. Lanza and W. L. Chick (eds.), Yearbook of Cell and Tissue Transplantation 1996/1997, 175–182.

Chapter 17

Neural transplantation: Spinal cord

Alan Tessler[1] & Marion Murray
Department of Anatomy and Neurobiology, Medical College of Pennsylvania and Hahnemann University, Philadelphia, Pennsylvania 19129 and [1] Philadelphia Veterans Administration Hospital, Philadelphia, PA 19104, U.S.A.

Several types of intraspinal transplant have effectively treated experimental models of human illnesses. Adrenal medullary allografts [1] and xenografts [2] placed into the subarachnoid space, for example, have reduced pain-related behavior in experimental peripheral neuropathy, presumably by releasing opioid peptides and catecholamines. Transplants of embryonic brainstem have also improved deficits in autonomic [3] and locomotor [4] function in experimental spinal cord injury, although the mechanism is unknown. The success of intraspinal transplants in treating spinal cord injury in laboratory animals has raised the possibility that transplants may one day be used as a treatment in humans.

Some features of spinal cord injury might be advantageous for treatment by transplants: (1) The success of transplantation replacement therapy for degenerative diseases may be limited by the progression of the disease. Spinal cord injury is an acute trauma that is unlikely to produce degenerative changes in transplanted tissue placed into the site of injury. (2) Most spinal cord injuries occur at cervical and thoracic levels and therefore spare the lumbar spinal cord which contains the circuitry that underlies hindlimb locomotion [5]. Treatment of these injuries will require preserving or enhancing the regeneration of axons originating from supraspinal neurons that initiate and modulate locomotion, but extensive reconstruction of intrinsic lumbar circuitry may not be necessary. (3) Most injuries are anatomically incomplete, so that even injuries that cause complete loss of hindlimb locomotion are likely to retain some anatomic continuity across the damaged segments [6]. Transplanted tissue may promote restoration of function using these spared systems. (4) There appears to be some redundancy in the populations of descending supraspinal and propriospinal axons that initiate and control locomotion [7, 8]. Therefore reconstructing anatomical continuity of all descending pathways may not be necessary to restore useful function.

Studies of intraspinal transplants have largely concentrated on demonstrating anatomic features of the integration of the tissue with the host, including survival, regeneration and synapse formation. Only limited evidence that transplants improve function after experimental injuries is available. The experience with transplants in brain, however, teaches that human subjects are likely to receive intraspinal transplants before the appropriate conditions or techniques have been fully defined. Spinal cord injured patients have received transplants of fetal spinal cord in countries outside the US, apparently without neurologic deterioration or increase in pain [9]. Because transplantation into the human spinal cord is likely to be imminent also in the US, an NIH-sponsored conference has been held to identify the syndromes most likely to benefit from transplantation and to discuss the assessment of the effects of transplants on locomotor function [9]. It is therefore timely to consider the mechanisms by which transplants might act to enhance recovery of function after spinal cord injuries and to evaluate some of the studies showing that transplants can enhance locomotor function in animals with spinal cord injury. Understanding the mechanisms by which transplants act should influence the choice of transplantation strategy in humans, particularly the choice of patients, timing of surgery and the most important characteristics of the tissue to be transplanted.

Transplant models

In addition to adrenal medulla and embryonic brainstem, three types of intraspinal transplant have been studied experimentally. All are designed to restore function by providing an environment that is enriched in neurotrophic and neurotropic substances which may promote neuron survival and regeneration.

(1) The peripheral nervous system (PNS) provides an appropriate environment for axonal regeneration. Grafts of peripheral nerve into the spinal cord in adults support axon growth within the grafts [10] for distances that can exceed the normal length of the axons [11]. Elongation is thought to be due to the presence of Schwann cells, which produce neurotrophic factors (reviewed in [12]) and synthesize and secrete elements of the extracellular matrix, including laminin and cell adhesion molecules upon which axons readily grow (reviewed in [13]). Grafts of Schwann cells [14] or implantation of guidance channels lined with Schwann cells [12] also support axon growth. The axons of retinal ganglion cells can extend through peripheral nerve grafts to their targets in the tectum, leave the grafts and establish functional synaptic contacts on postsynaptic tectal neurons [15]. Growth into host parenchyma beyond the graft is limited to 1 to 2 mm [16], however, and the grafts appear to support the long-term survival of only few axotomized neurons [17]. Furthermore, axons that do enter the graft may regress after a few months [18]. The environment provided by these PNS elements thus appears to be inadequate for permanent survival, and intraspinal growth is largely limited to the terrain provided by the Schwann cells. Long term survival seems to require formation of synapses upon adequate targets, perhaps because only appropriate targets produce adequate levels or specific types of trophic factors necessary for permanent survival. Recent evidence suggests that further outgrowth into host spinal cord can be promoted by administration of additional neurotrophic factors [19] or corticosteroids [20]. The importance of the peripheral nerve graft experiments is that they demonstrate conclusively that adult CNS axons are capable of robust regeneration when provided with an adequate environment. Better growth and improved survival appear to be promoted by additional interventions.

(2) A second approach is to transplant cells that provide specific factors which can support regeneration and survival of axotomized neurons. The technology has been developed that permits genetic engineering of cells and cell lines to produce specific factors. The feasibility of using genetically modified cells as intraspinal transplants has been demonstrated for fibroblasts genetically modified to express NGF [21] and other neurotrophic factors [22]. Fibroblasts genetically modified to produce neurotrophic factors and immortalized cell lines [23] will survive and integrate with adult host spinal cord. Cell lines derived from progenitor cells can develop into neurons and glial cells resembling those of the environment in which they are placed, including the spinal cord [24]. The use of genetically modified cells provides a very powerful approach toward reconstructing the injured spinal cord without the necessity of using fetal tissue, but this line of research is in an early stage of development. For example, the specific factors that are required need to be identified; not only is the number of these likely to be substantial, but both the temporal and dose requirements will also need to be elucidated.

(3) At present, fetal tissue has supplied the best evidence that spinal injury can be repaired so as to promote functional recovery. Transplants of fetal spinal cord (FSC) provide a more complex graft than either peripheral nerve or genetically modified cells. They integrate with host spinal cord and develop some of the morphological features of normal adult spinal cord [25] (Fig. 1). FSC transplant recipients have been shown to develop or recover locomotor performance to a degree that is superior to their control lesioned littermates. The following review emphasizes recent studies, centered on the use of FSC transplants, that consider the ways in which these transplants can repair injured cord and enhance useful function.

Mechanisms of action

At least four mechanisms have been proposed by which transplants may improve outcome after spinal cord injury. These are not mutually exclusive. One mechanism is by decreasing the astrocytic scarring that accompanies spinal cord injuries [26], either by preventing scar formation or by reducing an already established astrocytic scar [27]. Because the astrocytic scar has been postulated to act as a physical or chemical barrier to axon regeneration, reducing the scar might

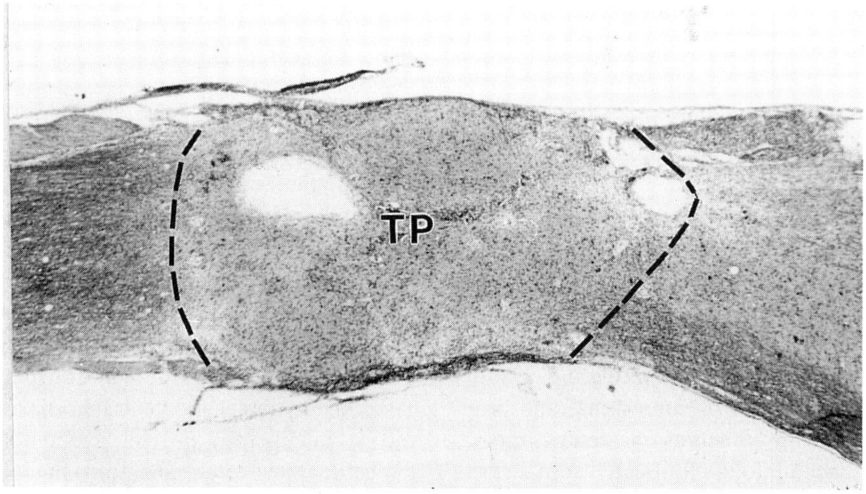

Fig. 1. Nissl myelin stained sagittal section through a fetal transplant (TP) placed in the site of a complete transection in an adult rat. Both rostral and caudal interfaces are seen (dotted lines). There is histological discontinuity between host and transplant and interruption of myelinated dorsal columns but good integration. The cellular organization within the transplant does not show the lamination characteristic of normal spinal cord, cysts may form and a modest astrocytic scar is present at the rostral and caudal interfaces.

improve outcome by reducing a feature of the environment that is not favorable for axon regeneration.

A second mechanism is by rescuing axotomized CNS neurons that would otherwise die. Retrograde cell death of axotomized neurons is thought to result at least in part from deprivation of target-derived neurotrophic factors. In many systems replacement of these factors allows axotomized neurons to survive [28], and the available evidence supports the idea that supply of neurotrophic factors plays a role in transplant-mediated cell rescue. Studies carried out on axotomized rubrospinal neurons in newborn rats provided the first evidence of rescue by fetal CNS transplants [29, 30]. Transplants of FSC rescued virtually all of the 50–60% of red nucleus neurons that normally die following axotomy at the midthoracic level [29]. Transplants of glia and transplants of regions of embryonic brain that are not normal targets of red nucleus neurons transiently maintain axotomized rubrospinal neurons after midthoracic section in newborns, but only FSC transplants effect permanent rescue [30]. This result is consistent with the idea generated from peripheral nerve graft studies that requirements for temporary rescue may be more general, but that permanent rescue depends on neurotrophic factors synthesized by the normal target of axotomized neurons. This notion receives additional support from the finding that brain-derived neurotrophic factor (BDNF) administered via gelfoam pledget at the site of injury also rescues newborn rubrospinal neurons in this model, although fewer neurons are rescued than after FSC transplant [31]. FSC transplants also effect substantial long-term rescue of axotomized red nucleus neurons in adults [32].

Studies on transplant-mediated rescue of Clarke's nucleus neurons support and extend the observations made on red nucleus neurons. Injury at the T8 level in newborn rats interrupts the dorsal spinocerebellar tract and causes 40% of ipsilateral Clarke's nucleus neurons at the L1 level to die, and the same lesion in adults kills 30% of the neurons [33]. Transplants of FSC, embryonic cerebellum and neocortex enable the neurons to survive in both adults and newborns, but transplants of embryonic striatum are ineffective [33]. The three types of transplants that rescue Clarke's nucleus neurons express high levels of neurotrophin-3 (NT-3) mRNA at the time of transplantation [34], whereas embryonic striatum does not. It seems likely, therefore, that a trophic factor common to these three tissues, such as NT-3, is responsible for the neuron rescue in Clarke's nucleus. Consistent with this formulation is our finding that NT-3 administered by gelfoam pledget to the lesion site in adults produces long-term rescue of about half of the Clarke's nucleus neurons otherwise destined to die [35]. For both red nucleus neurons in newborns and Clarke's nucleus neurons in adults, however, FSC transplants effect greater rescue than administration of a neurotrophic factor alone by gelfoam pledget. The superiority of FSC transplants may be due to their providing a single factor in larger amounts or for a longer time than does neurotrophic factor released by a gelfoam pledget or survival may depend on multiple factors or several conditions satisfied by a transplant. It will be of great interest to determine whether survival can be still further enhanced by the continuous supply of neurotrophic factors through an indwelling pump or by implantation of cells engineered to produce neurotrophic factors.

Transplants might also act by providing a substrate that allows the axons of supraspinal and propriospinal neurons to grow across an area of injury. FSC transplants placed into the transected spinal cord of newborns, for example, serve as bridges for the axons of corticospinal and raphespinal neurons, some of which continue to grow in host spinal cord caudal to the transection (reviewed in [36]). Axons regenerating into FSC transplants placed into adult hosts originate almost exclusively in neurons whose cell bodies are located within a few mm of the graft-host interface [37], and extension into caudal spinal cord has not been reported. Sparse regeneration in adults may be due to the limited regenerative capacity of adult CNS neurons. Grafts of peripheral nerve [38] and Schwann cells [12, 14] support axon regeneration from propriospinal neurons whose cell bodies are more distant than those that regenerate into FSC transplants. When Schwann cell grafts are supplemented by infusions of neurotrophins [19] or by systemic administration of methylprednisolone [20], brainstem neurons also extend axons into the grafts although the extent to which these axons regenerate into caudal host spinal cord remains to be determined. In addition, the fetal tissue continues to develop once transplanted into the host. The terrain provided by the maturing transplant may rapidly lose its ability to support regeneration and to synthesize molecules that promote, inhibit or repulse elongating axons [39, 40].

A fourth possible mechanism is that transplants serve as relay sites in which host neurons establish connections via transplant neurons. Although there is limited evidence that intraspinal neurons can establish connections with transplants [37], relay formation has been demonstrated in the dorsal root-transplant system. The cut central axons of dorsal root ganglion (DRG) neurons are unable to regenerate into spinal cord [41] but they do grow into FSC transplants and establish synapses with transplant neurons that resemble those formed in the dorsal horn of normal spinal cord and appear to be permanent [42–46]. Electrical stimulation of host dorsal roots elicits synaptically driven activity from transplant neurons, confirming that these connections are functional [47–49] (Fig. 2). Connections are formed by axons that originate from small and medium-sized DRG neurons and that extend only about 1 mm from the dorsal root into the transplant. These results confirm the idea that transplants can support the afferent limb of a functional relay between host neurons. Tracing studies with anatomical techniques have shown that the axons of transplant neurons extend up to 5 mm from the transplant and terminate in both the dorsal and ventral horns of host spinal cord [37], and thereby provide a potential efferent limb. Whether these axons also form functional connections has not been studied, but this will be the critical next step in determining whether host neurons can establish functional relays via transplant neurons.

Transplant-mediated enhancement of locomotion

There is a long history in neuroscience of poorly substantiated claims that interventions such as transplants can produce functional recovery after spinal injury. This reflects both the importance of the issue and the difficulty of the experimental analyses. The results of animal studies have shown that undamaged pathways in the CNS are capable of considerable plasticity that can promote recovery of function (reviewed in [50]). It is essential that the experimental studies be carefully executed with specific attention paid to the extent of the lesion and the nature of the recovery. Before the conclusions can be accepted, detailed anatomical and behavioral evidence must be provided that excludes alternate compensatory responses. Nevertheless, there are now several experimental models of spinal cord injury in which FSC transplants have been shown to enhance locomotor development in newborn recipients or promote locomotor recovery in adults. For example, when compared with newborn kittens [51] or newborn rats [52, 53] with spinal cord transections alone, transected animals with transplants exhibit superior weight support and postural stability, and superior coordination between forelimbs and hindlimbs during treadmill and overground locomotion.

Quantitative tests of locomotor function have shown that the performance of transected newborn rats with FSC transplants exceeds that of transected rats

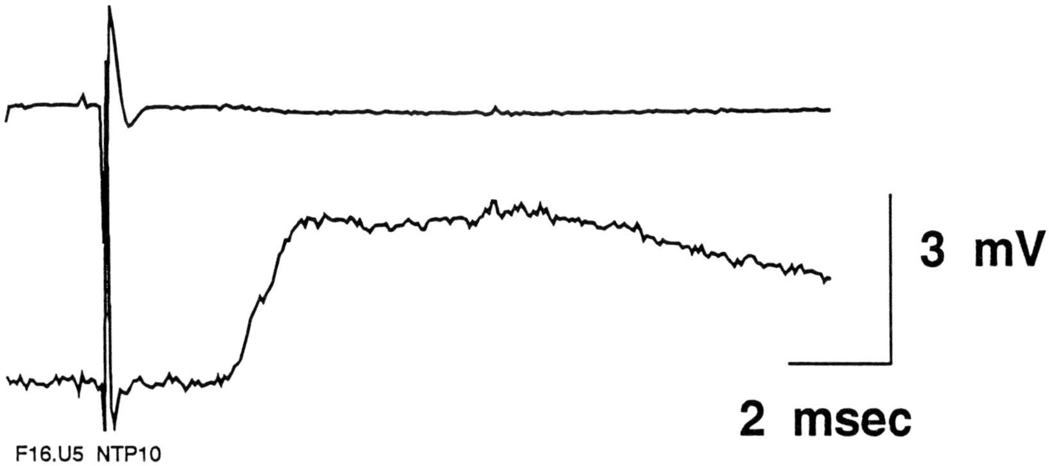

Fig. 2. Electrophysiological recordings from an animal that received a transplant about 2 months before the recording experiment. The upper trace shows the potential recorded at the junction between the juxtaposed root and the transplant following stimulation of the most peripheral part of the root at 50× threshold for the most excitable fibers in the root. While this stimulus intensity is high, the threshold for this EPSP was about 10 × threshold for the most excitable fibers, consistent with a small average fiber caliber. The lower trace shows the resulting EPSP recorded intracellularly from a neuron in the transplant. The initial upswing of the potential is the onset of the EPSP itself while the later, plateau-like potential probably represents currents initiated in the neuron as a result of the EPSP. The time calibration refers to both traces while the amplitude calibrations refer only to the intracellular trace.

without transplants. Enhanced performance is clearly apparent by 3 weeks of age and becomes further improved in the following weeks [52]. Differences are particularly apparent on the more demanding tests. For example, normal littermates and transected rats with transplants are able to make the large adjustments in postural control and forelimb-hindlimb coordination that are necessary to walk overground around a 90° corner, although the transected rats with transplants perform the task more slowly than their normal littermates. Transected rats without transplants rarely attempt the corner and fail when they do so because of their inability to support their hindlimbs and make the required postural adjustments. Rats with transplants seldom fall when challenged to walk down a stairway because their gait is stable, their postural adjustment adequate, most of the hindlimb steps are weight supported and the pattern of coordination between forelimbs and hindlimbs secure. In contrast, transected rats without transplants cannot use their hindlimbs to support their hindquarters and therefore drag their hindquarters down the stairway (Fig. 3). Neither the pathways nor the extent of regeneration required for the superior performance of transected rats with transplants is known in adequate detail. Immunocytochemical studies have shown serotonergic axons within the transplants and caudal host spinal cord. Because these axons originate in host brainstem neurons, this result raises the possibility that supraspinal or propriospinal pathways contribute to the enhanced locomotor development. Some results of our locomotor studies are consistent with this idea. For example, transected rats with transplants use different patterns of gait to cross wide (12″) and narrow (2″) runways and are able to adapt the precision of their gait to meet the more challenging conditions of the narrow runway. One interpretation of this behavioral result is that rats with transection + transplant rely on segmental spinal mechanisms to cross the wide runway but are able to call upon descending systems when necessary. Tracing studies are in progress to identify the pathways that regenerate into and through the transplants and that may be responsible for the enhanced locomotion in newborn rats with transection and transplant.

FSC transplants also promote limited recovery in adults. Stride length and base of support, for example, recover in adult rats whose spinal cord has been incompletely injured by an electromechanical impact device [54] or by hemisection [36, 55]. We have also observed transplant-mediated recovery in adult rats with complete lower thoracic spinal transection. These rats began training on a battery of quantitative locomotor tests at 3 weeks of age, and their spinal cords were transected at 8.5 weeks. Spinalized rats

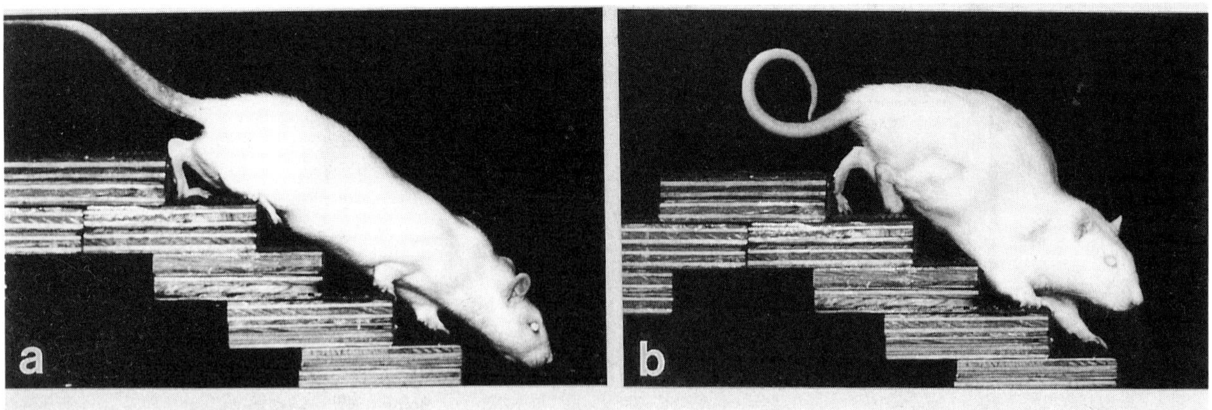

Fig. 3. a) Normal rat on stairs. Almost all steps by normal rats are weight supported during staircase descent. Two limbs are never supported on the same stair at the same time. b) Transplanted rat on stairs. Transplanted rats exhibit a majority of hindlimb weight supported steps during descent. There is considerable variability in their footfall pattern. In only some cases is there a diagonal footfall pattern similar to that used by a normal rat. The posture is hunched reflecting a shift of the center of mass forward.

with transplants recovered automatic bladder function within 2 days of surgery whereas spinalized rats without transplants did not recover bladder function until 10–13 days postoperative. Both groups of spinalized rats exhibited flaccid paralysis of the hindlimbs in the initial weeks after surgery, and the plantar surface of the foot was turned dorsal. During reflex walking on a treadmill supported by a sling at 2–3 weeks postoperative, rats with transplants began to place one or both hindfeet beneath their body with the toes flexed. Initially they exhibited this foot position only briefly but by the middle of the 4th week they maintained this posture most of the time. Transected rats without transplants do not recover either this foot placement or toe flexion. At 17 days postoperative both groups demonstrated vigorous kicking by the hindlimbs in response to intense tail pinch. By 1 month spinalized rats with transplants but not those without transplants made alternating stepping movements of the hindlimbs in response to mild tail pinch or to tail support. At this time rats with transplants began to make stepping movements which placed the plantar surface on the treadmill and no longer stepped with toes flexed. By the end of the 6th week most transplanted animals exhibited weight-supported steps during treadmill walking and some showed weight supported steps during overground locomotion. These results indicate that transplants contribute to locomotor recovery in rats with spinal transection as adults, although the extent of the recovery is limited. Alternating hindlimb movement and correct hindlimb posture recover, but weight support is only partial, postural adjustments are poor and there is little evidence for coordination between forelimbs and hindlimbs. In these animals, as in those operated as neonates, the transplants survived and integrated well with the host with minimal scarring. There is little evidence for regeneration through the transplant in adults, however, which may account for the more limited recovery of function in older animals. The role of the transplants in adult animals may lie in enhancing spinal reflexes sufficiently to provide weight support. These initial results suggest, however, that transplants are a promising approach toward treatment of spinal cord injury, perhaps as one component of a strategy that combines transplantation with training and pharmocological therapy (reviewed in [56]).

Acknowledgements

We are grateful to T. Eckenrode for preparation of figures and technical assistance. This research was supported by the Research Service of the Veterans Administration, NIH grant NS 24707, Paralyzed Veterans of America, American Paralysis Association and Allegheny Singer Research Institute.

References

1. Sagan J, Pappas GD and Perlow MJ (1986) Adrenal medullary tissue transplants in the rat spinal cord reduce pain sensitivity. Brain Res 384: 189–194.
2. Hama AT and Sagen JS (1994) Alleviation of neuropathic pain symptoms by xenogeneic chromaffin cell grafts in the spinal

subarachnoid space. Brain Res 651: 183–193.
3. Privat A, Mansour H and Geffard M (1988) Transplantation of fetal serotonin neurons into the transected spinal cord of adult rats: morphological development and functional influence. In: Gash DM and Sladek JR Jr (eds) Prog Brain Res, vol. 78, pp 155–166. New York: Elsevier.
4. Yakovleff A, Roby-Brami A, Guezard B *et al.* (1989) Locomotion in rats transplanted with noradrenergic neurons. Brain Res Bull 22: 115–121.
5. Calancie B, Needham-Shropshire B, Jacobs P *et al.* (1994) Involuntary stepping after chronic spinal cord injury. Brain 117: 1143–1159.
6. Kakulas BA (1988) The applied neurobiology of human spinal cord injury: a review. Paraplegia 26: 371–179.
7. Cheney PD, Fetz EE and Meives K (1991) Neural mechanisms underlying corticospinal and rubrospinal control of limb movement. Prog Brain Res 87: 213–252.
8. Houk JC, Keifer J and Barbo AG (1993) Distributed motor commands in the lumbar premotor network. TINS 16: 27–34.
9. Reier PJ, Anderson DK, Young W *et al.* (1994) Workshop on intraspinal transplantation and clinical application. J Neurotrauma 11: 369–377.
10. Aguayo AJ (1985) Axonal regeneration from injured neurons in the adult mammalian central nervous system. In: Cotman CW (ed) Synaptic plasticity, pp 457–484. New York: Guilford Press.
11. Aguayo AJ, Bjorklund A, Stenevi U *et al.* (1984) Fetal mesencephalic neurons survive and extend long axons across peripheral nervous system grafts inserted into the adult rat striatum. Neurosci Lett 45: 53–58.
12. Xu XM, Guenard V, Kleitman N *et al.* (1995) Axonal regeneration into Schwann cell-seeded guidance channels grafted into transected adult rat spinal cord. J Comp Neurol 351: 145–160.
13. Martini R (1994) Expression and functional roles of neural cell surface molecules and extracellular matrix components during development and regeneration of peripheral nerves. J Neurocytol 23: 1–28.
14. Martin D, Schoenen J, Delree P *et al.* (1993) Syngeneic grafting of adult rat DRG-derived Schwann cells to the injured spinal cord. Brain Res Bull 30: 507–514.
15. Sauve Y, Sawai H and Rasminsky M (1995) Functional synaptic connections made by regenerated retinal ganglion cell axons in the superior colliculus of adult hamsters. J Neurosci 15: 665–675.
16. Carter DA, Bray GM and Aguayo AJ (1994) Long term growth and remodeling of regenerated retino-collicular connections in adult hamsters. J Neurosci 14: 590–598.
17. Vidal-Sanz M, Bray GM and Aguayo AJ (1991) Regenerated synapses persist in the superior colliculus after the regrowth of retinal ganglion cell axons. J Neurocytol 20: 940–952.
18. Campbell G, Lieberman AR, Anderson PN *et al.* (1992) Regeneration of adult rat CNS axons into peripheral nerve autografts. J Neurocytol 21: 755–787.
19. Xu XM, Guenard V, Kleitman N *et al.* (1994) BDNF and NT-3 promote axonal regeneration of brainstem neurons into Schwann cell grafts in midthoracic spinal cord of adult rats. Soc Neurosci Abstr 20: 1111.
20. Chen A, Xu XM, Kleitman N *et al.* (1994) Methylprednisolone administration improves axonal regeneration into Schwann cell (SC) grafts in thoracic rat spinal cord. Soc Neurosci Abstr 20: 1111.
21. Tuszynski MH, Peterson DA, Ray J *et al.* (1994) Fibroblasts genetically modified to produce nerve growth factor induce robust neuritic ingrowth after grafting to the spinal cord. Exp Neurol 126: 1–14.
22. Tuszynski MH, Meyer S. Nakahara Y *et al.* (1994) Fibroblasts and Schwann cells genetically modified to secrete neurotrophic factors induce robust neuritic growth after transplantation to the spinal cord. Soc Neurosci Abst 20: 10.
23. Onifer SM, Whittemore SR and Holets VR (1993) Variable morphological differentiation of a raphe-derived neuronal cell line following transplantation into the adult rat CNS. Exp Neurol 122: 130–142.
24. Snyder EY (1994) Grafting immortalized neurons to the CNS. Curr Opin Neurobiol 4: 742–751.
25. Jakeman LB, Reier PJ, Bregman BS *et al.* (1989) Differentiation of substantia gelatinosa-like regions in intraspinal and intracerebral transplants of embryonic spinal cord tissue in the rat. Exp Neurol 103: 17–33.
26. Reier PJ, Houle JD, Tessler A *et al.* (1988) Astrogliosis and regeneration: new perspectives to an old hypothesis. In: Norenberg MD, Hertz L and Schousboe A (eds) The Biochemical Pathology of Astrocytes, pp 107–122. New York: Alan R Liss.
27. Houle JD and Reier PJ (1988) Transplantation of fetal spinal cord tissue into the chronically injured adult rat spinal cord. J Comp Neurol 269: 535–547.
28. Sofroniew MV, Cooper JD, Svendsen CN *et al.* (1993) Atrophy but not death of adult septal cholinergic neurons after ablation of target capacity to produce mRNAs for NGF, BDNF, and NT3. J Neurosci 13: 5263–5276.
29. Bregman BS and Reier PJ (1986) Neural tissue transplants rescue axotomized rubrospinal cells from retrograde death. J Comp Neurol 244: 86–95.
30. Bregman BS and Kunkel-Bagden E (1988) Effect of target and non-target transplants on neuronal survival and axonal elongation after injury to the developing spinal cord. Prog Brain Res 78: 205–211.
31. Diener PS and Bregman BS (1994) Neurotrophic factors prevent the death of CNS neurons after spinal cord lesions in newborn rats. NeuroReport 5: 1913–1917.
32. Mori F, Kowada M, Himes BT *et al.* (1995) Fetal spinal cord transplants rescue axotomized rubrospinal neurons from retrograde cell death in adult rats. Soc Neurosci Abstr 21: 1305.
33. Himes BT, Goldberger ME and Tessler A (1994) Grafts of fetal central nervous system tissue rescue axotomized Clarke's nucleus neurons in adult and neonatal operates. J Comp Neurol 339: 117–131.
34. Maisonpierre PC, Belluscio L, Friedman B *et al.* (1991) NT-3, BDNF, and NGF in the developing rat nervous system: parallel as well as reciprocal patterns of expression. Neuron 5: 501–509.
35. Shibayama M, Hattori S, Himes BT *et al.* (1995) Neurotrophin-3 (NT-3) prevents death of Clarke's nucleus neurons in adult rats. Soc Neurosci Abstr 21: 1057.
36. Bregman BS (1994) Recovery of function after spinal cord injury: transplantation strategies. In: Dunnett SB and Bjorklund A (eds) Functional Neural Transplantation, pp 489–529. New York: Raven.
37. Jakeman LB and Reier PJ (1991) Axonal projections between fetal spinal cord transplant and the adult rat spinal cord: A neuroanatomical tracing study of local interactions. J Comp Neurol 307: 311–34.
38. Houle JD (1991) Demonstration of the potential for chronically injured neurons to regenerate axons into intraspinal peripheral nerve grafts. Exp Neurol 113: 1–9.
39. Sugawara T, Itoh Y, Mori F *et al.* (1994) Fetal spinal cord transplants support dorsal root regeneration for the first few

days after implantation into adult spinal cord. Soc Neurosci Abstr 20: 469.
40. Sugawara T, Kowada M and Tessler A (1993) Relationship between extracellular matrix molecules and dorsal root axons regenerating into fetal spinal cord transplants. Soc Neurosci Abst 19: 679.
41. Pindzola RR, Doller C and Silver J (1993) Putative inhibitory extracellular matrix molecules at the dorsal root entry zone of the spinal cord during development and after root and sciatic nerve lesions. Dev Biol 156: 34–48.
42. Tessler A, Himes BT, Houle J *et al.* (1988) Regeneration of adult dorsal root axons into transplants of embryonic spinal cord. J Comp Neurol 270: 537–48.
43. Itoh Y and Tessler A (1990) Ultrastructural organization of regenerated adult dorsal root axons within transplants of fetal spinal cord. J Comp Neurol 292: 396–411.
44. Itoh Y and Tessler A (1990) Regeneration of adult dorsal root axons into transplants of fetal spinal cord and brain: a comparison of growth and synapse formation in appropriate and inappropriate targets. J Comp Neurol 302: 272–93.
45. Itoh Y, Sugawara T, Kowada M *et al.* (1992) Time course of dorsal root axon regeneration into transplants of fetal spinal cord: 1. A light microscopic study. J Comp Neurol 323: 198–208.
46. Itoh Y, Sugawara T, Kowada M *et al.* (1993) Time course of dorsal root axon regeneration into transplants of fetal spinal cord: an electron microscope study. Exp Neurol 123: 133–146.
47. Houle JD, Skinner RD and Turner KL (1991) Evoked synaptic potentials from regenerating dorsal root axons within intraspinal fetal spinal cord tissue transplants. Soc Neurosci Abstr 17: 568.
48. Reier PJ, Stokes BT, Thompson FJ *et al.* (1992) Fetal cell grafts into resection and contusion/compression injuries of the rat and cat spinal cord. Exp Neurol 115: 177–188.
49. Itoh Y, Tessler A, Kowada M *et al.* (1993) Electrophysiological responses in fetal spinal cord transplants evoked by regenerated dorsal root axons. Acta Neurochir Suppl 58: 24–26.
50. Goldberger ME, Murray M and Tessler A (1993) Sprouting and regeneration in the spinal cord: their roles in recovery of function after spinal injury. In: Gorio A (ed) Neuroregeneration, pp 241–264. New York: Raven.
51. Howland DR, Bregman BS, Tessler A *et al.* (1995) Transplants enhance locomotion in neonatal kittens whose spinal cords are transected. Exp Neurol 135: 123–145.
52. Miya DY, Clark B, Tessler A *et al.* (1994) Transplant mediated behavioral recovery of weight support & interlimb coordination in normal, spinal and spinal + transplant rats. Soc Neurosci Abstr 20: 1706.
53. Iwashita Y, Kawaguchi S and Murata M (1994) Restoration of function by replacement of spinal cord segments in the rat. Nature 367: 167–170.
54. Stokes BT and Reier PJ (1992) Fetal grafts alter chronic behavioral outcome after contusion damage to the adult rat spinal cord. Exp Neurol 116: 1–12.
55. Kunkel-Bagden E, Dai HN, Gao D *et al.* (1993) Fetal spinal cord transplants alter the recovery of locomotor function after spinal cord injury. Exp Neurol 123: 3–16.
56. Rossignol S and Barbeau H (1993) Pharmacology of locomotion: an account of studies in spinal cats and spinal cord injured subjects. J Amer Parapleg Soc 16: 190–196.

Section X:

Pituitary Cells

R. P. Lanza and W. L. Chick (eds.), Yearbook of Cell and Tissue Transplantation 1996/1997, 185–190.

Chapter 18

Pituitary cell transplantation

Zhong-ping Chen and Gérard Mohr
Division of Neurosurgery, and Lady Davis Institute for Medical Research, Sir Mortimer B. Davis – Jewish General Hospital, McGill University, Montréal, Quebec H3T 1E2, Canada

Although the transplantation of pituitary tissue dates back over 90 years, it was not until the 1950s that extensive studies were performed for the purpose of elucidating the gland's intrinsic physiology. During the last decade, however, interest in brain and nerve transplantation has reactualized the study of pituitary transplantation for therapeutic purposes. Recently, one of the most important findings was the demonstration that ectopic pituitary grafts could maintain their function *in vivo* [1–4]. Although functioning of the tissue was not totally physiological, numerous studies have demonstrated that pituitary tissue transplanted into extracerebral sites, such as under the skin, might provide a safe way to restore pituitary functions [1–6]. Alternatively, advances in pituitary hormone and hypothalamic peptide biosynthesis have improved treatment for hypophyseal insufficiency [7]. Unfortunately, long term observations revealed that treatment with human growth hormone (hGH) for dwarfism resulted in some cases in Creutzfeld-Jakob disease and leukemia [8–10]. Accordingly, pituitary cell transplantation may still prove to have therapeutic value, but, immune rejection remains a major obstacle. Successful immunoisolated cell transplantation has encouraged researchers to extend this technology to the field of pituitary cell transplantation [11–16], and may allow clinical applications in the future.

This chapter will review the latest contributions in the area of adenohypophyseal transplantation. Since the posterior pituitary lobe is mainly involved in the storage and release of oxytocin and vasopressin synthesized from the supraoptic and paraventricular nuclei, its transplantation is not feasible without a hypothalamic component [17] and will therefore not be included here.

The functioning pituitary graft

It has already been documented that pituitary tissue transplanted near the hypophysiotrophic area (HTA) can maintain normal function since hypothalamic peptides are able to diffuse to the gland [18, 19]. The prevailing view was that pituitary grafts transplanted to distant sites could not maintain normal hormonal secretory function without hypothalamic regulation. However, some investigators have recently reexamined the efficiency of ectopic pituitary grafts by directly evaluating pituitary hormone levels in the blood using radioimmunoassay (RIA). They found that pituitary tissue transplanted under the renal capsule promoted the growth of hypophysectomized rats with enhanced GH levels and were also helpful to restoring maternal behaviour [1]. Our group has implanted pituitary autografts subcutaneously in rats and observed the reestablishment of hormonal function as well as gain of body weight [4]. In recent years, clinical trails have been carried out in China using human fetal pituitary tissue. Intramuscular or subcutaneous implantation of the tissue induced growth in dwarf children and reduced the symptoms of hypopituitarism, at least for a short period (Bao Yaodong, personal communication 1995). Krieg *et al.* found that pituitary grafts implanted under the renal capsule responded to GH-releasing hormone (GRH) by increasing GH secretion [5]. Iturriza and Eberle reported that rats bearing kidney grafts of pituitary pars intermedia increased adrenocorticotropin (ACTH) secretion following chronic stress [2]. Van Dieten *et al.* found that pituitary grafts transplanted under the renal capsule were secreting luteinizing hormone (LH) influenced by administration of LH-releasing hormone (LHRH) [20]. However, whether the ectopic pituitary function is under the control of the hypothalamus, and if so, through which pathways remains largely unknown.

The levels of hypothalamic peptides are quite low in the systemic circulation of normal subjects and only after transport to the hypophysis via portal vessels can they modulate pituitary function. In hypophyseal insufficiency, such as in hypophysectomized animals, the hypophysiotrophic peptides are increased because of a lack of negative feedback. Nevertheless, it is unclear whether this enhancement of hypothalamic peptides in the circulation is high enough to regulate the function of ectopic pituitary grafts.

The pituitary gland itself may also adjust its own function since pituitary cells can secrete a wide variety of bioactive peptides including trophic peptides such as thyrotropin-releasing hormone (TRH) and glanin [21]. Recent studies on paracrine and autocrine interactions within the anterior pituitary evidenced a growing number of substances that may play a role in cell to cell communication [22]. It is possible that in pituitary hormone deficiency, due to a so-called ultrashort feedback [6], pituitary grafts secrete more trophic peptides which in turn stimulate the pituitary cells to secrete the appropriate hormones.

An increasing number of reports suggest that there are bidirectional interactions between the immune and the endocrine systems [23]. Immunocompetent cells can synthesize and secrete numerous pituitary hormones, including GH, prolactin (PRL), thyroid-stimulating hormone (TSH), ACTH, LH and follicle-stimulating hormone (FSH). However, the role of these hormones in influencing the function of other cells is controversial [24–26]. In 1982, Smith first reported that virus-infected leucocytes could release sufficient ACTH to stimulate steroidogenesis in hypophysectomized mice [24]. Others failed to demonstrate an increase in plasma corticosterone levels in hypophysectomized mice or rats after virus infection [25]. We hypothesize that pituitary grafts could also serve as a stimulator to the immune system and activate it to secrete pituitary hormones. If this is true, will it restore normal pituitary function?

Ectopic pituitary graft and hyperprolactinaemia

All pituitary hormones are secreted in response to specific releasing hormones from the hypothalamus, with the exception of PRL which is inhibited by a hypothalamic factor, likely dopamine [27]. It is obvious therefore, that ectopic pituitary grafts without direct hypothalamic input will secrete large amounts of PRL, so that this may even be used as a hyperprolactinaemia model [28]. Prolactin may also have some somatotrophic activities as it belongs to the so-called GH-PRL family with similar structural features [29]. Notwithstanding some evidence of hyperprolactinaemia related to weight gain [30], we do not know whether PRL secreted by ectopic pituitary grafts will be of benefit to dwarfism [31]. Hyperprolactinaemia will induce change of maternal behaviour. To solve this problem, we can either consider dopamine agonists, such as bromocriptine to inhibit secretion of PRL, or use selected pituitary cell transplantation as will be mentioned later in the chapter.

Donor selection

Age

Based on experience with central nerve system grafting, the optimal donor cells will be embryonic or those in an active stage of neurogenesis [32]. Although the pituitary has been named hypophysis cerebri, it is an endocrine gland rather than brain matter. Thus it functions only after its differentiation and can be transplanted at adult age. Studies on ontogenesis of anterior pituitary gland in the mouse showed that although five distinct secretory cells (corticotrophs: ACTH; thyrotrophs: TSH; gonadotrophs: LH and FSH; somatotrophs: GH and lactotrophs: PRL) can be detected at an early stage of the embryo [33, 34], they finally mutate at different stages of development. In humans, cells that produce GH and ACTH are identifiable at about nine weeks of gestation [35]. The pituitary is able to synthesize TSH from at least the 9th fetal week and its release into the fetal circulation can be demonstrated by 15 weeks of gestation [36]. The last cell type to become manifest is the lactotrope at about the fifth month of gestation [35]. During different stages of development the activity of various secretory cells is variable. For instance, GH secretion is comparatively high in early stage of life and decreases with aging [37]. Prolactin levels are low during neonatal life and gradually increase toward puberty. There is an age-related increase in the cell population engaged in PRL secretion [38]. FSH and LH secretion correlate with ovarian cycles beginning with puberty and will be disturbanced with aging [39]. Therefore, selection of pituitary donor will depend on which hormonal function is to be restored. For example, to reconstitute GH function, the choice of fetal pituitary gland will be preferred, while, for gonadotrophic function a mature pituitary will be favorable.

Nemeskeri *et al.* transplanted undifferentiated fetal pituitary cells under the renal capsules of rats in which the median eminence was removed; 28 days later, the presence of GH, PRL, ACTH, LH-β, FSH-β, TSH-β in pituitary cells was confirmed by immunohistological examination [6]. They concluded that undifferentiated fetal pituitary cells do not require hypothalamic hypophysiotrophic neurohormones for proliferation and cytodifferentiation, and that their development may be modulated by circulating trophic hormones. According to this finding, fetal pituitary could be used as a universal donor.

Sex

The most distinct differences of secretion manner of pituitary gland between male and female are PRL, FSH and LH [38]. Thus, in order to reconstitute gonadotrophic function, the donor should match the recipient with sex. Nevertheless, on the basis of Nemeskeri's observation that proliferation and cytodifferentiation of fetal pituitary cells can be modulated by circulating trophic hormones of the host [6], we could use embryonic pituitary irrespective of sex of the donor.

Sources

Pituitary donors are not easily obtained, particularly adult ones. Collection of human fetal tissue is fraught with ethical and social issues [40]. Can animal pituitary glands be used as donors for human transplantation? With respect to the molecular structure, there are only minor variations in many pituitary hormones among different species. It is not known, however, whether those variations will result in any change of biological activities. Hymer and his colleagues used human pituitary cells for transplantation in dwarf mice and showed an increase in body weight [11]. Although there is no evidence that animal pituitary hormones will be effective in humans, we may hypothesize that it works. The major problem to be solved is immune rejection and this will be discussed later.

Functional pituitary adenoma cells may also represent a potential source of pituitary hormones. Pituitary hormones secreted by adenoma cells have both biological function and molecular structure similar to normal hormones. We have encapsulated human GH secretory adenoma cells and transplanted them into hypophysectomized rats resulting in increase of body weight. It is not known however, whether these adenoma cells will develop a new adenoma after transplantation in the host? We hypothesize that this will not be the case since graft will be rejected by host immune system unless the adenoma cells undergo some structural immune modification. In any event, adenoma cells must be immunoisolated promptly in order to be used biologically as a hormonal source.

Recent progress in genetic engineering technology allows for the production of genetically transformed cells to secrete a given substance. The hGH gene has been successfully transfected in myoblasts and, both in vitro and in vivo studies confirmed effective hGH secretion [41, 42]. Furthermore, Chang *et al.* demonstrated prolonged release of hGH in dwarf rodents with recombinant murine fibroblasts [43]. Theoretically, once the gene for any pituitary hormone is sequenced, it can be incorporated into secretory cells. It is known that pituitary hormones should be secreted on their own specific dynamic fashion such as to contribute to normal function. For example, LH-FSH secretion should be corroborated by menstrual cycling, and ACTH secretion by stress. However, it is difficult to make a genetically engineered cell respond to various physical situations. Hopefully, GH, at least should be usable for therapeutic purpose even if its secretion is not pulsatile as normal physiological condition.

Selected pituitary cell transplantation

Although some patients suffer from panhypopituitarism, there are many cases with partial hypopituitarism [44]. For these patients, therefore, only certain pituitary hormones will be needed, and the so-called selected pituitary cell transplantation will be envisaged. To our knowledge, there is no ideal procedure to isolate different hormone secretory cells from the pituitary gland, although we can readily identify those cells and can even quantify hormone secretion from individual cells [45]. Nevertheless, it may be possible to achieve a similar goal by: (1) selecting fetal pituitary at a suitable stage of development, when only certain kinds of pituitary cells are differentiated; (2) selecting pituitary adenoma cells that secrete a particular hormone; and, (3) genetically engineered cells to secrete a specific hormone.

Preparation of donor cells

In order to allow sufficient time between procurement of a donor and transplantation, some preimplantation treatments are usually carried out and cell culture has been most commonly used. It does not only provide time, but also increases the number of cells and possibly reduces its immunogenic effect as well. However we cannot expect to obtain numerous cells by culture, as longer in vitro culture will eventually result in loss of secretory function. Cryogenic techniques have also used for maintaining donor cells [46, 47]. More recently, artificial cell techniques have also been introduced into pituitary cell transplantation. These will be discussed later.

The number and quality of pituitary cells grafted directly affects its function. It has been shown that at least 10% of the pituitary is needed to maintain its essential function [44]. The populations of distinct secretory cells in adenohypophysis are different, approximately 50% somatotrophs, 10–30% mammotrophs, 20% corticotrophs, 5% thyrotrophs, 15% gonadotrophs [48]. For the transplantation purpose, if only one kind of secretory cell is needed, less amounts of cells may be required. Nevertheless, it is the ability of cells to secrete hormones that is mostly relevant. The best method to assess how many cells are needed is to determine how many hormones a given number of cells will secrete in 24 hours, and extrapolate for the patient needs.

Immunoisolated pituitary cell transplantation

Immune rejection remains a major obstacle in tissue/cell transplantation. Physical isolation of the transplanted tissue from the host immune system by means of semipermeable membrane offers a promising solution to immunorejection [49]. The molecular weight (MW) of most pituitary hormones is around 20,000 Daltons which is far smaller than that of immunoglobulins (> 150,000). Therefore, a semipermeable membrane with a molecular weight cutoff of around 50,000 Daltons will prevent passage of immunoglobulins while still allowing diffusion of nutrients and pituitary hormones. There are now at least two types of membranes have been studied for pituitary cell transplantation.

The XM 50 hollow fiber, which is an acrylic copolymer with a nominal molecular weight cutoff of 50,000 Daltons, was first used in pancreatic islet cells transplantation and is now also applied to pituitary cell transplantation [11–14]. Hymer and his colleagues have loaded outbred mice pituitary cells in XM 50 hollow fiber and implanted them in dwarf mice intracranially resulting in a significant weight gain in the mice [11, 12]. Unfortunately, the hollow fiber became encapsulated by a layer of connective tissue [13] and the cells inside the capsules died within six to eight weeks from lack of oxygen and nutritions [50]. Based on the observation that local release of even small amounts of dexamethasone may control the tissue reaction, some researchers have tried to co-encapsulate graft cells with adrenocortical cells which will produce corticosteroids and reduce nonspecific inflammatory reaction [51].

Microencapsulation techniques have also been used to encapsulate pituitary cells [15]. We developed microcapsules with alginate and poly-L-lysine (PLL) which allow pituitary hormones such as GH, TSH and PRL freely to pass through [16]. Unfortunately, the capsules were not strong enough and most of them ruptured within four weeks after implantation into the peritoneal cavity of rats [52]. Chang *et al.* encapsulated recombinant hGH secretory cells and implanted them into the peritoneal cavity of dwarf mice. The microencapsulated cells secreted hGH for more than 100 days [43]. In order to improve the quality of microcapsules, we have tested several parameters, and selected an optimal combination of parameters allowing for development of very durable microcapsules for pituitary cell transplantation [53, 54]. An *in vivo* study with dwarf mice is currently being carried out.

The purpose of pituitary cell transplantation is to restore lost hormonal function. Previous studies indicated that immunosuppressive treatment is essential for pituitary allo-grafting [55], even when implanted intracranially, a site considered as immunologically privileged. In the clinical setting, if pituitary cell transplantation were to require combined immunosuppressive treatment, there would be no advantage compared to standard hormonal substitutional therapy. It therefore is obvious that immunoisolated pituitary cell transplantation would constitute a good alternative in clinical practice.

Functional evaluation for pituitary grafts

Before the era of the RIA, assessment of pituitary function mainly relied on a variety of end-organ effects and morphological criteria. As an endocrine gland, morphological survival of the pituitary graft does not always necessarily imply functional survival, and the end-organ function may be affected by many additional factors besides pituitary hormones [19]. It is now possible to measure all known pituitary hormones by RIA or enzyme-linked immunosorbent assay (ELISA).

It is well known that the secretion of pituitary hormones is a dynamic process. For instance, GH secretion follows a circadian cycle with the peak clearly related to the onset of sleep, particularly with stages three and four by electroencephalographic classification [56], while LH-FSH secretion is directly correlated with ovarian cycles. Thus, the time of blood sampling is very important, and when possible, hormone levels should be assessed dynamically.

Conclusion

Pituitary cell transplantation has enormous clinical potential. Further experimental studies should provide new knowledge about pituitary function and its regulation. Although many questions remain unanswered, we believe that in combination with immunoisolation, pituitary cell transplantation will be soon introduced into clinical practice, at least GH secretory cells for dwarfism.

References

1. Bridges RS and Millard WJ (1988) Growth hormone is secreted by ectopic pituitary grafts and stimulates maternal behavior in rats. Horm Behav 22: 194–206.
2. Iturriza FC and Eberle AN (1989) Secretion of melanocyte-stimulating hormone and adrenocorticotropin from transplanted pituitary pars intermedia in stressed and nonstressed rats. Neuroendocrinology 49: 610–616.
3. Esquifino AI, Agrasal C, Steger RW *et al.* (1989) Regulation of GH and TSH release from hyperplastic and ectopic pituitaries: Effects of dopamine in vitro. Life Sci 45: 199–206.
4. Chen Z, Hui G and Du Z (1990) An experimental study of pituitary transplantation. Transplantation 50: 513–515.
5. Krieg RJ, Johnson JH and Adler RA (1989) Growth hormone (GH) secretion in the pituitary grafted male rat: In vivo effects of GH-releasing hormone and isoproterenol and in vitro release by individual somatotropes. Endocrinology 125: 2273–2278.
6. Nemeskeri A, Setalo G, Kacsoh B *et al.* (1990) Fetal pituitary graft is capable of initiating hormone synthesis in median eminence removed adult rat. Endocrinol Exp 24: 283–292.
7. Jorgensen JO (1991) Human growth hormone replacement therapy: pharmacological and clinical aspects. Endocr Rev 12: 189–207.
8. Brown P (1988) The decline and fall of Creutzfeld-Jakob disease associated with human growth hormone therapy. Neurology 38: 1137.
9. Hara T, Komiyama A, Ono H *et al.* (1989) Acute lymphoblastic leukemia in a patient with pituitary dwarfism under treatment with growth hormone. Acta Paediatr Jpn Overseas Ed 31: 73–77.
10. Azagury M, Castaigne DH and Degos L (1995) Growth hormone and acute promyelocytic leukemia. Med Pediatr Oncol 24: 69–70.
11. Hymer WC, Wilbur DL, Page R *et al.* (1981) Pituitary hollow fiber units in vivo and vitro. Neuroendocrinology 32: 339–349.
12. Hymer WC, Harkness J, Bartke A *et al.* (1981) Pituitary hollow fiber units in the dwarf mouse. Neuroendocrinology 32: 350–354.
13. Vasilatos-Younken R and Hibbard E (1986) Hollow fiber-encapsulated pituitary cells for the study of adenohypophyseal regulation of growth in poultry: 1. Preparation and use. Poult Sci 66: 891–898.
14. Vasilatos-Younken R (1986) Hollow fiber-encapsulated pituitary cells for the study of adenohypophyseal regulation of growth in poultry: 2. Recipient growth responses. Poult Sci 66: 899–903.
15. Dupuy B, Cadic C, Gin H *et al.* (1991) Microencapsulation of isolated pituitary cells by polyacrylamide microlatex coagulation on agarose beads. Biomaterials 12: 493–496.
16. Chen Z and Bao Y (1994) Study of microencapsulated pituitary transplantation: Preparation of the capsule and its property. Chin Med J [Eng] 107: 200–204.
17. Charlton HM (1992) Hypothalamic transplantation. Ciba Found Symp: 268–275.
18. Halasz B, Pupp L, Uhlarik S *et al.* (1965) Further studies on the hormone secretion of the anterior pituitary transplantation into the hypophysiotrophic area of the rat hypothalamus. Endocrinology 77: 343–355.
19. Tulipan NB, Zacur HA and Allen GS (1985) Pituitary transplantation: Part 1. Successful reconstitution of pituitary-dependent hormone levels. Neurosurgery 16: 331–335.
20. Van Dieten JA, de Koning J and Van Rees GP (1989) Regulation by ovarian factors of the LHRH-induced LH response in pituitary glands in situ or grafted under the kindey capsule in intact and ovriectomized rats. J Endocrinol 123: 41–45.
21. Houben H and Denef C (1994) Bioactive peptides in anterior pituitary cells. Peptides 15: 547–582.
22. Schwartz J and Cherny R (1992) Intercellular communication within the anterior pituitary influencing the secretion of hypophyseal hormones. Endocr Rev 13: 453–475.
23. Gaillard RC (1994) Neuroendocrine-immune system interactions. The immune-hypothalamo-pituitary-adrenal axis. TEM 5: 303–309.
24. Smith EM, Meyer WJ and Blalock JE (1982) Virus-induced corticosterone in hypophysectomized mice: a possible lymphoid adrenal axis. Science 218: 1311–1312.
25. Olsen NJ, Nicholson WE, DeBold CR *et al.* (1992) Lymphocyte-derived adrenocorticotropin is insufficient to stimulate adrenal steroidogenesis in hypophysectomized rats. Endocrinology 130: 2113–2119.
26. Weigent DA and Blalock JE (1994) Effect of the administration of growth hormone-producing lymphocytes on weight gain and immune function in dwarf mice. Neuroimmunomodulation 1: 50–58.
27. Ben-Jonathan N (1985) Dopamine: a prolactin inhibiting hormone. Endocr Rev 6: 564–589.
28. Adler RA (1986) The anterior pituitary-grafted rat: Avalid model of chronic hyperprolactinaemia. Endocr Rev 7: 302–313.

29. Nicoll CA, Mayer GL and Russell SM (1986) Structure features of prolactin and growth hormone that can be related to their biological properties. Endocr Rev 7: 169–203.
30. Ferreira MF, Pires JS and Sobrinho LG (1993) Prolactin, weight gain and psychgeny. Molecular and Clinical Advances in Pituitary Disorders. Proceedings of the 3rd Internatinal pituitary Congress: 243–247.
31. Bouchon R (1994) Pituitary grafts and behavior in mice. Effects of sex of donor and number of pituitary grafts on exploratory and learning abilities. Physiol Behav 55: 711–715.
32. Fisher LJ and Gage FH (1993) Grafting in the mammalian central nervous system. Physiol Rev 73: 583–616.
33. Voss JW and Rosenfeld MG (1992) Anterior pituitary development: short tales from dwarf mice. Cell 70: 527–530.
34. Andersen B and Rosenfeld MG (1994) Pit-1 determines cell types during development of the anterior pituitary gland. A model for trancriptional regulation of cell phenotypes in mammalian organogenesis. J Biol Chem 269: 29335–29338.
35. Thorner MO, Vance ML, Horvath E *et al.* (1992) The anterior pituitary. In: Wilson JD and Foster DW (eds) Williams Textbook of Endocrinology, 8th ed, pp 221–224. Philadeiphia: W.B. Saunders Company.
36. Beck-Peccoz P, Cortelazzi D, Persani L *et al.* (1993) Hypothalamic-pituitary-thyroid axis during fetal development. Molecular and Clinical Advances in Pituitary Disorders. Proceedings of the 3rd International pituitary Congress: 67–72.
37. Rudman D, Kutner MH, Rogers CM *et al.* (1981) Impaired growth hormone secretion in the adult population-relation to age and adiposity. J Clin Invest 67: 1361.
38. Becu-Villalobos D, Lacau-Mengido IM, Diaz-Torga GS *et al.* (1992) Ontogenic studies of the neural control of adenohypophyseal hormones in the rat. II. Prolactin. Cell Mol Neurobiol 12: 1–19.
39. Wise PM (1994) Nathan Shock Memorial Lecture 1991. Changing neuroendocrine function during aging: impact on diurnal and pulsatile rhythms. Exp Gerontol 29: 13–19.
40. Hurd RE (1992) Ethical issues surrounding the transplantation of human fetal tissue. Clin Res 40: 661–666.
41. Barr E and Leiden JM (1991) Systemic delivery of recombinant proteins by genetically modified myoblasts. Science 254: 1507–1509.
42. Dhawan J, Pan LC, Pavlath GK *et al.* (1991) Systemic delivery of human growth hormone by injection of genetically engineered myoblasts. Science 254: 1509–1512.
43. Chang PL, Shen N and Westcott AJ (1993) Delivery of recombinant gene products with microencapsulated cells in vivo. Human Gene Therapy 4: 433–440.
44. Frohman LA (1987) Diseases of the anterior pituitary. In: Felig P, Baxter JD, Broadus AE and Frohman LA (eds) Endocrinology and Metabolism, Second ed, pp 247–279. New York: McGraw-Hill Book Company.
45. Arita J (1993) Analysis of the secretion from single anterior pituitary cells by cell immunoblot assay. Endocr J 40: 1–15.
46. Maganto P, Cienfuegos JA, Santamaria L *et al.* (1988) Cryopreservation and transplantation of hepatocytes: an approach for culture and clinical application. Cryobiology 25: 311.
47. Dixit V, Darvasi R, Arthur M *et al.* (1993) Cryopreserved microencapsulated hepatocytes transplantation studies in gunn rats. Transplantation 55: 616–622.
48. Horvath E and Kovacs K (1995) Anatomy and histology of the normal and abnormal pituitary gland. In: DeGroot JL, Besser M, Burger HG, Jameson JL, Loriaux DL, Marshall JC, Odell WD, Potts JT and Rubenstein AH (eds) Endocrinology, Third ed, pp 160–165. Philadelphia: W.B. Saunders Company.
49. Lysaght MJ, Frydel B, Gentile F *et al.* (1994) Recent progress in immunoisolated cell therapy. J Cell Biochem 56: 196–203.
50. Chen Z, Chen G and Bao Y (1993) XM 50 hollow fiber-encapsulated pituitary adenoma cells for the study of transplantation: 1. Morphological observation on grafts. ACTA Academiae Medicinae Suzhou 13: 90–92.
51. Cadic-Amadeuf CM, Vitiello S, Baquey CV *et al.* (1992) Inflamatory reaction induced by agarose implants reduced by adding adrenal cells to the polymer. ASAIO J 38: M386–M389.
52. Chen Z and Bao Y (1995) Study of microencapsulated pituitary transplantation. ACTA Academiae Medicinae Sinicae: in press.
53. Chen Z, Bao Y, Gorczyca W *et al.* (1995) Study of microencapsulation for pituitary transplantation: Capsule preparation and *in vitro* study. JACSIB 23: 597–604.
54. Chen Z and Mohr G (1995) Microencapsulation for cell implantation into central nervous system: The importance of alginate viscosity and related factors. 1995 Quadrennial Meeting of the American Society for Stereotactic and Functional Neurosurgery. Marina del Rey, CA. March 8–11.
55. Tulipan NB, Huang S and Allen GS (1986) Pituitary transplantation: Cyclosporine enables transplantation across a minor histocompatibility barrier. Neurosurgery 18: 316–320.
56. Mendelson WB (1982) Studies of human growth hormone secretion in sleep and waking. Int Rev Neurobiol 23: 367.

Section XI:

Retinal Cells

R. P. Lanza and W. L. Chick (eds.), Yearbook of Cell and Tissue Transplantation 1996/1997, 193–201.

Chapter 19

Retinal cell transplantation

Robert B. Aramant and Magdalene J. Seiler
Department Ophthalmology and Visual Sciences and Department Anatomy and Neurobiology, University of Louisville School of Medicine, KY 40292, U.S.A.

Transplantation of embryonic neural retina may some day be a treatment for retinal diseases such as macular degeneration and retinitis pigmentosa, but only after strategies are developed that result in appropriate functional connections of the transplant with the host retina and in restored visual function. We have been developing and testing a transplant system which provides information that may help solve these problems. Our basic experimental approach has been to use embryonic donor tissue which has the greatest potential for developing appropriate synaptic connections and elicits little or no immune responses when transplanted within the central nervous system of the same species. We have shown that long-term embryonic retinal transplants grow neuronal processes that form synapses with the host retina. Our latest approach (unpublished) has been to transplant intact sheets of embryonic retina that can develop normal retinal layers.

Introduction

Many incurable retinal diseases such as macular degeneration or retinitis pigmentosa, involve the degeneration of photoreceptors and retinal pigment epithelium (RPE) cells but leave the optic nerve and ganglion cells intact. Retinal transplantation has the potential to prevent further retinal degeneration and to restore eyesight by replacing damaged cells. However, there is no convincing evidence yet that retinal transplants can restore visual function in a damaged retina.

Retinal transplantation studies have involved either RPE cells [1–6] or cells of the neural retina [7–12]. There are different techniques for transplanting the donor tissue: as dissociated cells [2, 13, 14], cell aggregates [7, 8, 15, 26] or sheets [4, 6, 12].

Donor tissue

Our working hypothesis has been that the best donor tissue is fresh tissue from embryos, transplanted as nearly intact as possible. In contrast to dopaminergic brain transplants for Parkinson's disease which have to function as dopamine-producing minipumps [17], transplanted retinal or pigment epithelial cells have to maintain cellular interaction, organization and polarity to communicate in a functional way with the host. Primordial cell contacts are established early [18] and will later develop the intercellular microcircuitry. When these contacts are disrupted by cell dissociation or in the formation of cell aggregates, the transplanted cells have to reestablish these contacts. Such transplants are therefore unlikely to function optimally. In addition, dissociated cells need to have a certain density to survive as transplants [19].

Cells and tissue can lose important properties in culture, such as polarity, differentation, and organization (e.g. [20]). Fresh donor cells from embryos have many advantages over cultured or mature donor cells. If attention is given to optimal donor parameters (e.g., donor age, storage medium, donor tissue preparation), embryonic cells are in general well-tolerated immuno-

---

Abbreviations: B – Bruch's membrane; BrdU – bromodeoxyuridine; ChAT – choline acetyl transferase; CNS – central nervous system; CRALBP – cellular retinal binding protein; DMSO – dimethyl sulfoxide; E – embryonic day (day of gestation); GABA – γ-amino-butyric acid; GAD – glutamic acid decarboxylase; GC – ganglion cell layer; GFAP – glial fibrillary acidic protein; HRP – horseradish peroxidase; IN – inner nuclear layer; IP – inner plexiform layer; IPM – interphotoreceptor matrix; IS – photoreceptor inner segments; MAP – microtubule-associated protein; NF – neurofilament; NSE – neuron-specific enolase; ON – outer nuclear layer; OP – outer plexiform layer; OS – photoreceptor outer segments; P – post-natal day (after birth); PKC – protein kinase C; RCS – Royal College of Surgeons; RPE – retinal pigment epithelial cells; TH – tyrosine hydroxylase.

logically by a host of the same species when placed into the CNS or the eye, because they do not yet carry antigenic sites [21, 22]. In addition, embryonic donor cells have a higher capacity to proliferate, to differentiate and to integrate with the host than cells of more mature donor tissue [7, 23]. Cryopreserved embryonic retinal cells, in contrast to older retinal cells, can be successfully transplanted even after 8 months of storage [23].

Transplantation of RPE cells – rescue of photoreceptors

Transplantation of RPE cells is aimed at arresting retinal degeneration. Restoration of vision would require in addition successful transplantation of photoreceptors. The retinal degeneration in the RCS rat is due to a RPE defect [24] (review: Voaden [25]). The progression of retinal degeneration can be halted by transplantation of normal RPE cells between postnatal days 10 and 26 [13, 14] (review: Sheedlo *et al.* [2]). Injection of PBS, insertion of gelatin [26], or injection of basic fibroblast growth factor (bFGF) [27] can have similar transient effects up to two months after surgery. On the other hand, long-term survival of RCS photoreceptors can only be achieved by RPE transplants [19, 28]; however, the rescue effect decreases over time between 6 and 12 months [19]. The rescue effect is related to the age of the donor cells, but only neonatal RPE cells support long-term survival [2, 29]. Fetal donor RPE have not been tested in the RCS rat.

There have been several attempts to test the visual function of RPE-transplanted RCS rats [5, 30], with conflicting results. The differences appear to relate to differences in transplantation technique (trypsinized dissociated cells appear to be not as good as untrypsinized cell aggregates) and the retinal areas covered by the transplanted RPE cells. Corneal electroretinograms (ERGs) could be detected from eyes with multiple RPE transplant patches arising from cell aggregates [5], but not from eyes with single RPE transplant patches arising from dissociated cells [30]. The ERG responses decreased with age of the host [5].

The success of RPE transplants in RCS rats and the availability of human fetal tissue has lead to the first clinical trials of human fetal RPE transplants to patients with age-related macular degeneration [4]. The transplants were not rejected [4]. However, up to now it has not been reported that RPE transplants can prevent further retinal degeneration or improve vision.

Transplantation of neural retina

RPE transplantation can temporarily rescue existing photoreceptor cells. However, when photoreceptors are irreversibly lost, they need to be replaced. Relatively few research groups have investigated retina-to-retina transplantation [7,10–12, 16, 28, 31, 32].

Prerequisite for functional interaction is the development of neuronal fibers and synaptic connections between retinal transplant and host retina. Photoreceptors in retinal transplants presumably would need to develop near normal morphology with axon terminals and with inner and outer segments in contact with functional RPE (either from host or transplant). In addition, the layers should be parallel so that useful visual information can be produced.

Transplants in the retina would need to grow long distances through the optic nerve to make connections with target regions in the brain. Results so far make this an unlikely expectation [33, 34]. However, it might be sufficient for transplant cells to connect to existing host ganglion cells, thereby permitting transmission of light-generated responses from retinal transplants to the CNS. The connections could be established either directly, or through host interneurons in the inner nuclear layer.

Functional testing of retinal transplants

Convincing demonstrations that retinal transplants can mediate light-induced "behaviors" in mammals have been limited to intracranially transplanted retinas which do provide a model to study the developmental cues of target-directed outgrowth and innervation (review: Lund *et al.* [34]). These transplants, placed in the brain far away from the host eye, develop connections with nearby subcortical visual structures [35]. They are relatively easy to test for their ability to mediate visual function, because light can be shone directly on the transplant, and normal (eye-mediated) visual stimulation can be avoided by covering the eyes or cutting the optic nerve [36]. Lund and colleagues (review: Lund *et al.* [37]; Lund and Coffey [38]) have shown that stimulation of intracranial retinal transplants in rats can produce pupillary constriction [36, 39] and conditioned suppression [40]. Furthermore, they demonstrated that light stimulation of intracranial transplants will produce open-field photophobia (i.e., light avoidance) [40].

However, functional testing of retina-to-retina transplants involves several difficult methodological problems. (1) The visual responses mediated by the transplant need to be distinguished from those of the remaining (undamaged) photoreceptor cells in the host eye. (2) With present techniques, the transplants only cover a small area of the host retina, thus their effects may be very small. (3) It is difficult to determine which sub-cortical visual structures are used by the transplanted cells, assuming connections with intact host ganglion cells.

To detect the transplant response against the background of the host retina, normal host photoreceptors need to be eliminated as much as possible by disease or environmentally induced damage (constant light). The measures of visual function need to be very sensitive and particularly appropriate. Unfortunately, there is no clear evidence so far that retina-to-retina transplants are functionally effective.

Most efforts have been directed at electrophysiological assessment of transplant function. Such studies have included measurement of the electroretinogram (ERG), ganglion cell recordings, and the visually evoked cortical potential (VECP), with mixed results. For example, very small intra-retinal ERGs (5% of normal amplitude) were recorded from RCS rats with RPE transplants [30], but ganglion cell recordings from degenerate C3H mouse retinas following retinal transplants 1–2 months after surgery yielded no light-driven activity [41]. Encouraging results have been reported by Silverman *et al.*, [11] who measured light-driven VECPs presumably mediated by photoreceptor transplants in RCS or light-damaged rats. Nevertheless, even if transplant-mediated neural activity can be demonstrated, it does not prove that vision has been restored. For this, one needs to go to behavioral models.

"Behavioral" testing of visual function of retina-to-retina transplants has been reported by Del Cerro *et al.* [32] and by Silverman *et al.* [11] However, the paradigms used by each group were not appropriate to demonstrate the type of visual function needed for clinical applications of retinal transplants. The light-inhibited acoustic startle reflex studied by Del Cerro *et al.* [32] is an unconscious response. In addition to being inappropriate for demonstrating vision, the results obtained in the transplant group were atypical and difficult to interpret. The pupillary reflex studied by Silverman *et al.* [11] is also an involuntary, subcortical response.

Our group is currently developing a behavioral test, in which the rats are first trained in a threshold light detection task, to establish visual thresholds of light-damaged animals before and after transplantation.

Our work in retinal transplantation

Our research has evolved from experiments in transplantation of retinal cell aggregates into lesioned adult retinas [15]. In these early studies, the embryonic donor tissue was disrupted during transplantation, but rearranged itself into folded sheets and rosettes, and fused with the host retina. However, our current approach is to transplant pieces of embryonic retina in intact sheets to the subretinal space, using an anterior approach. We have always used embryonic retinal tissue from either mouse, rat, rabbit or human donors. Host animals were rats [42] or rabbits [43]. Early on, one of our studies showed that it was possible to transplant from one species to another when the host was immunosuppressed [44] which later lead to studies with human embryonic retinal transplants to immunosuppressed rats [45].

Our current research is focused on the investigation of neuronal connections between graft and host, and on improving graft organization by transplanting intact sheets of embryonic retina.

Cell types in retinal transplants

The neuroblastic donor cells develop into the different retinal cell types after transplantation [16, 31, 46, 47]. The sequence of development has been followed by immunohistochemistry. So far this has been studied mostly in aggregate, not in intact-sheet transplants.

Using antibodies against MAP 1A (microtubule associated protein) and NSE (neuron-specific enolase), only a few ganglion-cell like cells have been identified [16, 48]. Other ganglion-specific markers such as neurofilament, an intermediate filament protein, and the cell surface marker THY-1 failed to detect ganglion cells in the transplant. Different types of amacrine cells, (cholinergic, dopaminergic, GABA-ergic and also somatostatin-like staining cells) could be found in the transplants [42], but transplant inner plexiform layers, as seen with different antibodies, were not normally laminated. Bipolar cells were immunoreactive for PKC (protein kinase C) [16, 47, 48]. Graft horizontal cells were stained with neurofilament, MAP 1A, HPC-1 and vimentin antibodies. Rod photoreceptors were immunoreactive for rhodopsin, S-antigen and α-transducin-specific antibodies [31, 46, 47] (see also Fig. 1b). Cones have been identified in human transplants with antibodies for cone-specific red/green and blue opsins, synaptophysin and NSE (and some with S-antigen) [31, 46]. Müller cells and astrocytes have been identified in the transplants using the glia-specific vimentin, S-100 and GFAP (glial fibrillary acidic protein) antibodies [46, 49, 50]. Müller cells are also immunoreactive for CRALBP (cellular retinaldehyde binding protein) [31, 46].

Later, additional immunohistochemical studies have been done on rabbit retinal transplants [51–53].

The outer limiting membrane of transplant rosettes is formed by adherent junctions between photoreceptors and Müller cells. In the rosette lumen, photoreceptor outer segments develop, and Müller cells extend microvilli into the lumen. Typical photoreceptor-ribbon synapses are found in the outer plexiform layers. In spite of some abnormal synapses, all synaptic types (including bipolar ribbon synapses) can be identified in the inner plexiform layers [31, 54, 55].

To sum up: Retinal transplants contain most cell types and layers found in normal retina. The photoreceptors contain several substances important for phototransduction. This makes it likely that transplant photoreceptors have the capacity to respond to light. The abnormal lamination of aggregate transplants is probably due to the disruption of the tissue during transplantation. However, transplantation of pieces of embryonic retina in intact sheets can improve the organization of retinal grafts (see below).

Human retinal transplants

For an eventual future clinical application, it would be important to know how human embryonic retinal cells develop in a transplant situation. This could also offer a model for the study of the normal human retinal development. Accordingly, human embryonic retinal cell aggregates were transplanted to the eyes of immunosuppressed rats. The donor age ranged from 5–10 weeks postconception, and the transplants were studied to a total age of almost 9 months postconception. (Total age = donor age + survival time after transplantation.)

These transplants developed according to their intrinsic human retinal developmental timetable [45, 46]. Rosettes had formed in the transplant at the earliest age studied (13 weeks total age). At this stage, the transplants contained a ganglion cell layer and an outer layer of mostly neuroblastic cells. Immature cone cells could be seen close to the outer limiting membrane of rosettes. At 20 weeks, inner and outer nuclear layers were separated by an outer plexiform layer [45]. At 20 weeks, several cell markers (S-antigen, rhodopsin, α-transducin, cellular retinal binding protein) were expressed for the first time [46]. In photoreceptor layers, both cones and rods had developed inner segments and beginning outer segments. In older transplants (32–41 weeks), photoreceptor outer segments were fully developed [45, 46, 54].

The development of human embryonic retinal transplants appears to approximately parallel normal *in utero* development. Transplant cones, rods and Müller cells all express their cell-specific proteins. The photoreceptors develop both inner and outer segments and contain several essential proteins for processing light. The transplants can reach a degree of maturity comparable to newborn retina.

Another group has reported human second trimester xenografts to the anterior chamber. However, these experiments were not comparable because of the short survival times (up to 1 month), the graft placement and the donor age [56].

We have done an additional study with human transplants to immunodeficient "nude" rats [31]. In this rat model, human transplants can be studied long-term, up to one year, with no immunosuppression. The success rate of long-term transplants to "nude" rats was very high, in contrast to human transplants to immunosuppressed rats. The development of the transplants was characterized with antibodies for synaptophysin, NSE, red-green or blue opsin, S-antigen, rhodopsin, vimentin, CRALBP, and GFAP. Immunohistochemistry for rat MHC-II and microglia/macrophages revealed a negligible immune response from the host. "Nude" rats offer an excellent model for the study of human retinal xenografts without the negative effects of immunosuppression. Compared to immunosuppressed rats, transplantation to nude rats gives consistent results and superior long-term survival of hosts and transplants.

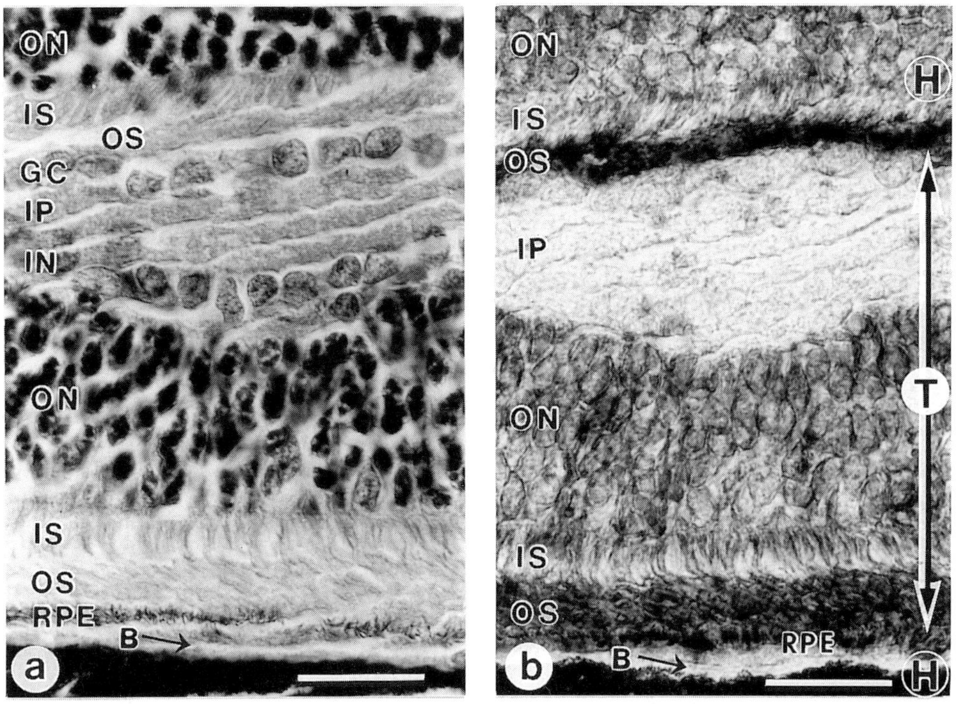

Fig. 1. Intact-sheet rat transplant (T), E20 donor, 2 wks after transplantation in subretinal space between host (H) outer nuclear layer (ON) and RPE. a) Hematoxylin-eosin staining. b) Adjacent section, stained for rhodopsin (dark), lightly counterstained with hematoxylin. The outer segments of transplant photoreceptors are intensely stained. The host RPE are unstained, but appear dark because of their melanin content. Bruch's membrane (B) is unstained, the underlying choroid is black. Note that the transplant outer segments in contact with host RPE are much longer than the host outer segments that have been separated from their RPE.

Cografts of retina and RPE [57]

The retinal pigment epithelium (RPE), important for the development of the neuronal retina [58], may also influence the differentiation of retinal grafts. Rabbit E16 retinal cells, with and without RPE, were grafted to adult rabbit eyes. For the first time in our studies with aggregate transplants, a complete layering was observed in some of the retinal cografts. Sometimes near clusters of RPE, an inner limiting membrane with Müller cell endfeet and, occasionally, an apparent ganglion cell layer could be seen. In grafts with retina alone, no inner limiting membrane was seen, and most photoreceptors were degenerated at 12 weeks. Cografting of rabbit embryonic RPE cells and retina apparently induced the formation of an inner limiting membrane that was not seen in rabbit grafts of retina alone. In addition, cografted RPE cells can promote the long-term survival of transplant photoreceptors.

Donor cell label

A recurring question in our different studies is how to distinguish donor cells from host cells. The problem with most cytoplasmic dyes is diffusion of the dye from donor cells to host cells. To overcome this problem, we have successfully used genetically or DNA-marked donor cells [59]. We have applied two strategies: transplantation of bromodeoxyuridine (BrdU)-labelled cells and of transgenic cells carrying the *E. coli* β- galactosidase gene. Transplanted BrdU-labelled retinal donor cells are distinctly identified immunohistochemically several months after transplantation (example after 2 weeks in Fig. 2b). A completely different approach is the use of transgenic NSE-lacZ mouse [60] donor cells (Dr. Sutcliffe, La Jolla, CA). These cells can be identified several months after transplantation in the host because they express a bacterial enzyme. This study also showed transplant-derived fibers in the host retina, indicating neuronal outgrowth [59].

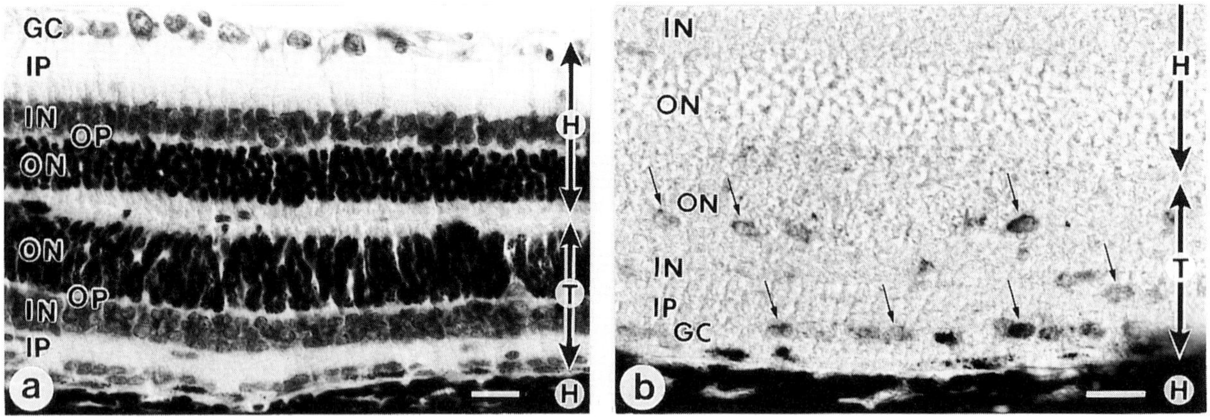

Fig. 2. Intact-sheet rat transplant (T), E19 donor, 2 wks after transplantation in subretinal space between the photoreceptors and RPE of the host (H). The orientation of this transplant is reversed relative to the host retina. a) Hematoxylin-eosin staining. b) Adjacent section. Prelabelled donor cells, stained by immunohistochemistry for BrdU (bromodeoxyuridine). Arrows point to specifically labelled cells.

Connections between transplants and host retina

Do embryonic retinal transplants grow neuronal fibers and make synaptic connections with the host retina? It seems possible because, often, transplant and host can be completely integrated, i.e., there are no glial barriers [49]. This integration is seen especially well with embryonic brain transplants to adult retina which grow large and fuse exceptionally well with the host retina [61]. We therefore used embryonic brain transplants to develop tracing techniques in retinal wholemounts. Horseradish peroxidase (HRP) tracing in wholemounts *in vitro* disclosed extensive fiber outgrowth from brain transplants into the host retina [62].

In another pilot study, fluorogold was injected into the superior colliculus of host rats with long-term retinal transplants to investigate whether there were fiber connections between host retinal ganglion cells and transplants. Fluorogold retrogradely labelled the host retinal ganglion cells. In some experiments, specifically labelled fibers were found in the transplants (see Aramant *et al.* [16], Seiler *et al.* [63]). Fluorogold labelling, which can also be detected by immunohistochemistry, is now routinely used in most of our experiments, simultaneously with BrdU-label of donor cells and DiI* tracing (see below).

We have used the carbocyanine DiI [64] to trace neuronal fibers from transplants in fixed tissue. This was done with embryonic rat (E16–22) or human (9–13 wks) transplants to adult rats. Normal Long-Evans rats received rat transplants. The hosts for human transplants were athymic nude rats. Three to eleven months after surgery, the eyes were fixed with 4% paraformaldehyde (sometimes with added 0.1% glutaraldehyde). DiI-coated glass microneedles (0.5–1 mm long) were inserted into the transplants, which were then stored at room temperature for 3–15 months. This filled cells that had processes in the area near the needle. DiI-labelled transplant cells showed fiber outgrowth into the host retina. After photoconversion of the dye to an electron-dense precipitate, these neuronal processes could be followed with better resolution than with fluorescence (examples in Figs 3a and b). Selected photoconverted sections were embedded for electron microscopy. Synapses could be found along transplant processes that had grown into the host inner plexiform layer. These results indicate that neuronal fibers originating from embryonic retinal transplants form synapses in the host retina [65].

* DiI = 1,1′-dioctadecyl-3,3,3′,3′-tetramethylindo-carbocyanine perchlorate, Molecular Probes, OR.

Transplantation of embryonic retina in intact sheets

Although transplantation of embryonic retinal aggregates resulted in the development of almost all cell types and layers, the organization of these transplants was far from normal. Especially the formation of rosettes made it inconceivable that such transplants could be used to restore visual function. We have therefore developed a technique to transplant intact

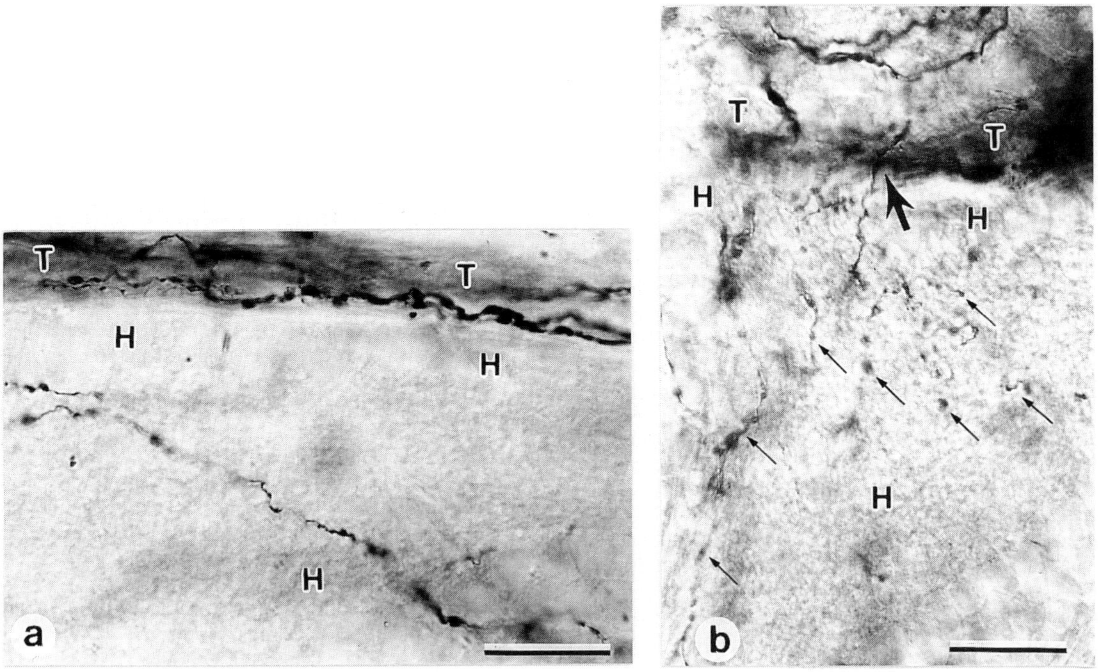

Fig. 3. Examples of DiI tracing from retinal transplants (rat E16, both fixed 4 months after transplantation). Photoconverted vibratome sections. a) Detail of transplant (T) and host (H). Fibers in transplant along the interface close to host, and transplant fibers in the inner plexiform layer of the host retina. b) Transplant (T) on top of host retina (H). Labelled fiber crossing from transplant into host (thick arrow) and branching in host inner plexiform layer. Small arrows point to labelled fibers and varicosities which could indicate synaptic areas.

pieces of embryonic retina to the subretinal space [66]. Transplants placed with the right polarity, with their photoreceptors towards the host RPE, developed parallel layers with photoreceptor inner and outer segments (see Fig. 1a). Transplant outer segments stain intensely for rhodopsin and are in contact with the host RPE (Fig. 1b). However, transplants placed in a reversed orientation, with their photoreceptors facing the host photoreceptors or the inner nuclear layer, did not develop outer segments (Figs. 2a,b). Optimal transplants with fully differentiated photoreceptors will be the basis for our continuing research. The next step will be to see whether such transplants increase visual sensitivity in hosts with degenerated photoreceptors.

Acknowledgements

The authors wish to express their appreciation to Ms. Ann M. Potter and Ms. Carla J. Porter for their excellent technical assistance, and to Dr. James Longley for his editorial help which has been very valuable. We are grateful to the following researchers for providing us with antibodies: Dr. C.J. Barnstable, New Haven, CT (HPC-1); Dr. L.A. Donoso, Philadelphia, PA (monoclonal antibody against S-antigen); Dr. K.R. Fry, Houston, TX (monoclonal antibody AB-5); Dr. B.K.K. Fung, Los Angeles, CA (monoclonal antibody against rod α-transducin); Drs. G. Garwin, J. Saari, Seattle, WA (rabbit antiserum against CRALBP); Dr. R.S. Molday, Vancouver, Canada (monoclonal antibodies against rhodopsin: rho 4D2, 1D4); Dr. W. Oertel, Munich, F.R.G. (sheep antiserum against GAD); Drs. J.J. Plantner, E. Kean, Cleveland, OH (rabbit antiserum against rhodopsin); Dr. P. Salvaterra, Duarte, CA (monoclonal antibody against ChAT); and Dr. P.R. Vulliet, Davis, CA (monoclonal antibody against TH). We thank Drs. G. Sutcliffe and S. Forss-Petter, La Jolla, CA, for providing us with the transgenic NSE-lacZ mouse strain. The different studies presented in this paper have been supported by the Schepens Eye Research Institute, Boston; the Kentucky Lions Eye Foundation; the Vitreoretinal Research Foundation, Louisville; an unrestricted grant from the Research To Prevent Blindness, Inc.; the Jewish Hospital Foundation, Louisville, and the National Institute of Health grant EY08519.

References

1. El Dirini AA, Wang H, Odgen TE *et al.* (1992) Retinal pigment epithelium implantation in the rabbit: technique and morphology. Graefe's Arch Clin Exp Ophthalmol 230: 292–300.
2. Sheedlo HJ, Li L, Gaur VP *et al.* (1992) Photoreceptor rescue in the dystrophic retina by transplantation of retinal pigment epithelium. Int Rev Cytol 138: 1–49.
3. Wongpidachai S, Weiter JJ, Weber P *et al.* (1992) Comparison of external and internal approaches for transplantation of autologous retinal pigment epithelium. Invest Ophthalmol Vis Sci 33: 3341–3352.
4. Algvere PV, Berglin L, Gouras P *et al.* (1994) Transplantation of fetal retinal pigment epithelium in age-related macular degeneration with subfoveal neovascularization. Greafe's Arch Clin Exp Ophthalmol 232: 707–716.
5. Jiang LQ and Hamasaki D (1994) Corneal electroretinographic function rescued by normal retinal pigment epithelial grafts in retinal degenerative Royal College of Surgeons rats. Invest Ophthalmol Vis Sci 35: 4300–4309.
6. Sheng Y, Gouras P, Cao H *et al.* (1995) Patch transplants of human fetal retinal pigment epithelium in rabbit and monkey retina. Invest Ophthalmol Vis Sci 36: 381–390.
7. Aramant R, Seiler M and Turner JE (1988) Donor age influences on the success of retinal transplants to adult rat retina. Invest Ophthalmol Vis Sci 29: 498–503.
8. Gouras P, Du J, Gelanze M *et al.* (1991) Transplantation of photoreceptors labeled with tritiated thymidine into RCS rats. Invest Ophthalmol Vis Sci 32: 1704–1707.
9. Silverman MS, Hughes SE, Valentino TL *et al.* (1991) Photoreceptor transplantation to dystrophic retina. In: Anderson RE, Hollyfield JG and LaVail MM (eds) Retinal Degenerations, pp 321–335. Boca Raton FL: CRC Press.
10. Gouras P, Du J, Kjeldbye H *et al.* (1992) Reconstruction of degenerate rd mouse retina by transplantation of transgenic mouse photoreceptors. Invest Ophthalmol Vis Sci 33: 2579–2586.
11. Silverman MS, Hughes SE, Valentino T *et al.* (1992) Photoreceptor transplantation: Anatomic, electrophysiologic, and behavioral evidence for the functional reconstruction of retinas lacking photoreceptors. Exp Neurol 115: 87–94.
12. Silverman MS, Ogilvie JM, Lett J *et al.* (1994) Photoreceptor transplantation: potential for recovery of visual function. In: Christen Y, Doly M and Droy-Lefaix MT (eds) Retina, Aging and Transplantation. Proceedings of "Seminaires Ophthalmologiques IPSEN", pp 43–59. Paris: Elsevier.
13. Li L-X and Turner JE (1988) Inherited retinal dystrophy in the RCS rat: prevention of photoreceptor degeneration by pigment epithelial cell transplantation. Exp Eye Res 47: 911–917.
14. Lopez R, Gouras P, Kjeldbye H *et al.* (1989) Transplanted retinal pigment epithelium modifies the retinal degeneration in the RCS rat. Invest Ophthalmol Vis Sci 30: 586–588.
15. Turner JE and Blair JR (1986) Newborn rat retinal cells transplanted into a retinal lesion site in adult host eyes. Dev Brain Res 26: 91–104.
16. Seiler M, Aramant R, Ehinger B *et al.* (1991) Characteristics of embryonic retina transplanted to rat and rabbit retina. Neuro-ophthalmology 11: 263–279.
17. Björklund A (1992) Dopaminergic transplants in experimental Parkinsonism: cellular mechanisms of graft induced functional recovery. Curr Opin Neurobiol 2: 683–689.
18. Fisher SK and Linberg KA (1975) Intercellular junctions in the early human retina. J Ultrastructure Res 51: 69–78.
19. Li L and Turner JE (1991) Optimal conditions for long-term photoreceptor cell rescue in RCS rats: The necessity for healthy RPE transplants. Exp Eye Res 52: 669–679.
20. Gundersen D, Powell SK and Rodriguez-Boulan E (1993) Apical polarization of N-CAM in retinal pigment epithelium is dependent on contact with the neural retina. J Cell Biol 121: 335–343.
21. Widner H and Brundin P (1988) Immunological aspects of grafting in the mammalian central nervous system. A review and speculative synthesis. Brain Res Rev 13: 287–324.
22. Dunnett SB (1990) Neural transplantation in animal models of dementia. Eur J Neurosci 2: 567–587.
23. Aramant R and Seiler M (1991) Cryopreservation and transplantation of immature rat retina into adult rat retina. Dev Brain Res 61: 151–159.
24. Mullen RJ and LaVail MM (1976) Inherited retinal dystrophy: primary defect in pigment epithelium determined with experimental rat chimeras. Science 192: 799–801.
25. Voaden MJ (1990) Retinitis pigmentosa and its models. Prog Retinal Res 10: 293–331.
26. Silverman MS and Hughes SE (1990) Photoreceptor rescue in the RCS rat without pigment epithelium transplantation. Curr Eye Res 9: 183–191.
27. Faktorovich EG, Steinberg RH, Yasumura D *et al.* (1990) Photoreceptor degeneration in inherited retinal dystrophy delayed by basic fibroblast growth factor. Nature 347: 83–86.
28. Gouras P, Lopez R, Du J *et al.* (1990) Transplantation of retinal cells. Neuro-ophthalmology 10: 165–176.
29. Sheedlo HJ, Li L and Turner JE (1993) Effects of RPE age and culture conditions on support of photoreceptor cell survival in transplanted RCS dystrophic rats. Exp Eye Res 57: 753–761.
30. Yamamoto S, Du J, Gouras P *et al.* (1993) Retinal pigment epithelial transplants and retinal function in RCS rats. Invest Ophthalmol Vis Sci 34: 3068–3075.
31. Aramant RB and Seiler MJ (1994) Human embryonic retinal transplants in athymic immunodeficient rat hosts. Cell Transplantation 3 (6): 461–474.
32. Del Cerro M, Ison JR, Bowen GP *et al.* (1991) Intraretinal grafting restores function in light-blinded rats. NeuroReport 2: 529–532.
33. Matthews MA and West LC (1982) Optic fiber development between dual transplants of retina and superior colliculus placed in the occipital cortex. Anat Embryol 163: 417–433.
34. Lund RD, Hankin MH, Sefton AJ *et al.* (1988) Conditions for optic axon outgrowth. Brain Behav Evol 31: 218–226.
35. Craner SJ, Radel JR, Jen LS *et al.* (1989) Light-evoked cortical activity produced by illumination of intracranial retinal transplants: experimental studies in rats. Exp Neurol 104: 93–100.
36. Radel JD, Das S and Lund RD (1992) Development of light-activated pupilloconstriction in rats as mediated by normal and transplanted retinae. Eur J Neurosci 4: 603–616.
37. Lund RD, Radel JD and Coffey PJ (1991) The impact of retinal transplants on types of behavior exhibited by host rats. TINS 14: 358–362.
38. Lund RD and Coffey PJ (1994) Visual information processing by intracerebral retinal transplants in rats. Eye 8: 263–268.
39. Klassen H and Lund RD (1990) Retinal graft-mediated pupillary responses in rats: restoration of a reflex function in the mature mammalian brain. J Neurosci 10: 578–587.
40. Coffey PJ, Lund RD and Rawlins N (1990) Detecting the world through a retinal implant. Prog Brain Res 82: 269–275.
41. Gouras P, Du J, Yamamoto S *et al.* (1993) Anatomy and physiology of photoreceptor transplants in degenerate C3H mouse retina. Invest Ophthalmol Vis Sci 34 (Suppl.): 1096.

42. Aramant R, Seiler M, Ehinger B *et al.* (1990) Neuronal markers in rat retinal grafts. Dev Brain Res 53: 47–61.
43. Seiler M, Aramant R and Ehinger B (1990) Transplantation of embryonic retina to adult retina in rabbits. Exp Eye Res 51: 225–228.
44. Aramant R and Turner JE (1988) Cross-species grafting of embryonic mouse and grafting of older postnatal rat retinas into the lesioned adult rat eye: the importance of Cyclosporin A for survival. Dev Brain Res 41: 303–307.
45. Aramant R, Seiler M, Ehinger B *et al.* (1990) Transplantation of human embryonic retina to adult rat retina. Restor Neurol Neurosci 2: 9–22.
46. Seiler MJ and Aramant RB (1994) Photoreceptor and glial markers in human embryonic retina and in human embryonic retinal transplants to rat retina. Dev Brain Res 80: 81–95.
47. Aramant RB and Seiler MJ (1994) Embryonic retinal cell transplantation to adult retina. In: Christen Y, Doly M and Droy-Lefaix MT (eds) Retina, Aging and Transplantation. Proceedings of "Seminaires Ophthalmologiques IPSEN", pp 31–42. Paris: Elsevier.
48. Seiler M, Aramant R, Bergström A *et al.* (1991) Organization of embryonic rabbit retinal transplants of aggregated donor tissue. Soc Neurosci Abstr 17: 1137.
49. Seiler M and Turner JE (1988) The activities of host and graft glial cells following retinal transplantation into the lesioned adult rat eye: developmental expression of glial markers. Dev Brain Res 43: 111–122.
50. Seiler M and Turner JE (1989) Host and graft glial cell activities following retinal transplantation to the adult rat eye. In: Weiler R and Osborne NN (eds) Neurobiology of the Inner Retina, pp 481–486. Berlin: Springer.
51. Bergström A, Ehinger B, Wilke K *et al.* (1994) Development of cell markers in subretinal rabbit retinal transplants. Exp Eye Res 58: 301–314.
52. Juliusson B, Mieziewska K, Bergström A *et al.* (1994) Interphotoreceptor matrix components in retinal cell transplants. Exp Eye Res 58: 615–622.
53. Szél Á, Juliusson B, Bergström A *et al.* (1994) Reversed ratio of color-specific cones in rabbit retinal cell transplants. Dev Brain Res 81: 1–9.
54. Ehinger B, Bergström A, Seiler M *et al.* (1991) Ultrastructure of human retinal cell transplants with long survival times in rats. Exp Eye Res 53: 447–460.
55. Zucker CL, Ehinger B, Seiler M *et al.* (1994) Ultrastructural circuitry of retinal cell transplants to rat retina. J Neur Transplant Plast 5 (1): 1–12.
56. Epstein LG, Cvetkovich TA, Lazar E *et al.* (1992) Successful xenografts of second trimester human fetal brain and retinal tissue in the anterior chamber of the eye of adult immunosuppressed rats. J Neural Transplant Plast 3: 151–158.
57. Seiler MJ, Aramant RB and Bergström A (1995) Co-Transplantation of embryonic retina and retinal pigment epithelial cells to rabbit retina. Curr Eye Res 14: 199–207.
58. Layer PG and Willbold E (1989) Embryonic chicken retinal cells can regenerate all cell layers in vitro, but ciliary pigmented cells induce their correct polarity. Cell Tissue Res 258: 233–242.
59. Seiler MJ and Aramant RB (1995) Transplantation of embryonic retinal donor cells, labelled with BrdU or carrying a genetic marker, to adult retina. Exp Brain Res 105: 59–66.
60. Forss-Petter S, Danielson PE, Catsicas S *et al.* (1990) Transgenic mice expressing β-galactosidase in mature neurons under neuron-specific enolase promoter control. Neuron 5: 187–197.
61. Aramant R, Seiler M and Adolph AR (1988) Cografts of retina and tectum or cerebellum to adult rat retina. Soc Neurosci Abstr 14: 1276.
62. Stirling RV, Aramant R, Seiler M *et al.* (1989) In vitro labelling to demonstrate connection between host retina and fetal tectum or cerebellum grafts in adult rats. Soc Neurosci Abstr 15: 308.
63. Aramant R, Seiler M, Ehinger B *et al.* (1991) Transplanting embryonic retina to the retina of adult animals. In: Anderson RE, Hollyfield JG and LaVail MM (eds) Retinal Degenerations, pp 275–288. Boca Raton FL: CRC Press.
64. Godement P, Vanselow J, Thanos S and Bonhoeffer F (1987) A study in developing visual systems with a new method of staining neurons and their processes in fixed tissue. Development 101: 697–713.
65. Aramant RB and Seiler MJ (1995) Fiber and synaptic connections between embryonic retinal transplants and host retina. Experimental Neurology 133: 1–12.
66. Aramant RB and Seiler MJ (1995) Organized embryonic retinal transplants to normal or light-damaged rats. Soc Neurosci Abstr 21: 1308.

Section XII:

Skin

R. P. Lanza and W. L. Chick (eds.), Yearbook of Cell and Tissue Transplantation 1996/1997, 205–219.

Chapter 20

Cultured epidermal and dermal skin replacements

John F. Hansbrough
Department of Surgery, The University of California, San Diego Medical Center, San Diego, CA 92103, U.S.A.

Overview: Advances in burn treatment

The development of centralized burn centers in most geographic areas has played an important role in improving patient survival following extensive burn injuries. Through advances in the treatment of shock, nutritional support, ventilatory support and many other aspects of critical care we have developed great expertise in treating burn injuries, and survival of extensively burned patients has markedly improved in the past several decades. However, technologies are now needed to improve functional and cosmetic outcomes following severe burns. In large part, future improvements will depend upon advances in our ability to control and accelerate various aspects of wound coverage and wound healing.

Surgical excision of the burn wound (Fig. 1) probably represents the greatest single refinement in burn care which has increased patient surviva [1–5]. Removal of burned and devitalized skin removes the major source for infection for burn patients, and metabolic benefits may occur since the necrotic tissue serves as a massive inflammatory focus. Although blood losses may be severe during the operative procedures, increased survival has been shown in those burn centers which practice early surgical excision of burn wounds [2]. Meshing and expansion of split-thickness skin grafts permits larger areas of autologous skin coverage with a given amount of skin, thus increasing the area of wound closure with each surgical procedure and accelerating the time required for permanent wound closure in patients who require repeated surgery. However, the cosmetic and functional results of utilizing meshed skin grafts for wound closure are frequently unsatisfactory, since the mesh interstices must heal by epithelialization which is frequently accompanied by fibrosis. The "mesh pattern" is usually grossly evident and is unsightly (Fig. 2), but more importantly, hypertrophic scarring and contraction may be severe when meshed grafts are utilized for wound coverage.

Now that survival after severe burn injury has almost become routine except for the most severe cases, we must develop technologies and expertise to decrease scar formation and wound contraction and improve the quality of life of burn patients. We hope that wound research and the biotechnology industry can help us to achieve these goals.

Development of dermal replacements

Full-thickness burns, by definition, are characterized by destruction of both the epidermal and dermal layers of the skin. In an effort to develop a dermal replacement for full-thickness wounds, we have worked with Advanced Tissue Sciences, LaJolla CA, to developments a tissue replacement composed of human neonatal fibroblasts cultured on a biodegradable, synthetic matrix (Dermagraft^{TM}). Polyglycolic acid (PGA) and polyglactin-910 (PGL) fibers are in common surgical usage as resorbable sutures [6]. PGA suture materials are known as Dexon^{TM}, and are supplied by Davis and Geck Inc., Danbury, CT. PGL suture materials are known as Vicryl^{TM} and are supplied by Ethicon Inc., Somerville, NJ. Various surgical specialties have also employed these fibers in woven or knitted meshes, as they generate a relatively limited inflammatory reaction and are primarily biodegraded by hydrolysis, rather than by enzymatic degradation. In subcutaneous tissues, absorption is complete within 60 to 90 days [6].

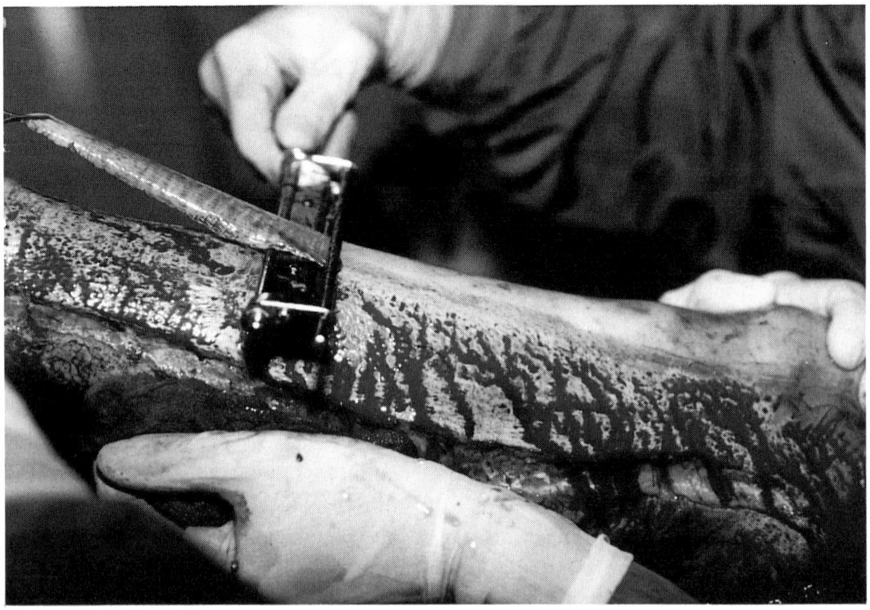

Fig. 1. Surgical excision of the burn wound has marked a major advance in burn care. Under anesthesia, burned skin is removed utilizing a variety of techniques. Excision must be followed by coverage of the wound with the patient's own skin, or with an acceptable temporary skin substitute. The use of standard dressings on excised wounds will quickly lead to maceration and infection.

Fig. 2. The raised, unsightly appearance of full-thickness wounds following closure with meshed and expanded skin grafts is evident on these wounds. Hypertrophic scarring is most severe in the interstices, or the openings in the skin created by the meshing procedure.

These materials are used as tubes in nerve reconstruction [7], to limit scar formation in laminectomies [8], to replace dura [9], to re-create the pelvic floor [10], and for many procedures in the abdomen, particularly in repairing various types of hernias [11–13]. Of particular relevance, related to application to potentially contaminated burn wounds, it has been suggested that bacterial growth may be partially inhibited in regions of absorbing PGA/PGL sutures, possibly via a local tissue pH effect [14, 15].

Human fibroblasts, in contrast with epidermal cells, are relatively nonantigenic. Thus they can be utilized to construct dermal tissue replacements which can theoretically be stored and used "off the shelf" as soon as the patient is in condition to receive grafts. In addition, the donor cells as well as the mother can be extensively screened for infectious agents, to insure the sterility of the transplanted cells. In particular, sensitive viral assays can be performed while the cells are "quarantined" using frozen storage.

To construct this dermal tissue [16–22], human dermal fibroblasts are isolated from neonatal foreskin obtained aseptically after circumcision. Epidermis and dermis are separated by incubation in 0.25% trypsin/0.2% EDTA for one to two hours at 37 °C. Dermis is minced and digested with collagenase B, and the tissue digest is filtered through sterile gauze to remove debris. Fibroblasts are maintained in Dulbecco's modified Eagle's medium (DMEM), and are passaged when cells reach 80% to 90% confluence. Cells are removed from flasks and resuspended for seeding at a concentration of 4×10^6 cells/ml. Dermal grafts are prepared by seeding 4×10^5 viable fibroblasts, determined by trypan blue exclusion, in a minimum volume of DMEM onto each 4 cm^2 area of PGA or PGL mesh. The fibroblasts readily attach *in vitro* to the mesh fibers and they become confluent in 2 to 3 weeks (i.e., all mesh openings are covered by cells and tissue matrix, assayed by inverted phase microscopy). A representative cross-section of this tissue construct, which has been termed Dermagraft™, is shown in Fig. 3.

As they grow in the mesh, the fibroblasts secrete multiple matrix proteins, resulting in formation of an extracellular dermal matrix which fills the mesh interstices. Immunofluorescence studies have shown that this matrix contains dermal collagens types 1,2, 3 and 6, elastin, fibronectin, tenascin, and decorin. Levels of fibronectin are high, and resemble the high levels which are seen in fetal skin, as opposed to the lower levels seen in adult skin. High levels of fibronectin may be particularly important for this dermal replacement, since fibronectin promotes cellular migration and may enhance basement membrane assembly [23–26]. Decorin is a high molecular weight, core protein of dermal proteoglycans. Basic fibroblast growth factor is detectable in the *in vitro* matrix by Western blot. Polymerase chain reaction (PCR) analysis identified mRNA for insulin-like growth factors 1 and 2, and platelet-derived growth factor (PDGF). Tearing strength of the grafts is high [27], which predicts that the grafts will accept sutures or staples without disruption.

Fleischmajer *et al.* [28] reported the structure of this three-dimensional dermal fibroblast model, grown on a nylon mesh. Adherence of fibroblasts was seen within 3 hours and fibroblasts began to stretch across the mesh opening within 5–7 days of the initial inoculation. After 2 weeks, the mesh squares were occupied by large numbers of fibroblasts with large rough endoplasmic reticulum and Golgi's, and numerous pinocytotic vesicles. The cells appeared elongated with long pseudopods and had a tendency to arrange in a parallel fashion, thereby delineating well-demarcated spaces occupied by numerous collagen fibrils and microfibrils. The presence of collagen fibrils and microfibrils was detected in cultures as early as 5 days after seeding, and they increased in numbers to eventually fill all of the intercellular spaces.

Microfibrils are distributed through various connective tissues and are characterized by a diameter of less than 20 nm, but lack the 67 nm periodicity of collagen. Two major subgroups of microfibrils have been described, namely, those with diameters of 3–5 nm, and those with diameters of 6–18 nm [29]. The first group has been identified as type VII collagen [30, 31]. The second group includes 10–13 nm microfibrils, which are hollow on cross-section and have a beaded appearance with some periodicity. Frequently, 10-nm microfibrils of this group are seen in proximity to basal lamina and may be anchored to these structures. Microfibrils may play a role in certain diseases of connective tissues. A significant reduction in microfibrils has been reported in cultured fibroblasts derived from patients with unilateral Marfan's syndrome, and this deficiency may be responsible for changes noted in the skin and blood vessels of these patients [32, 33].

The studies of Fleischmajer *et al.* showed that 10 nm microfibrils were formed in the three dimensional fibroblast cultures, and in structure they appeared very similar to the 10 nm microfibrils which are seen in normal tissues. They were 10–12 nm in diameter, had a beaded appearance, and labelled with fibrillin at 67-nm periodicity. Since microfibrils of this size may

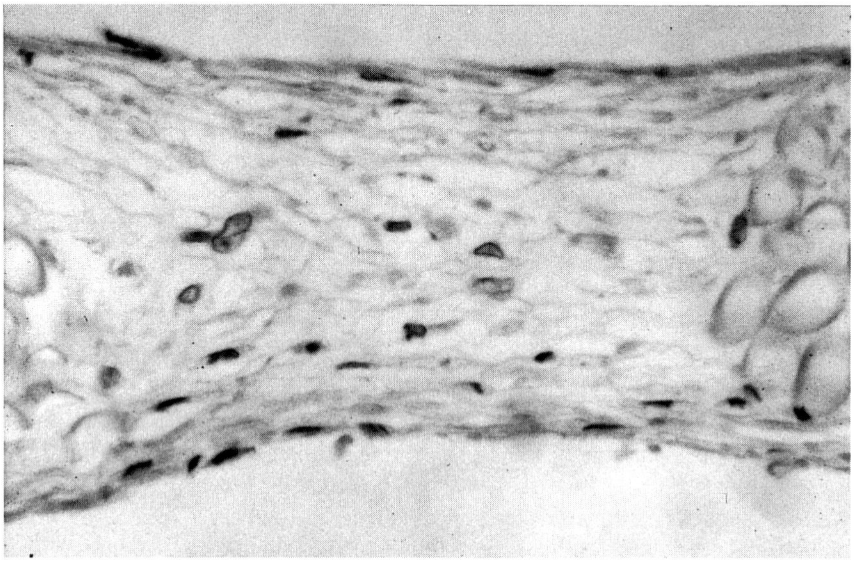

Fig. 3. Histologic cross-section of Dermagraft®, which is composed of human neonatal fibroblasts which are cultured in a biodegradable mesh composed of polyglactin fibers (Vicryl®. During the several week culture process the fibroblasts secrete multiple matrix proteins which accumulate in the mesh, as well as numerous cytokines.

play an important role in determining the elasticity or distensibility of tissues such as lungs, skin, and the cardiovascular system, the identification of microfibrils in this dermal tissue construct is exciting. Thus, this dermal model system allows the fibroblasts to form pseudopods and to deposit collagen and microfibrils in well demarcated intercellular spaces, resembling the *in vivo* process. In contrast, monolayer fibroblast cultures do not form these intercellular chambers.

We have utilized Dermagraft in an attempt to furnish a continuous dermal replacement layer beneath meshed skin grafts, with the thought that this would improve the cosmetic and functional outcomes following mesh grafting. In initial experiments in our laboratory, this living graft was first placed over 2 × 2 cm full-thickness wound defects in athymic mice. The wounds were excised to the level of the panniculus carnosus. Rapid incorporation of Dermagraft into the wound was observed, followed by partial epithelialization from the wound edges [21]. In contrast, when PGA or PGL meshes without fibroblasts were placed in the wound, the meshes rapidly separated from the wound and no fibrovascular ingrowth was seen. In subsequent animal studies [22], we placed meshed, split-thickness human skin grafts, 8 to 12 thousandths of an inch in thickness, on top of the dermal replacements. Skin was used within 48 hours of harvesting from cadaver donors and was maximally stretched (expanded) over the surface of the Dermagraft and the edges of the meshed skin were sutured to the wound edges. The PGA\PGL-fibroblast grafts routinely supported "take" of the overlying skin grafts. Vascularization of the meshed skin occurred rapidly with subsequent closure of the mesh graft interstices by epithelialization, within 10 days, across the surface of the dermal replacement. Histologic examination revealed complete incorporation of the tissue replacement into the wounds. Occasional mononuclear cell infiltrates with a rare multinucleated (granulomatous) giant cell were seen, but most areas showed little inflammatory response. Immunofluorescent microscopy revealed deposition of basement membrane proteins laminin and type IV collagen beneath the epithelialized interstices. Progressive dissolution of the mesh occurred *in vivo*, and by 4 weeks after grafting no evidence remained of the PGA or PGL mesh fibers. No inflammatory reaction was seen in the later stages of resorption. By gross and histologic observations on various days after grafting, no differences were noted between grafts composed of PGA (Dexon) or PGL (Vicryl) meshes. Control grafts containing PGA or PGL mesh without cultured cells did not support "take" and epithelialization of overlying meshed skin grafts. No gross or histologic differences were noted between fresh and cryopreserved grafts. The PGA/PGL meshes containing cultured human fibroblasts (Dermagraft®) maintain viability after controlled-rate freezing and subsequent thawing. Clinical trials of this dermal replacement are in progress at our Burn Center [34] and other burn centers in the United States.

The use of a living dermal replacement which can be applied immediately to the wound following excision may offer an advantage over other techniques which require a delay in the application of a skin or dermal replacement. It has been shown that delayed application of skin grafts does not inhibit wound contraction as effectively as does immediate grafting. In open wounds, immediate application of split-thickness skin grafts will moderately reduce wound contraction [35]. If, in contrast, the split-thickness grafts are not applied until the wound has actively begun to granulate and contract, the inhibitory effect of the split-thickness skin graft is lost. On the other hand, delayed full-thickness skin grafting of open wounds in rabbits inhibited contraction [36]. Further trials will determine if the early application of Dermagraft to full-thickness wounds can result in less contraction and scar formation, and if Dermagraft can impart qualities to meshed grafts so that they perform more like full-thickness grafts than partial-thickness grafts. Clearly, wounds which are covered with full-thickness grafts heal with far superior cosmetic and functional qualities than wounds which are closed with split-thickness skin grafts [37].

The development of effective temporary skin replacements

As discussed above, if we examine the various refinements in burn care over the past thirty to fifty years, it appears that excisional therapy has been associated with the most significant increase in survival of seriously burned patients [1–5]. Increasingly early after injury and after shock resuscitation and stabilization in most burn centers, patients are transported to the operating room where their wounds are surgically excised under general anesthesia (1). Wound excision has important benefits, probably most importantly in the reduction of the load of necrotic tissue which can serve as a culture medium for microorganisms [38]. Early burn wound excision may also decrease some of the metabolic responses to burn injury [39]. However, following surgical excision we are left with massive open wounds which in many patients cannot be immediately closed with autologous skin due to the lack of available "donor sites" for skin graft harvesting. We need to develop improved technologies for closing these wounds with both temporary and permanent skin and skin replacements.

There are numerous temporary skin replacements either currently available for clinical use, or in clinical testing [40–42]. These materials include temporary and permanent skin replacements, epidermal and dermal skin replacements, and synthetic/biosynthetic and biologic skin replacements. This chapter will focus on skin replacements which consist of, or include, cultured cells. Several acellular materials which have been designed and tested as temporary skin replacements: Biobrane®, Dow Hickam Inc., Sugar Land, TX [43, 44] and Integra®, Integra Life Sciences Inc., Plainsboro, NJ [45–47] will not be discussed extensively in this chapter, since they do not contain cultured cells. Biobrane, composed of a thin layer of silicone rubber bound to nylon mesh, is currently in clinical use as a dressing for partial-thickness wounds and as a temporary skin replacement for excised wounds. Integra, a cross-linked collagen matrix which is bound to a silicone rubber layer, is designed to function as a temporary skin replacement and to permit fibrovascular ingrowth into the matrix, has not been approved for use by the FDA. With both materials, the silicone rubber layer is designed to function as an "epidermal" barrier.

The "gold standard" for temporary wound coverage, cadaver allograft skin, is utilized by most burn centers in this country for achieving temporary wound closure in excised wounds. Cadaver skin is effective for temporary closure of these wounds and it can be life-saving for patients with massive burn injuries. It is the wound covering material of choice for closure of excised burn wounds [48]. Cadaver skin closes the wound and prepares the wounds for definitive autografting. When viable allograft skin has adhered to the wound and is subsequently removed, a well-vascularized wound base usually remains which can then accept skin autografts. However, there are significant problems associated with the use of cadaver skin. There is high demand for allograft skin and it frequently is unavailable for use at individual burn centers. Since the U.S. Food and Drug Administration has recently increased its surveillance of tissue banking operations, the supply of cadaver skin has become even more limited. In addition the quality of cadaver skin is variable, since donors may be quite elderly and frequently limited areas of the body are able to be harvested. Prolonged postmortem time periods until skin is harvested may decrease the viability of the skin; in addition prolonged skin storage at refrigerated temperatures, and cryopreservation procedures, may decrease skin viability and thus limit its performance on the excised wound bed. Perhaps most importantly, the potential of cadaver skin for transmitting infection, both bacterial and viral, is always present. It will probably never be possible to insure that allograft skin is free of contaminating viruses, since the donor may be infected but seronegative at the time of harvesting, and the donor of skin, unlike the blood donor, cannot be retested at a later date.

Allograft skin is rejected within several weeks of placement on the wound. 20 years ago, allograft skin routinely survived a month or even longer. However, as overall care for burn patients has improved, particularly in the areas of early excision and nutritional support, immune function has substantially improved so that allograft skin is rejected much more quickly. An equally important effect is the separation of epidermis from epidermis following placement of cadaver skin on the wound, a process which appears to be related to cryopreservation procedures.

Utilizing our experience with cultured fibroblasts and Dermagraft®, we are attempting to develop a laboratory-grown temporary replacement for human skin. An important aspect of human fibroblasts is that they are relatively nonimmunogenic, so they do not stimulate a rejection response in the host. Thus, allogeneic cells can be utilized to produce a storable tissue product which can be used immediately after injury on the patient. We utilized a biosynthetic dressing, Biobrane® (Dow Hickam Inc., SugarLand, TX), which had been moderately successful as a temporary skin replacement [43], and have attempted to improve its performance as a biologic skin replacement. Biobrane is composed of a nylon mesh which is bonded to a thin silicone rubber membrane; the membrane provides a barrier against water-vapor transmission from the wound and against bacterial invasion from the environment. Although Biobrane has been utilized as a temporary skin replacement by some burn centers, it does not encourage the excellent fibrovascular ingrowth which results following the application of viable human skin. As a result, Biobrane has not been universally accepted as a temporary skin replacement for excised wounds. We attempted to improve the biologic and clinical performance of Biobrane by seeding human neonatal fibroblasts in the nylon mesh. We found that the cells proliferated rapidly within the mesh and after several weeks a densely cellular "tissue" formed surrounding the nylon fibers, which contained high levels of secreted human matrix proteins as well as multiple growth factors and in these respects resembled Dermagraft, described above. Animal testing showed that this skin replacement could be used either fresh or cryopreserved (viable), or as a non-viable product, to close full-thickness wounds [49]. We believe that the matrix proteins and growth factors contained in the "tissue" encourage fibrovascular ingrowth from the wound bed and decrease the inflammatory response to the foreign materials, nylon and silicone.

These grafts were successful in closing full-thickness excised wounds in athymic mice [49]. Even though the human fibroblasts are non-immunogenetic in the allogeneic situation, they would quickly incite a rejection response in the heterograft situation. Thus we utilized athymic mice for preclinical testing of the grafts, since these animals lack a cell-mediated immune response. The grafts stimulated excellent fibrovascular ingrowth and also appeared to completely inhibit the wound contract process. In *in vivo* sections you can see the dense array of fibroblasts and blood vessels, ten days after placement of grafts on the wound (Fig. 4). We quantified the adherence, working with our bioengineers, and found that this material, which is now termed Dermagraft-Transitional Covering™ or DG-TC™, was actually more adherent to the wound over a two week period compared to cryopreserved human allograft skin.

When we measured messenger RNAs for various growth factors in the DG-TC, high levels of message were found for multiple cytokines including acidic and basic FGFs, keratinocyte growth factor, and TGFα and TGFβ [49] (Table 1). Levels of cytokine-specific message were far higher than those found in specimens of adult skin. In fact, there was very little message for cytokines in adult skin, but interestingly, neonatal skin contained very high levels of mRNAs for multiple trophic cytokines. Thus the cytokine profile of DG-TC parallels that of neonatal skin, rather than adult skin. We believe that the high levels of growth factors which are reflected by this mRNA profile will improve the physiologic performance of this temporary skin replacement.

Handling of skin replacements is very important for the burn surgeons, who must work with extensive wounds and close them rapidly. Many of these wounds are irregular in shape and configuration. DG-TC has good handling characteristics, can be moved on the wound easily and sutured or stapled quickly. Very importantly, the material is safe for human use since

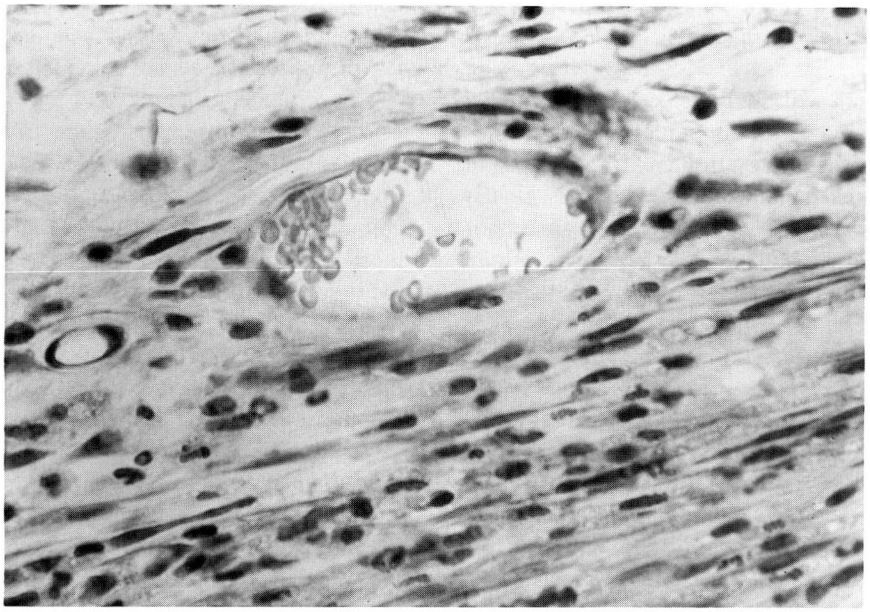

Fig. 4. Histologic section of Dermagraft-TC, two weeks following placement of the temporary skin replacement on full-thickness wounds on athymic mice. Dense numbers of fibroblasts are seen between the nylon fibers, and the wound tissue is penetrated by numerous blood vessels. Hematoxylin and eosin, × 200.

Table 1. Quantification* of cytokine-specific messenger RNA in skin and a tissue construct containing cultured fibroblasts

	Actin	aFGF	bFGF	KGF	TGFα
Neonatal foreskin	+5	+++	+++	++++	++++
Human Adult Skin	+5	0	0	0	+
*Dermagraft-TC®	+5	++	++	++++	++++

(* Human neonatal fibroblasts cultured in Biobrane® dressing.)

Amounts of cytokine-specific mRNA in the different groups were quantified by polymerase chain reaction (PCR) and levels were compared by the densities of the gel bands. aFGF, acidic fibroblast growth factor; bFGF, basic fibroblast growth factor; KGF, keratinocyte growth factor; TGFα, transforming growth factor alpha. Each (+) represents an approximate 10-fold increase in band density with a single (+) corresponding to a visible but barely discernible band. Levels of actin mRNA served as an internal control to standardize the quantity of tissue analyzed in each experiment. mRNA analysis was performed by Johnnie Underwood Ph.D., University of California, San Francisco.

the mother, the baby and the cultured cells can be screened over a period of time for multiple pathogenic agents including viruses. Since the donor and mother can be retested for viral infections after the cells are obtained, there should be essentially zero incidence of viral contamination with these grafts. This is very important, since as discussed above the potential for viral contamination with human cadaver skin is always present.

A material such as this can be stored "on the shelf" and utilized immediately to close excised wounds. Ready availability is very important for these patients because it has been shown that patients who are excised soon after admission have a better outcome than the former technique of waiting for the eschar to separate over time [1]. It is critically important to have a temporary graft material readily available for wound coverage. It would be a further advantage to have a skin replacement which will not be rejected. Since human fibroblasts appear to be non-immunogenic in the allogeneic situation, this offers a further advantage to skin replacements which contain human fibroblasts. On the other hand, human keratinocytes are highly immunogenic. Although early studies suggested that the culture process would decrease their antigenicity, perhaps by decreasing expression of surface transplantation antigens, in clinical practice the cultured allogeneic epithelium appears to be rapidly rejected.

Pilot clinical studies in ten patients with this temporary skin replacement, Dermagraft-TC, have been successfully completed, under the direction of Advanced Tissue Sciences, Inc. [50]. Seven of the patients were treated at our own burn center, and three were treated at other burn centers. An additional patient with 60% full-thickness burns was treated at our Burn Center under compassionate use protocol. He was covered with greater than one square foot of DG-TC (Fig. 5). The grafts functioned well in all patients studied, achieving successful wound closure in all patients until autograft skin could be applied. In several patients the DG-TC grafts outperformed adjacent areas of cryopreserved cadaver skin, which developed early sloughing of the epidermal layer either from immunologic rejection or as a result of damage to the dermal-epidermal junction from the cryopreservation process. A multicenter pivotal trial of DG-TC in currently in progress in the United States.

Development of permanent cultured skin replacements

A method which holds great promise to permanently close full-thickness wounds utilizes cultured epithelial cells. After separation of epidermis from the dermis using enzymatic treatment, keratinocytes can be induced to proliferate and to expand greatly in culture. Successful epidermal culture technologies were developed over 15 years ago [51, 52], and cultured keratinocyte sheets were soon applied to burn wounds [53]. Numerous trials have been subsequently performed to determine the ability of cultured epithelium to permanently close full-thickness wounds [42, 54–65]. Reports of clinical trials from multiple burn centers are contained in the January/February 1992 issue of the Journal of Burn Care and Rehabilitation, and the results were summarized by Odessey [66]. Although some success has been achieved, the cultured epithelial sheets are extremely fragile, and they can be difficult to handle and apply to the wound. The subsequent wound care must be meticulous, and cultured grafts appear to be highly susceptible to infection. For these reasons the general success of cultured epithelium for closing wounds is not dependable in most centers. Rives from France has reported perhaps the best results [67]; however, the intensity of wound care which he reports appears extraordinary in terms of time and personnel. In addition, several studies have reported that the healed epithelium can be quite fragile and the skin is prone to breakdown and contractions [56–59]. Clinical results following grafting of cultured epithelium appear to be optimized when the grafts are placed over cadaver dermis [60, 66]. However, the timing for achieving this aim may prove difficult, and if cadaver skin is rejected and lost prior to the availability of cultured epidermal grafts for the burn patient, the opportunity to utilize cadaver dermis is lost.

One of the largest clinical series of the use of epidermal cultured grafts, reported in 1993 from the U.S. Army Burn Center at San Antonio, showed disappointingly poor take of cultured epithelial autograft [55]. As little as 13 to 20 percent take was found in deeper wounds. Other reports describe adequate initial take of cultured epithelium but later wound problems such as shearing and wound contraction [56]. Obviously it would be a medical and economic disaster to wait three to four weeks and spend many thousands of dollars for cultured epithelium, only to achieve unsatisfactory take, persistence and long-term performance of the grafts.

Subsequent reports have noted that histologically, full-thickness burn wounds resurfaced by CEA (cultured epithelial autografts) contain abnormal features, both clinically and histologically. All reports seem to agree that the epidermis developed from CEA grafts develops fully differentiated basal, spinous, granular, and cornified layers, but lacks rete ridges at one week after grafting [60], two weeks [61], four weeks [62], 14 weeks [53], six months [61, 63], and 11 months [64]. Woodley *et al.* [65] studied biopsies of CEA which were placed on four severely burned patients 14 to 35 days after injury. All of the patients were excised to fascia prior to grafting. The grafted wounds were examined eight to 148 days after transplantation, and the sites were compared to control specimens of normal skin which were obtained from elective surgical procedures. In all four patients the reconstituted basement

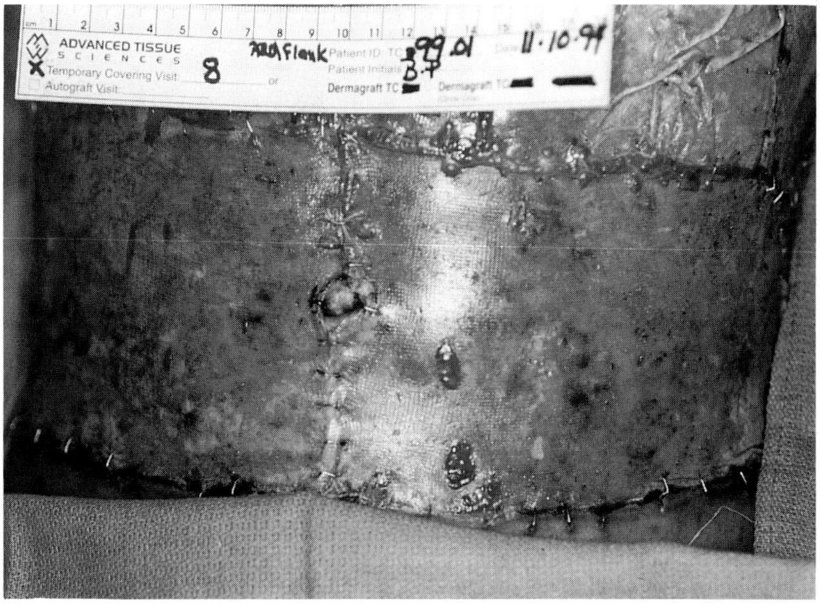

Fig. 5. Following full-thickness burns over 60% of the body surface, some of the surgically excised wounds on the abdomen of this 16 year old boy were covered with Dermagraft-TC (DG-TC). 8 days after placement of the grafts the DG-TC is shown tightly adherent to the wounds, serving as an effective tempoary wound covering until permanent skin grafting can be performed.

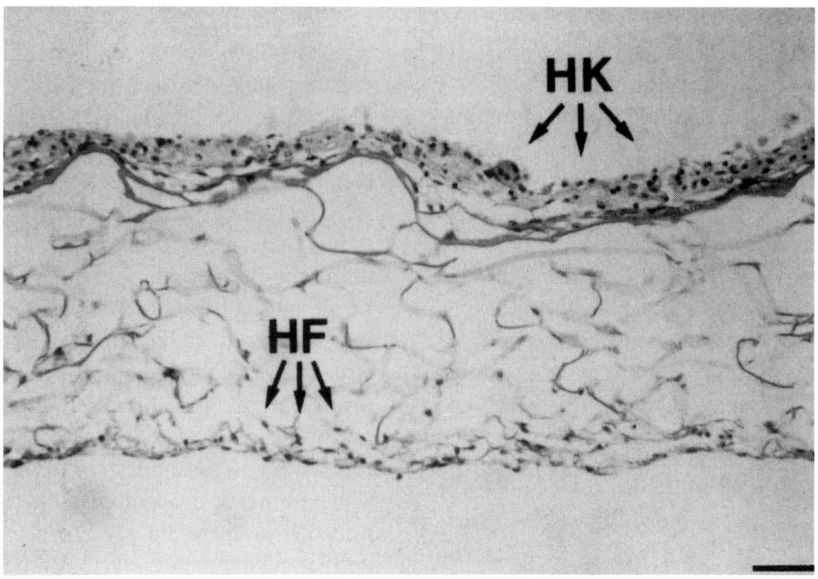

Fig. 6. A composite skin graft developed in our laboratory is composed of human fibroblasts and keratinocytes which are cultured on a cross-linked collagen membrane. Inclusion of fibroblasts in the grafts was necessary in order to achieve confluent growth of keratinocytes on the surface of the membrane. HK, human keratinocytes; HF, human fibroblasts. Hematoxylin and eosin stain, × 25.

membrane zone beneath the autografts was incomplete and lacked type IV (basement membrane) collagen and anchoring fibrils. Anchoring fibril collagen (type VII) was present, demonstrated by immunofluorescence, at the junction between the CEA and the underlying dermis. However, since mature anchoring fibrils were not

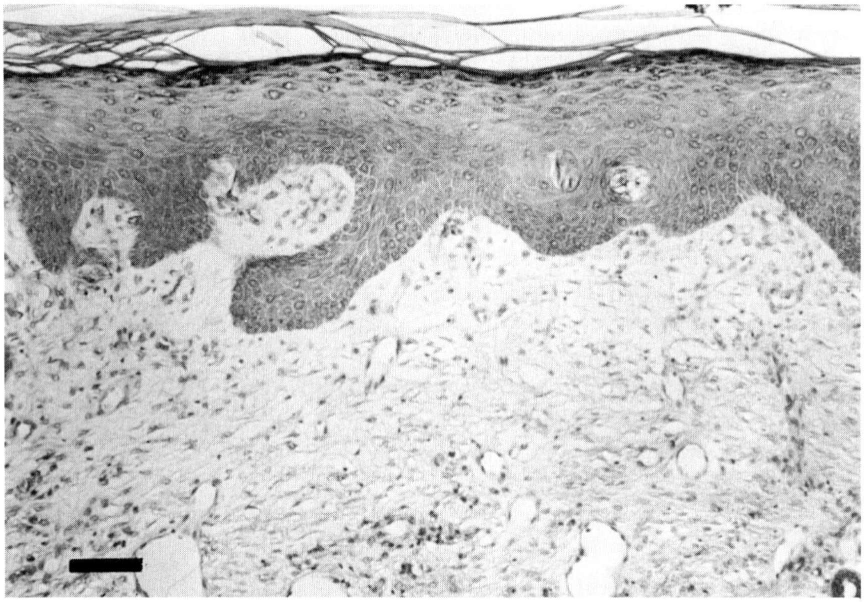

Fig. 7. Histologic section from patient's wound which was closed 26 days previously by the composite graft shown in the previous figure. Extensive interdigitations are seen connecting the epidermal and dermal layers; the projections resemble rete ridges but are not as regular in pattern. Hematoxylin and eosin, × 25.

found, it was suggested that the anchoring fibril collagen was unable to become organized into structurally intact and functional anchoring fibrils.

Culturing keratinocytes on flexible, biocompatible and biodegradable membranes may be able to facilitate the transfer of cell sheets to the wound bed, by eliminating the need to cleave the attachments between cells and the culture surface with proteolytic enzymes prior to transfer of grafts. In addition, the handling of delicate cultured epithelium is difficult and labor-intensive, and the simple transfer of cell-coated membranes to the wound would greatly simplify graft transfer, operative surgery and wound care, particularly if the membrane could be left in place postoperatively to secure the cells to the wound surface. These experiments are underway in our own laboratory.

Development of epidermal-dermal (composite) cultured grafts

When cultured epithelial sheets are placed on the full-thickness wound which by definition lacks a dermis, there is delayed basement membrane formation with initial development of a flat dermal-epidermal junction. This may have very important consequences to wound healing. In normal skin, rete ridges are present which result in interdigitation of epidermis and dermis; this markedly increases the surface area of the dermal-epidermal junction which furnishes strength and durability of the skin [68]. On a molecular level, the dermis is important for epithelial development and maturation [69], including the development of ultrastructural elements such as anchoring fibrils which help attach epidermis to the dermis. Since the full-thickness wound by definition lacks dermis, and dermis does not or is very slow to regenerate, developing of attachment structures is delayed [34]. This may account for the disappointing take of epithelial sheet grafts. For these reasons it appears that it is necessary to replace the dermis, as well as the epidermis, in full-thickness wounds. The best means to accomplish this remains to be determined. Several years ago our laboratory developed a dermal analogue composed of human fibroblasts cultured in a collagen-glycosaminoglycan membrane [71, 72]. We found that inclusion of fibroblasts in the membrane was necessary in order to obtain confluent growth of keratinocytes on the surface of the dermal substrate (Fig. 6). Without fibroblasts, we were not able to achieve satisfactory epithelial growth on the surface. We subsequently showed that levels of collagen type IV, a structural protein component of basement

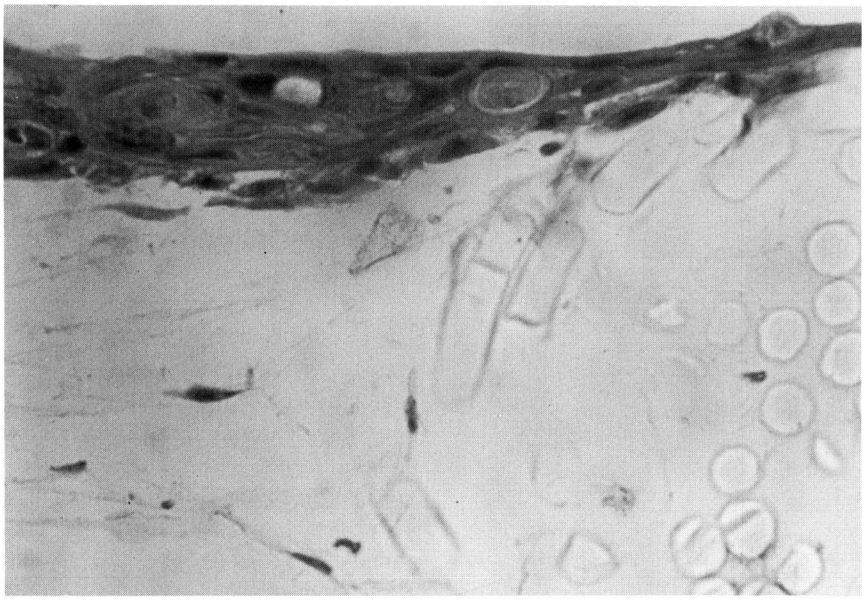

Fig. 8. Human keratinocytes were cultured on the surface of Dermagraft®, reaching confluence after 4 to 6 days of culture. At this point the composite grafts are ready for transfer to wounds. Hematoxylin and eosin, × 100.

membrane, were far higher in the culture medium when keratinocytes were cultured together compared to cultures of the individual cell types [73]. Keratinocytes cultured alone produced little laminin, another structural protein component of basement membrane, while both fibroblasts and composite cultures produced high amounts of laminin. Clearly, fibroblasts play an important role in the development of the basement membrane and associated structures.

These composite grafts successfully closed wounds in athymic mice [74]. When we placed the composite grafts on patients [75], we had some areas of excellent wound take with good skin appearance and minimal hypertrophic scar formation. As early as 10 days post-grafting, we found extensive rete ridge interdigitation (Fig. 7), far different from the flat dermal-epidermal junction which occurs when cultured epithelial sheets are applied to wounds. This was a very uniform finding in our seven patients. Basement membrane proteins laminin and type IV collagen were identified immunohistochemically in all patients. We believe that early formation of normal anatomic structures is facilitated in these composite grafts which contain both epidermal and dermal components.

However, overall take of this composite skin was approximately 50 percent, not successful enough to make this a routinely acceptable skin replacement. We believe that graft take was limited for several reasons. The wounds which receive autologous cultured grafts are really chronic wounds, since we must wait three to four weeks after injury to produce the cultured grafts. The wounds contain a variety of proteases [76], which undoubtedly readily attack the collagen-GAG membrane. In addition, many of the wounds have high bacterial counts [77] which also limits the success of fragile grafts.

As described above, we have evaluated a dermal replacement composed of fibroblasts cultured on biodegradable polyglyactin mesh [21, 78]. Since the polyglactin fibers undergo hydrolysis when placed in the wound, rather than attack and degradation by proteases and inflammatory cells, we hypothesized that these cultured grafts will be noninflammatory when placed on the wound. The cultured fibroblasts secrete matrix proteins which are deposited in the dermal analogue during the culture process. This material, Dermagraft®, is undergoing clinical evaluation as a replacement for human dermis [79]. The studies indicate that Dermagraft can support adherence and vascularization of overlying meshed skin grafts in both animals and humans, with an absence of immune responses and minimal or no inflammatory respons-

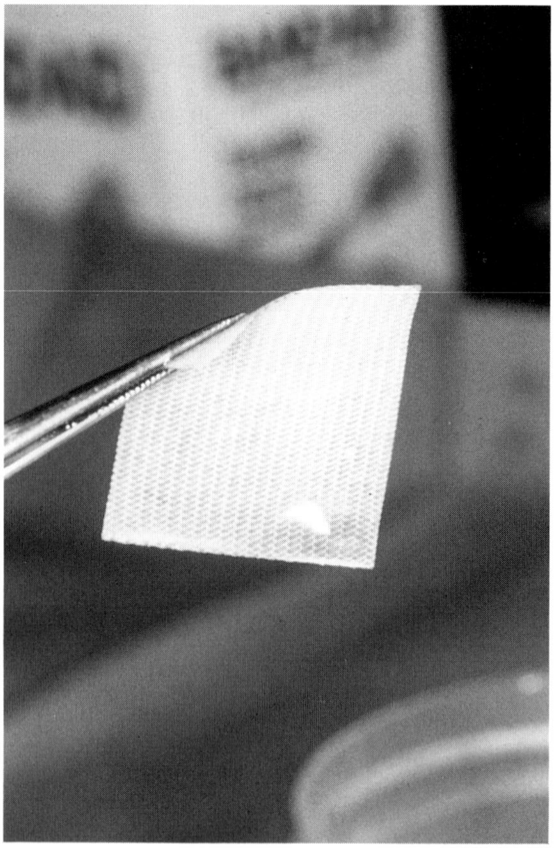

Fig. 9. The Dermagraft-keratinocyte skin replacement has excellent structural stability and handling qualities, due to the physical properties of the resorbable polyglactin mesh.

es. We and others are also investigating the attributes of other biodegradable matrix fibers and supports.

We have cultured adult human keratinocytes on the surface of Dermagraft, and have achieved confluent growth of the keratinocytes resulting in a multi-layered epithelium (Fig. 8). These composite grafts successfully closed wounds on athymic mice, and persistence of human epidermis, an important aspect of these types of studies, was demonstrated by positive staining for human involucrin, an epidermal differentiation antigen. Transplantation of the grafts also resulted in early formation of basement membrane structures including appearance of laminin and type IV collagen, by immunostaining [80].

Very importantly, these composite grafts are easy to handle because of the excellent structural integrity of the polyglactin mesh (Fig. 9). This benefits the cell culture technician as well as the surgeon. The grafts are easy to remove from the culture flask or dish by simply lifting them out, since they do not require enzymatic treatment to release the epithelium from the culture vessel surface. The grafts drape easily and can be moved around on the wound bed, they have good resistance to tearing, and they can be easily sutured or stapled to the woun. Thus an important advantage of using a dermal replacement in conjunction with the cultured epithelium is that graft handling and placement are greatly facilitated. Clinical trials of these composite grafts are planned for the future.

Bell *et al.* developed one of the first composite skin replacements, comprised of fibroblasts cast in collagen lattices which were allowed to contract and then seeded with dispersed, non-cultured epidermal cells [81]. A multilayered keratinizing epidermis with desmosomes, tonofilaments, and hemidesmosomes developed with formation of basement membrane structures within 2 weeks. It was suggested that the dermal replacement could provide better interactions between fibroblasts and other cell types *in vitro*, compared to cells cultured on a plastic or glass surface. Organogenesis Inc. (Canton, MA) has further developed this composite skin replacement (Graftskin®). In our laboratory [82, 83], Graftskin placed on full-thickness excised wounds on athymic mice persisted for the 60 day study period with minimal contraction and resulted in well-differentiated epidermis with stratum corneum. Basement membrane proteins were identified by immunohistochemical staining and electron microscopy within 2 weeks of graft transplantation. Graftskin® is currently undergoing evaluation in clinical trials as a skin replacement for burn wounds and cutaneous ulcers [84].

Summary

There is much progress to be made to optimize the development of laboratory-grown temporary and permanent skin replacements. It seems clear that replacement of both epidermal and dermal layers will be important for achieving optimal "take" of cultured grafts and for optimizing the quality of wound healing. Although the use of retained cadaver allodermis on the wound bed may improve the performance of cultured epithelium [60, 61], the development of successful complete dermal-epidermal skin replacements ("composite grafts") would greatly simplify burn management in the future, handling and stability of the cultured grafts should be improved, and clinical outcomes would be expected to be superior. Unfortunately, funding for this type of "applied research" has not achieved high priority from the federal government granting agencies, despite the great clinical need for improved technology, and future progress will depend largely upon commercial support.

References

1. Sheridan RL, Tompkins RG and Burke JF (1993) Management of the burn wound by prompt excision and immediate closure. J Intensive Care Med 9: 6–19.
2. Feller I, Tholen D and Cornell RG (1980) Improvements in burn care, 1965 to 1979. JAMA 244: 2074–8.
3. Burke JF, Bondon CC and Quinby WC (1974) Primary burn excision and immediate grafting: A method of shortening illness. J Trauma 14: 389–95.
4. Gray DT, Pine RW, Harnar TJ *et al.* (1982) Early surgical excision versus conventional therapy in patients with 20 to 40 percent burns. Am J Surg 144: 76–80.
5. Tompkins RG, Burke JF, Schoenfield DA *et al.* (1986) Prompt eschar excision: a treatment system contributing to reduced burn mortality. Ann Surg 204: 272–81.
6. Laufman H and Rubel T (1977) Synthetic absorbable sutures. Surg Gynecol Obstet 145: 597–608.
7. MacKinnon SE and Dellon AL (1990) Clinical nerve reconstruction with a bioabsorbable polyglycolic acid tube. Plast Reconstr Surg 85: 419–24.
8. Nussbaum CE, McDonald JV and Baggs RB (1990) Use of vicryl (polyglactin-910) mesh to limit epidural scar formation after laminectomy. Neurosurg 26: 649–54.
9. Nussbaum CE, Maurer PK and McDonald JV (1989) Vicryl (polyglactin-910 mesh) as a dural substitute in the presence of Pia arachnoid injury. J Neurosurg 71: 124–7.
10. Hoffman MS, Roberts WS, Lapolla JF *et al.* (1989) Use of Vicryl mesh in the reconstruction of the pelvic floor following exenteration. Gynecol Oncol 35: 170–1.
11. Tyrell J, Silberman H, Chandrasoma P, Niland J and Shull J (1989) Absorbable versus permanent mesh in abdominal operations. Surg Gynecol Obstet 168: 227–32.
12. Dayton MT, Buchele BA, Shirazi SS and Hunt LB (1988) Use of an absorbable mesh to repair contaminated abdominal-wall defects. Arch Surg 121: 954–60.
13. Patsner B, Mann WJ Jr, Chaias E, Orr JW Jr (1990) Intestinal complications associated with use of the dexon mesh sling in gynecologic oncology patients. Gynecol Oncol 38: 146–8.
14. Edlich RF, Panek PH, Rodeheaver GT *et al.* (1973) Physical and chemical configuration of sutures in the development of surgical infection. Ann Surg 177: 679–88.
15. Lilly GE, Cutcher JL, Jones JC and Armstrong JH (1972) Reaction of oral tissues to suture material. IV. Oral Surg 33: 152–61.
16. Triglia D, Braa SS, Yonan C and Naughton GK (1991) Cytotoxicity testing using neutral red and MTT assays on a three-dimensional human skin substrate. Toxic in Vitro 5: 573–8.
17. Slivka SR, Landeen L, Zimber MP and Bartel RL (1991) Biochemical characterization, barrier function and drug metabolism in an *in vitro* skin model. J Cell Biol 115A: 1370.
18. Slivka SR, Zeigler F and Bartel RL (1991) An *in vitro* skin model for the study of keratinocyte responses to irritants. J Cell Biol 115A: 2072.
19. Slivka SR, Halberstadt CR, Landeen L *et al.* (1991) Characterization of a three-dimensional cultured human skin model. American Inst Chem Engineers Ann Mtg, Los Angeles, CA.
20. Slivka SR, Landeen L, Zimber MP and Bartel RL (1991A) Characterization of a three-dimensional human skin culture model for *in vitro* percutaneous absorption studies. Pharm Research 8S: 143.
21. Cooper ML, Hansbrough JF, Spielvogel RL *et al.* (1991) *In vivo* optimization of a living dermal substitute employing cultured human fibroblasts on a biodegradable polyglycolic acid or polyglactin mesh. Biomaterials 12: 243–8.
22. Hansbrough JF, Cooper ML, Greenleaf G *et al.* (1992) Evaluation of a biodegradable matrix containing cultured human fibroblasts as a dermal replacement beneath meshed split-thickness skin grafts. Surgery 111: 438–46.
23. Clark RAF (1990) Fibronectin matrix deposition and fibronectin receptor expression in healing and normal skin. J Invest Dermatol 94S: 128–34.
24. Clark RAF, Lanigan JM, DellaPelle P *et al.* (1982) Fibronectin and fibrin provide a provisional matrix for epidermal cell migration during wound epithelialization. J Invest Dermatol 79: 264–9.
25. Brownell AG, Bessem CC and Slavkin HC (1981) Possible functions of mesenchyme cell-derived fibronectin during formation of basal lamina. PNAS USA: 3711–5.
26. Couchman JR, Austria MR and Woods A (1990) Fibronectin-cell interactions. J Invest Dermatol 94S: 7–14.
27. Cohen R, Zimber M, Hansbrough JF *et al.* (1991) Tear strength properties of a novel cultured dermal tissue model. Annals Biomed Engineering 19: 600.
28. Fleischmajer R, Contard P, Schwartz E *et al.* (1991) Elastin-associated microfibrils (10 nm) in a three-dimensional fibroblast culture. J Invest Dermatol 97: 638–43.
29. Maddox BK, Sakai LY, Keene DR and Glanville RW (1989) Connective tissue microfibrils. J Biol Chem 264: 21381–5.
30. Bruns RR, Press W, Engvall E *et al.* (1986) Type VI collagen in extracellular, 100 nm periodic filaments and fibrils: identification by immunoelectron microscopy. J Cell Biol 103: 393–404.
31. Keene DR, Engvall E and Glanville RW (1988) Ultrastructure of type VI collagen in human skin and cartilage suggests an anchoring function for this filamentous network. J Cell Biol 107: 1995–2006.
32. Godfrey M, Olson S, Burgio RG *et al.* (1990) Unilateral microfibrillar abnormalities in a case of asymmetric Marfan's syndrome. Am J Hum Genet 46: 661–71.
33. Godfrey M, Menashe V, Weleber RG *et al.* (1990) Cosegration of elastin-associated microfibrillar abnormalities with the Marfan phenotype on families. Am J Hum Genet 46: 652–60.
34. Hansbrough JF, Dore' C and Hansbrough WB (1992) Clinical trials of a living dermal tissue replacement beneath meshed, split-thickness skin grafts on excised burn wounds. J Burn Care Rehab 13: 519–29.
35. Stone PA and Madden JW (1974) Effect of primary and delayed split skin grafting on wound contraction. Surg Forum 25: 41–4.
36. Donoff RB and Grillo HC (1975) The effects of skin grafting on healing open wounds in rabbits. J Surg Res 19: 163–7.
37. Rudolph R, Vande Berg J and Ehrlich HP (1992) Wound contraction and scar contracture. In: Cohen IK, Diegelmann RF and Lindblad WJ (eds) Wound Healing; Biochemical and Clinical Aspects, pp 96–114. Philadelphia: WB Saunders Co.
38. Hansbrough JF (1987) Burn wound sepsis. J Intensive Care Med 2: 313–27.

39. LaLonde C and Demling RH (1987) The effect of complete burn wound excision and closure on postburn oxygen consumption. Surgery 102: 862–8.
40. Pruitt BA and Levine NS (1984) Characteristics and uses of biologic dressings and skin substitutes. Arch Surg 119: 312–22.
41. Atnip RG and Burke JF (1983) Skin coverage. Curr Prob Surg 20: 623–86.
42. Hansbrough JF (1992) Wound Coverage with Biologic Dressings and Cultured Skin Substitutes. Austin: RG Landes.
43. Purdue GF, Hunt JL, Gillespie RW *et al.* (1987) Biosynthetic skin substitute versus frozen human cadaver allograft for temporary coverage of excised burn wounds. J Trauma 27: 155–7.
44. Tavis MN, Thornton NW, Bartlett RH *et al.* (1980) A new composite skin prosthesis. Burns 7: 123–30.
45. Yannas IV and Burke JF (1980) Design of an artificial skin. I. Basic design principles. J Biomed Mat Res 14: 65–81.
46. Yannas IV, Burke JF, Gordon PL *et al.* (1980) Design of an artificial skin. II. Control of chemical composition. J Biomed Mat Res 14: 107–31.
47. Heimbach D, Luterman A, Burke J *et al.* (1988) Artificial dermis for major burns. A multicenter randomized clinical trial. Ann Surg 208: 313–20.
48. Greenleaf G and Hansbrough JF (1995) Current trends in the use of allograft skin for burn patients and reflections on the future of skin banking in the United States. J Burn Care Rehab 15: 428–31.
49. Hansbrough JF, Morgan J, Greenleaf G and Underwood J (1994) Development of a temporary living skin replacement composed of human fibroblasts cultured in Biobrane[R], a synthetic dressing material. Surgery 115: 633–44.
50. Davis M, Hansbrough J, Mozingo D *et al.* (1994) Dermagraft-TC, a human dermal tissue for treatment of burns requiring temporary coverings. Wound Healing Society, Minneapolis MN, April (abstract).
51. Rheinwald JG and Green H (1975) Serial cultivation of strains of human epidermal keratinocytes: the formation of keratinizing colonies from single cells. Cell 6: 331–43.
52. Green H, Kehinde O and Thomas J (1979) Growth of cultured human epidermal cells into multiple epithelia suitable for grafting. PNAS USA 76: 5665–8.
53. Gallico GG, O'Connor NE, Compton CC *et al.* (1984) Permanent coverage of large burn wounds with autologous cultured human epithelium. NEJM 311: 448–51.
54. Compton CC (1993) Wound healing potential of cultured epithelium. Wounds 5: 97–111.
55. Rue LW III, Cioffi WG, McManus WF and Pruitt BA Jr (1993) Wound closure and outcome in extensively burned patients treated with cultured autologous keratinocytes. J Trauma 34: 662–8.
56. Desai MH, Mlakar JM, McCauley RL *et al.* (1991) Lack of long-term durability of cultured keratinocyte burn-wound coverage: A case report. J Burn Care Rehab 12: 540–5.
57. Kumagai N, Nishina H, Tanabe H *et al.* (1988) Clinical application of autologous cultured epithelia for the treatment of burn wounds and burn scars. Plast Reconstr Surg 82: 99–108.
58. Herzog SR, Meyer A, Woodley D and Peterson HD (1988) Wound coverage with cultured autologous keratinocytes: use after burn wound excision, including biopsy followup. J Trauma 28: 195–8.
59. Clugston PA, Snelling CFT, Macdonald IB *et al.* (1991) Cultured epithelial autografts: Three years of clinical experience with eighteen patients. J Burn Care Rehab 12: 533–9.
60. Cuono C, Langdon R and McGuire J (1986) Use of cultured autografts and dermal allografts as skin replacement after burn injury. Lancet 1: 1123–4.
61. O'Connor NE, Gallico GG, Kehinde O *et al.* (1984) Grafting of burns with cultured epithelium prepared from autologous epidermal cells: II. Intermediate term results on three pediatric patients. In: Hunt TK, Heppenstall HBS, Pines E *et al.* (eds) Soft and Hard Tissue Repair: Biological and Clinical Aspects, pp 283–92. New York: Praeger Publishers.
62. Pittelkow MR and Scott RE (1986) New techniques for the *in vitro* culture of human keratinocytes and perspectives on their use for grafting of patients with extensive burns. Mayo Clin Proc 61: 771–7.
63. Eldad A, Burt A, Clarke JA and Gusterson B (1987) Cultured epithelium as a skin substitute. Burns 13: 173–80.
64. Cuono CB, Langdon R, Birchall N *et al.* (1987) Composite autologous-allogeneic skin replacement: development and clinical application. Plast Reconstr Surg 80: 626–35.
65. Woodley DT, Peterson HD, Herzog SR *et al.* (1988) Burn wounds resurfaced by cultured epidermaltografts show abnormal reconstitution of anchoring fibrils. JAMA 259: 2566–71.
66. Odessey R (1992) Addendum: Multicenter experience with cultured epidermal autograft for treatment of burns. J Burn Care Rehab 13: 174–80.
67. Rives JM, Sellam P, Karcenty B *et al.* (1994) Cultured epithelial autografts (CEA) management and colonization control in extensive burn injuries: "Thoughts on local dressing". 9th Congress, Internat Soc Burn Inj, Paris: 152.
68. Briggaman RA and Wheeler CE Jr (1975) The epidermal-dermal junction. J Invest Dermatol 65: 71–84.
69. Coulomb B, Lebreton C and Dubertret L (1989) Influence of human dermal fibroblasts on epidermalization. J Invest Dermatol 92: 122–5.
70. Woodley DT, Peterson HD, Herzog SR *et al.* (1988) Burn wounds resurfaced by cultured epidermal autografts show abnormal reconstitution of anchoring fibrils. JAMA 259: 2566–71.
71. Boyce S and Hansbrough JF (1988) A skin autograft substitute: biological attachment and growth of cultured human keratinocytes on a graftable collagen and chondroitin-6-sulfate substrate. Surgery 103: 421–31.
72. Boyce S, Christianson D and Hansbrough J (1988) Structure of a collagen-GAG skin substitute optimized for cultured human epidermal keratinocytes. J Biomat Res 22: 939-57.
73. Cooper ML, Andree C, Hansbrough JF *et al.* (1993) Direct comparison of a cultured composite skin substitute containing human keratinocytes to an epidermal sheet graft containing human keratinocytes on athymic mice. J Invest Dermatol 101: 811–9.
74. Cooper ML and Hansbrough JF (1991) Use of a composite skin graft composed of cultured human keratinocytes and fibroblasts and a collagen-GAG matrix to cover full-thickness wounds on athymic mice. Surgery 109: 198–207.
75. Hansbrough JF, Boyce ST, Cooper ML and Foreman TJ (1989) Burn wound closure with cultured autologous keratinocytes and fibroblasts attached to a collagen-glycosaminoglycan substrate. JAMA 262: 2125–30.
76. Jeffrey JJ (1992) Collagen degradation. In: Cohen IK, Diegelmann RF and Lindblad WJ (eds) Wound Healing: Biochemical and Clinical Aspects, pp 177–194. Philadelphia: WB Saunders Inc.
77. Greenleaf G, Cooper ML and Hansbrough JF (1991) Microbial contamination in allografted wound beds in burn patients. J Burn Care Rehab 12: 442–5.

78. Hansbrough JF, Cooper ML, Cohen R *et al.* (1992) Evaluation of a biodegradable matrix containing cultured human fibroblasts as a dermal replacement beneath meshed skin grafts on athymic mice. Surgery 111: 438–46.
79. Hansbrough JF, Doré C and Hansbrough WN (1992) Clinical trials of a living dermal tissue repiacement placed beneath meshed, split-thickness skin grafts on excised burn wounds. J Burn Care Rehab 13: 519–29.
80. Hansbrough JF, Morgan JL, Greenleaf GE and Bartel R (1993) Composite grafts of human keratinocytes grown on a polyglactin mesh-cultured fibroblast dermal substitute function as a bilayer skin replacement in full-thickness wounds on athymic mice. J Burn Care Rahab 14: 485–94.
81. Bell E, Ehrlich HP, Buttle DJ and Nakatsuji T (1981) Living tissue formed *in vitro* and accepted as skin-equivalent tissue. Science 211: 1052–4.
82. Hansbrough JF, Morgan J, Greenleaf G *et al.* (1994) Evaluation of Graftskin composite grafts on full-thickness wounds on athymic mice. J Burn Care Rehab 15: 346–53.
83. Nolte CJM, Oleson MA, Hansbrough JF *et al.* (1994) Ultrastructural features of bilayered skin cultures grafted onto athymic mice. J Anatomy 185: 325–33.
84. Sabolinski ML, Rovee DT, Parenteau NL *et al.* (1995) The efficacy and safety of GRAFTSKIN for the treatment of chronic venous ulcers. Abstract, Wound Healing Soc, April 1995, Minneapolis MN; and Symposium on Advanced Wound Care, San Diego, CA, May (abstracts).

R. P. Lanza and W. L. Chick (eds.), Yearbook of Cell and Tissue Transplantation 1996/1997, 221–226.

Chapter 21

The transplantation of endothelial cells

Stuart K. Williams
Department of Surgery, University of Arizona, Health Sciences Center, Tucson, AZ 85724, U.S.A.

Early development of endothelial cell transplantation technology

The need for viable endothelial cells on the lumenal surface of vascular implants (e.g. coronary artery bypass grafts, saphenous vein grafts, polymeric grafts) has been realized only with the understanding of the complex physiology of these cells. The development and first clinical experiences with polymeric vascular grafts provided the initial evidence of the importance a functional endothelium on the lumenal surface of blood vessels. These synthetic blood vessel performed adequately in large diameter positions (e.g. abdominal aortic replacements) however as the diameter of the synthetic grafts needed became smaller long term function was compromised. The best evidence for the need of luminal endothelial cells is derived from the use of synthetic grafts with diameters less then 6 mm. These grafts exhibit extremely poor, clinically unacceptable, long term patency due predominantly to the formation of blood clots resulting in lost patency.

The bypass of occluded native vessels with diameters less then 6 mm has been limited to the use of autologous native blood vessels. While these vessels have provided acceptable long term patency clinicians have realized the importance of maintaining the integrity of the lumenal lining of endothelial cells [1–4]. Autologous vessels function not only as a proteinaceous, collagen tube but function adequately due to the antithrombogenic nature of the endothelial cell lining. Procedures which disrupt the endothelium, such as angioplasty and atherectomy provide additional evidence for the need of a functional endothelium.

The first report of the successful isolation and transplantation of endothelial cells was a report by Dr. Malcolm Herring in 1978 [5]. His methods involved the use of large vessel derived endothelium obtained by scraping the lumenal surface of vein segments. These cells were subsequently transplanted onto the lumenal surface of polymeric graft materials a process which was termed seeding. The methods used for seeding include mixing the isolated cells with autologous blood and subsequently using this blood-cell inoculum to preclot porous grafts. The lumenal blood contacting surface was observed to be predominantly fibrin and red cells with endothelial cells present predominantly within the resulting clot. The hypothesis behind these studies was that seeded endothelial cells would migrate from within the clot to the lumenal blood flow surface, proliferate and finally form a continuous monolayer through the formation of typical inter-endothelial cell junctions. Subsequent work by a large number of investigators including work from the laboratories of Schmidt [6], Stanley [7] and Graham [8] established the ability to accelerate the formation of monolayers on prosthetic grafts using large vessel endothelial cells as a source of cells for seeding.

Publication of these studies established the feasibility of performing endothelial cell transplantation and resulted in extensive work to optimize seeding technology. Numerous questions were identified which needed to be addressed before endothelial cell transplantation was expected to be found clinically acceptable.

Vessel source for transplanted endothelium

As described above, Herring first reported the use of venous derived endothelium for subsequent transplantation. The yield of cells from a suitable piece of vein segment was approximately 1×10^4 cells [9]. Thus, based on the largest piece of vein which could be obtained from a patient using an acceptable technique, only 10,000 endothelial cells could be obtained. Under even the most ideal conditions this cell number would provide a relatively sparse coating of cells requiring extensive cell population doubling to achieve a monolayer. Kempzinski *et al.* [10] reported that even if cells could be isolated and deposited on the surface of polymeric grafts with optimal efficiency a large number of these cells would be lost from the lumenal surface of the graft due to the effects of blood flow induced shear on the lumenal surface. Endothelial cells seeded with this technology would be required to undergo even more extensive proliferation or alternatively a new source of endothelial cells obtainable in large numbers must be found.

Cultured endothelium for transplantation

Methods for the isolation and culture of non-human endothelial cells were first reported in the 1920s providing methods for the expansion of endothelial cell numbers by in vitro proliferation [11]. However the same methods used for non-human endothelial cell proliferation had no stimulatory effect on human endothelial cell proliferation. During the 1970s methods for human endothelial cell isolation, based on the pioneering work of Lewis in 1922, provided human endothelial cells in culture, however, these cells exhibited minimal proliferative capacity. A number of endothelial cell growth factors were discovered that slightly stimulated human endothelial cell proliferation. The major breakthrough in human endothelial cell growth in culture was work performed by Dr. Susan Thornton, working with Dr. Elliot Levine, which identified heparin as an integral co-factor with endothelial cell growth factor for the stimulation of human umbilical vein endothelial cell growth in culture [12]. Jarrell and co-workers [13] subsequently reported that heparin and ECGF would stimulate human adult endothelial cell growth in culture. For the first time cells from human artery and vein segments from numerous anatomical positions could be grown in large quantities. Large numbers of human endothelium could be produced from single endothelial cells providing nearly unlimited supplies of human adult endothelium. These breakthroughs in culture methods provided a means to produce large quantities of autologous endothelial cells by obtaining a small segment of a patients blood vessel, isolating and culturing the endothelium in heparin and ECGF and transplanting these cells at very high seeding densities onto prosthetic grafts.

Large vessel derived endothelial cell seeding at high density was achieved in vitro, however, additional questions were raised delaying clinical utilization of this technique. For instance the derivation of cells from large vessel sources would require a separate surgical procedure most likely from patients with an already compromised circulatory system. This increased risk of a separate vascular complication had to be weighed against the need for endothelial cell transplantation. If cultured cells were used for transplantation the conditions of culturing must be evaluated to determine the effect on subsequent cell function. Bovine derived serum components such as viruses and growth factors such as heparin and ECGF would be carried with transplanted cells into patients requiring characterization of their effects. Considering that endothelial cells have an extremely low mitotic index, the stimulation of endothelial cell growth must be considered abnormal requiring evaluation of how cell growth may alter endothelial cell function. Finally, little data is available concerning the appropriateness of transplanting vein derived endothelial cells into an arterial circulation with respect to the ability of these cells to differentiate and function under arterial conditions.

Given the small numbers of cells available if autologous vein derived endothelium are used without culture, and concerns about culturing effects, investigators have been pursuing alternate sources of endothelial cells for transplantation. In 1986 Jarrell and Williams [14] reported methods for the isolation of autologous microvessel endothelial cells from adipose tissue for use in cell transplantation. The source of fat for this EC isolation was initially fat deposits associated with omentum. This fat is well vascularized with a density of endothelial cells in excess of 10^6 endothelial cells per gram of fat isolated. The need to place cells in culture to increase number is obviated by the large number of endothelium available per gram of omental associated fat. However, access to omental associated fat still would necessitate a separate surgical procedure with related complications.

The use of omental associated fat derived endothelial cells was questioned by Visser *et al.* [15] who reported that omental tissue derived from humans contains predominantly mesothelial cells and not endothelium. This report raised a significant controversy concerning the use of omental associated fat as a source of endothelial cells for transplantation. Since this initial report by Visser *et al.* subsequent reports have clarified the controversy and provided additional options for obtaining microvessel endothelial cells [16]. First it was established that investigators have erroneously used the terms omentum and omental fat interchangeably. These two tissues are histologically and anatomically distinct. While omentum as used by Visser *et al.* is a vascularized tissue with a predominant number of mesothelial cells present, omental associated fat, as used by Jarrell and Williams is composed of predominantly endothelial cells and adipocytes.

Alternate anatomic sources for microvascularized autologous human fat for endothelial cell transplantation have been reported since these original reports. A significant improvement was the identification of subcutaneous fat as a source of endothelial cells and the use of liposuction to derive this fat from patients [17]. A patient undergoing endothelial cell transplantation therefore does not need to undergo a laparotomy to remove omental fat or undergo a separate vascular procedure to remove a segment of vein. Rather a small amount (50 cc) of fat can be removed using a hand held syringe cannula device. This procedure requires less then 5 minutes and requires a small skin incision or trocar puncture to insert the liposuction cannula.

The use of liposuction fat also is advantageous since the tissue is effectively minced during the liposuction removal process. Scanning electron microscopic evaluation of liposuction fat (Fig. 1) illustrates the morphological characteristics of this tissue. The predominant cell type present at this magnification is adipocytes. Recently, a complete characterization of cells present in liposuction fat was reported and established that the major cell type present in this tissue is endothelium [16].

The methods for the isolation of endothelial cells from fat are essentially modifications of the methods reported by Wagner in 1972 for the isolation of endothelial cells from rat epididymal fat pads [18]. Human fat microvessel endothelial cells are isolated by first digesting the fat with a proteolytic enzyme mixture composed predominantly of collagenase and trypsin. The characteristics of this enzyme are critical for successful isolation of endothelial cells from fat [19]. Following collagenase digestion the slurry is centrifuged resulting in the separation of buoyant adipocytes from more dense endothelium. The endothelial cell rich pellet is then used for cell transplantation procedures.

Numerous sources of endothelial cells have been evaluated for cell transplantation including both large blood vessel and microvessel endothelial cells. In addition, culture techniques have been optimized for the long term cultivation of human endothelial cells. As a result of these studies endothelial cell transplantation has been evaluated in both preclinical animal trials and human trials.

Animal models of endothelial cell transplantation

The availability of methods for endothelial cell isolation and culture as well as methods to deposit cells onto graft surfaces has resulted in numerous investigations to evaluate the efficacy of endothelial cell transplantation [20–28]. The earliest animal models were developed to evaluate the ability to accelerate the formation of endothelial cell linings on polymeric grafts. Most of these studies have been performed in the canine model due primarily to the acceptance of this animal model as a predictor of graft function in humans. Earliest studies established the ability to accelerate the formation of a continuous monolayer of endothelial cells on the lumenal surface of vascular grafts (Fig. 2). In general, monolayer formation is highly dependant upon the density of cells used to treat the graft surface. When cells are transplanted at confluent densities or greater, monolayer formation appears to be complete in under three weeks. Subconfluent densities require more extended periods of time.

Subsequent studies using animal models have focussed on establishing the effects of endothelial cell transplantation on graft patency. Complete review of the literature concerning endothelial cell transplantation and graft patency in animal models identifies a large variation in the types and size (both i.d. and length) of graft materials tested, graft placement sites, anticoagulation used, types of animals (dog, pig, rabbit, rat, baboon) and the type of endothelial cells transplanted. Although there is very little consistency between author's methods, when graft patency is evaluate using more stringent statistical analysis a consistent positive benefit of endothelial cell seeding on graft patency has been observed. These results have prompted a number of researchers to evaluate endothelial cell transplantation in human trials.

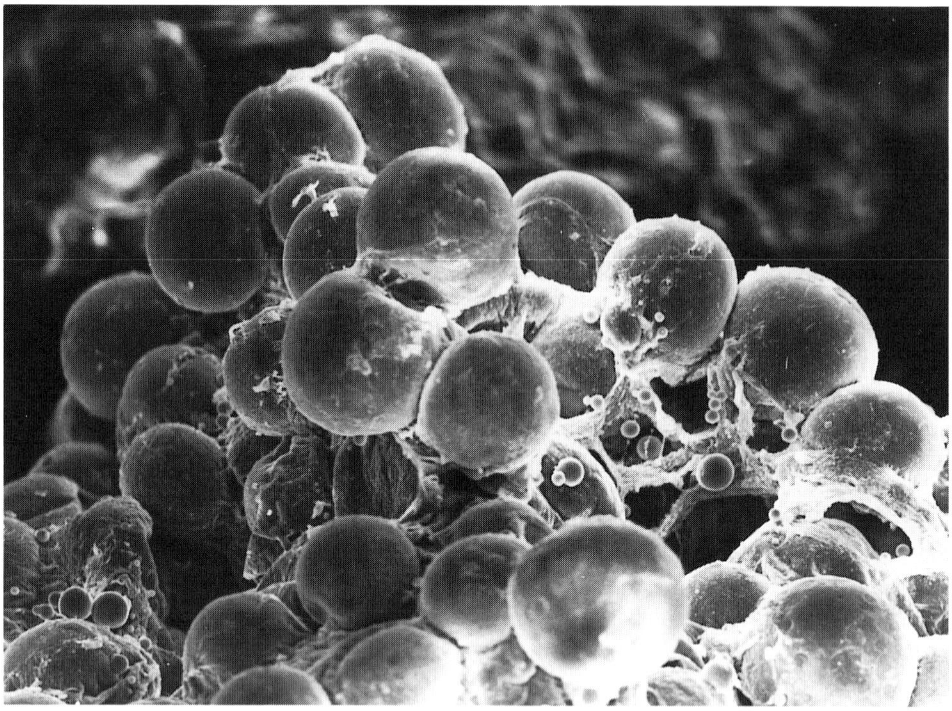

Fig. 1. Scanning electron micrograph of human liposuction-derived subcutaneous fat illustrating the predominance of adipocytes.

Human studies of endothelial cell transplantation

Just as Malcolm Herring was the first to report successful endothelial cell transplantation in animal models, he was the first to report results of a significantly large trial of human endothelial cell transplantation [29]. Numerous other studies involving smaller numbers of patients have been reported permitting some early conclusions to be drawn. First, Dr. Herring's report using large vessel (vein) endothelial cell transplantation concluded that large vessel endothelial cell seeding, that is cell transplantation using a very low density of endothelial cells, does not result in improved patency of synthetic grafts used in peripheral bypass procedures. These results suggest that the earliest methods developed for endothelial cell transplantation do not significantly improve graft function. Trials using more optimized cell isolation and deposition techniques as well as trials using microvessel endothelial cell transplantation are ongoing. Results from these studies, and more definitive conclusions on the efficacy of endothelial cell transplantation will hopefully emerge as the results of these trials are compiled.

Future of endothelial cell transplantation

While the original goal of endothelial cell transplantation studies has focussed on small diameter vascular grafts the availability of operating room compatible methods for endothelial cell transplantation has expanded the possible application of these cells. The development of gene therapy methodologies has begun to target the endothelia cell as a cellular vehicle for gene therapy. Endothelial cells reside at the interface between blood and tissue and therefore maintain a strategic position for the production and release of therapeutic materials. Molecular biologists have begun to take advantage of the extensive methodology established for endothelial cell transplantation. Reports of genetic modification of endothelial cells have exploded recently as the attractiveness of this cell is realized. The use of genetically modified endothelial cells has moved beyond prosthetic devices and studies are ongoing to establish the use of these cells in almost every anatomic site under consideration for gene therapies. Again, the availability of methods to easily isolate and transplant endothelial cells, especially subcutaneous fat derived

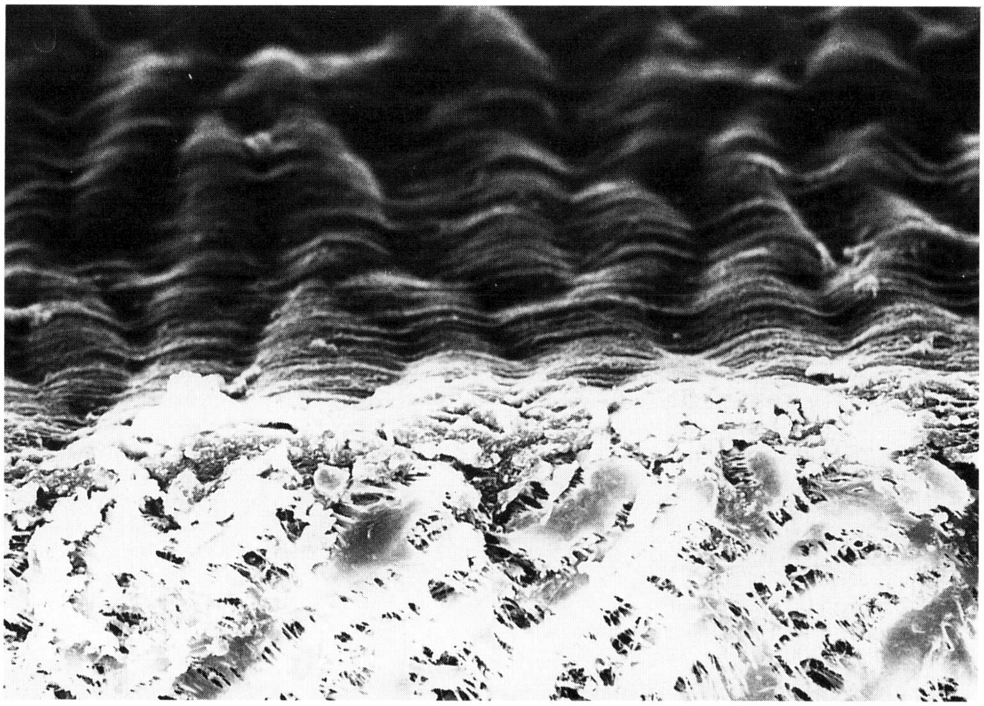

Fig. 2. Endothelial cell monolayer established on an ePTFE graft using autologous microvascular endothelial cell transplantation.

microvessel endothelial cells, provides an exciting new use for endothelial cell transplantation.

References

1. Quist WC, Haudenshield CC and Logerfo FW (1992) Qualitative microscopy of implanted vein grafts: Effects of graft integrity on morphological fate. J Thorac Cardiovasc Surg 103: 671–677.
2. Richardson JV, Wright CB and Hiratzka LF (1980) The role of endothelium in the patency of small venous substitutes. J Surg Res 28: 556–562.
3. Cambria RP, Megerman J and Abbot WM (1985) Endothelial preservation in reversed and in situ autogenous vein grafts. Ann Surg 202 (1): 50–55.
4. Angelini GD, Breckenridge IA, Psaila JV, Williams HM, Henderson AH and Newby AC (1987) Preparation of human saphenous vein for coronary artery bypass grafting impairs its capacity to produce prostacyclin. Cardiovasc Res 21: 28–33.
5. Herring M, Gardner A and Glover J (1978) A single staged technique for seeding vascular grafts with autogenous endothelium. Surg 84: 498–504.
6. Schmidt SP, Hunter TJ, Sharp WV, Malindzak GS and Evancho MM (1984) Endothelial cell seeded four millimeter Dacron vascular grafts. J Vasc Surg 1: 434–441.
7. Stanley JC, Burkel WE and Ford JW (1982) Enanced patency of small diameter, externally supported Dacron iliofemoral grafts seeded with endothelial cells. Surg 92: 994.
8. Graham LM, Vinter DW and Ford JW (1980) Immediate seeding of enzymatically derived endothelium in Dacron vascular grafts. Early studies with autologous canine cells. Arch Surg 115: 1289–1294.
9. Kesler KA, Herring MB, Arnold MP, Glover JL, Park HM, Helmus MN and Bendick PJ (1986) Enhanced strength of endothelial attachment on polyester elastomer and polytetrafluoroethylene graft surfaces with fibronectin substrate. J Vasc Surg 3: 58–64.
10. Kempczinski RF, Rosenman JE, Pearce WH, Rodersheimer LR, Berlatzky Y and Ramalanjaona GR (1985) Endothelial cell seeding of new PTFE vascular prostheses. J Vasc Surg 2: 424–429.
11. Lewis WH (1922) Endothelium in tissue culture. Am J Anat 30: 39–59.
12. Thornton SC, Mueller SN and Levine EM (1983) Human endothelial cells: Cloning and long term serial cultivation employing heparin. Science 222: 623–624.
13. Jarrell BE, Levine EM, Shapiro SS, Williams SK, Carabasi RA, Mueller SN and Thornton SC (1984) Human adult endothelial cell growth in culture. J Vasc Surg 1: 757–764.
14. Jarrell BE, Williams SK, Stokes G, Hubbard FA, Carabasi RA, Koolpe E, Greener D, Pratt K, Moritz MJ, Radomski J and Speicher L (1986) Use of freshly isolated capillary endothelial cells for the immediate establishment of a monolayer on a vascular graft at surgery. Surg 100 (2): 392–399.

15. Visser MJP, Van Bockel JJ, Van Muijen GNP and Van Hinsbergh VWM (1991) Cells derived from omental fat tissue and used for seeding vascular prostheses are not endothelial in origin: A study on the origin of epitheloid cells derived from human omentum. J Vasc Surg 13: 373–381.
16. Williams SK, Wang TF, Castrillo R and Jarrell BE (1994) Liposuction derived human fat used for vascular graft sodding contains endothelial cells and not mesothelial cells as the major cell type. J Vasc Surg 19: 916–923.
17. Williams SK, Jarrell BE, Rose DG, Pontell J, Kapelan BA, Park PK and Carter TL (1989) Human microvessel endothelial cell isolation and vascular graft sodding in the operating room. Ann Vasc Surg 3(2): 146–152.
18. Wagner RC and Matthews MA (1975) The isolation and culture of capillary endothelium from epididymal fat. Microvasc Res 10: 286–297.
19. Williams SK, McKenney S and Jarrell BE (1995) Collagenase lot selection and purification for adipose tissue digestion. Cell Transplantation 4: 281–289.
20. Allen BT, Long JA, Clark RE, Sicard GA, Hopkins KY and Welch MJ (1984) Influence of endothelial cell seeding on platelet deposition and patency in small-diameter Dacron arterial grafts. J Vasc Surg 1: 224–232.
21. Belsen TA, Schmidt SP, Falkow LJ and Sharp WV (1982) Endothelial cell seeding of small-diameter vascular grafts. Trans Am Soc Artif Intern Organs 28: 173.
22. Douville EC, Kempczinski RF, Birinyi LK and Ramalanjaona GR (1987) Impact of endothelial cell seeding on long term patency and subendothelial proliferation in a small-caliber highly porous polytetrafluoroethylene graft. J Vasc Surg 5: 544.
23. Schmidt SP, Hunter TJ and Hirko M (1985) Small diameter vascular prostheses: Two designs of PTFE and endothelial cell seeded and nonseeded Dacron. J Vasc Surg 2: 292–297.
24. Shepard AD, Eldrup-Jorgensen J and Keough EM (1986) Endothelial cell seeding of small caliber synthetic grafts in the baboon. Surg 99: 318.
25. Shindo S, Takagi A and Whittemore AD (1987) Improved patency of collagen impregnated grafts after in vitro autogenous endothelial cell seeding. J Vasc Surg 6: 325.
26. Stanley JC, Burkel WE and Linbald B (1985) Endothelial cell seeding of synthetic vascular prostheses. Acta Chir Scand Suppl 529: 17–27.
27. Tannenbaum G, Ahlborn T, Benvenisty A, Reemstma K and Nowygrod R (1983) High density seeding of cultured endothelial cells leads to rapid coverage of PTFE grafts. Curr Surg 222: 623.
28. Williams SK, Rose DG and Jarrell BE (1994) Microvascular endothelial cell sodding of ePTFE vascular grafts: Improved patency and stability of the cellular lining. J Biomed Mater Res 28: 203–212.
29. Herring M, Compton RS, Gardner AL and LeGrand DR (1987) Clinical experiences with endothelial seeding in Indianapolis. In: Zilla P, Fasol R and Deutsch M (eds) Endothelialization of Vascular Grafts, pp 218–224. Basel: Karger.

R. P. Lanza and W. L. Chick (eds.), Yearbook of Cell and Tissue Transplantation 1996/1997, 227–231.

Chapter 22

Epidermal cell transplantation

David Lawlor and Basil A. Pruitt
United States Army Institute of Surgical Research, Fort Sam Houston, TX 78234, U.S.A.

The current interest in epidermal transplantion is largely the result of recently developed methods of cultivating large quantities of keratinocytes. These cells have the potential to cover large open wounds with an epidermis that possesses many of the properties of skin. The greatest impetus for transplanting epidermal cells is to effect wound closure in patients with extensive burns and a paucity of donor sites. Other potential clinical applications of epidermal transplants include acceleration of wound healing in chronic venous stasis ulcers [1], coverage of deficits produced by soft tissue transfer in limb reconstruction, closure of defects produced by the debridement of infected tissue in patients with necrotizing fasciitis and gas gangrene, repair of traumatic degloving injuries, and promotion of healing in exfoliative skin disorders [2]. More recently cultivated keratinocytes have been proposed as delivery vehicles for gene therapy [3].

Transplanted epidermis must satisfy several requirements in order to function as a clinically effective skin substitute. The properties of an ideal epidermal graft include the following: (1) nonantigenicity; (2) absence of local and systemic toxicity; (3) tissue compatibility; (4) pliability and flexibility to permit conformation to irregular surfaces; (5) elasticity to permit body motion and sufficient tensile strength to resist linear and shear stresses; (6) resistance to microbial colonization or subgraft suppuration; (7) impermeability to microorganisms and water; (8) acceptable water vapor transmission; (9) minimal requirements for storage; (10) indefinite shelf life; (11) cosmetic acceptability with minimal need for revision; and (12) reasonable cost [4]. A clinically effective skin substitute should quickly and persistently adhere to the wound bed; reduce water, protein and electrolyte loss from the wound surface; reduce wound pain; and promote joint motion. In practice, neither standard split-thickness skin grafts nor cultured keratinocytes fulfill these goals completely. Desirable features which have been elusive to epidermal transplants include resistance to ulceration, blistering and infection, and the reconstitution of a basement membrane and subcutaneous tissue that mimics dermis. The development of dermal analogues represents an attempt to address these issues.

The Ebert papyrus (1500 BC) contains the earliest known description of covering wounds with skin flaps. Susruta of India, in 500 BC, utilized free skin grafts "by slapping the skin of the buttock with a wooden shoe until it was quite congested, and then, with a leaf cut to proper shape as a pattern, cutting out a piece of skin with its subcutaneous fat, transplanting it and sewing it into place, uniting it to the freshened edges of the defect" [5, 6]. In 1867, Tigri of Siena performed the first pure epidermal autograft when he successfully transplanted the elevated layer of a blister [7]. Reverdin is credited with the application of split-thickness skin autograft in 1869 when he transferred thinly cut squares of skin to the granulating wound of a patient's thumb [5]. In 1871, he presented his work using allograft and xenograft skin to cover wounds [8]. Pollock, applying the principles of Reverdin's work, performed the first skin graft for a burn injury using both autologous and homologous tissue [9]. He not only described the closure of the wounds with autograft skin but subsequent ulceration of the small area of allograft skin as well. Soon after this, allograft skin was reported as a useful wound dressing and its eventual loss became well recognized [10]. The immunology of allogeneic skin loss was not understood until the 1940s when Medawar established the accelerated rejection of second-set allograft skin in man [11, 12].

In recent years the closure of extensive skin defects has required expansion of cutaneous autografts to obtain wound coverage. Such methods include the Meek-Wall method which achieves a 1 : 9 expansion by separating cut squares of autograft skin on pre-folded gauze [13]. Tanner's method of meshing split-thickness skin grafts [14] has become widely popular today although expansion remains limited to 1 : 4 or at most 1 : 6 for acceptable results. Overlaying widely meshed autograft skin with allograft skin using a sandwich technique [15] or applying sheets of allograft skin studded with minced autograft skin [16] enhances success by providing a more favorable wound environment for re-epithelialization. The results of a clinical study of epidermal sheets applied in "sandwich" fashion over 1 : 4 expanded meshed autograft skin have been interpreted as showing that treatment to be effective in accelerating closure of graft interstices. However, the final graft take was not improved over that of cultured autologous keratinocytes alone [17].

Amnion, which exerts an angiogenic effect and increases the capillary density of the underlying wound bed, has limited practical utility because it lacks an epidermis and is therefore susceptible to desiccation [18]. Both amnion and allograft skin carry the risk of viral transmission which for human immunodeficiency virus (HIV) is the same as for a unit of blood. Cutaneous xenografts, primarily porcine, are widely used as temporary dressings to promote healing of clean superficial partial-thickness burns. Xenograft skin does not engraft on immunocompetent hosts but rather adheres to the wound bed as a consequence of fibrin deposition or fibrovascular ingrowth [19].

By the 1970s, meshed and sheet autograft skin was established as the most successful means of attaining permanent epidermal replacement and allograft skin was considered to be the best available temporary wound dressing. For the last two decades, cultured keratinocytes have been the main focus of research in epidermal transplantation as improvements in resuscitation, ventilatory support, and topical antimicrobial therapy have brought about the survival of patients with greater disparity between open burn wounds and available donor sites.

The history of keratinocyte cultivation has been well documented [20]. In 1898, Ljunggren was able to preserve human skin fragments in ascitic fluid and then transplant them back into the donors. Subsequently, by using saline solutions enriched with biological fluids and occasionally glucose other researchers were able to produce an outgrowth of epithelial and dermal cells from skin explants (Carrel and Burrows, 1910; Hadda, 1912; Kreibich, 1914). Medawar was able to separate the epidermis from the dermis in 1941 and reimplant cultured epidermis on the donor in 1948. Fundamental contributions to the field of keratinocyte cultivation included evidence that keratinocytes maintained their viability after treatment with trypsin to separate individual cells (Billingham and Reynolds, 1952), the development of fibroblast-free epidermal cultures (Prunieras, 1965), and the demonstration of enhanced proliferation of cultured keratinocytes on collagen gels (Karasek and Charlton, 1971).

In 1975 Rheinwald and Green developed techniques to expand relatively pure populations of keratinocytes in culture [21] which eventually permitted the production of sheets of cultured autologous epidermis which could be applied to burn wounds in patients with limited donor sites [22]. These investigators demonstrated that a feeder layer of lethally irradiated mouse 3T3 mesenchymal cells in an appropriately supplemented medium permitted the clonal expansion of keratinocytes in culture while suppressing fibroblast overgrowth. Epidermal growth factor (EGF) 10 ng/ml is used to promote cloning efficiency. EGF antagonizes the effect of contact inhibition of growth in the center of the keratinocyte colonies by stimulating the migration of cells while both EGF and cholera toxin are felt to oppose the onset of terminal differentiation [23]. More recently, Tsao *et al.* described a serum-free keratinocyte growth medium containing bovine pituitary extract 140 μg/ml and a lower calcium concentration 0.03 mM [24]. After subculturing the keratinocytes obtained by either method by exposing them to trypsin for three to four minutes and then centrifuging, resuspending and replating the cells, they are grown to confluence. Under optimal conditions the keratinocytes are allowed to stratify after a 10,000 fold expansion over 21 days. The epidermal cells can then be released as thin sheets from flasks with the enzyme dispase and mounted on petrolatum impregnated gauze to facilitate application to a prepared recipient wound bed.

The three week delay between harvesting skin from a patient and the availability of sufficient quantities of cultured autologous epidermis makes the production of a "universal" skin substitute desirable. The reported research experience with cultured epidermal allograft was well summarized by Phillips in 1991 [25]. Since the Langerhans cells were absent in keratinocyte preparations after seven days culture time it was thought that cultured epidermal allografts might exhibit prolonged or even permanent survival. In fact, experimental work by Hefton and others was initially interpreted as confirming the permanent survival of cultured allogeneic keratinocytes on partial thickness wounds. However, subsequent animal and human studies utilizing monoclonal antibodies against major histocompatibility class I antigens, *in situ* DNA hybridization with Y chromosome probes in sex-mismatched grafts, DNA restriction fragment length polymorphism analysis, and blood group antigen mismatching have demonstrated that cultured allogeneic keratinocytes are ultimately rejected and do not survive indefinitely.

Cultured epidermal allografts have been observed to promote the healing and re-epithelialization of partial thickness wounds such as venous stasis ulcers, superficial burns and split-thickness skin graft donor sites. Possible mechanisms of enhanced healing after application of cultured allogeneic keratinocytes include the following: (1) release of mediators such as epidermal cell-derived thymocyte activating factor, fibronectin and transforming growth factor alpha which may stimulate host keratinocyte proliferation and migration; (2) stimulation of host immune cells to populate the wound and produce growth factors that hasten re-epithelialization; (3) production of basement membrane components facilitating wound closure; and, finally, (4) prevention of wound desiccation by providing an effective vapor barrier.

Phillips *et al.* showed, in patients with a mean age of 71 years, that split thickness donor site healing was shortened from 15.3 days in patients with standard nonadherent dressings to only 8.4 days with cultured epidermal allografts [26]. These observations have led investigators to view cultured allogeneic keratinocytes as a convenient biologic system for the delivery of growth factors to an optimized wound environment. However, Cairns *et al.* have recently described accelerated second set rejection and enhanced cytotoxic lymphocyte alloreactivity with cultured mouse keratinocyte allografts. These observations raise concerns that multiple applications of allogeneic keratinocytes may elicit a second set rejection and that the inflammatory response may delay, rather than promote, wound healing [27].

Cultured autologous keratinocytes were first used for wound coverage in two thermally injured patients in 1981 [22]. Only a small fraction of the burns were actually covered successfully by the cultured cells, however, the authors were able to conclude that the main cause of graft failure seemed to be infection and that freshly excised wounds provided superior wound beds relative to chronic granulation tissue. These observations have been validated during a decade of clinical experience with cultured epidermis. Enthusiasm for cultured epidermis blossomed after Gallico's success in grafting half the body surface of two children with 97 and 98 percent total body surface area burns using cultured autologous keratinocytes. The authors estimated that the "take rate" of these cultured cells was 60–80% [28].

During the past decade clinical experience with cultured autologous keratinocytes in burn patients has produced variable and unpredictable results with estimated graft "take" ranging from 0 to 85% [6, 29, 30, 31, 32]. Odessey, after reviewing a multicenter experience with cultured epidermal autografts, has reported an average final "take" of 60%, usually assessed three to four weeks after graft placement, in the 104 of over 240 patients for whom complete data were available [33]. In these patients early eschar excision ($\leq$ 10 days postburn), successful temporary wound coverage with allograft and the absence of clinical wound infection were associated with improved "take". 90% "take" was achieved in 14 patients in whom the epidermis of engrafted allogeneic skin was removed by tangential excision or dermabrasion and the allodermis left intact before applying the cultured autologous keratinocytes. The multicenter European experience with cultured autologous keratinocytes for burn wound coverage noted substantial differences in engraftment between patients older than 18 years of age (28%) and those less than 18 years of age (47%) [34].

The report by Rue *et al.* has tempered enthusiasm for cultured epidermal autografts. These investigators described mean body surface area of definitive wound coverage as a more meaningful assessment of the contribution of cultured autologous keratinocytes to wound closure [35]. In 16 patients only 4.7% of the burn wound was definitively covered with cultured keratinocytes at a cost of $9,300/% body surface area covered despite an overall graft "take"of 47%. In those patients whose wounds were excised to fascia, eliminating the possible contribution of residual viable skin appendage keratinocytes to re-epithelialization, only 2.8% of the burn wound was definitively covered despite application to 21% of the body surface area. Their final graft "take" was only 32.5% while those patients who underwent burn wound excision to deep dermis had 57.7% final engraftment. However, in the dermal excision group it is possible that the proliferation of residual keratinocytes from the remaining skin appendages may have partially repopulated the wound surface. Late graft loss was identified as a significant problem particularly in the larger burns in the absence of demonstrable wound bed infection. In the ten patients whose burn size was $\geq 70\%$ of the total body surface area, the graft "take" at the time of discharge was only 30.6% despite initial engraftment of 59.8% when the first dressing was removed. In contrast, those patients with smaller burns had 73.5% graft "take" when discharged from the hospital. Review of 19 patients who received 31 applications of cultured autologous keratinocytes at the United States Army Institute of Surgical Research confirm prior observations that engraftment was best on dermis, poorer on fascia, and worst on granulation tissue. The recovery of fungi and gram negative bacilli from cultures of the wound bed was associated with decreased final graft "take" and less definitive wound coverage as compared to those wound beds from which gram positive organisms were recovered.

Mechanical instability, due to the lack of a mature dermo-epidermal junction, an incomplete basal lamina, a paucity of anchoring fibrils and a lack of discrete rete ridges [36] may contribute to blistering, shearing, and ulceration. The possibility of rejection, mediated by the host's immune response to the antigens of exogenous xenogeneic serum supplements [37, 38, 39] or persistent xenogeneic feeder layer fibroblasts [40, 41], may also contribute to graft loss and has driven efforts to develop an improved "defined" keratinocyte growth medium which will support both keratinocyte proliferation and differentiation. Finally, the "melting away" of cultured keratinocytes without inflammatory cell infiltration suggests that culturing techniques may not adequately support the retention or expansion of the progenitor keratinocyte subpopulation and that graft loss is merely a reflection of senescence or even apoptosis of terminally differentiated keratinocytes incapable of producing the progeny necessary to maintain wound closure.

The possibility of overcoming the current limitations of cultured skin heralds a future of exciting research. Chimeric allogeneic-autologous cultured epithelium has been shown to promote epidermal regeneration in Fusenig transplantation chambers on the dorsum of mice. If this approach is successful the delay in therapy while waiting for the clonal expansion of cultured autologous keratinocytes may be decreased substantially [42, 43, 44]. Ongoing research in dermal replacement in combination with the application of cultured autologous keratinocytes to produce a bilaminate construct which better mimics normal skin promises to address the mechanical instability of cultured epidermal sheets, optimize engraftment and improve the functional outcome. The details of this work are outlined elsewhere. Finally, recent work with autologous keratinocyte suspensions in pigs produced rates of re-epithelialization comparable to cultured autologous epidermal sheets although the necessary maintenance of a liquid wound environment seems tedious and perhaps impracticable for patient care [45].

In summary, while autologous epidermal transplants are necessary for definitive wound coverage, xenogeneic and allogeneic epidermis will remain valuable as temporary biologic wound dressings until autologous tissue becomes available. The clonal expansion of keratinocytes in culture represents a great step forward in understanding the biology of the epidermis and providing the means to achieve timely wound closure of massive skin defects. Initial reports of clinical success with cultured keratinocytes have been tempered by the problems of reproducibility of results, late graft loss, assessment of graft take versus percent of wound covered, and the functional limitations of grafts that lack dermal replacement. Additional work to develop methods of maintaining a nonimmunogenic keratinocyte stem cell population in culture complements ongoing efforts to provide a dermal substitute to create a skin analogue which will more reliably engraft and duplicate the function of normal skin.

References

1. Hefton JM, Cald D, Biozes DG *et al.* (1986) Grafting of skin ulcers with cultured autologous epidermal cells. J Am Acad Dermatol 14: 399–405.
2. Carter DM, Lin AN, Varghese MC *et al.* (1987) Treatment of junctional epidermolysis bullosa with epidermal autografts. J Am Acad Dermatol 17: 246–250.
3. Morgan JR, Barrandon Y, Green H *et al.* (1987) Expression of an exogenous growth hormone gene by transplantable human epidermal cells. Science 237: 1476–1479.
4. Pruitt BA Jr and Levine NS (1984) Characteristics and uses of biologic dressings and skin substitutes. Arch Surg 119: 312–322.
5. Ehrenfried A (1909) Reverdin and other methods of skin grafting. Boston Med Surg J 161: 911–927.
6. Gallico GG and O'Connor NE (1985) Cultured epithelium as a skin substitute. Clin Plast Surg 12: 149–157.
7. Arons JA, Wainwright DJ and Jordon RE (1992) The surgical applications and implications of cultured human epidermis: a comprehensive review. Surgery 111: 4–11.
8. Reverdin ML (1871) Sur la greffe epidermique. C R Acad Sci (Paris) 73: 1280–1282.

9. Haynes BW Jr (1987) The history of burn care. In: Boswick JA (ed) The Art and Science of Burn Care, pp 3–9. Rockville, Md: Aspen Publishers Inc.
10. Alsbjorn BF (1992) Biologic wound coverings in burn treatment. World J Surg 16: 43–46.
11. Gibson T and Medawar PB (1943) The fate of skin homografts in man. J Anat 77: 299–310.
12. Medawar PB (1944) The behaviour and fate of skin autografts and skin homografts in rabbits. J Anat 78: 176–199.
13. Meek CP (1958) Successful microdermagrafting using the Meek-Wall microdermatome. Am J Surg 96: 557–558.
14. Tanner JC Jr, Vandeput J and Olley JF (1964) The mesh skin graft. Plast Reconstr Surg 34: 287–292.
15. Alexander JW, MacMillan BG, Law E *et al.* (1981) Treatment of severe burns with widely meshed skin autograft and meshed skin allograft overlay, J Trauma 21: 433–438.
16. Yang CC, Shih TS, Chu TA *et al.* (1980) The intermingled transplantation of auto- and homografts in severe burns. Burns 6: 141–145.
17. Teepe RGC, Kreis RW, Koebrugge EJ *et al.* (1990) The use of cultured autologous epidermis in the treatment of extensive burn wounds. J Trauma 30: 269–275.
18. Faulk WP, Matthews R, Stevens PJ *et al.* (1980) Human amnion as an adjunct in wound healing. Lancet 1: 1156–1158.
19. Silverstein P, Curreri PW and Munster AM (1971) Evaluation of fresh viable porcine cutaneous xenografts as temporary burn wound cover. Annual Research Progress Report, US Army Institute of Surgical Research, Brooke Army Medical Center, June 30, section 51: 1–5.
20. Andreassi L (1992) History of keratinocyte cultivation. Burns 18 (Suppl 1): S2–4.
21. Rheinwald JG and Green H (1975) Serial cultivation of strains of human epidermal keratinocytes: the formation of keratinizing colonies from single cells. Cell 6: 331–344.
22. O'Connor NE, Mulliken JB, Banks-Schlegel S *et al.* (1981) Grafting of burns with cultured epithelium prepared from autologous epidermal cells. Lancet 1: 75–78.
23. Parkinson EK and Yeudall WA (1992) The epidermis. In: Freshney RI (ed) Culture of Epithelial Cells, pp 59–80. New York: Wiley-Liss, Inc.
24. Tsao MC, Walthall BJ and Ham RG (1982) Clonal growth of normal human epidermal keratinocytes in a defined medium. J Cell Physiol 110: 219–229.
25. Phillips TJ (1991) Cultured epidermal allografts-a temporary or permanent solution? Transplantation 51: 937–941.
26. Phillips TJ, Provan A, Colbert D *et al.* (1993) A randomized single-treatment of split-thickness skin graft donor sites. Arch Dermatol 129: 879–882.
27. Cairns BA, deSerres S, Matsui M *et al.* (1994) Cultured mouse keratinocyte allografts prime for accelerated second set rejection and enhanced cytotoxic lymphocyte response. Transplantation 58: 67–72.
28. Gallico GG, O'Connor NE, Compton CC *et al.* (1984) Permanent coverage of large burn wounds with autologous cultured human epithelium. N Engl J Med 311: 448–451.
29. Pittelkow MR and Scott RE (1986) New techniques for the in vitro culture of human skin keratinocytes and perspectives on their use for grafting of patients with extensive burns. Mayo Clin Proc 61: 771–777.
30. Eldad A, Burt A and Clarke JA (1987) Cultured epithelium as a skin substitute. Burns 13: 173–180.
31. Latarjet J, Gangolphe M, Hezez G *et al.* (1987) The grafting of burns with cultured epidermis as autografts in man. Scand J Plast Reconstr Surg 21: 241–244.
32. Herzog SR, Meyer A, Woodley D *et al.* (1988) Wound coverage with cultured autologous keratinocytes: use after burn wound excision, including biopsy followup. J Trauma 28: 195–198.
33. Odessey R (1992) Addendum: Multicenter experience with cultured epidermal autograft for treatment of burns. J Burn Care 13: 174–180.
34. De Luca M, Albanese E, Bondanza S *et al.* (1989) Multicentre experience in the treatment of burns with autologous and allogeneic cultured epithelium, fresh or preserved in frozen state. Burns 15: 303–309.
35. Rue LW, Cioffi WG, McManus WF *et al.* (1993) Wound closure and outcome in extensively burned patients treated with cultured autologous keratinocytes. J Trauma 34: 662–668.
36. Woodley DT, Peterson HD, Herzog SR *et al.* (1988) Burn wounds resurfaced by cultured epidermal autografts show abnormal reconstitution of anchoring fibrils. J Am Med Assoc 259: 2566–2571.
37. Johnson MC, Meyer AA, deSerres S *et al.* (1990) Persistence of fetal bovine serum proteins in human keratinocytes. J Burn Care Rehabil 11: 504–509.
38. Johnson LF, deSerres S, Herzog SR *et al.* (1991) Antigenic cross-reactivity between media supplements for cultured keratinocyte grafts. J Burn Care Rehabil 12: 306–312.
39. Meyer AA, Manktelow A, Johnson M *et al.* (1988) Antibody response to xenogeneic proteins in burn patients receiving cultured keratinocyte grafts. J Trauma 28: 1054–1059.
40. Cairns BA, deSerres S, Brady LA *et al.* (1993) Fibroblasts from feeder layer persist in cultured keratinocyte grafts. Proceedings of the American Burn Association 25: 6.
41. Cairns BA, deSerres S, Brady LA *et al.* (1994) Viable, immunologically active, mouse 3T3 fibroblasts persist in human cultured epidermal autografts used in burn wound coverage. Presented at the American Association for the Surgery of Trauma Meeting: 74.
42. Rouabhia M, Germain L, Bergeron J *et al.* (1994) Successful transplantation of chimeric allogeneic-autologous cultured epithelium. Transplantation Proceedings 26: 3361–3362.
43. Rouabhia M, Germain L, Bergeron J *et al.* (1995) Allogeneic-syngeneic cultured epithelia-a successful therapeutic option for skin regeneration. Transplantation 59: 1229–1235.
44. Suzuki T, Ui K, Shioya N *et al.* (1995) Mixed cultures comprising syngeneic and allogeneic mouse keratinocytes as a graftable skin substitute. Transplantation 59: 1236–1241.
45. Andree C, Page C, Vogt P *et al.* (1994) Direct comparison of cultured epidermal sheet grafts and keratinocyte suspensions in a liquid environment in pigs. Proceedings of the American Burn Association 26: 102.

Section XIII:

Tissue Engineering/Hybrid Tissues

R. P. Lanza and W. L. Chick (eds.), Yearbook of Cell and Tissue Transplantation 1996/1997, 235–245.

Chapter 23

Tissue engineering: Cartilage, bone and muscle

Clemente Ibarra[1], Robert Langer[1,2] and Joseph P. Vacanti[1]
[1] *Department of Surgery, Children's Hospital and Harvard Medical School, Boston, MA 02115, U.S.A.;* [2] *Department of Chemical Engineering, Massachusetts Institute of Technology, Cambridge, MA 02319, U.S.A.*

As a very rapidly developing field of research, *tissue engineering* has been defined differently by many authors but one common aspect of its application is the study of the utilization of living cells to restore or create a lost or absent tissue structure or function (Fig. 1) [1, 2, 3].

Perhaps one of the most successful areas of tissue engineering has been that of "structural" tissues (cartilage, bone, tendon, muscle). The use of isolated cultured chondrocytes as means of cell therapy has started to find clinical application [4]. Experimentally, tissue engineered cartilage has shown potential for application in different fields of surgery such as orthopaedics, plastic and reconstructive surgery, urology and other medical specialties [2, 5, 6, 7, 8, 9]. Tissue engineering of bone also offers a very attractive potential alternative in the reconstruction of large bone defects that present either as a consequence of trauma, congenital or neoplastic diseases. The isolation and in vitro culture of bone-related cells has lately found a an important application in the study of the biologic effects of naturally or synthetically obtained growth factors such as bone morphogenetic protein (BMP) and other peptides [10]. The study of the biology of muscle transplantation and the use of genetically altered muscle cells in the treatment of muscle injury and myopathies has been the main focus of researchers in the field of skeletal muscle-related tissue engineering [2].

In this chapter we will attempt to give an overview of the current advances and some of the possible future perspectives of tissue engineering in relation to cartilage, bone and muscle.

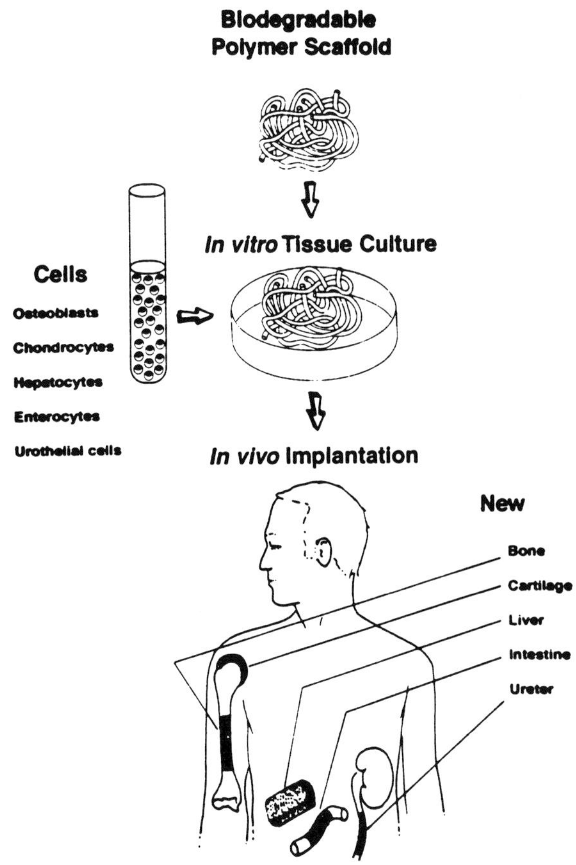

Fig. 1. Schematic representation of *Tissue Engineering*. Cells obtained from the desired tissue are cultured *in vitro*, seeded onto synthetic, biocompatible, biodegradable polymer scaffolds and new tissue is created *in vivo* or *in vitro*. (Reproduced with permission of the authors from: Langer R, Vacanti JP (1993) Science 260: 920–26.)

Tissue engineering of cartilage

Cartilage has very limited capacity for spontaneous repair after injury or disease [11, 12, 13, 14]. Injury to articular cartilage leads to the development of early osteoarthritis (OA) [15]. Deep articular cartilage lesions that reach vascular subchondral bone are ultimately repaired by the formation of fibrocartilaginous tissue. The new repair tissue can sometimes resemble hyaline cartilage grossly and histologically, but biochemically and biomechanically it behaves in a different manner. The newly formed cartilage is thought to originate from bone marrow-derived mesenchymal cells which under appropriate conditions turn into bone through endochondral ossification pathways. The process is stopped at some point during this process and the tissue remains as hyaline-like cartilage. However, the cells are not articular cartilage chondrocytes and do not produce a normal hyaline cartilage matrix. This causes the new cartilaginous repair tissue to suffer extensive degenerative changes. Fibrillation and fissuring usually occur by six months with eventual loss of such hyaline-like cartilage [16, 17]. Wakitani *et al.* reported similar results after attempting to repair articular cartilage defects in rabbits with bone marrow and periosteum-derived mesenchymal cells [18]. Vacanti *et al.* reported to have demonstrated in preliminary experiments that chondrocytes harvested from articular surfaces, rib and ear fibrocartilage and perichondrium differentiate into cartilage whereas cells obtained from periosteum form new bone after passing through a phase in which they resembled cartilage grossly and microscopically [19].

Different arthroscopic techniques and prosthetic joint replacement are the current treatment modalities for large cartilaginous lesions and severe degeneration of articular cartilage. Prosthetic implants made of alloplastic materials are frequently used in plastic and reconstructive surgery to replace or mold an absent or defective body structure. All of these treatment methods present with several short-term and long-term potential complications that can limit their use [1, 17, 19] specially in younger, active patients. Matsusue described an arthroscopic technique for transplantation of autologous osteochondral fragments to treat chondral defects in the knee [20]. The potential of this technique is limited by scarcity of donor tissue and thus can only be used to repair small cartilaginous defects. This is one of the problems that face other areas of reconstructive and transplantation surgery [2]. Autografts [21] and allografts [22, 23] which rely on creeping substitution by host cells for integration, are currently used to reconstruct bone defects. Nevertheless, articular cartilage does not seem to show the same success as bone when transplanted. Even though chondrocytes seem to survive transplantation better, cartilage allografts usually undergo degenerative changes due to mechanical and biochemical factors [24]. The use of a biologic implant created from the patient's own tissues or cells could be a more reasonable and lasting answer to these problems. Brittberg recently described the use of isolated autologous chondrocytes in suspension to repair deep cartilage defects in patients' knees [3]. Chondrocytes suspended in culture media were injected into a chondral defect covered with a periosteal patch. Similar techniques using chondrocytes in suspension had previously been described in experimental studies by different authors [25, 26, 27, 28, 29, 30, 31]. Two questions regarding this technique can arise: Will the cells in suspension attach to the surrounding tissue consistently? Is the newly-formed cartilage derived from the injected cultured chondrocytes or from cells in turn derived from the periosteum or subchondral bone after the cells in suspension had been washed away into the joint space by joint motion and synovial fluid? If the latter were the case, the new cartilage would have been formed from cells other than articular chondrocytes and thus could eventually break-down as hyaline-like fibrocartilaginous repair tissue usually does.

Considering the natural difficulties that the use of isolated cultured cells in suspension carried for transplantation, researchers initiated the use of systems that could allow cell survival and delivery to the desired site.

The use of naturally occurring matrices as cell delivery devices to produce cartilage with varying rates of success, has been described. Wakitani [32] and Kimura [33] have used collagen gels and isolated chondrocytes to create cartilage. Itay reported the use of fibrin glue as a vehicle in chondrocyte transplantation with limited ability to repair cartilage lesions [31]. Cell behavior and survival in fibrin glue can limit its use [34]. Perichondrium has also been used to grow new cartilage [35]. Bruns reported the formation of hyaline-like cartilage by using perichondrium and collagen sponges or fibrin glue with poor integration of the new tissue to the host's cartilage [36]. Tsai *et al.* described the use of fibrin glue and autologous periosteal grafts in a rabbit model [37]. Agarose gels have been used as scaffolds to support chondrocytes for culture in vitro [38] and peptide stimulation of cells has also been studied as another means of creating new cartilage [39, 40].

Based on techniques originally intended to create parenchymal tissue for transplantation [41, 42] Vacanti *et al.* introduced a novel approach for the creation of new cartilage by *tissue engineering* [43]. These techniques combined cell isolation and culture procedures, already used by different researchers and the use of synthetic, biocompatible, biodegradable materials. After unsuccessful attempts to create cartilage

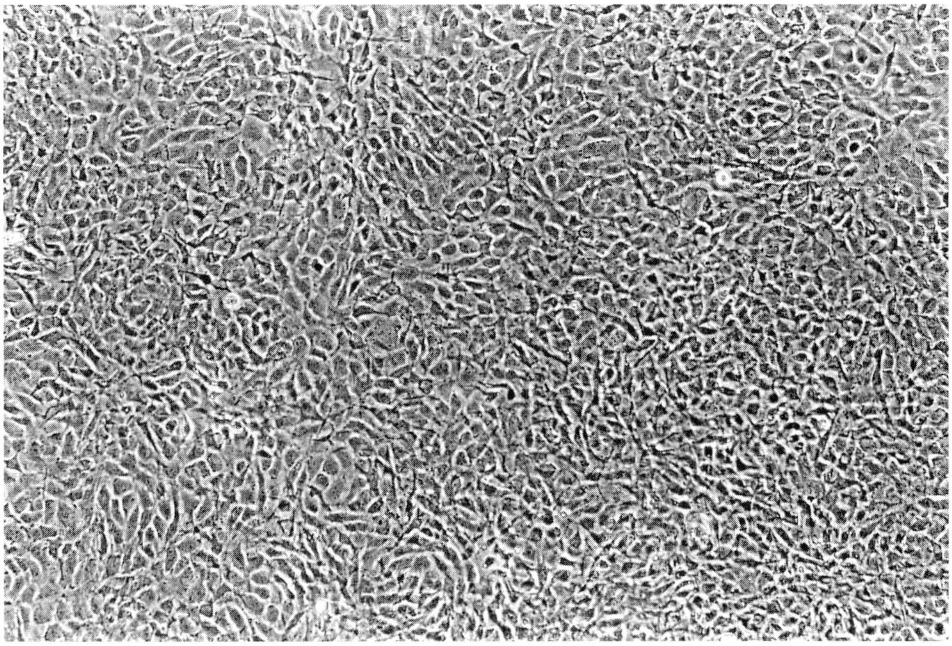

Fig. 2a.

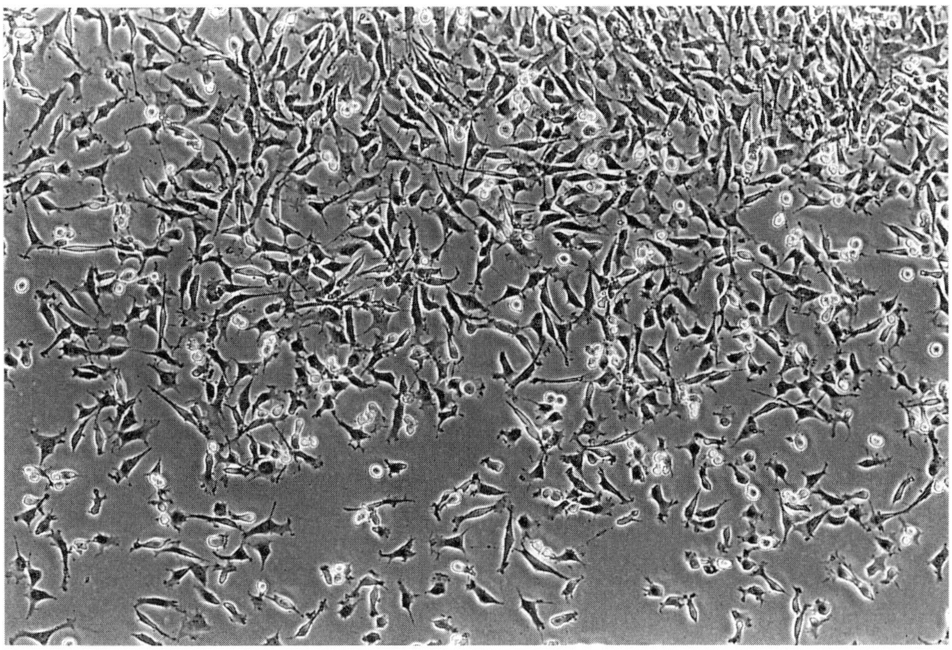

Fig. 2b.

Figs. 2(a)–(b). Chondrocytes forming a monolayer during culture *in vitro*. A. Chondrocytes forming a monolayer in 2-dimension culture *in vitro*. B. Monolayer of chondrocytes in 2-dimension culture *in vitro*.

by seeding isolated chondrocytes on demineralized bone, Green predicted in the 1970s that the advent of new synthetic biocompatible materials might aid in the transplantation of chondrocytes [44]. In 1988 Vacanti *et al.* were able to create new hyaline cartilage in nude athymic mice by using isolated bovine chondrocytes (Fig. 2), and biodegradable suture materials [43] (Fig. 3,4). The unbraided suture fibers provided a temporary scaffold onto which cells could attach until they created their own supporting matrix. This scaffolding structure allowed for diffusion of nutrients and waste products and at the same time allowed the cells to attach to the polymer, to each other and to bridge the interfiber spaces with extracellular matrix (Fig. 5). The cell-polymer constructs were then implanted subcutaneously in nude mice. The generation of cartilage was assessed at different time points grossly and histologically using hematoxylin and eosin (H&E) and other specific stains for cartilage. Cells were also labeled in vitro with BrdU or fluorescein (chloromethyl derivative of amino and hydroxycoumarin) before implantation to determine whether the tissue obtained after in vivo culture of the specimens had been formed by the cells seeded onto the polymers. After the first successful attempts using unbraided polyglactin 910 sutures (Vicryl, Ethicon Somerville, NJ) and polyglycolic acid (PGA) (Dexon, Davis & Geck, Danbury, CT), different polymer configurations were utilized with similar success. Non-woven mesh of PGA approximately 100 μm thick, with a fiber diameter of 14–15 μm and interfiber spaces of 150–200 μ and co-polymeric structures combining PGA and poly-L-lactic acid (PLLA) were designed to successfully create cartilage constructs with predetermined shapes [45]. According to the design of the polymer, its physical characteristics could be controlled in terms of shape, size, time of degradation and affinity for cell attachment. The optimal cell concentrations to create cartilage from cell-polymer constructs were determined [46] and cell survival studies were performed demonstrating the capacity of chondrocytes to maintain functional activity after being stored in a refrigerator at 4 °C in an appropriate culture media for up to 30 days [47]. Injectable polymers also proved effective to create new cartilage [48]. Calcium alginate gels were used to suspend chondrocytes in culture. The cell-containing gel was then injected through a syringe into plastic molds to form discs which were implanted into subcutaneous pockets in athymic mice. The chondrocyte-calcium alginate gel was also directly injected subcutaneously in the mice. Cartilage was formed after 8 weeks of implantation. Finally, very specific applications of the new tissue-engineered cartilage started to be tested. Nasoseptal implants and a temporomandibular joint disc have been designed using tissue engineered cartilage [7, 8]. Cartilage in the shape of a human ear has been successfully obtained [6] and and a similar model of vascularized autologous cartilage covered with skin is currently being developed for use as a pedicled graft for ear reconstruction in a rabbit model [49]. Cartilaginous tubes lined with respiratory epithelium have been egineered for tracheal replacement [9] and successful joint resurfacing has been accomplished in a rabbit model with tissue engineered cartilage [5]. It is evident that the spectrum for application of tissue engineered cartilage is a very broad one. If the combination of the use of synthetic biodegradable polymers with cell culture enhancement by growth factors is successfully achieved in vitro, perhaps aided by the use of bioreactors, the feasibility of the utilization of tissue engineered cartilage as a common treatment of injured or diseases articular cartilage or in the reconstruction of cartilaginous structures using autologous chondrocytes, could very soon be a reality.

Tissue engineering of bone

Replacement of bone lost to injury or disease has been attempted for many years through different approaches. The use of bone autografts [21] and allografts [22, 23] has become common practice in orthopaedics and reconstructive surgery. However, the sources of autologous bone for use as grafts are very limited and harvesting always produces secondary morbid sites. Scarcity of donor tissue for allograft use [2], the biology of allograft integration to host tissue [24] and the possible risk of transmission of infectious diseases [50] limit their use. Bone substitutes and alloplastic materials have been used alone [51, 52, 53, 54, 55, 56] and in conjunction with demineralized bone [57] or autogenous bone grafts [58, 59]. Growth factors have also been used to stimulate bone development from isolated mesenchymal cells [60].

Based on the experience gained by engineering new cartilage, Vacanti *et al.* were able to obtain bone by seeding bovine periosteal cells onto sheets of non-woven PGA mesh. The cells were isolated by in vitro culture of periosteum explants obtained under sterile conditions from newborn calf forelimbs. Once the cells had formed a monolayer on the tissue culture dish, the cells were seeded onto the polymer by scraping the bot-

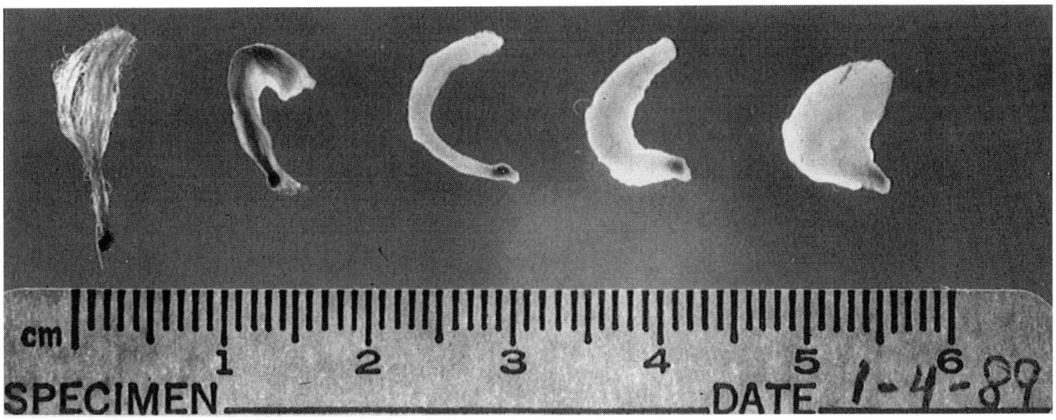

Fig. 3. Gross experimental specimens of new cartilage. Note the appearance of polymer fibers prior to seeding with chondrocytes (*far left*). The polymer fiber is an unbraided, 17 mm in length, of suture material "0" Vicryl (Ethicon, Somerville, NJ), that has been knotted at one end. To the right of the unseeded polymer fibers are representative specimens of different implants excised (*from left to right*) at days 8, 18, 28, and 49. With time, the polymer fibers began to dissolve as the cartilage developed into a homogenous plate of cartilage. (Reproduced with permission from Vacanti CA, Langer R, Schloo B *et al.* (1991) Plast Reconst Surg 87: 753–59).

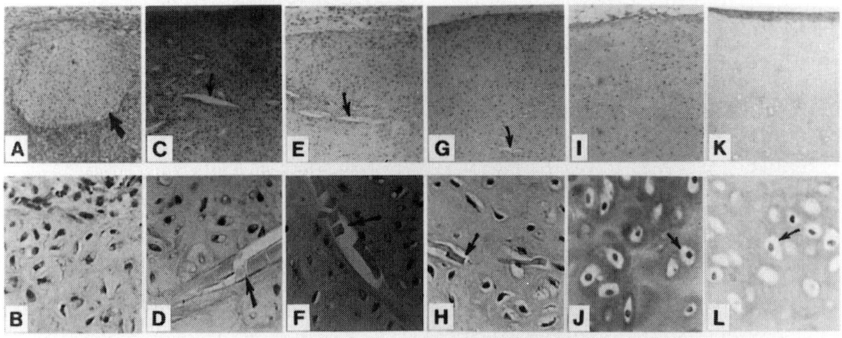

Fig. 4. Photomicrographs (taken with a Nikon 35 mm camera) demonstrating H&E staining of experimental explants excised (*from left to right*) at 8, 18, 28, 49, 81, and 168 days. (Upper row original magnification, ×4; lower row original magnification, ×20). *A*, At day, isolated "nests" of cartilage (*arrow*) were found embedded in fibrous tissue. There was some evidence of a mild inflammatory response consisting of infiltrates of polymorphonuclear leukocytes and giant cells. *B*, Higher magnification of the nests of cartilage. Note the presence of polymer fibers (*arrows*) at days 18 (*C* and *D*) and 28 (*E* and *F*) after which time progressively less evidence of residual fibers is seen. Specimens (*C–L*) appear to be homogeneous plates of cartilage that mature histologically with time. There is increased deposition of basophilic matrix associated with increasing intercellular distances. Also, the formation of lacunae (*arrows*) in which the chondrocytes are enclosed is observed in the day 81 and day 168 specimens (*J* and *L*). This parallels the normal histogenesis of cartilage. There is no evidence of neovascularization in the day 49 (*G* and *H*), day 81 (*I* and *J*), and day 168 (*K* and *L*) implants. Aldehyde fuschin-alcian staining was highly suggestive of the presence of chondroitin sulfate in the specimens form days 18, 28, 49, 81, and 168. (Reproduced with permission from Vacanti CA, Langer R, Schloo B *et al.* (1991) Plast Reconst Surg 87: 753–59).

tom of the dish with sterile sheets of PGA. The cells attached readily to the polymer as assessed by phase contrast microscopy. The cell-polymer constructs were cultured in vitro for an additional week until the polymer was coated with several layers of periosteal cells. Metabolic activity of the cells was determined before subcutaneous implantation of the cell-polymer constructs into athymic mice. Positive immunohistochemical staining for osteocalcin, a bone-specific protein, in the supernatant from the culture media, confirmed the presence of functioning osteoblasts. The constructs were then implanted in subcutaneous pockets in the dorsum the nude mice. The samples were harvested at different time points for gross and histologic examination. By 6 weeks, the constructs showed gross and microscopic appearance of cartilage with focal areas of

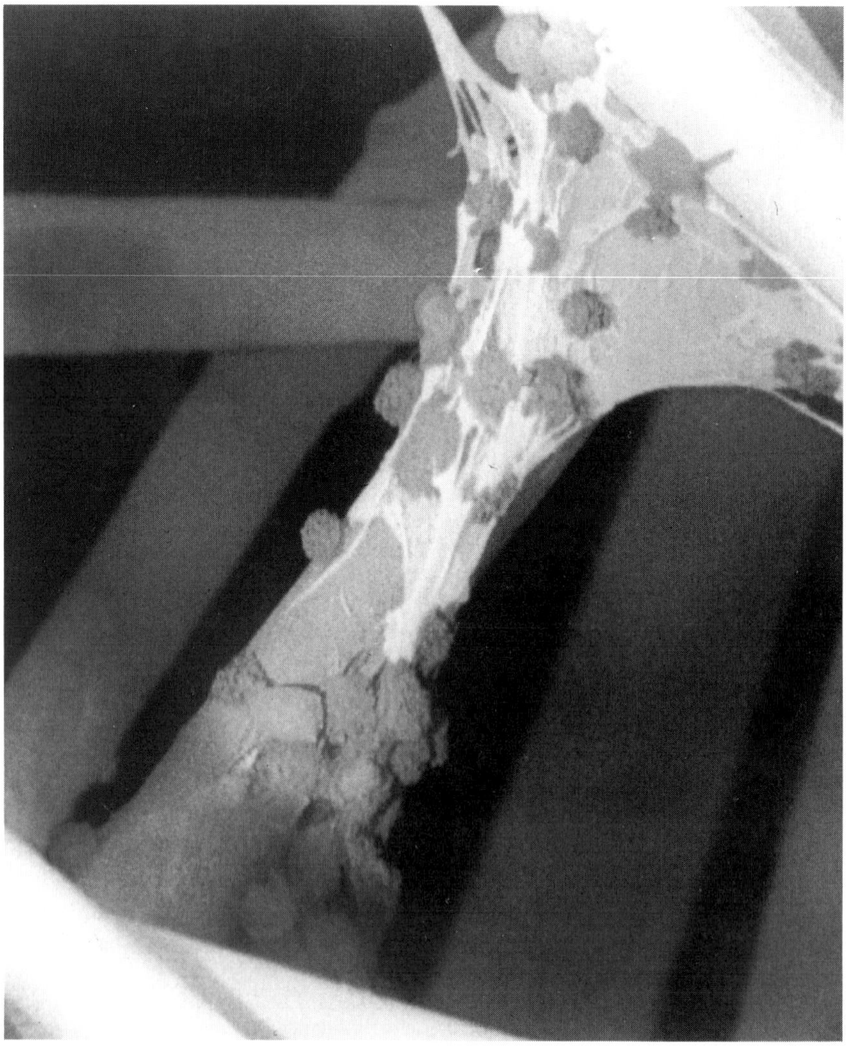

Fig. 5. Scanning electron micrograph of chondrocytes attached to PGA fibers. Notice extracellular matrix being laid down by cells between polymer fibers (Reproduced with permission of David Mooney, Ph.D and the publisher, from the cover picture of Tissue Engineering; 2, 1995).

vascular invasion and bone formation. After 10 weeks of in vivo culture, most of the specimens had turned into bone with marked vascular proliferation. Areas of tissue undergoing endochondral ossification and small islands of cartilage were still present. The newly formed bone contained cellular elements of bone marrow. Polymers seeded with chondrocytes isolated by enzymatic digestion from bovine articular cartilage and plain polymers used as controls confirmed the observations that bone was only obtained when periosteal cells were used. Regardless of implantation time or location, polymers seeded with chondrocytes turned into cartilage but never progressed into bone. However, when periosteal cells were used, cartilage formation was observed during the first few weeks of in vivo implantation with subsequent transformation into mature, organized bone as determined by gross examination and histologic evaluation using H&E stains (Fig. 6).

Having created tissue-engineered bone, the next logical step was to delineate its possible applications. In this manner, polymers seeded with periosteal cells and articular cartilage chondrocytes were used to fill cranial bone defects created in the parietal, frontal and temporal bones of nude rats [46, 61]. Similar cranial defects covered with plain polymer and with nothing at

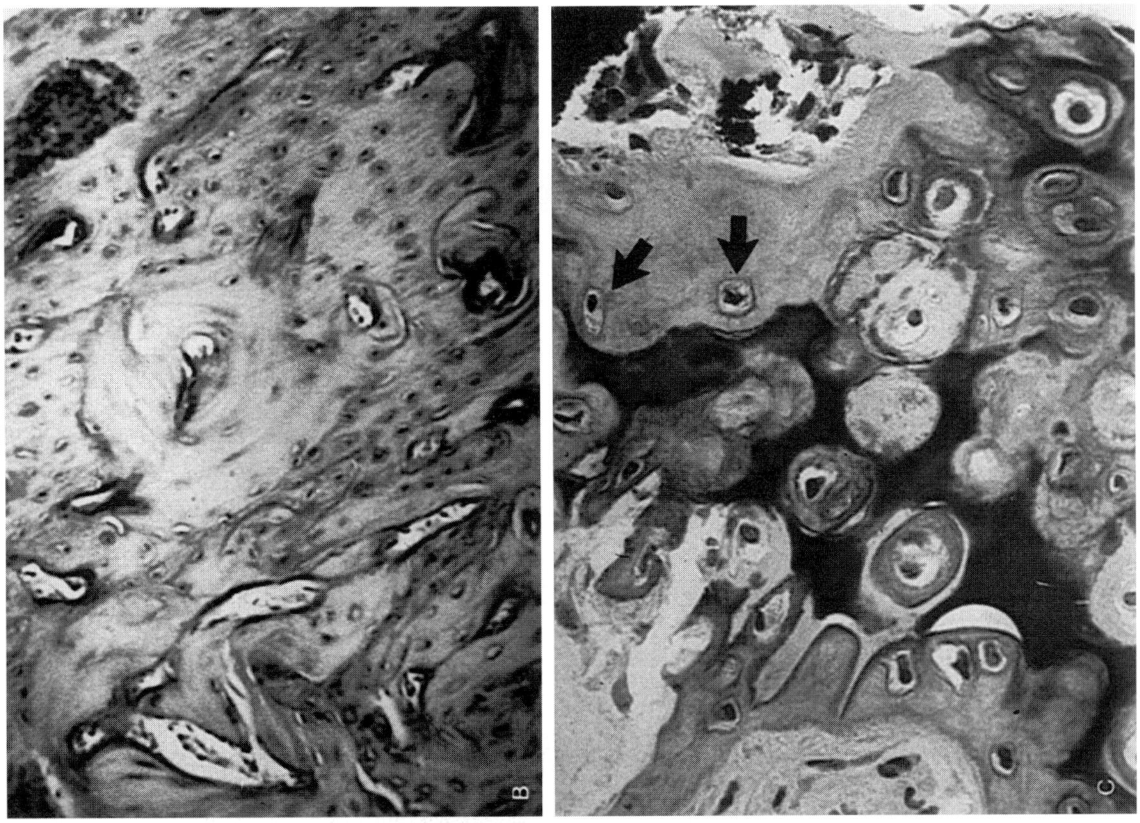

Fig. 6. Histologic section of tissue-engineered bone (H&E stain). Note the transition from cartilage to bone 8 weeks after implantation. Cells in the cartilage, seen centrally, which appear to hypertrophy and then atrophy, are seen as osteocytes (*arrows*) in the new bone. (Photomicrograph taken with a Nikon 35 mm camera; original magnification, ×40).

all were used as controls. Samples harvested at different time points were studied grossly and histologically with H&E stains. Previous results were reproduced in this model. Early samples showed new cartilage covering the defects that had been filled with both chondrocytes and periosteal cells on polymer, whereas specimens harvested after 9 and 12 weeks demonstrated bone formation in the cranial defects that had been filled with periosteal cells on polymer and cartilage covering the defects that had been filled with chondrocytes on polymer. Defects cover with plain polymer or not filled with anything, remained as empty bone defects. Mid shaft bone defects created in the femur of nude rats and fixed with miniplates and screws to maintain the bone gap were filled with cell-polymer constructs in similar fashion. Radiographic controls showed evidence of bone formation and signs of bone healing in the periosteal cell-polymer group. Animals in which chondrocytes on polymer, plain polymer or nothing had been used to fill the bone gap, did not show any radiographic signs of bone formation. After 24 weeks, the miniplates were surgically removed. The bone defects filled with periosteal cells showed complete bone healing with exuberant callus formation covering the miniplates. Bone defects filled with chondrocytes on polymer showed cartilage formation filling up the gap. This explained the difference in radiographic appearance in the animals of these two groups. Finally, the bone gaps that had been filled with plain polymer or nothing showed atrophic non-unions. In some of the animals of these two groups the miniplates had broken.

Mid shaft bone defects similarly created in the femur of nude rats have also been repaired with tissue engineered vascularized bone grafts. PGA mesh seeded with bovine periosteal cells in similar fashion was wrapped around the saphenous neurovascular bundle of nude rats. Organized bone engrafted on a vascular pedicle was obtained. Mid shaft bone defects were then created in the ipsilateral femur of the rat, and the new vascularized bone grafts was transferred as a pedicled graft. Grafts created in this manner demonstrated earlier engraftment and healing of the bone defect than the previous method described [62].

Tissue engineered composites of bone and cartilage

Composite structures of bone and cartilage have also been engineered by suturing together cell-polymer scaffolds seeded with either chondrocytes or periosteal cells. In every construct either one of the cell types was labeled with a flourescent dye in vitro before implantation into animals for up to 16 weeks [61]. Early specimens showed formation of cartilage only. Over time, new bone and cartilage was demonstrated grossly and histologically. Bone developed exclusively on the side of the polymer construct originally seeded with periosteal cells and cartilage formed on the opposite side of the construct with no evidence of bone formation. This was confirmed by detection of the specific labeled cells confined to one side of the specimen. A distinct bone-cartilage interface was formed and could clearly be identified by the presence of nylon suture material used to mark the area. This experiment clearly demonstrated once again that periosteal or osteoblastic cells cultured in vivo can eventually give rise to mature bone through and endochondral ossification-like pathway, whereas chondrocytes implanted in similar fashion will evolve into mature cartilage [19].

Having reached this point, one of the paths to follow could be to attempt combining some of the strategies used by different researchers such as the use of new synthetic biodegradable polymer scaffolds with improved biomechanical properties [63] and the use of autologous cells cultured and expanded in the presence of growth factors. It was necessary to try to understand the normal biology of the isolated cells without other stimuli than that their own intrinsic signals to set up a base line with which to compare the biologic effect of such peptides when applied to tissue engineering. The stimulation and enhancement of cell growth and tissue formation with growth factors in vitro, with the aid of bioreactors [64, 38] will definitely allow for a more controlled study and development of tissues and organs for transplantation.

Tissue engineering of muscle

Faulkner described skeletal muscle-related *tissue engineering* as the rational modification of the structure and function of skeletal muscles by manipulation of molecules, cells or constituent structures of the whole tissue to restore the structure and/or function of injured or diseased muscles [3]. According to a modified description of the Bioengineering Program at the University of Michigan in August, 1992, skeletal muscles may be engineered by utilizing the natural adaptive, or restorative capacities, by introduction of gene sequences into the genome of the animal, or by transplantation of cells, whole tissues or biologic substitutes. Following these principles several researchers have attempted different approaches to find alternatives to the treatment of injured and diseased muscle.

Current treatment modalities to restore the lost or altered function of injured muscles are the transposition or transfer and transplantation of complete muscles or muscle segments. This can be accomplished by using muscles either as free grafts [65] or microsurgically vascularized whole muscles with or without simultaneous repair of its nerve supply . Nevertheless, transposition and transplantation of muscles invariably result in structural and functional deficits (33% loss of muscle mass and 40% decrease in maximum force). Tenotomy and repair are perhaps the most important factors related to such a decrease in function. In spite of these deficits, transplanted or transferred muscles can usually develop enough force and power to perform their desired function (maintenance of posture or patency of sphincters [66], movement of a limb, development of facial symmetry and facial expression [67], function as a cardiac assist pump [3], etc.). However, secondary morbid sites can originate as a consequence of removing a functioning muscle from its original site of function. This again is one of the current problems of reconstructive and transplant surgery [2]. A possible solution to these problems could be the creation of a whole muscle unit, by combining the strategies of different fields of research. The isolation, culture, characterizationand biology of skeletal muscle cells has been widely described in the literature [68, 69, 70]. In 1991 Vandenburgh *et al.* described the creation of three-dimentional artificial muscle organs in vitro by applying computer-aided mechanical forces to differentiating skeletal muscle myoblasts. Development of tendon seemed to be developing and muscle activity could apparently be directed [71]. Cao and Vacanti on the other hand, recently described the development of structurally sound tissue engineered tendons by using tendon fibroblasts (tenocytes) and synthetic, biocompatible, biodegradable polymer scaffolds [72]. Current studies on the "engineering" of innervated smooth muscle structures point towards a promising field of research (Vacanti, unpublished data). The ideal scenario would be the creation of a complete innervated muscle-tendon unit with its own vascular supply in vivo or in vitro, using autologous myocytes and tenocytes. In this manner a complete functional structure could be readily obtained and implanted in the required site without causing a secondary morbid site and without rejection or foreign-body inflammatory reaction problems.

Myoblast transfer and gene therapy are other modalities of tissue engineering [73]. By combining the current knowledge of genetic engineering and tissue culture techniques, cells with a normal genome have started to be used to try to correct genetic deficiencies or alterations present in myopathies such as Duchenne muscular dystrophy (DMD). The gene that encodes for the protein dystrophin is normally expressed in skeletal muscle cells [74]. Dystrophin is absent form the muscles of patients with DMD [75] due to an X-linked mutation in the dystrophin gene in the form of a deletion, a duplication or a point mutation. Morgan demonstrated that after implantation of normal cells from newborn mice partial restoration of the dystrophin deficiency could be achieved in a dystrophic mdx mouse model [76]. Diseased muscle cells that lacking normal expression of dystrophin have been fused with normally functioning cells by intramuscular injection of normal myoblasts (satellite cells) obtained from a healthy related donor and multiplied in culture [77]. Dystrophin expression in muscle biopsy specimens was documented one month after transplantation by identification of dystrophin messenger RNA (mRNA) after reverse transcription and polymerase chain reaction (PCR). The goal of myoblast transfer in the treatment of DMD patients is to create muscle mosaics with functioning and non-functioning cells, simulating what naturally occurs in carriers whom are usually asymptomatic. Although results of myoblast transfer are encouraging the efficiency of the procedure has to be improved before it offers a real therapeutic alternative to DMD patients. Further studies have to be undertaken to understand better the effect of variables such as age and immune response of the host towards the transplanted cells to clearly delineate the potential applicability of myoblast transfer as therapy for DMD patients.

Another approach that has been attempted experimentally is the transfer of a functional dystrophin gene directly into skeletal muscle tissue. Transgenic mdx mice have been successfully created by microinjection of a full-length murine dystrophin complementary DNA (cDNA) vector [78]. Dystrophin was overexpressed by such transgenic mice. A better understanding of the integration and regulation of the recombinant DNA is required before this procedure can be applied successfully in the treatment of myopathies. Direct intramuscular injection of human dystrophin plasmid DNA has been successfully used to obtain dystrophin expression in rodent skeletal muscle in vivo and in vitro [79, 80, 81, 82]. Before this method can be clinically effective, larger numbers of transfected myofibers must be created and it should be clearly determined whether human cells will incorporate and express the injected genetic material. Retrovirus-mediated and adenovirus-mediated gene transfer have also been attempted experimentally with varying rates of success [73, 83, 84, 85, 86]. Nevertheless, for retroviral gene delivery to be effective there must be active cell division taking place and differentiated skeletal muscle is in a permanent post-mitotic state. The use of retroviruses is also limited by the size of genes they can deliver and the dystrophin gene exceeds such a size [75, 82]. Adenoviruses can elicit an immune response from the host, can inhibit protein synthesis in host's cells and present with the potential risk of cancer development [75].

The knowledge that has been gained through the different approaches aimed at the generation of sound and functional skeletal muscle has led to very encouraging experimental results that might lead in a near future to the creation of complete skeletal muscle units in vitro or to cure severeley diseased muscle groups in patients that currently face very incapacitating and nearly untreatable muscle injuries or diseases.

In summary, the techniques of tissue engineering have shown promise in the cration of the structural tissues of cartilae, bone, tendon, and muscle. With continued refinement of the concepts and techniques, experimental human application will be possible.

References

1. Vacanti CA and Mikos AG (1995) Letter from the editors. Tissue Engineering 1: 1–2.
2. Langer R and Vacanti JP (1993) Tissue Engineering. Science 260: 920–6.
3. Faulkner JA, Carlson BM and Kadhiresan VA (1994) Review: whole skeletal muscle transplantation: mechanisms responsi-

ble for functional deficits. Biotech and Bioeng 43: 757–63.
4. Brittberg M, Lindahl A, Nilsson A *et al.* (1994) Treatment of deep cartilage defects in the knee with autologous condrocyte transplantation. N Engl J Med 331: 889–95.
5. Vacanti CA, Kim WS, Schloo B *et al.* (1994) Joint resurfacing with cartilage grown in situ from cell-polymer structures. Am J Sports Med 22: 485–88.
6. Vacanti CA, Cima LG, Ratkowski D *et al.* (1992) Tissue engineered growth of new cartilage in the shape of a human ear using synthetic polymers seeded with chondrocytes. Mat Res Soc Symp Proc 252: 367–73.
7. Puelacher WC, Mooney D, Langer R *et al.* (1994) Design of nasoseptal cartilage replacements synthesized from biodegradable polymers and chondrocytes. Biomaterials 15: 774–78.
8. Puelacher, WC, Wisser J, Vacanti CA *et al.* (1994) Temporomandibular joint disc replacement made by tissue-engineered growth of cartilage. J Oral Maxillofac Surg 52: 1172–77.
9. Vacanti CA, Paige KT, Kim WS *et al.* (1994) Experimental tracheal replacement using tissue engineered cartilage. J Pediatr Surg 29: 201–205.
10. Kataoka H H and Urist MR (1993) Transplant of bone marrow and muscle-derived connective tissue cultures in diffusion chambers for bioassay of bone morphogenetic protein. Clin Orthop 286: 262–70.
11. Fuller JA and Chadially FN (1972) Ultrastructural observations on surgically produced partial-thickness defects in articular cartilage. Clin Orthop 86: 193–205.
12. Ghadially FN, Thomas I, Oryschak AF *et al.* (1977) Long-term results of superficial defects in articular cartilage: a scanning electron-microscope study. J Pathol 121: 213–17.
13. Mankin HJ (1982) Current Concepts Review. The response of articular cartilage to mechanical injury. J Bone Joint Surg: 64A: 460–66.
14. Meachim G (1963) The effect of scarification on articular cartilage in the rabbit. J Bone Joint Surg: 45B: 150–161.
15. Davis MA, Ettinger WH, Neuhaus JM *et al.* (1989) The association of knee injury and obesity with unilateral and bilateral osteoarthritis of the knee. Am J Epidemiol 130: 278–88.
16. Colettti JM Jr, Akeson WH, Woo SLY (1972) A comparison of the physical behavior of normal articular cartilage and the arthroplasty surface. J Bone Joint Surg 54A: 147–160.
17. Furukawa T, Eyre DR, Koide S *et al.* (1980) Biochemical studies on repair cartilage resurfacing experimental defects in the rabbit knee. J Bone Joint Surg 62A: 79–89.
18. Wakitani S, Goto T, Pineda SJ *et al.* (1994) Mesenchymal cell-based repair of large, full-thickness defects of articular cartilage. J Bone Joint Surg 76A: 579–92.
19. Vacanti CA and Upton J (1994) Tissue engineered morphogenesis of cartilage and bone by means of cell transplantation using synthetic biodegradable polymer matrices. Clinics in Plastic Surgery 21: 445–62.
20. Matsusue Y, Yamamuro T and Hama H (1993) Arthroscopic multiple osteochondral transplantation to the chondral defect in the knee associated with anterior cruciate ligament disruption. Arthroscopy 9: 318–21.
21. Springfield DA (1987) Massive autogenous bone grafts. Orthop Clin North AM 18: 249–56.
22. Friedlander GE and Mankin HJ (1984) Transplantation of osteochondral allografts. Ann Rev Med 35: 311–24.
23. Mankin HJ, Gebhardt MC and Tomford WW (1987) The use of frozen cadaveric allografts in the management of patients with bone tumors of the extremities. Orthop Clin North Am 18: 275–89.
24. Ostrum RF, Chao EYS, Bassett CAL *et al.* (1994) Bone injury, regeneration and repair. In: Simon SR (ed) Orthopaedic Basic Science, pp 277–323. American Academy of Orthopaedic Surgeons.
25. Chesterman PJ and Smith AU (1968) Homotransplantation of articular cartilage and isolated chondrocytes: an experimental study in rabbits. J Bone Joint Surg 50B: 184–97.
26. Bentley G and Greer RG III (1971) Homotransplantation of isolated epiphyseal and articular cartilage chondrocytes into joint surfaces of rabbits. Nature 230: 385.
27. Green WT Jr (1977) Articular cartilage repair: Behavior of rabbit chondrocytes during tissue culture and subsequent allografting. Clin Orthop 124: 237.
28. Lipman JM, McDevitt CA and Sokoloff A (1983) Xenografts of articular chondrocytes in the nude mouse. Calcif Tissue Int 35: 767.
29. Takigawa M, Shirai E, Fukuo K *et al.* (1987) Chondrocytes dedifferentiated by serial monolayer culture form cartilage nodules in nude mice. Bone Mineral 2: 449–62.
30. Grande D, Pitman MI, Peterson L *et al.* (1989) The repair of experimentally produced defects in rabbit articular cartilage by autologous chondrocyte transplantation. J Orthop Res 7: 208–18.
31. Moskalewski S (1991) Transplantation of isolated chondrocytes. Clin Orthop 272: 16–20.
32. Wakitani S, Kimura T, Hirooka A *et al.* (1989) Repair of rabbit articular cartilage surfaces with allograft chondrocytes embedded in collagen gel. J Bone Joint Surg 63B: 529.
33. Kimura T, Yasui N, Oshawa S *et al.* (1983) Chondrocytes embedded in collagen gels maintain cartilage phenotype during long-term cultures. Clin Orthop 186: 231–39.
34. Homminga GN, Buma P, Koot HWJ *et al.* (1993) Chondrocyte behavior in fibrin glue. Acta Orthop Scand 64: 441–45.
35. Upton J, Sohn SA and Glowacki J (1981) New cartilage derived from transplanted perichondrium: What is it? Plast Reconstr Surg 68: 166–74.
36. Bruns J, Kersten P, Lierse W *et al.* (1992) Autologous rib perichondrial grafts in experimentally induced osteochondral lesions in the sheep knee joint: morphological results. Virchows Archiv A Pathol Anat 421: 1–8.
37. Tsai CL, Liu TK, Fu SL *et al.* (1992) Preliminary study of cartilage repair with autologous periosteum and fibrin adhesive system. J Formosa Med Assoc 91: S239–45.
38. Sittinger M, Bujia J, Minuth WW *et al.* (1994) Engineering of cartilage tissue using bioresorbable polymer carriers in perfusion culture. Biomaterials 15: 451–56.
39. Wozney JM (1988) Novel regulators of bone formation. Molecular clones and activities. Science 242: 1528.
40. Tesch GH, Handley CJ, Cornell HJ *et al.* (1992) Effect of free and bound insuline-like growth factors on proteoglycan metabolism in articular cartilage explants. J Orthop Res 10: 14–22.
41. Vacanti JP (1988) Beyond transplantation. Arch Surg 123: 545–49.
42. Vacanti JP, Morse MA and Saltzman WM (1988) Selective cell transplantation using bioabsorbable artificial polymers as matrices. J Pediatr Surg 23: 3.
43. Vacanti CA, Langer R, Schloo B *et al.* (1991) Synthetic polymers seeded with chondrocytes provide a template for new cartilage formation. Plast Reconstr Surg 87: 753–59.
44. Green WR Jr (1977) Articular cartilage repair: Behavior of rabbit chondrocytes during tissue culture and subsequent allografting. Clin Orthop 124: 237.

45. Kim WS, Vacanti JP, Cima L *et al.* (1994) Cartilage engineered in predetermined shapes employing cell transplantation on synthetic biodegradable polymers. Plast Reconstr Surg 94: 233–37.
46. Puelacher WC, Kim SW and Vacanti JP (1994) Tissue engineered growth of cartilage: the effect of varying the concentration of chondrocytes seeded onto synthetic polymer matrices. Oral Maxillofac Surg 23: 49–53.
47. Kim WS, Vacanti JP, Upton J *et al.* (1993) Potential of cold-preserved chondrocytes for cartilage reconstruction. Plastic Surgery Research Council.
48. Paige KT, Cima I and Yaremchul MJ (in press) Injectable cartilage.
49. Cao YL and Vacanti JP. Unpublished data.
50. Asselmeier MA, Caspari RB and Bottenfield S (1993) A review of allograft processing and sterilization techniques and their role in transmission of the human immunodeficiency virus. Am J Sports Med 21: 170–175.
51. Gatti AM, Zaffe D and Poli GP (1990) Behaviour of tricalcium phosphate and hydroxiapatite granules in sheep bone defects. Biomaterials 11: 513–17.
52. Krukowski M, Shively RA, Osdoby P *et al.* (1990) Stimulation of craniofacial and intramedullary bone formation by negatively charged beads. J Oral Maxillofac Surg 48: 468–75.
53. Roux FX, Brasnu D, Loty B *et al.* (1988) Madreporic coral: A new bone graft substitute for cranial surgery. J. Neurosurg 69: 510–13.
54. Rozema FR, Bos RR, Pennings AJ *et al.* (1990) Poly(L-lactide) implants in repair of defects of the orbital floor: An animal study. J Oral Maxillofac Surg 48: 305–9.
55. Thaller SR, Hoyt J, Borjeson K *et al.* (1993) Reconstruction of calvarial defects with anorganic bovine bone mineral (Bio-Oss) in a rabbit model. J Craniofac Surg 4: 9–84.
56. Yukna RA (1990) Polymer grafts in human periodontal osseous defects. J Periodont 61: 633–642.
57. Mulliken JB and Gowacki J (1980) Induced osteogenesis for repair and construction in the craniofacial region. Plast Reconstr Surg 65: 553–60.
58. Costantino PD, Friedman CD, Jones K *et al.* (1992) Experimental Hydroxyapatite cement cranioplasty. Plast Reconstr Surg 90: 174–91.
59. Dahlin C, Alberius P and Linde A (1991) Osteopromotion for cranioplasty: An experimental study in rats using a membrane technique. J Neurosurg 74: 487–91.
60. Kataoka H and Urist MR (1993) Transplant of bone marrow and muscle-derived connective tissue cultures in diffusion chambers for bioassay of bone morphogenetic protein. Clin Orthop 286: 262–270.
61. Vacanti CA, Kim WS and Mooney D (1993) Tissue engineered composites of bone and cartilage using synthetic polymers seeded with two cell types. Orthopaedic Transactions 18: 276.
62. Cao Y and Vacanti CA. Unpublished data.
63. Yaszemski MJ, Payne RG and Hayes WC (1995) The ingrowth of new bone tissue and initial mechanical properties of a degrading polymeric coposite scaffold. Tissue Engineering 1: 41–52.
64. Freed LE, Vunjak-Novakovic G and Langer R (1993) Cultivation of cell-polymer cartilage implants in bioreactors. J Cell Biochem 51: 257–64.
65. Faulkner JA and Cote C (1986) Functional deficits in skeletal muscle grafts. Fed Proc 45: 1466–69.
66. Freilinger G and Deutinger M (1992) Third Vienna muscle symposium. Blackwell MVZ, Vienna, Austria.
67. Freilinger G (1975) A new technique to correct facial paralysis. Plast Reconstr Surg 56: 44–8.
68. Yasin R, Van Beers G, Nurse KCE *et al.* (1977) A quantitative technique for growing human adult skeletal muscle in culture starting from mononucleated cells. J Neurol Sci 32: 347–60.
69. Yasin R, Kundu D and Thompson E (1981) Growth of adult human cells in culture at clonal densities. Cell Differ 10: 131–37.
70. Florini J and Magri K (1989) Effect of growth factors on myogenic differentiation. Am J Physiol 256: C701–11.
71. Vandenburgh HH, Swasdison S and Karlisch P (1991) Computer-aided mechanogenesis of skeletal muscle organs from single cells in vitro. FASEB Journal 3: 2860–67.
72. Cao Y, Vacanti JP, Ma X *et al.* (1994) Generation of neo-tendon using synthetic polymers seeded with tenocytes. Transpl Proc 26: 3390–92.
73. Brooks SV, Cole NM and Faulkner JA (in press) Tissue engineering of skeletal muscle.
74. Hoffman EP, Brown RH and Kunkel LM (1987) Dystrophin: the protein product of Duchenne muscular dystrophy locus. Cell 51: 919–28.
75. Hoffman EP, Fishbeck KH, Brown RH *et al.* (1988) Characterization of dystrophin in muscle-biopsy specimens from patients with Duchenne's or Becker's muscular dystrophy. N Engl J Med 318: 1363–68.
76. Morgan JE, Hoffman EP and Partridge TA (1990) Normal myogenic cells from newborn mice restore normal histology to degenerating muscles of the mdx mouse. J Cell Biol 111: 2437–49.
77. Gussoni E, Pavlath GH, Lanctot AM *et al.* (1992) Normal dystrophin transcripts detected in Duchenne muscular dystrophy patients after myoblast transplantation. Nature 356: 435–38.
78. Cox GA, Cole NM, Matsumura K *et al.* (1993) Overexpression of dystrophin in transgenic mdx mice eliminates dystrophic symptoms without toxicity. Nature 364: 725–29.
79. Lin H, Parmacek MS, Morle G *et al.* (1990) Expression of recombinant genes in myocardium in vivo after direct injection of DNA. Circulation 82: 2217–21.
80. Wolff JA, Malone RW, Williams P *et al.* (1990) Direct gene transfer into mouse muscles in vivo. Science 247: 1465–68.
81. Acsadi G, Dickson G, Love DR *et al.* (1991) Human dystrophin expression in mdx mice after intramuscular injection of DNA constructs. Nature 352: 815–18.
82. Ono T, Ono K, Mizukawa K *et al.* (1994) Limited diffusability of gene products directed by a single nucleus in the cytoplasm of multinucleated myofibers. FEBS Lett 337: 18–22.
83. Dunckley MG, Wells DJ, Walsh FS *et al.* (1993) Direct retroviral-mediated transfer of a dystrophin minigene into mdx mouse muscle in vivo. Hum Mol Genet 2: 717–23.
84. Stratford-Perricaudet LD, Makeh I, Perricaudet M *et al.* (1992) Widespread long-term gene transfer to mouse skeletal muscle and heart. J Clin Invest 90: 626–630.
85. Ragot T, Vincent N, Chafey P *et al.* (1993) Efficient adenovirus-mediated transfer of a human minidystrophin gene to skeletal muscle of mdx mice. Nature 361: 647–50.
86. Vincent N, Ragot T, Gilgenkrantz H *et al.* (1993) Long term correction of mouse dystrophic degeneration by adenovirus-mediated transfer of a minidystrophin gene. Nature Genetics 5: 130–34.

R. P. Lanza and W. L. Chick (eds.), Yearbook of Cell and Tissue Transplantation 1996/1997, 247–252.

Chapter 24

Tissue engineering: Liver

Susumu Eguchi, Steve C. Chen, Jacek Rozga and Achilles A. Demetriou
Department of Surgery and the Liver Support Unit, Cedars-Sinai Medical Center, Los Angeles, CA 90048, U.S.A.

Background

Treatment of liver failure

Various attempts have been made to support animals and patients with liver insufficiency, utilizing various extracorporeal support systems including cross-circulation, whole liver blood perfusion, hemadsorption, hemodialysis, plasma exchange, total body washout, use of microsomal enzymes bound to artificial carriers and others [1]. However, none of these therapeutic modalities succeeded in gaining wide clinical acceptance.

Charcoal hemoperfusion has been extensively studied in the laboratory and was used to treat severe acute liver failure clinically without conclusive evidence of efficacy [2]. In general, this and other techniques which rely primarily on blood detoxification, have not succeeded in improving patient survival in fulminant hepatic failure.

It would appear that whole liver perfusion would be more beneficial because it combines synthetic liver function with detoxifying capacity. From a practical standpoint, however, human liver perfusion cannot be carried out because of lack of organ availability; if a suitable organ becomes available, it is used for transplantation. Xenogeneic liver perfusion is cumbersome because it involves maintaining animal colonies at each treatment site and because several organs are usually needed to treat a single patient over a short period of time due to development of perfused liver parenchymal hemorrhage and thrombosis. In the past twenty years, several reports of whole liver blood perfusions using human, primate, bovine, and porcine livers have been published [1, 3, 4]. None of these reports succeeded in convincingly demonstrating that the treatments had a beneficial effect on patient survival, when compared to standard supportive therapeutic measures. In some instances in which whole blood *ex vivo* liver perfusion was carried out, there was a transient improvement in encephalopathy with some patients “waking up” from coma. However, significant complications were associated with these liver perfusions, especially hemolysis and thrombo-cytopenia. Thus it appears that, at least for now, whole organ transplantation remains the only method with clinically-proven efficacy for treating severe acute liver failure.

Extracorporeal liver support

In attempting to develop systems for temporarily supporting patients until an organ becomes available for transplantation, and because of the complexity and vast number of metabolic and other physiologic functions provided by the liver, it was felt that to provide effective *ex vivo* liver support construction of a liver support system utilizing intact, viable, functioning isolated liver cells will be needed. The two major technical advantages of use of hepatocyte based systems over whole organ perfusion, are the ability to cryopreserve cells and use them as needed and the elimination of highly antigenic endothelial cells and macrophages present in the intact liver.

Sorrentino was the first to coin the term “artificial liver” and used fresh liver homogenates which metabolized salicylic and barbituric acids, ketone bodies and produced urea from ammonia [5]. Hori described the use of cross-species hemodialysis [6]. Use of liver cells to construct liver assist systems was extensively investigated by Eiseman *et al.* [7–10], Uchino *et al.* [11]

and more recently by our group [12–15]; these experimental animal studies demonstrated that a number of metabolic functions which are either impaired or lost as a result of severe liver failure, can be replaced, to some degree, by these systems.

Clinical experience with hepatocyte-based systems

Several important advances in the experimental laboratory made possible the development of hepatocyte liver support systems for clinical use. These include:

1) Development of simple methods of hepatocyte isolation from animals [16–18].
2) Demonstration that cultured hepatocytes function better when attached to a "physiologic" matrix [19, 20].
3) Improvement of hepatocyte culture conditions.

A hepatocyte-based device was first used by Matsumura [21], who converted a renal dialyzer to an "artificial" liver by adding a cryopreserved rabbit liver cell suspension to a dialysis chamber. He replaced the usual cuprophane membrane with a cellulose membrane which was permeable to middle-range molecules but not proteins. The device was used to treat a 45-year-old man in hepatic failure due to inoperable cholangiocarcinoma who underwent hemodialysis for five hours; total bilirubin was reduced from 25.0 to 16.8 mg/dl. Three days later, after a second 4 1/2 hour treatment, the total bilirubin level decreased from 18.0 to 8.0 mg/dl. There was no evidence that the device affected the course of the disease and there have been no further reports of the use of this system.

A group of 59 patients with liver failure was treated in Latvia with daily 6-hour hemoperfusions through a 20 ml polychlorovinyl capsule filled with activated charcoal and less than 0.4 g of porcine hepatocytes [22]. A control group of 67 patients received standard medical therapy. The authors noted improved survival in the hepatocyte-treated group. No evidence of physiologic, biochemical and metabolic improvement in the treatment group was presented.

Sussman *et al.* [23, 24] utilized a cell line (C3A) derived from human hepatoblastoma. Patients underwent whole blood perfusion continuously for relatively long periods of time. This method required administration of heparin to prevent blood coagulation in the system. In the initial group of ten patients, no significant effect on disease outcome was noted and only one patient survived. Subsequently, a preliminary report described the results of a prospective controlled clinical trial carried out at Kings College Hospital in London. No significant effect on patient survival was noted in the group treated with the cell support system compared to the untreated control group [25].

Several other systems are currently in various stages of laboratory development with clinical trials planned in the future. Problems facing the field include: 1) difficulty in scaling up systems tested in small animals for clinical use; 2) proving system efficacy by demonstrating evidence of detoxification and synthetic function; 3) technical problems (hemolysis, coagulation, thrombo-cytopenia, need for heparin administration); and 4) defining the test patient population, quantitating the degree of liver failure, taking into account the underlying disease etiology. The above limitations make meaningful comparisons among various devices difficult.

Bioartificial liver

System design

We have developed a Bioartificial Liver (BAL) which combines biological and artificial mechanical components [12, 13, 26]. Blood is removed from a patient through a double-lumen catheter in either the saphenous or superficial femoral vein at a rate of 60–90 ml/minute, and is run through a plasma separator, a transmission reservoir, a hollow fiber module inoculated with microcarrier-attached porcine hepatocytes, an oxygenated water bath maintained at 37 °C, four roller pumps and oxygen-permeable silicone tubing. The primary mechanism of solute transport in the hollow fiber module is fluid convection (Starling flow) driven across the membrane by a trans-fiber pressure gradient. The faster the axial flow rate, the faster the Starling flow. A transmission reservoir is used to enhance system-efficiency by allowing plasma recirculation through the BAL circuit at 200 ml/min. Following recirculation through the BAL, plasma and red blood cells are reconstituted and returned to the patient *via* the venous cannula.

System components

Porcine hepatocytes are isolated aseptically from adult pigs as previously described [12, 13]. Isolated hepatocytes are attached to collagen-coated dextran microcarriers as previously described [12, 13].

Approximately 1.0×10^9 cells are added to 1.6 g (dry weight) of hydrated microcarriers. Microcarrier-attached porcine hepatocytes are inoculated into the extra-fiber compartment of hollow fiber modules. Each hollow fiber module consists of a polycarbonate cylinder (29.1 mm I.D., 31.2 mm O.D.) containing 670 cellulose nitrate/cellulose acetate fibers (635 μm I.D., 760 μm O.D., wall thickness 62.5 μm; 510 mm overall length, 445 mm potted length), with an extra-fiber volume of 177 ml. Total fiber internal surface area is 5,850 cm^2, external surface area is 7,010 cm^2 and the pore diameter in the semi-permeable fiber wall is 0.2 μm. Approximately 6×10^9 viable microcarrier-attached porcine hepatocytes are inoculated into the extra-fiber chamber of each hollow-fiber module.

Clinical experience

Nineteen patients were treated with the BAL. Patients belonged to one of two groups. Group I patients ($n = 11$) were candidates for liver transplantation at the time of presentation with acute liver failure; ten patients in this group had classic fulminant hepatic failure without history of chronic underlying liver disease and one patient had primary non-function of a transplanted liver. Ten Group I patients were in deep stage 4 coma and one was in stage 2. Group II ($n = 8$) consisted of patients with chronic underlying disease who presented with an acute exacerbation; these patients were not candidates for transplantation at the time of presentation. Patients were treated with the BAL for 7-hour periods one or more times.

All patients tolerated BAL treatment(s) well. They remained hemodynamically stable and did not experience significant adverse reactions. As expected, use of citrate to prevent clotting in the plasma separation system, resulted in ionized hypocalcemia which was treated with calcium chloride intravenous infusion; serum calcium levels were monitored hourly during each treatment. Patients did not experience any hypersensitivity reactions. The effects of BAL treatments on relevant plasma biochemical parameters for both groups of patients are summarized in Table 1.

All patients in Group I were successfully "bridged" to transplantation. Patients in this group were treated 1–3 times and the mean "bridge" interval from the initiation of the first BAL treatment to transplantation was 39 hours. Patients in this group had significant neurologic impairment with a Glasgow Coma Scale of 7.2 ± 0.8 and a Comprehensive level of Consciousness Score of 25.2 ± 2.6. The latter neurologic assessment system appears to be a better indicator of brain stem dysfunction. The most significant observed clinical effect was improvement in the patients' neurologic status with reversal of the decerebrate status. This was accompanied by a significant decrease in intracranial pressure (from 19.1 ± 2.2 to 9.0 ± 1.2 mm Hg; $p < 0.005$) accompanied by a significant increase in cerebral perfusion pressure (from 71.0 ± 4.3 to 85.1 ± 2.7 mm Hg; $p < 0.008$). An improvement in the CLOC score was also noted (29.8 ± 2.1; $p < 0.01$). All patients underwent successful liver transplantation, survived, and were discharged from the hospital neurologically intact. There was no evidence that pretreatment with the BAL had an adverse effect on subsequent liver allograft survival and function.

Group II patients with chronic liver disease were not candidates for transplantation. Two patients in this group recovered from their acute illness, later underwent successful liver transplantation and recovered. The remaining six patients clinically experienced transient beneficial effects but eventually they succumbed because of lack of adequate synthetic function and development of sepsis, multiple organ failure, and inability of their livers to regenerate and recover.

Discussion

Charcoal hemoperfusion has been used to treat severe acute liver failure with mixed results [27–30]. Although there is clear experimental evidence that the technique has some beneficial effects, a controlled prospective clinical study has failed to demonstrate significant clinical advantages [2]. Most methods that relied primarily upon blood detoxification showed limited success as well [27–30]. Due to the complexity and vast number of metabolic and other physiologic functions provided by the liver and the need for broad metabolic support in acute liver failure, it was felt that construction of an extracorporeal liver support system would require utilization of viable isolated hepatocytes rather than either specific cell components or enzymes. We have previously demonstrated that attachment of a hollow fiber module inoculated with normal hepatocytes to Gunn rats (which are unable to conjugate bilirubin) *via* arterial and venous cannulas, resulted in the appearance of bilirubin conjugates in their bile [31]. In subsequent experiments, utilizing dogs with severe irreversible acute ischemic liver failure, we demonstrated that treatment with a hollow fiber module inoculated with cryopreserved matrix-attached allogeneic

Table 1. Changes in serum biochemical parameters following BAL treatment

Group I	Bili. (totl.) mg/dl	Bili. (dir.) mg/dl	NH_3 μmol/l	Glucose mg/dl	AST IU/l	ALT IU/l
Pre-BAL	20.8 ± 3.0	9.9 ± 1.8	157.8 ± 14.3	117.6 ± 12.1	1725 ± 542	1378 ± 375
Post-BAL	17.7 ± 2.4	8.0 ± 1.3	121.7 ± 10.6	171.4 ± 23.2	1089 ± 288	913 ± 250
p value*	0.003	0.005	0.02	0.009	0.02	0.006
Group II						
Pre-BAL	22.8 ± 5.1	11.5 ± 2.9	201.3 ± 46.6	144.2 ± 12.71	1196 ± 816	364 ± 219
Post-BAL	19.5 ± 4.3	9.4 ± 2.4	143.0 ± 25.3	166.0 ± 17.5	1278 ± 897	347 ± 219
p value*	0.01	0.02	0.05	0.09	0.4	0.5

* Paired Student *t* Test.

and xenogeneic (porcine) hepatocytes, resulted in significant beneficial effects. These effects included lower serum ammonia, lactate and pH levels and higher serum glucose levels and systolic blood pressure in hepatocyte-treated animals compared to controls [12]. In later studies, we utilized plasma separation, high-performance sequential perfusion through a column loaded with activated cellulose-coated charcoal particles and through a hollow-fiber module inoculated with viable matrix-anchored porcine hepatocytes, to treat dogs with ischemic liver failure with significant improvement [13]. Our experimental animal data suggest that such a hybrid system can provide both detoxifying and synthetic liver functions and is superior to charcoal plasma perfusion alone.

Use of plasma perfusion of isolated porcine hepatocytes, offers several distinct advantages over use of whole organ perfusion:

a) Hepatocytes can be easily isolated and cryopreserved.
b) The method is logistically rapid and simple.
c) Plasmapheresis is an established clinical procedure, familiar to nursing and other hospital personnel and it does not require heparin anticoagulation.
d) The procedure is cost-effective.
e) Decreased antigenicity.
f) The procedure is safe.
g) Potential for either immuno-isolation or modification of hepatocytes to render them less immunogenic.

With the ability to isolate and maintain hepatocytes functional *in vitro* for relatively long periods of time, various systems of extra corporeal liver support can be designed. In general, for a system to be useful it should be simple, cost-effective, easily accessible and safe. As indicated above, use of plasma separation and plasma perfusion in the extracorporeal liver support module, eliminates the need for systemic heparin anticoagulation and is not associated with the mechanical blood cell destruction often seen with whole blood organ perfusion. Use of hollow-fiber technology for designing an extracorporeal liver support system, takes advantage of experience gained from the industrial use of hollow-fiber bioreactors for tissue culture and vaccine production, in system design and determination of material biocompatibility. This experience has resulted in the development of several commercially available hollow-fiber modules. In addition, a large volume of engineering data has been obtained from the study of industrial production bioreactors, which can be utilized in designing systems for clinical use.

Various types of matrix can be used for hepatocyte attachment. It has been demonstrated that isolated hepatocyte viability and expression of differentiated function *in vitro*, was enhanced when hepatocytes were maintained as a monolayer on a collagen biomatrix; several types of biomatrix have been described for maintaining primary hepatocyte cultures for long periods of time. We believe that "anchored" hepatocytes attached to a collagen matrix, are better able to express differentiated liver functions in an extracorporeal liver support system, because of a more physiologic environment which allows cell-cell and cell-matrix interactions. Various systems can be developed utilizing many types of matrix for hepatocyte attachment. In general, system design should allow effective transport from the cellular to the plasma-flow compartment and adequate perfusion of the cells. We chose a commercially available type of cell matrix, dextran microcarriers coated with type I collagen.

Successful *in vivo* use of a particular experimental device which utilizes matrix-attached hepatocytes, requires definition of the appropriate conditions (flow, transport, bio-compatibility, cell mass and other) which will allow optimal hepatocyte function in an adverse environment (perfusion with plasma from patients with liver failure). The system can be further optimized by complementing hepatocyte function by the addition of various columns (i.e. charcoal, specific affinity columns) on-line with the hepatocyte-containing module.

Management of patients with acute severe liver failure is a major clinical challenge. We need to better understand normal and abnormal liver physiology, develop methods of assessing the degree of liver dysfunction, standardize methods of intervention and develop rational procedures to support the acutely failing liver until it either recovers or is replaced. We believe that our approach offers several distinct advantages over use of whole organ perfusion:

1) Hepatocytes can be easily isolated and cryopreserved.
2) Harvesting large numbers of hepatocytes can be easily accomplished.
3) Methods of hepatocyte cryopreservation, allowing cell storage for later use, have been developed. On short notice, cells can be thawed, washed and inoculated into the support system device.
4) Plasmapheresis is an established clinical procedure, familiar to nursing and other hospital personnel. Citrate regional anticoagulation is used without a need for systemic heparin anticoagulation.
5) The procedure is cost-effective. Cell harvesting can be carried out cost-effectively because of use of EDTA and only a very small amount of recirculated collagenase during the isolation process.
6) There is no need for maintaining animal colonies at hand for possible clinical use and no need for emergency surgical liver removals, both of which are costly.
7) Decreased antigenicity. The process of hepatocyte harvesting and preparation, results in a 98% enriched (pure) hepatocyte population. Use of purified hepatocytes eliminates endothelial and other types of highly antigenic liver cells.
8) The procedure is safe. In animal experiments, porcine hepatocytes were used to treat dogs with severe ischemic liver failure. There was no evidence of hemolysis and thrombocytopenia in the treated animals; both of these complications are usually seen following whole organ perfusion.
   In addition, treated animals did not experience hypersensitivity and other reactions following treatment. Porcine hepatocytes in the extra-corporeal liver support system remained viable for the duration of the treatment.
9) Potential for either immuno-isolation or modification of hepatocytes to render them less immunogenic. It may be possible to immuno-isolate hepatocytes by using selective molecular weight cut-off membranes, encapsulation techniques, pretreatment with *UV* B and specific antisera, cryopreservation and other modalities used to reduce cell immunogenicity. In addition, genetic manipulation of hepatocytes, may result in the development of low-antigenicity, immortalized, differentiated "liver cell" lines for use in liver support systems.

We used a BAL to treat a small number of patients with fulminant hepatic failure. Various significant beneficial effects were noted. However, the most important clinical effect was the neurologic improvement and the accompanied reduction in intracranial pressure. Most patients with fulminant liver failure die because of cerebral edema and brain herniation. For a liver support device to have significant clinical impact on survival in this patient group, it should achieve reduction in intracranial pressure and cerebral edema. If this beneficial effect can be seen consistently, it will represent a significant advance in this field.

Use of a liver support system for the treatment of chronic liver disease and acute exacerbation of chronic disease will require early aggressive intervention and longer periods of support to allow time for the diseased liver to recover/regenerate. Prospective controlled studies of liver support systems are needed to determine and fully establish their efficacy in the treatment of these severely ill patients.

References

1. Brunner G, Mito M (eds) (1992) Artificial Liver Support. Berlin: Springer-Verlag.
2. O'Grady JG, Gimson AES, O'Brien CJ *et al.* (1988) Controlled trials of charcoal hemoperfusion and prognostic factors in fulminant hepatic failure. Gastroenterology 94: 1186–1192.
3. Parbhoo SP, James IM, Adjukiewicz A *et al.* (1971) Extracorporeal pig liver perfusion in treatment of hepatic coma due to fulminant hepatitis. Lancet ii: 659–665.
4. Abouna GM, Cook JS, Fisher LMA *et al.* (1972) Treatment of acute hepatic coma by *ex vivo* baboon and human liver perfusions. Surgery 71: 537–546.
5. Sorrentino F (1956) Prime ricerche per la realizzatione di un fegato artificiale. Chir Patol Sper 4: 1401–14.

6. Hori M (1982) Artificial liver: The concept and working hypothesis of hybrid organs: from a 25-year-old anecdote to the 21st century model. Trans Am Soc Artif Intern Organs 28: 639–41.
7. Eiseman B, Norton L and Kralios NC (1976) Hepatocyte perfusion within a centrifuge. Surg Gynecol Obst 142: 21–8.
8. Olumide F, Eliashiv A, Kralios N *et al.* (1977) Hepatic support with hepatocyte suspensions in a permeable membrane dialyzer. Surgery 82: 599–606.
9. Soyer T, Lempinen M, Walker JE *et al.* (1973) Extracorporeal assist of anhepatic animals with liver slice perfusion. Am J Surg 126: 20–24.
10. Eiseman B, Lien DS and Raffucci F (1965) Heterologous liver perfusion in treatment of hepatic failure. Ann Surg 162: 329–345.
11. Uchino J, Tsuburaya T, Kumagai F *et al.* (1988) A hybrid bioartificial liver composed of multiplated hepatocyte monolayers. ASAIO Trans 34: 972–77.
12. Rozga J, Williams F, Ro M-S *et al.* (1993) Development of a Bioartificial Liver: Properties and function of a hollow-fiber module inoculated with liver cells. Hepatology 17: 258–65.
13. Rozga J, Holzman MD, Ro M-S *et al.* (1993) Hybrid Bioartificial Liver Support treatment of animals with severe ischemic liver failure. Ann Surg 217: 502–11.
14. Rozga J, Podesta L, LePage E *et al.* (1993) Control of cerebral oedema by total hepatectomy and extracorporeal liver support in fulminant hepatic failure. Lancet 342: 898–99.
15. Rozga J, LePage E, Moscioni AD *et al.* (1993) Clinical Use of a Bioartificial Liver to treat fulminant hepatic failure. Ann Surg 219: 538–546.
16. Berry MN and Friend DS (1969) High yield preparation of isolated rat liver parenchymal cells. A biochemical and fine structural study. J Cell Biol 43: 466–68.
17. Wang S, Renaud G, Infante J *et al.* (1985) Isolation of rat hepatocytes with EDTA and their metabolic function in primary culture. In Vitro Cell Dev Biol 21: 526–30.
18. Kraemer BL, Staecker JL, Sawada N *et al.* (1986) Use of a low-speed, iso-density percoll centrifugation method to increase the viability of isolated rat hepatocyte preparations. In Vitro Cell Dev Biol 22: 201–11.
19. Reid LM and Rojkind M (1989) New techniques for culturing differentiated cells: reconstituted basement membrane rafts. Methods Enzymol 58: 263–78.
20. Caron JM (1990) Induction of albumin gene transcription in hepatocytes by extracellular matrix proteins. Mol Cell Biol 10: 1239–43.
21. Matsumura KN, Guevara GR, Huston H *et al.* (1987) Hybrid bioartificial liver in hepatic failure: Preliminary clinical report. Surgery 101: 151–57.
22. Margulis MS, Erukhimov EA, Andreiman LA and Viksna LM (1989) Temporary organ substitution by hemoperfusion through suspension of active donor hepatocytes in a total complex of intensive therapy in patients with acute hepatic insufficiency. Resuscitation 18: 85–94.
23. Sussman NL and Kelly JH (1993) Improved liver function following treatment with an extracorporeal liver assist device. Artif Organs 17: 27–30.
24. Sussman NL, Chong MG, Koussayir T, He DE, Shong TA, Whisennand HH and Kelly JH (1992) Reversal of fulminant hepatic failure using an extracorporeal liver assist device. Hepatology 16: 60–5.
25. Wendon J, Hughes R, Langley P *et al.* (1994) A controlled trial of the Hepatix extracorporeal liver assist device (ELAD) in acute liver failure. Hepatology 20: 140A (abstract).
26. Neuzil D, Rozga J, Moscioni AD *et al.* (1993) Use of a Bioartificial Liver in a patient with acute liver insufficiency. Surgery 113: 340–343.
27. Chang T, Williams R and Nose Y (1983) A multifaceted approach to artificial liver support. Trans Am Soc Artif Intern Organs 29: 795–799.
28. Berk PD and Goldberg JH (1988) Charcoal Haemoperfusion. Plus Ca Change, Plus C'est La Meme Chose. Gastroenterology 94: 1228–1230.
29. Gimson AES, Mellon PJ, Braude S, Canalese J and Williams R (1982) Earlier charcoal haemoperfusions in fulminant hepatic failure. Lancet ii: 681–683.
30. Bihari D, Hughes RD, Gimson AES *et al.* (1983) Effects of serial resin hemoperfusion in fulminant hepatic failure. Int J Artif Organs 6: 299–302.
31. Arnaout WS, Moscioni AD, Barbour RL and Demetriou AA (1990) Development of bioartificial liver: Bilirubin conjugation in Gunn rats. J Surg Res 48: 379–382.

R. P. Lanza and W. L. Chick (eds.), Yearbook of Cell and Tissue Transplantation 1996/1997, 253–264.

Chapter 25

Pancreas

Robert P. Lanza and William L. Chick
BioHybrid Technologies, Inc., Shrewsbury, MA 01545, U.S.A.

The restoration of normal glucose metabolism has only recently been achieved in patients with Type I diabetes mellitus by the transplantation of islets of Langerhans. Pancreatic islets, which comprise only 1–2% of the human adult pancreatic volume, can be transplanted without extensive surgery. However, despite the promise of this therapy for the treatment of diabetes, the clinical success of islet transplantation has been sporadic, with only 28 documented cases of insulin independence occurring from 242 attempts at clinical islet allotransplantation since 1983 [1]. There are multiple reasons for this low success rate. Studies indicate that transplanted islets are exquisitely sensitive to both conventional rejection and to damage by autoimmune activity specifically directed against the beta cells. Local cytokine release is toxic to islet cells and inhibits normal insulin secretion by those cells not lethally affected. In addition, 'naked' islet transplants require a life-long regimen of high dose immunosuppressive drugs to prevent immune rejection. Unfortunately, use of these agents is associated with a variety of problems. Interestingly, many of these immunosuppressive drugs, including glucocorticoids and cyclosporine, also have dose dependent delererious effects on glucose homeostasis and beta cell function [2–5].

Ultimately, the goal of islet transplantation is to treat patients without generalized immunosuppression, and early enough in the course of the disease to prevent or retard the development of complications. Immunoisolation of islets in biohybrid devices offers a distinct advantage in this respect. These devices are based upon the principle of isolating living tissue from the immune system of the host by selectively permeable membranes. Low molecular weight substances such as nutrients, electrolytes, oxygen, and bioactive secretory products are exchanged across the membrane while immunocytes, and other transplant rejection effector mechanisms, are excluded [6,7]. This approach has the potential not only to allo allogeneic transplantation of islets without immunosuppression, but also to permit the use of xenografts (pancreatic islets can also be isolated from porcine, bovine and canine glands; annimal insulins are fully active in man and have been used to treat diabetics for over 70 years). In the form of a perfused vascular inplant, the islets can be distributed in a chamber surrounding the membrane, and the device inplanted as a shunt in the vascular system [8–10]. Alternatively, the islets can be encapsulated (i.e. immunoisolated) within membrane diffusion chambers [11–15] or within microcapsules [16–20] and placed intraperitoneally [11–20], subcutaneously [21,22] or in other sites [23].

Intravascular pancreas devices

Although much of the immunoisolation work to date has been carried out with membrane diffusion chambers and microcapsules, the modern era of biohybrid device development began approximately 20 years ago with the introduction of islet-containing perfusion devices implanted as arteriovenous shunts. These devices offer certain obvious advantages. The islets are distributed in a chamber surrounding a selectively permeable membrane, and the device implanted as a shunt in the cardiovascular system. The islets have direct vascular access, and are supplied with oxygen by diffusion from the arterial circulation at a partial pressure of about 100 torr (mm Hg). By contrast, extravascular devices implanted intraperitoneally, intramuscularly, or in other tissues must exchange oxygen and nutrients over larger diffusion distances, and with a microvasculature that normally delivers oxygen at a partial pressure of only 40 torr [24,25]. In this regard, there are data that suggest that islet viability and/or insulin secretory function are detrimentally influenced at low oxygen tensions [26–28]. In addition, it is possible to access the cell chamber of vasular devies for removal and replacement of nonfunctioning islets once the device is implanted. The seeding ports can be sedigned to be accessible for reseeding using a needle and syringe, and can be positioned just under the surface of the skin.

The original perfusion devices were developed in the mid-1970s by William Chick *et al.* [29,30]. These devices utilized bundles of capillary fibers seeded on their outside surfaces with isolated islet cells. Tissue culture medium was circulated through the lumen of the fibers and the islets secreted insulin in response to stimulatory glucose concentrations. However, the use of these small diameter fibers (ID $\leq$ 1 mm) as vascular inplants was limited to short-term, *ex vivo* studies because of clotting [29,31–36]. Experiments in which tubular membranes with an inner diameter of 2.7 mm were used resulted in the first demonstration of extended *in vivo* patency [37]. These larger diameter fibers remained patent as AV shunts for seven weeks in dogs which did not receive any systemic anticoagulation.

An artificial pancreas device which utilizes a single, coiled, tubular membrane with an inner diameter of 5–6 mm was investigated by our group for the past several years [8–10]. This device design incorporates a poly acrylonitrile-polyvinyl chloride (PAN-PVC) membrane within an acrylic housing. The islet chamber is created by the space between the membrane and the housing. The PAN-PVC membrane can not be sutured so it is connected to standard polytetrafluorethylene graft material of the same diameter which extends beyond the housing and is used for anastomosis to the vascular system.

In vivo studies indicate that this design has excellent biocompatibility with respect to blood clotting. Four dogs which received devices without islet tissue are still ongoing after three years. The function of this perfused artificial pancreas was evaluated by inplanting devices seeded with canine islet allografts into severely diabetic pancreatectomized dogs using a protocol that had been shown to optimize long-term insulin secretion in devices maintained *in vitro* [9]. These experiments indicate that the devices can significantly improve glucose homeostatis and can function for more than a year [8]. In a series of 17 dogs which received two of these devices containing canine islets, 11 animals no longer required insulin for control of fasting blood levels. Four dogs were terminated due to sepsis or device thrombosis. The 3, 6, and 12 month graft survival rates for the remaining animals were 100, 71, and 29%, respectively. Histologic evaluation of the sevices after removal revealed viable islets with granulated β-cells. No evidence of infiltration of immune cells was observed suggesting that the membrane was indeed immunoprotective. In two animals, the devices were removed one year after implantation. In both cases, the exogenous insulin required to control fasting blood glucose concentrations increased by more than 20 U/day after device removal. In addition, approximately 25% of the islets remained viable after one year *in vivo* (Figure 1). One of the devices was perfused with tissue culture medium *in vitro* following removal, and showed continued release of insulin.

Data from the implantation of devices containing xenogeneic islets are limited but does indicate that discordant xenografts should also be feasible [8,38]. One dog which received devices containing islets demonstrated excellent control of fasting glucose levels for almost two months without exogenous insulin. The results using porcine islets are even more preliminary, but in at least two dogs devices containing porcine islets substantially decreased the exogenous insulin requirement for more than eight months [39].

However, a number of issues remain which may ultimately limits the therapeutic potential of this approach. Perhaps most importantly, our data suggest that the size and geometry of the current perfusion devices imposes a critical limitation on the amount of islet tissue that can be transplanted into a patient. At present, two perfusion devices would be required to treat a patient with an insulin requirement of approximately 30 units/day. Attempts to lengthen the coiled, tubular membrane, thereby increasing insulin secretion, have failed because of clotting. In addition, the glycemic control provided by the perfusion device design clearly was not optimal. Nevertheless, much has been learned from these studies. They represent an important first step toward developing a newer, simpler, more viably for transplanting islets using microreactors.

Extravascular pancreas devices

Diffusion devices

Numerous devices of this type have been evaluated by our own and other groups [40]. These are typically tubular or planar designs. Membrane materials used to fabricate these devices include PAN-PVC, polypropylene, polycarbonate, and cellulose nitrate

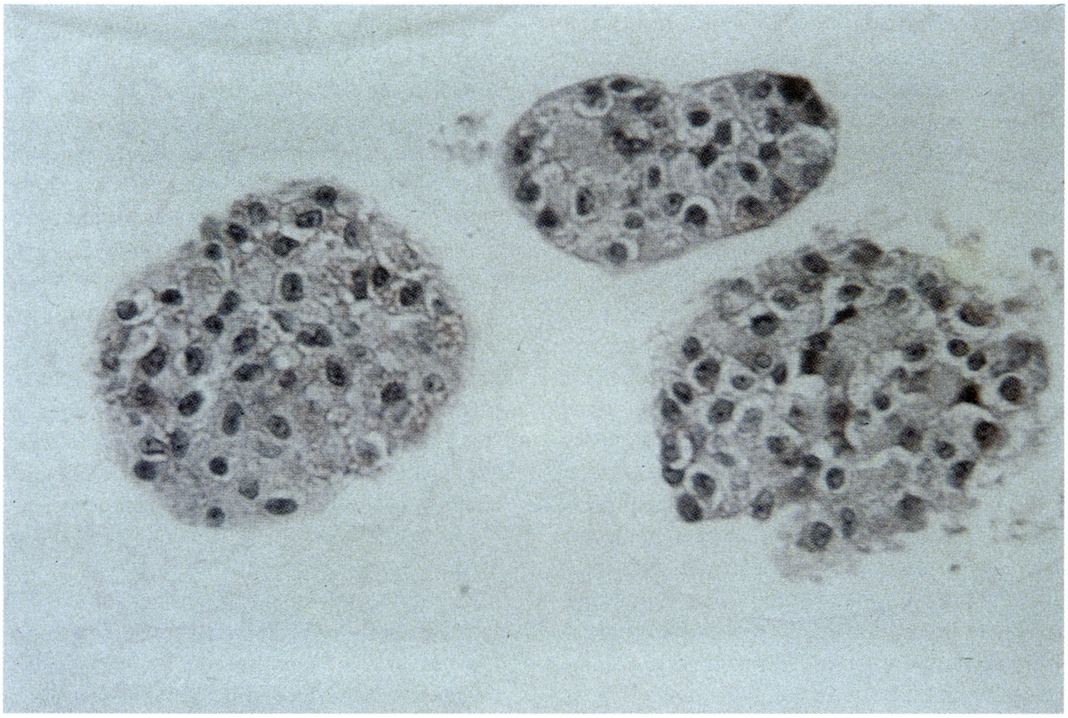

Fig. 1. Canine islets retrieved from a vascular device after >1 year in a diabetic dog without any immunosuppression. After device removal, the exogenous insulin requirement of the animal increased >20 U/day. Reprinted with permission from Lanza *et al.* Diabetes 1992; 41: 1503–1510.

[39,41]. These studies have been reviewed elsewhere [42–45]; only those systems which have shown significant progress to date will be discussed further.

In 1982, Woodward *et al.* [46] examined the influence of geometry on the occurrence of fibrosis. He demonstrated that a disc shaped chamber induced a zone of collagen-rich connective tissue between the implant and the vascular system, whereas a hollow-fiber elicited only a minimal response. A number of investigations have been carried out to assess the potential of tubular chambers as an artificial pancreas. Others found that islet and insulinoma tissue seeded within hollow fibers restored normoglycemia in diabetic rodents when implanted intraperitoneally [47–49]. Histologic analysis of recovered implants, however, revealed a fibrous tissue layer surrounding the membranes. The same type of fiber was also observed to elicit an inflammatory pericapsular response in the pig [48]. This tissue reaction was generally more intense, though qualitatively similar to that seen in the rat, except for lymphoid clusters with giant and pseudo-epithelioid cells that were observed only in the pigs. The reaction consisted of several layers of fibroblasts and collagen with polymorphonuclear leukocytes, macrophages, histiocytes, and small lymphocytes. The fenestrated outer wall of the tubular membrane was always infiltrated by collagen, fibroblasts, and macrophages.

Recent reports from our laboratory [11,12,50–52] described a series of experiments using wider-bore tubular chambers with a diameter of 1.6 to 4.8 mm. These studies were carried out with PAN-PVC membranes having a smooth outer skin. Porcine, bovine, and canine islets placed within these chambers restored normoglycemia in streptozotocin (STZ)-induced diabetic rats for more than a year without immunosuppression [52]. Only minimal tissue reactivity was observed. The external membrane surfaces were generally free of fibrotic overgrowth and exhibited only occasional host cell adherence. Encapsulated canine xenografts implanted in spontaneously diabetic BB rats also had the same success, resulting in fasting normoglycemia for periods of several weeks to more than eight months [51] (Figure 2). Intravenous glucose tolerance test (IVGTT) K-values (decline in glucose levels, percent per minute) after implantation in spontaneously diabetic BB and STZ rats were 2.3±0.4 and 2.6±0.2 to 3.5±0.3, respectively, compared with 3.1±0.1 and 3.3±0.1 for normal control groups. In constrast, the K-values for untreated diabetic rats were <1. Both light and electron microscopy of long-term functioning grafts revealed well preserved islets (Figure 3), with hormone-producing alpha, beta and delta cells.

While these wider-bore PAN-PVC membranes solved many of the problems associated with diffusion chambers (e.g. fibrosis, abscess formation, adhesions) [41,53], studies in large animals closer to man will likely be required before trials can be contemplated. Experiments in totally pancreatectomized, severely diabetic dogs have in fact already been performed in our laboratory [50]. They indicated that canine islet inplants can provide long-term correction of hyperglycemia without the use of immunosuppressive and/or antiinflammatory drugs. Insulin independence was achieved for >10 weeks in dogs with preimplantation insulin requirements of 30–40 units per day (a dosage in the range of what most human patients require) (Figure 4). Little or no fibrosis was observed for periods as long as 30 weeks.

In view of these encouraging results, a number of unsolved issues critical to the wide scale clinical success of these devices must be addressed. These include (1) long-term biocompatibility (with risk of fibrosis, peritonitis, intestinal adhesions, and abscess formation), (2) membrane breakage, (3) suitability for retrieval (or, alternatively, the use of the peritoneum or other implantation site as a 'dumping ground'), (4) further improvements in glycemic control, and (5) potential limitations imposed by the size and geometry of these chambers.

Biocompatibility

Experiments performed in our laboratory have demonstrated the feasibility of long-term immunoisolation if islets by artificial [PAN-PVC] membranes and the long-term biocompatibility of the membrane versus the graft and versus the recipient [11,12,50–52]. These data indicate that islet implants can provide correction of hyperglycemia in dogs and rodents for periods of several months to more than a year without the use of immunosuppressive drugs. Diffusion chambers fabricated from permselective acrylic membranes showed little or no evidence of an inflammatory response when implanted intraperitoneally in either spontaneously or streptozotocin-induced diabetic rats [51,52]. Complications such as abscess formation or intestinal adhesions, which have been observed with other technologies [41,53] were not observed with these implants. These studies have recently been extended to implantation in the peritoneal cavity of a large animal, the dog, with surgically induced diabetes [50]. Histological examination of the chambers revealed that they were biocompatible. The outer surface showed only scattered foci of macrophages and lymphocytes. Intactness and sterility of the chambers, however, was a crucial factor in the success of the inplants. In addition to loss of islet viability, damaged or contaminated membranes were often encapsulated by fibrous tissue which exhibited an interstitial acute and/or chronic inflammatory reaction and development of granulation tissue was observed. Before testing can be undertaken in diabetic patients, it will be important to determine the cause(s) of this peritoneal tissue reaction.

Membrane breakage

Most of the transplants described above ultimately failed because of membrane breakage. Under stress, the tubular chambers can bend, leading to fracture of the membrane walls and subsequent destruction of the encapsulated islet tissue. By 5–7 months postimplantation, 80–90% of the membrane chambers in dogs had broken. The tubular membranes used in most of these studies had a wall thickness of only 69–105 μm. The chambers fabricated from these membranes were relatively fragile, and susceptible to breakage. An increase in the membrane wall thickness, or the use of small spherical chambers or microreactors may minimize this problem. Indeed, our results indicate that such chambers are extremely durable and difficult to mechanically disrupt.

Methods for device removal and/or sampling

It is uncertain whether implantation of extravascular devices such as membrane chambers or microcapsules will require localization and removal. If necessary, and depending upon the site of implantation, the devices could be removed by laparoscopy or needly aspiration with minimal patient discomfort and morbidity. However, surgical excision could theoretically be necessary if the devices were to become fibroencapsulated. Open surgery, of course, carries risk of infection and would be a more extensive surgical procedure.

Blood glucose control

The motivation for islet transplantation is to provide physiologic control of blood glucose concentration. *In*

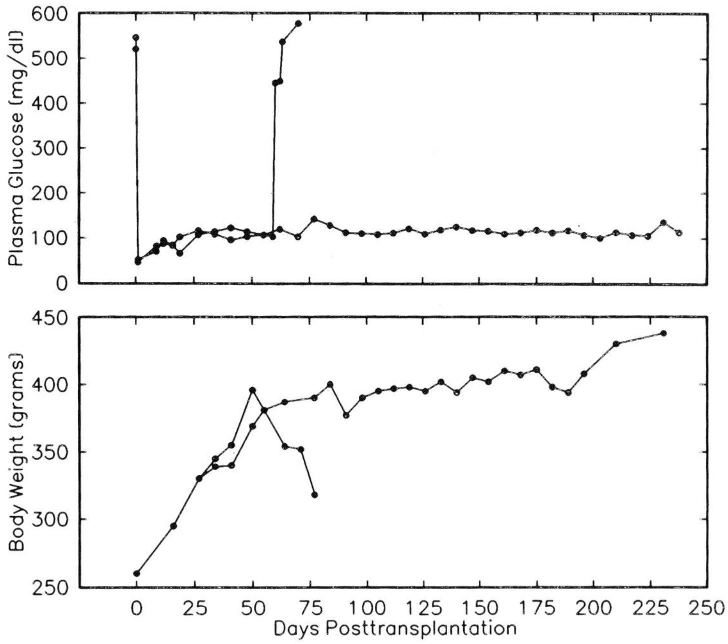

Fig. 2. Fasting plasma glucose concentrations (top) and body weight (bottom) in two spontaneously diabetic BB/Wor rats that received discordant canine islet xenografts encapsulated in permselective acrylic membranes (i.e., hollow fibers). The membranes were removed from one rat (●) at 2 months. The other animal (○) continued to maintain normoglycemia after more more than 8 months. No immunosuppression was used in this study. Reprinted with permission from Lanza *et al.* Endocrinology 1992; 131: 637–647.

vitro and *in vivo* experiments with tubular chambers have demonstrated only moderately delayed changes (lagtime <10 minutes) in insulin secretion in response to changes in glucose concentration. Perifusion of encapsulated canine islets with glucose elicited an approximately fourfold average increase from the basal insulin secretion [11]. There was a delay of only 7±1 minutes before the insulin concentration in the perfusate began to increase. Although this response is well within a time frame compatible with closed-loop insulin delivery (pharmacokinetic modeling of glucose homeostasis in man suggests that the lag time of the increase in insulin delivery by an artificial pancreas must be <15 minutes to avoid the overexcursion of postprandial blood glucose) [54], a reduction in the volume of the islet cell compartment would further improve the transmission of the glycemic signal from the blood to the islets, and of insulin from the islets to the recipient.

In a set of dog experiments [50,55], wider-bore chambers were implanted into the peritoneum of six totally pancretectomized dogs, and the animals monitored for glycemic control by fasting and post-prandial blood-glucose determinations, and by responses to both intravenous glucose and oral glucose. All of the dogs had varying degrees of reduced insulin requirements for control of fasting blood glucose levels. Implantation of the chambers completely supplanted exogenous insulin therapy in three animals for 51 to >90 days (each of these implants continued to maintain blood glucose control for >20 weeks). The fasting glucose concentrations averaged 81±6 mg/dl for these three animals during the first month. This was lower than the fasting glucose levels prior to pancreatectomy, which averaged 91±3 mg/dl. The reason for these slightly lower levels is presently unclear.

Diffusion limitations

Immunoisolated islets lack intimate vascular access, and must be supplied with oxygen and nutrients by diffusion from the nearest blood vessels over distances greater than those normally encountered. In wider-bore membrane chambers, the problem of cell death or dysfunction as a result of oxygen supply limitations, or accumulation of wastes or other agents is likely to be more severe. Our observations with 4.8 mm ID chambers is consistent with this. Chambers retrieved from the peritoneal cavity of dogs several months after

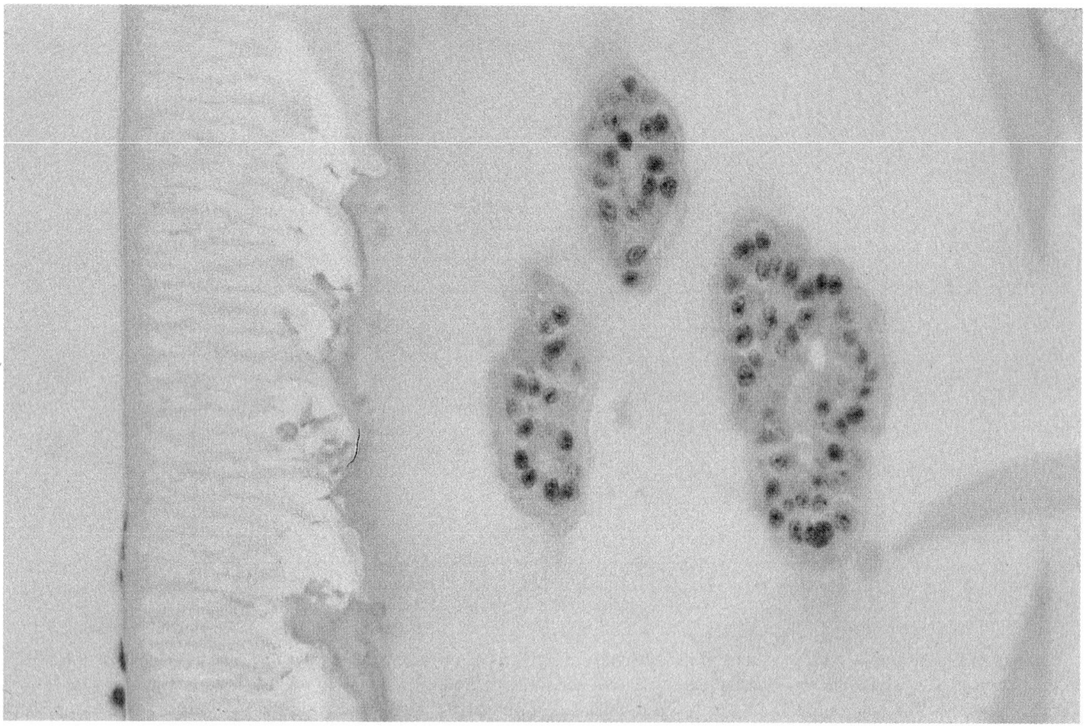

Fig. 3. Encapsulated porcine islets retrieved from the peritoneal cavity of diabetic rats 307 days after xenotransplantation. Reprinted with permission from Lanza *et al.* Transplantation 1993; 56: 1067–1072.

allotransplantation contained a central necrotic core. Only a rim of islets remained viable within approximately 0.5–1 mm of the inner membrane wall. Similar results were obtained with canine islet implants into rats. These findings may also explain the surprisingly large number of islets required to achieve blood glucose control. Clearly, careful attention must be paid to the diffusion distances and transport properties of the membranes. The relatively small volume (and short distances) associated with smaller spherical devices would maximize the transfer of oxygen and nutrients to the islets, and minimize the suppression of function by any islet hormonal or waste products which could accumulate in the islets compartment.

Microcapsules

Over the past decade, several methods for microencapsulating islets have been investigated [56]. Microcapsules offer a number of distinct advantages over the use of other biohybrid devices, including (1) a greater surface to volume ratio, (2) can simply be injected, and (3) are retrievable by lavage and needle aspiration (or alternatively, can be fabricated from biodegradable materials). However, problems severely limit the usefulness of most of these approaches, and only the alginate-poly-L-lysine (PLL) system has resulted in long-term function in larger animals. This type of microencapsulation system is widely used to immobilize microbial cells for industrial applications [57,58]. The procedure involves extruding a mixture of cells and sodium alginate using a droplet generation device into a $CaCl_2$ solution.

Using this procedure to entrap islets in calcium alginate hydrogels, Lim and Sun [16] then coated these negatively charged gelled droplets with positively charged PLL. However, these capsules were unstable and produced an inflammatory response when implanted into the pertioneal cavity of animals [59,58]. Modifications in the encapsulation procedure have improved the biocompatibility of the capsules, resulting in a dramatic increase in the duration of islet allograft function in diabetic rodents to more than a year [61]. Implanted rat-to-mouse islet xenografts produced blood glucose control for a shorter period of time. Although prolongation of survival of canine islets has also been achieved

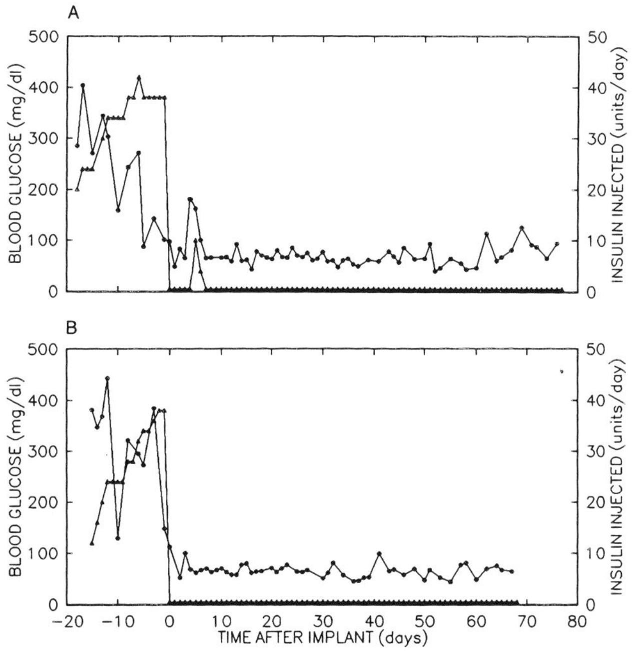

Fig. 4. Successful treatment of experimental diabetes in a large animal with encapsulated islets. First description of the ability of an extravascular device to sustain normogyclemia in diabetic dogs without any immunosuppression. Exogenous insulin requirements (△) and fasting blood glucose concentrations (○) in two dogs before and after device implantation. Reprinted with permission from Lanza *et al.* Diabetes 1992; 41: 886–889.

with the alginate-PLL technique, these studies have been performed in mice, and have usually required adjunctive treatment with immunosuppressive agents [62–64]. Weber *et al.* [65] found that alginate-PLL microcapsules containing canine islets functioned for <2 weeks in diabetic NOD mice. With anti-CD4 monoclonal antibody treatment, however, long-term functional survival was observed in many of the recipients.

Recently, prolongation of discordant xenograft survival in diabetic mice has been achieved in our laboratory without immunosuppression using alginate microspheres without the synthetic poly (L-lysine) membrane [66]. Uncoated alginate spheres containing porcine and bovine islets routinely reversed hyperglycemia after injection into diabetic mice. All but one (13/14) of the grafts functioned for at least 1 month, and 11 animals for at least 10 weeks. In comparison, nonencapsulated islet transplants failed to function, or sustained euglycemia for <4 days.

The ability of uncoated microspheres to achieve marked prolongation of discordant xenograft survival is surprising, as the destruction of the grafts might have been expected to occur based on the presence of circulating preformed natural antibodies in the recipients that are reactive to cells of the donor [67]. The proteins of the complement system are smaller in size than thyroglobulin and presumably would have had similar acess to the encapsulated islet graft. Naturally produced antibody and activation of the complement cascade are generally thought to target the donor endothelium [68,69]. The cytotoxic antibodies in man against pig xenoantigens are believed to be directed against α-galactosyl epitopes, which are expressed on vascular endothelium but are not expressed on many parenchymal cells [70,71]. These antibodies are believed to trigger the hemorrhagic necrosis that leads to solid organ rejection [72–75]. However, the role of xenoreactive natural antibodies in islet rejection is still uncertain. An explanation for the immunoprotective effect of the alginate spheres also needs to accommodate the fact that cytokines (M_r of 10–30 kD), nitric oxide and other toxic moieties are small enough to diffuse readily into the gel matrix, yet did not induce dysfunction or destruction of the islet graft.

Our data suggest that the alginate matrix served as a physical barrier to prevent direct effector cell contact with the donor tissue. However, we cannot rule out the possibility that immunological effector molecules were excluded from the negativelyy charged alginate gels based on chemical characteristics. The destruction of discordant islet xenografts may depend upon proteins hat carry charges under physiological conditions. For example, one or more of the components of the complement system with its cascading series of plasma enzymes, regulatory proteins, and proteins capable of cell lysis could be affected by the electrostatic interaction between gel support and protein charges. Other phenomena may also be involved in the mechanism(s) of ptotection. Recently, Zekorn *et al.* [76] have shown that encapsulated islets are protected form high doses of IL-1 (M_r of 17 kD) inside hollow fiber membranes with a molecular weight cutoff of 50 kD. The authors suggested that an nonspecific coating of the membranes by serum proteins may have caused the protective effect.

Although our results using uncoated alginate microspheres in mice are encouraging, it will be important to study the nature and extent of this immunoprotectivity in other animal models. Preliminary experiments in diabetic rats suggest that uncoated alginate spheres containing porcine and bovine islets can reverse hyperglycemia for periods of up to more than 175 days [77] (Figure 5). However, these results have been inconsistent and usually have required adjunctive treatment with immunosuppressive agents. Whether this immunosuppression can be eliminated, or the alginate microspheres modified to provide more complete immunoprotection requires further assessment and study.

Experiments in spontaneously diabetic dogs have also been performed in our laboratory. Our results indicate that long-term survival of canine islet allografts can be achieved by islet encapsulation inside uncoated alginate spheres (Figure 6). Although low-dose CsA was also administered, by three weeks portimplantation the whole blood trough levels of the drug were below readable limits by HPLC. Implantation of the microspheres completely supplanted exogenous insulin therapy in the dogs for 60 to >150 days (three of the five implants continue to maintain function). These results may have important implications in assessing the potential role for this type of microreactor as therapy for human insulin-dependent diabetes. Moreover, by using recombinant methods and encapsulating other tissues, it may also prove possible to treat patients suffering frova wide variety of other disorders requiring formone or enzyme replacement therapy.

Soon-Shiong *et al.* [78] have also reported successful long-term implantations of microencapsulated allografts in larger animals. His group treated spontaneous diabetes in dogs administered subtherapeutic doses of cyclosporine therapy.. However, these PLL-coated microspheres only maintained euglycemia for 63 to 140 days (comparable to the results obtained in our laboratory without the syntheetic poly L-lysine membrane). Using the same technology, Soon-Shiong has also achieved insulin-independence in a patient for approximately 1-month. However, this patient was on a regimen of immunosuppressive drugs, and subsequently required exogenous insulin therapy. Calafiore *et al.* [80] have also achieved insulin independence in one of three alloxan-induced diabetic dogs and, transient-

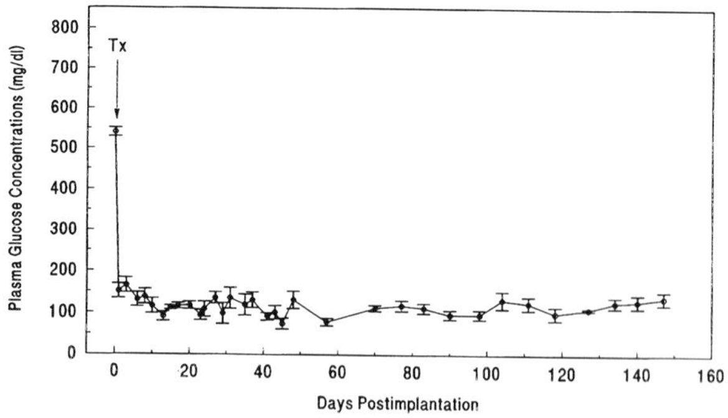

Fig. 5. Mean nonfasting plasma glucose levels in four STZ-induced diabetic rats that received intraperitoneal implants of alginate-encapsulated bovine islets. All of the animals received low-dose CsA (10–20 mg/kg/day s.c.). Immunohistochemical staining of the grafts (>100–150 days) revealed healthy viable islets (80–100%, comparable to the day zero control specimens), with well granulated alpha, beta, and delta cells. This method for transplanting discordant islets into rats is simple and inexpensive, and may also be a useful procedure for transplanting other cells and tissues. These spheres can be formed simply by extruding a mixture of cells and sodium alginate with a 16-gauge angiocatheter into a $CaCl_2$ solution. Reprinted with permission from Lanza *et al.* Transplantation 1995; 59: 1485–1487.

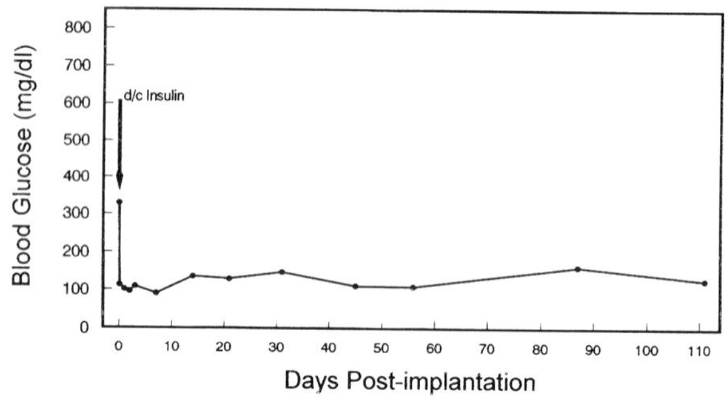

Fig. 6. Fasting blood glucose concentrations of a spontaneously diabetic dog that received an intraperitoneal implants of alginate-encapsulated canine islets. Reprinted with permission from Lanza *et al.* Tissue Engineering 1995; 1: 181–196.

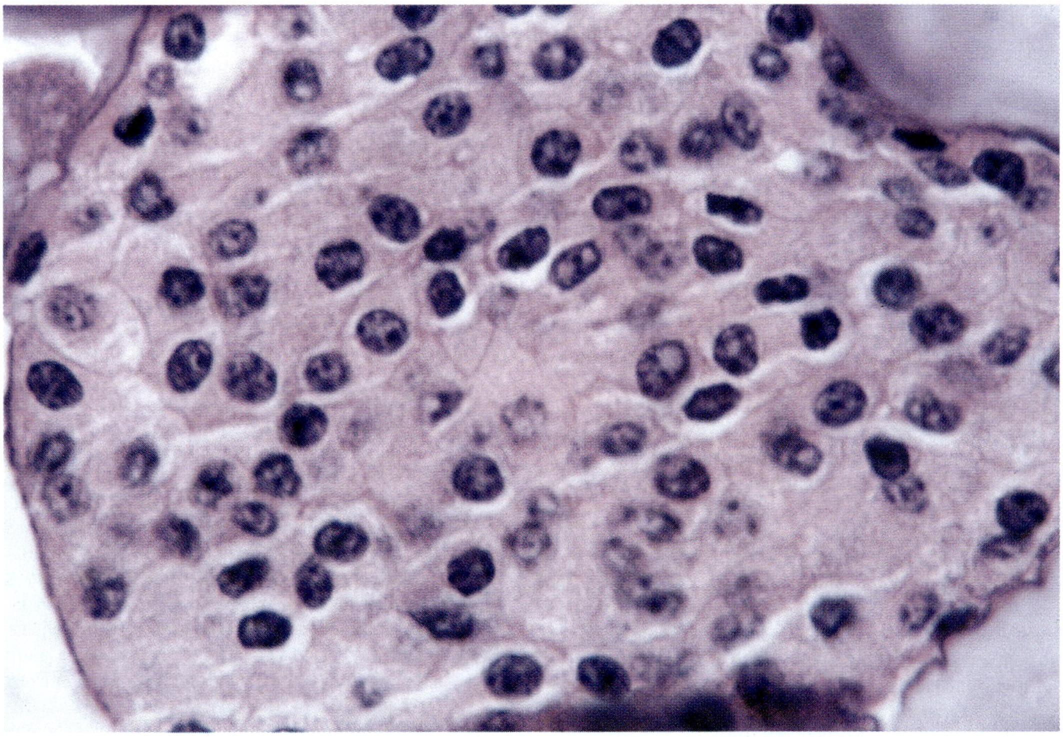

Fig. 7. Encapsulated bovine islet retreived from the peritoneal cavity of a dog 6 weeks after xenotransplantation. These new biodegradable 'microreactors' are injectable and do not appear to require the use of immunosuppressive drugs to prevent rejection.

ly, in one of two patients without any pharmocologic immunosuppression. However, these microcapsules were coated with PLL and were deposited in artificial prostheses directly anastomosed to blood vessels.

More recently, our laboratory has successfully tested a new type of microcapsule in dogs using discordant pig and cow islets [81]. In a set of experiments in dogs, coated and uncoated islets were implanted into the peritoneum of nomal dogs for periods of several weeks. Few or no islets survived in the uncoated microcapsules-even with the use of triple immunosuppressive therapy. However, when the islets were immobilized inside the new type of coated microcapsule, the islet tissue remained viable both with and without low-dose immunosuppression (Figure 7). Immunohistochemical staining revealed well-granulated alpha, beta, and delta cells consistent with functionally active hormone synthesis and secretion. To further test the secretory function of the islets, the explanted microcapsules were incubated in either basal (50 mg/dl) or stimulatory (300 mg/dl) glucose. In both the immunosuppressed and non-immunosuppressed dog experiments, the explanted islets responded with an approximately four- to sixfold average increase above basal insulin secretion. We hope to bring this new approach to xenotransplantation to clinical reality within the next year.

Conclusion

The pace af advancement in encapsulation technology during the past decade strongly suggests that transplantation of cells and tissues with differentiated functions will play an important therapeutic role in medicine in the future. It appears likely that by the year 2000 clinical trials utilizing encapsulated islet xenografts to treat diabetic patients will become a reality. This in turn will serve as an important starting point for developing living drug delivery for treating a wide range of additional disorders. For patients with diabetes the coming years hold promise for greatly improved therapy.

References

1. Hering BJ, Geier C, Schultz AO, Bretzel RG, Federlin K. International Islet Transplant Registry, 14th AIDSPIT Meeting, Igls, Austria, 1995.
2. Gunnarsson R, Klintmalm G, Lundgren G *et al.* Deterioration in glucose metabolism in pancreatic transplant recipients given cyclosporin. Lancet 1983, 2: 571.
3. Alejandro R, Feldman EC, Bloom AD *et al.* Effects of cyclosporin on insulin and C-peptide secretion in healthy beagles. Diabetes 1989, 38: 698.
4. Schlumpf R, Largiader F, Uhlschmid GK *et al.* Is cyclosporine toxic for transplanted pancreatic islets? Transplant Proc 1986, 28: 1169.
5. Van Schilfgaarde R, van der Burg MPM, van Suylichem HG *et al.* Does cyclosporin influence beta cell function? Transplant Proc 1986, 28: 1175.
6. Soon-Shiong P, Lu ZN, Lanza RP *et al.* An *in vitro* method of assessing the immunoprotective properties of microcapsule membranes using pancreatic and tumor cell targets. Translant Proc 1990, 22: 754.
7. Darquy S, Reach G. Immunoisolation of pancreatic β cells by microencapsulation. Diabetologia 1985, 28: 776.
8. Lanza RP, Solomon BA, Monaco AP *et al.* Devices implanted as AV shunts. In: Lanza RP, Chick WL (eds) Pancreatic Islet Transplantation: Volume III Immunoisolation of Pancreatic Islets. Austin, Landes/CRC Press, 1994, p 154.
9. Sullivan SJ, Maki T, Borland KM *et al.* Biohybrid artificial pancreas: Long-term implantation studies in diabetic, pancreatectomized dogs. *Science* 252: 718, 1991.
10. Monaco AP, Maki T, Ozato H *et al.* Transplantation of islet aalografts and xenografts in totally pancreatectomized diabetic dogs using the hybrid artificial pancreas. *Ann Surg* 214: 339, 1991.
11. Lanza RP, Butler DH, Borland KM *et al.* Xenotransplantation of canine, bovine, and porcine islets in diabetic rats without immunosuppression. *Proc Natl Acad Sci USA* 88: 11100, 1991.
12. Lanza RP, Butler DH, Borland KM *et al.* Successful xenotransplantation of a diffusion-based biohybrid artificial pancreas: A study using canine, bovine, and porcine islets. *Transplant Proc* 24: 669, 1992.
13. Lanza RP, Borland KM, Lodge P *et al.* Treatment of severely diabetic, pancreatectomized dogs using a diffusion-based hybrid pancreas. *Diabetes* 41: 886, 1992.
14. Race JM, LeGrelle M, Bethoux JP *et al.* Macroencapsulation in small animals. In: Lanza RP, Chick WL (eds): Immunoisolation of Pancreatic Islets. Austin, Landes/CRC Press, 1994, p 133.
15. Lanza RP. Macroencapsulation in large animals. In: Lanza RP, Chick WL (eds) Immunoisolation of Pancreatic Islets. Austin, Landes/CRC Press, 1994, p 139.
16. Lim F, Sun AM. Microencapsulated islets as bioartificial endocrine pancreas. *Science* 210: 908, 1980.
17. Norton J, Weber C, Reemtsma K. Microencapsulation: prevention of islet graft rejection. In: Van Schilgaarde R, Hardy M, eds. *Transplantation of the Endocrine Pancreas in Diabetes Mellitus*. Amsterdam, Elsevier, 1988.
18. Lanza RP, Soon-Shiong P. Experimental xenotransplantation of encapsulated islets. In: Cooper DKC, Kemp E, Reemtsma K, White DJK eds. *Xenotransplantation: The Transplantation of Organs and Tissues Between Species*. Heidelberg, Springer-Verlag, 1991, p 297.
19. Weber CJ, Reemstma K. Microencapsulation in small animals-xenografts. In: Lanza RP, Chick WL (eds) of Immunoisolation of Pancreatic Islets. Austin, Landes/CRC Press, 1994, p 59.
20. Soon-Shiong P. Encapsulated islet transplantation: pathway to human clinical trials. In: Lanza RP, Chick WL (eds) Immunoisolation of Pancreatic Islets. Austin, Landes/CRC Press, 1994, p 81.
21. Lacy PE, Hegre OD, Gerasimidi-Vazeou A *et al.* Maintenance of normoglycemia in diabetic mice by subcutaneous xenografts of encapsulated islets. *Science* 254: 1282, 1991.
22. Valente U, Ferro M, Campisi C *et al.* Allogeneic pancreatic islet transplantation by means of artificial membrane chambers in 13 diabetic recipients. *Transplant Proc* 12: 223, 1980.
23. Icard P, Penfornis F, Gotheil C *et al.* Tissue reaction to implanted bioartificial pancreas in pigs. *Transplant Proc* 22: 724, 1990.
24. Spokane RB, Clark LC, Bhargava HK *et al.* An implanted peritoneal oxygen tonometer that can be calibrated *in situ*. Trans Am Soc Artif Intern Organs 1990, 36: M719.
25. Colton CK, Avgoustiniatos. Bioengineering in development of the hybrid artificial pancreas. J Biomed Eng 1991, 113: 152.
26. Dionne KE, Colton CK, Yarmush ML. Effect of hypoxia on insulin secretion by isolated rat and canine islets of Langerhans. Diabetes 1993, 42: 12.
27. Kuhtreiber WM, Lanza RP, Beyer AM *et al.* Relationship between insulin secretion and oxygen tension in hybrid diffusion chambers. ASAIO Journal 1993, 39: M247.
28. Ohta M, Nelson D, Nelson J *et al.* Oxygen and temperature dependence of stimulated insulin secretion in isolated rat islets of Langerhans. J. Biol. Chem 1990, 265: 17525.
29. Chick WL, Perna JJ, Lauris V *et al.* Artificial pancreas using live β-cells: Effects of glucose homeostasis in diabetic rats. Science 1977, 197: 780.
30. Chick WL, Like AA, Lauris V. Beta cell culture on synthetic capillaries: An artificial endocrine pancreas. Science 1975, 187: 847.
31. Tze WJ, Wong FC, Chen IM *et al.* Implantable artificial endocrine pancreas unit used to restore normoglycemia in the diabetic rat. Nature 1976, 264: 466.
32. Whittemore AD, Chick WL, Galletti PM *et al.* Effects of the hybrid artificial pancreas in diabetic rats. Trans Am Soc Artif Intern Organs 1977, 13: 336.
33. Sun AM, Parisius W, Healy GM *et al.* The use in diabetic rats and monkeys of artificial capillary units containing cultured islets of Langerhans. Diabetes 1977, 26: 1136.
34. Feldman S, Dodi G, Haid K *et al.* Artificial hybrid pancreas. Surg Forum 1977, 28: 439.
35. Tze WJ, Wong FC, Chen LM. Implantable artificial capillary unit for pancreatic islet allograft and xenograft. Diabetologia 1979, 16: 247.
36. Sun AM, Parisius W, MacMorine HG *et al.* An artificial endocrine pancreas containing cultured islets of Langerhans. Artif Organs 1980, 4: 275.
37. Galletti PM, Trudell LA, Panol G *et al.* Feasibility of small bore AV shunts for hybrid artificial organs in nonheparinized beagle dogs. Trans Am Soc Artif Organs 1981, 17: 185.
38. Lanza RP, Sullivan SJ, Chick WL. Islet transplantation with immunoisolation. Diabetes 1992, 41: 1503.
39. Maki T, Otsu I, O'Neil JJ *et al.* Treatment of diabetes wih a bioartificial pancreas containing porcine islets. Diabetes 1995, 44 (Suppl 1): 129A.
40. Lanza RP, Chick WL. Introduction Lanza RP, Chick WL (eds). In: Immunoisolation of Pancreatic Islets. Austin, Landes/CRC Press, 1994, p1.

41. Theodorou NA, Vrbova H, Tyhurst M *et al.* An assessment of diffusion chambers for use in pancreatic islet cell transplantation. Transplantation 1979, 27: 350
42. Lanza RP, Soon-Shiong P. Experimental xenotransplantation of encapsulated islets. In: Cooper DKC, Kemp E, Reemtsma K, White DJK (eds) Xenotransplantation: The Transplantation of Organd and Tissues Between Species. Heidelberg, Springer-Verlag, 1991, 297.
43. Colton CK, Avgoustiniatos ES. Bioengineering in development of the hybrid artificial pancreas. J. Biomech Eng 1991, 113: 152.
44. Scharp DW, Mason NS, Sparks RE. Islet immuno-isolation: The use of hybrid artificial organs to prevent islet tissue rejection. World J Surg 1984, 8: 221.
45. Scharp DW, Mason NS, Sparks RE. The use of hybrid artificial organs tp provide immunoisolation to endocrine grafts in transplanntation. In: Slavin S (ed) Bone Marrow & Organ Transplantation. Elsevier Science Publishers, Israel, 1984, p 601.
46. Woodward CL. How fibroblasts and giant cells encapsulate inplants: Considerations in design of glucose sensors. Diabetes Care 1982, 5: 278.
47. Altman JJ, Houlbert D, Bruzzo F *et al* Implantation of semipermeable hollow fibers to prevent immune rejection of transplanted pancreatic islets. In: Federlin K, Pfeiffer E, Raptis S (eds) Islet-Panccreas-Transplantation and Artificial Pancreas. New York, Thieme-Stratton, 1982.
48. Altman JJ, Houlbert D, Callard P *et al.* Long-term plasma glucose normalization in experimental diabetic rats with microencapsulated implants of benign human insulinomas. Diabetes 1986, 35: 625.
49. Archer J, Kaye R, Matter G. Control of steptozotocin diabetes in Chinese hamsters by cultured mouse islet cells without immunosuppression. J Surg Res 1980, 28: 77.
50. Lanza RP, Borland KM, Lodge P *et al.* Treatment of severely diabetic, pancreatectomized dogs using a diffusion-based hybrid pancreas. Diabetes 1992, 41: 886.
51. Lanza RP, Borland KM, Staruk JE *et al.* Transplanation of encapsulated canine islets into spontaneously diabetic BB/Wor rats without immunosuppression. Emdocrinology 1992, 131: 637.
52. Lanza RP, Beyer AM, Staruk JE, Chick WL. Biohybrid artificial pancreas: longterm function of discordant islet xenografts iun streptozotocin diabetic rats. Transplantation 1993, 56: 1067.
53. Andersson A. Survival of pancreatic islet allografts. Lancet 1979, 2: 585.
54. Kraegen EW, Chisholm DJ, MacNamara ME. Timing of insulin delivery with meals. Horm Metab Res 1981, 13: 365.
55. Lanza RP, Lodge P, Borland KM *et al.* Transplantation of islet allografts using a diffusion-based biohybrid artificial pancreas: long-term studies in diabetic, pancreatectomized dogs. Transplant Proc 1993, 25: 978.
56. Lanza RP, Chick WL (eds), Pancreatic Islet Transplantation: Volume III Immunoisolation of pancreatic islets. Austin, Landes/CRC Press, 1994.
57. Hackel V, Klein J, Megret R *et al.* Immobilization of microbial cells in polymeric matrices. Eur J Appl Microbiol 1975, 1: 291.
58. Kierstan M, Bucke C. The immobilization of microbial cells, subsellular organelles and enzymes in calcium alginate gels. Biotechnol Bioeng 1977, 9: 387.
59. Sun AM, O'Shea G, Van Rooy H *et al.* Microencapsulation d'îlots de Langerhans et pancréas artificiel. J Ann Diabetol (Hotel-Dieu) 1982, 20: 161.
60. Sun AM, O'Shea GM, Goosen MFA. Injectable microencapsulated islet cells as a bioartificial pancreas. Appl Biochem Biotechnol 1984, 10: 87.
61. O'Shea GM, Goosen MFA, Sun AM. Prolonged survival of transplanted islets of Langerhans encapsulated in a biocompatible membrane. Biochim Biophys Acta 1984, 804: 133.
62. Wang Y, Hao L, Gill R *et al.* Autoimmune diabetes in NOD mouse is L3T4 T-lymphocyte dependent. Diabetes 1987, 36: 535.
63. Califiore R, Janjic D, Koh N *et al.* Transplantation of microencapsulated canine islets inro NOD mice: prolongation of survival with superoxide dismutase and catalase. Clin Res 1987, 35: 499A.
64. Ricker A, Bhati V, Bonner-Weir S *et al.* Microenvapsulated xenogeneic islet grafts in NOD mouse; dexamethasone and inflammatory response. Cold Spring Harbor Symposium, October 1987: 53A.
65. Weber CJ, Zabinski S, Koschitzky T *et al.* The role of CD4+ helper T cells in the destruction of microencapsulated islet xenografts in NOD mice. Transplantation 1990, 49: 396.
66. Lanza RP, Kuhtreiber WM, Ecker D *et al.* Xenotransplantation of porcine and bovine islets without immunosuppression using incoated alginate microspheres. Transplantation 1995, 59: 1377.
67. Platt JL, Bach FH. The barrier to xenotransplantation. Transplantation 1991, 52: 937.
68. Platt JL, Vercellotti GM, Dalmasso AP *et al.* Transplantation of discordant xenografts: A review of progress. Immunol Today 1990, 11: 450.
69. Platt JL, Bach FH. Mechanisms of tissue injury in hyperacute xenograft rejection. In: Cooper DKC, Kemp E, Reemtsma K *et al.* (eds) Xenotransplantation: The Transplantation of Organs and Tissues Between Species. Heidelberg, Germany, Springer-Verlag, 1991, p 69.
70. Cooper DKC, Good AH, Malcolm AJ *et al.* Identification of α-galactosyl and other carbohydrate epitopes that are bound by human anti-pig antibodies: relevance to discordant xenografting in humans. Transplant Immunol 1992, 1: 198.
71. Good AH, Cooper KKC, Malcolm AJ *et al.* Identification structutures that bind human antiporcine antibodies: implications for discordant xenografting in humans. Transplant Proc 1992, 24: 559.
72. Mozes M, Gewurz H, Gunnarsson A *et al.* Xenograft rejection by dog and man: isolated kidney perfusions with blood plasma. Transplant Proc 1971, 3: 531.
73. Cooper DKC. Depletion of natural antibodies in nonhuman primates–a step towards successful discordant xenografting in humans. Clin Transplant 1992, 6: 178.
74. Platt JL, Fishel RJ, Matas AJ *et al.* Immunopathology of hyperacute xenograft rejection in a seine-to-primate model. Transplantation 1991, 52: 214.
75. Linn BS, Jensen JA, Portal P *et al.* Renal xenograft prolongation by suppression of natural antibody. J Surg Res 1968, 8: 211.
76. Zekorn T, Siebers U, Bretzel RG *et al.* Protection of islets of Langerhans from interleukin-1 toxicity by artificial membranes. Transplantation 1990, 50: 391.
77. Lanza RP, Ecker D, Kuhtreiber WM *et al.* A simple method for transplanting discordant islets into rats using alginate gel spheres. Transplantation 1995, 59: 1485.
78. Soon-Shiong P, Feldman E, Nelson R *et al.* Long-term reversal of diabetes by the injection of immunoprotected islets. Proc Natl Acad Sci USA 1993, 90: 5843.

79. Soon-Shiong P, Heintz RE, Merideth N *et al.* Insulin independence in a type 1 diabetic patient after encapsulated islet transplantation. Lancet 1994, 143: 950.
80. Calafiore R. Transplantation of microencapsulated pancreatic human islets for therapy of diabetes mellitus. ASAIO Journal 1992, 38: 34.
81. Lanza RP, Chick WL. Encapsulated cell therapy. Scientific American Science and medicine 1995, 2: 32.

R. P. Lanza and W. L. Chick (eds.), Yearbook of Cell and Tissue Transplantation 1996/1997, 265–274.

Chapter 26

Tissue Engineering: Skin

Gail K. Naughton and William R. Tolbert
Advanced Tissue Sciences, La Jolla, CA92037, U.S.A.

Tissue engineering is moving rapidly from potential to accepted commercial and medical applications. As an emerging technology, it holds the promise of an entirely new approach to the repair and reconstruction of tissues and organs damaged by disease, accident, or congenital abnormalities and defects. Advantages of using normal human tissues produced by *in vitro* culture techniques include safety, efficacy of the product, and availability of a consistent material. The human cell source for these tissue products can be fully characterized and safety tested for possible contaminants by standard, well recognized cell banking procedures [1]. Normal human tissues may be more efficacious in that they supply a balanced natural mixture of growth factors and both structural and interactive functional proteins. These tissues can communicate with the body's own repair mechanisms to stimulate angiogenesis, remodeling and other steps to restore complete function. The ability to efficiently produce these tissues in culture and store them in a frozen state, allows an "off-the-shelf" availability for physicians, as contrasted to tissue and organs currently restricted to limited cadaveric and/or surgical discard sources.

Products first to move from tissue engineering potential to commercial application are those based on the skin as cell source and as the target for therapeutic intervention. The skin, as the largest body organ, provides the separation and barrier between the organism and his environment. It is a complex, highly organized structure of a continually proliferating epidermis, consisting of several cell types including keratinocytes responsible for differentiating into the stratum corneum. Fibroblasts, the main cell component of the dermis, secret various macromolecules including collagen, elastin, fibronectin, decorin, tenascin, laminin and proteoglycans to form the connective tissue underlayment that provides the skin with its high tensile strength [2, 3]. When the skin is damaged or breaks down due to environmental insult or disease, the results range from minor to life threatening. Therefore, the use of skin models to measure the level of environmental toxic and damaging factors such as chemicals or solar radiation *in vitro*, as well as development and use of skin biological components for therapeutic intervention have been the subject of tissue engineering research.

Advanced Tissue Sciences' core tissue engineering technology creates a three dimensional living stroma and resultant tissue *in vitro*. Traditional culture of non-transformed animal and human cells involves the attachment and growth of the cells into two dimensional flat sheets in containers such as petri dishes, flasks and roller bottles. These monolayer cultures do not reflect the normal cellular environment and potentiate artificial effects such as dedifferentiation [4, 5]. In contrast, a three dimensional matrix culture allows cells to develop and assemble into tissues that more closely resemble their *in vivo* counter parts [6]. In normal growth and development, specialized connective tissue cells form stroma and this living framework provides the three-dimensional structure for each tissue and organ type. In addition to structural components, the stroma contains various attachment and growth factors that are required for full differentiation and function of the tissue. For each application specific cells are seeded onto a polymer mesh framework in an optimum environment to stimulate differentiated tissue development. The cells attach, divide and secrete extracellular matrix proteins and growth factors, using the hydrolyzable polymer mesh as a scaffolding [7]. The resultant human tissue construct may then be transplanted to treat human medical applications [8, 9] or used in an *in vitro* model to replace and improve toxicology testing currently performed on animal or human clinical subjects [10, 11].

In vitro toxicology testing

The skin serves as a complex barrier protecting the internal environment against external factors, including bacteria, viruses, environmental toxins and UV light as well as physical damage. For the epidermis to produce the most effective toxicological, immunological and biochemical barrier, the major cell types of the epidermis, keratinocytes, and in the dermis, fibroblasts, must function together in a dynamic integrated fashion [12, 13]. The epidermal-dermal intercellular biochemical signals such as interleukins (IL), cytokines and other factors provide the skin with local homeostatic signals to ensure that skin's integrity in response to a variety of environmental insults.

The progressive and compelling need to reduce or replace the use of animals as a predicative model for assessing human toxicity to environmental factors has encouraged the development of reliable and well standardized, *in vitro* alternative methods. Using its proprietary technology, Advanced Tissue Sciences (La Jolla, CA) has successfully developed a three dimensional human skin tissue model called Skin^{2TM}. The *in vitro* model was tissue engineered to mimic the skin both biochemically and morphologically (Fig. 1). This skin tissue system is composed entirely of metabolically and mitotically active dermal and epidermal layers supported by a scaffold of nylon mesh. These layers are produced in culture by initial seeding with neonatal foreskin fibroblasts that secrete functionally active extracellular matrix proteins; followed by over-seeding with neonatal foreskin keratinocytes to produce a functional multilayered epidermis overlaid with stratum corneum [2, 14]. This skin model has been used to study the development of various supramolecular structures including elastin-associated microfibrils and collagen fibrillogenesis. The system allows the cells to form an architecture at a molecular level that resembles the *in vivo* tissue [15, 16].

These structural and biochemical similarities to normal human tissue are the basis for the success of Skin^2 (Advanced Tissue Science) in a wide variety of *in vitro* testing protocols [2, 12, 17, 18, 19]. Small pieces of tissue are provided which can be challenged by topical applications of substances whose irritant potential is measured by the time course of cellular enzyme release as well as by residual viability. These materials have been used commercially by a large number of companies in development of their finished products before release to the public. Estimates of more than 10,000 new chemical entities per year make the development of effective and economical safety testing procedures imperative.

Validation studies in blinded, side by side comparisons with animal methods such as the Draize ocular irritation assay have been conducted in the US and Europe. An example of such a formal validation process is the 3-year study being conducted by the European Commission and the European Cosmetic Association on *in vitro* alternatives including Skin^{2}to evaluate the phototoxic potential of personal and skin care products. In this study, various concentrations of test chemicals are applied to the epidermal side of the test tissue for contact times of 1 or 24 hours followed by exposure to 2.9 J/cm^2 of ultraviolet A radiation. After exposure the test substrates are rinsed free of test chemicals and allowed to recover for 24 hrs. Cytotoxicity is measured by the MTT mitochondrial assay to determine which chemicals can mask the phototoxic reaction. Current results have shown that data from Skin^2 studies have a high degree of correlation with data from human and animal models now used to evaluate this phototoxic potential of chemicals [19].

Recently the US Department of Transportation, Transport Canada and the Swiss Government have approved the use of Skin^2 as an *in vitro* laboratory test alternative to live animal testing for potentially corrosive materials. The testing requirement was mandated by a United Nations directive for transportation of hazardous materials. This recognition is an important milestone as the first US government regulatory approval for a human tissue engineered product. Extensive studies are on-going worldwide for further regulatory acceptance on these alternative testing approaches.

Therapeutic applications for dermal replacement tissue

For medical applications, the dermal or lower layer of the skin has significant advantages which makes it suitable as an initial target for tissue engineered therapeutics. The human diploid fibroblast cell type, its main constituent, grows well in culture with a proliferative

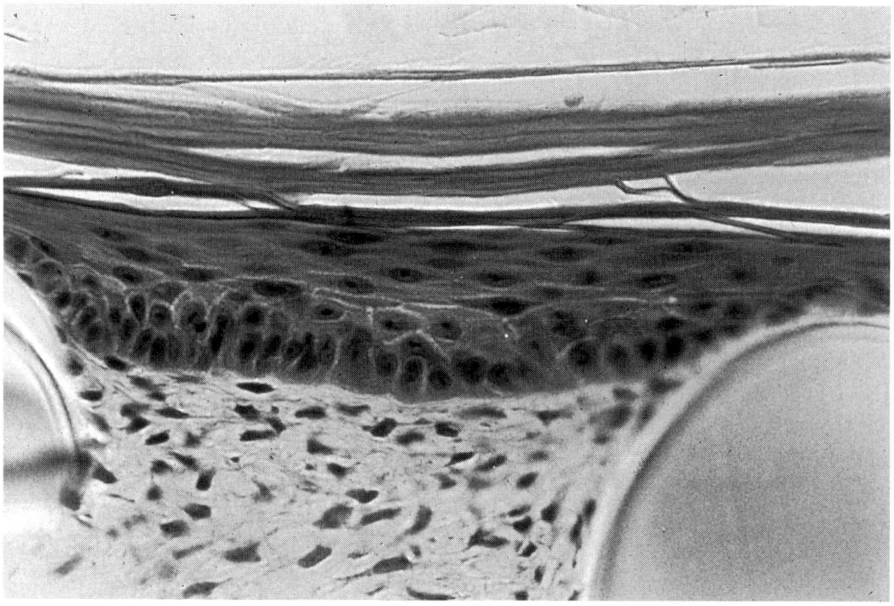

Fig. 1. Cross section of human skin *vs* Skin^{2TM} ZK1300 model: hematoxylin-eosin staining.

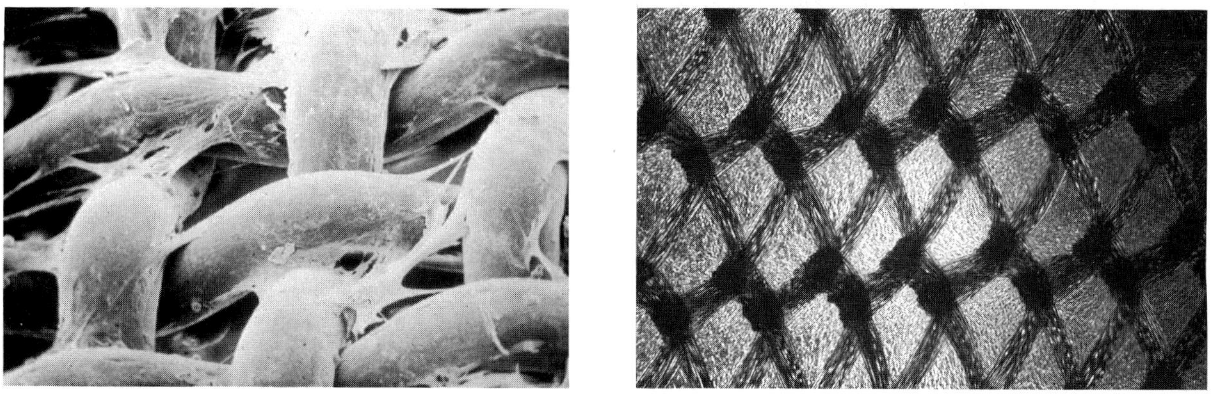

Fig. 2. (A) Scanning electron micrograph (S.E.M) of fibroblasts stretching across the mesh substrate 1 to 2 days post seeding. (B) Photomicrograph (40×) of confluent cells on the mesh substrate 4 to 5 days post seeding.

potential of approximately 60 population doublings (10^{18} fold increase over the original cell isolation) and has long been used for the production of human biologicals [20, 21, 22]. Unlike keratinocytes which carry HLA-DR surface antigens that may cause allograft rejection phenomena [23]; implantation of allogenic human fibroblasts does not stimulate an immune response [24, 25]. We have implanted over 400 patients in human trials with Dermagraft™ products of cultured human foreskin fibroblasts from two individual neonatal donors without any clinical evidence of immune response or rejection.

Dermal replacement is also a major medical need as the dermal tissue, unlike the epidermis, does not regenerate *in vivo* after serious burns and instead, potentiates the formation of scar tissue. "Loss of the dermis in extensive full thickness burn wounds, therefore poses a serious problem which is not completely solved by the application of meshed, split-thickness autograft skin"[23]. The meshing of the autograft in treatment of these burned patients allows the available unburned skin to provide the maximum coverage in as short a time as possible and is a life saving procedure. The epidermis quickly spreads to fill in the mesh openings

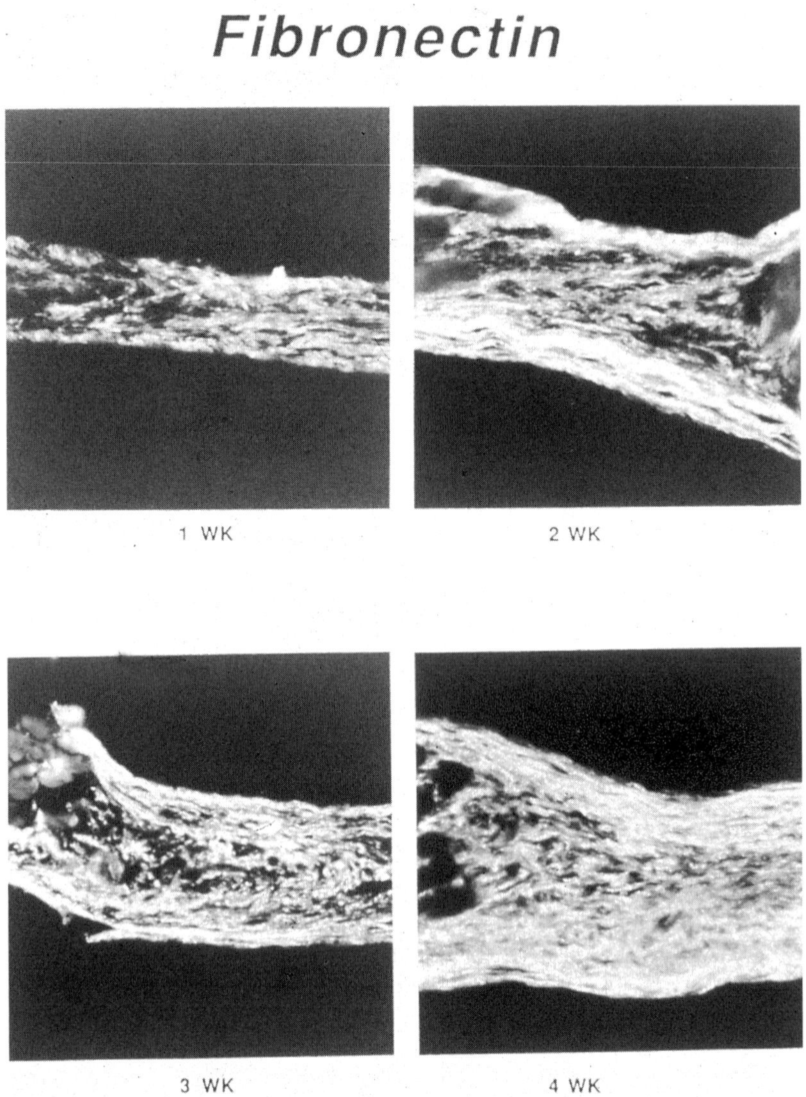

Fig. 3. Photomicrographs (400×) showing increase in fibronectin immuno-fluorescence of dermal replacement tissue during culture: (A) 1 week, (B) 2 weeks, (C) 3 weeks and (D) 4 weeks.

for "graft take" from 85% to 100%. In the long term, however, the lack of normal underlying vascularized dermis can result in skin fragility, blistering, and formation of hypertrophic scar. Animal experiments as well as clinical results have demonstrated this dependence on dermal integrity [8, 9, 13, 23, 26]. This natural failure of the dermis to regenerate may also play a role in other wound situations, which have shown clinical response to dermal replacement as a therapeutic approach.

The Dermagraft family of products is designed to replace this dermal layer and potentiate a normal wound healing process. Dermagraft products have been evaluated in pre-clinical studies and in pilot and pivotal human clinical trials for use in treating severe burn wounds as a dermal replacement under autograft [27, 28, 29, 30] and as a temporary covering graft to replace cadaveric allogenic skin in management of extensive burns prior to permanent autografting [9, 31, 32, 33, 34, 35]. Dermagraft product has also been used in clinical studies of chronic skin ulcers [36, 37]. These products are based on the three dimensional growth of human diploid fibroblast cells grown on a polymer scaffold (Fig. 2) and secreting a mixture of growth

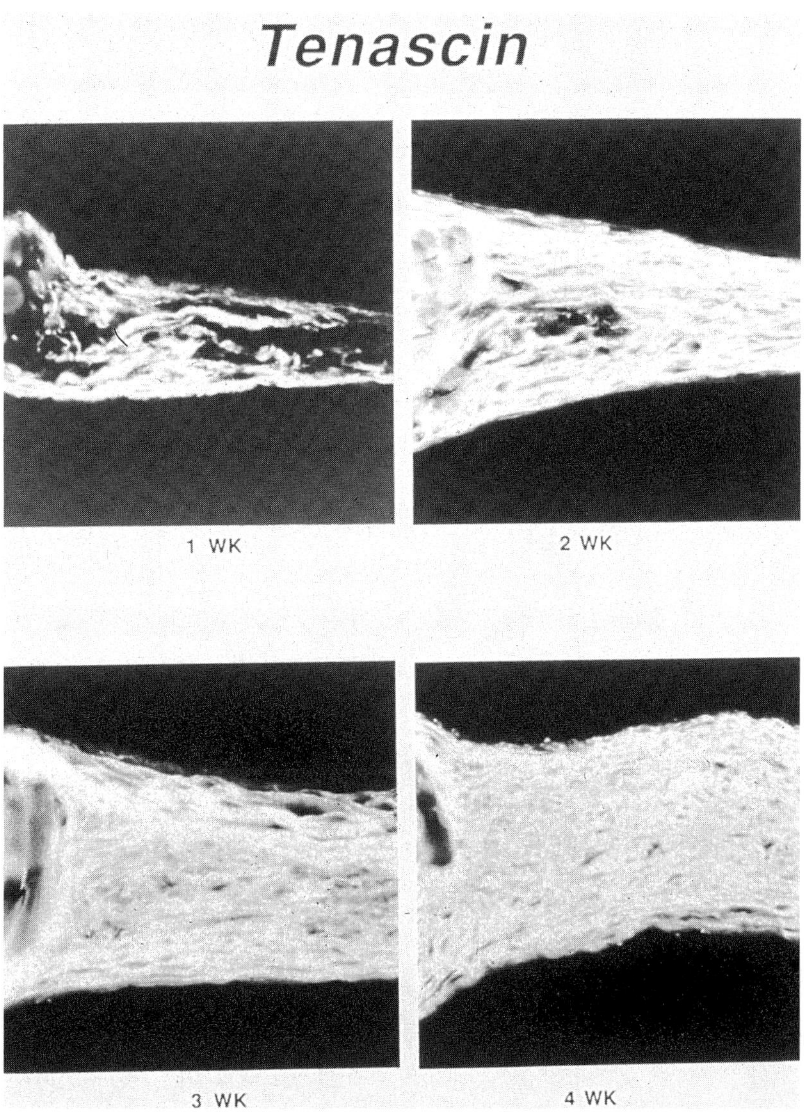

Fig. 4. Photomicrographs (400×) showing increase in tenascin immuno-fluorescence of dermal replacement tissue during culture: (A) 1 week, (B) 2 weeks, (C) 3 weeks and (D) 4 weeks.

factors and matrix proteins (Figs 3, 4) [3].

Human fibroblast cell strains were established from surgically removed neonatal foreskins and cultured by standard methodology [20, 38]. Maternal blood samples were tested for potential infectious disease exposure including HIV, HTLV, CMV and hepatitis. An initial screen was made of the cultured cells for sterility, mycoplasma, and for 8 human viruses: Adeno, HSV 1&2, CMV, HIV-1&2, and HTLV I&II. Master cell banks (MCB) were created at passage 3 and manufacturer's working cell banks (MWCB) at passage 5. Extensive additional testing of the cell banks (Table 1) was done according applicable sections of the FDA "Points to Consider" [1] and the European Union Committee for Proprietary Medicinal Products (CPMP) guidelines [39]. The Dermagraft products are produced by seeding the polymer scaffolds at passage 8 at a population doubling level of approximately 30 (which is at about half the life span for this cell type). Using this procedure a single foreskin donor provides sufficient cell seed to produce a quarter of a million square feet of finished Dermagraft product. This accepted procedure for producing biologic products for medical application, derived from human cells

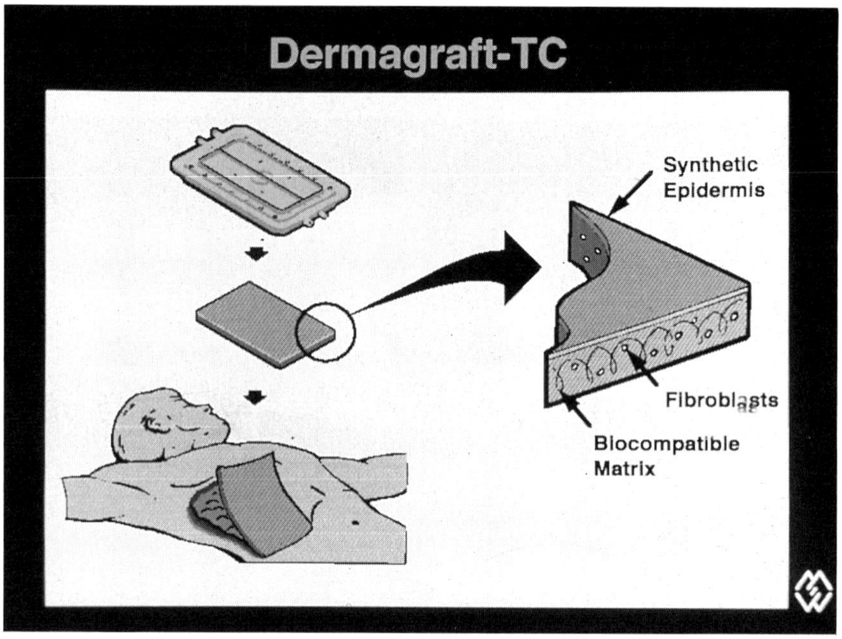

Fig. 5. Cartoon of Dermagraft-TC as it is applied to the excised wound bed to stimulate neovascularization.

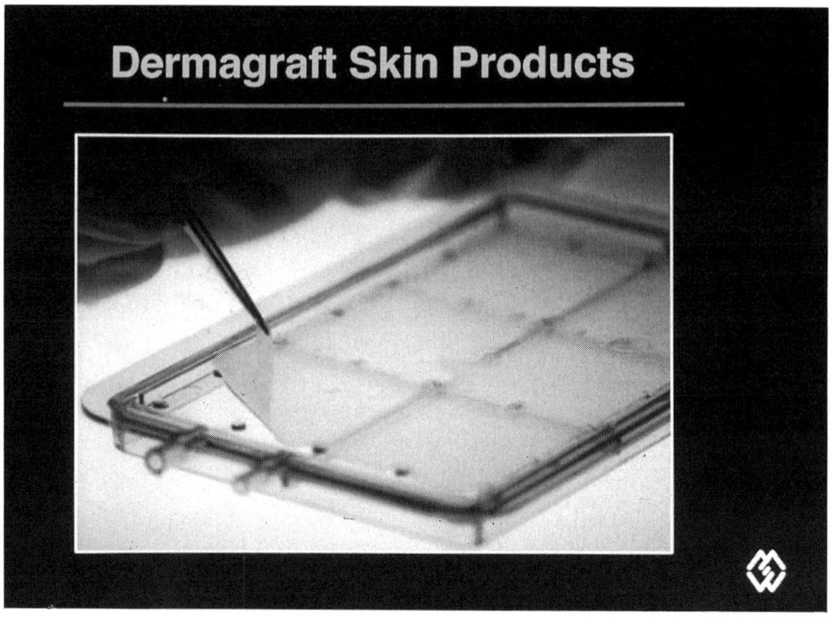

Fig. 6. Photograph of Dermagraft-TC as it is being removed from the bioreactor cassette where it is grown, frozen, stored and shipped to physicians.

[21, 22], provides a much higher level of safety, uniformity and availability than is possible with the minimally controlled and tested cadaveric tissues from multiple donors currently used for implantation [40, 41, 42].

Current medical practice is to use human cadaveric skin, either fresh or cryopreserved, as a biologic dressing for severely burned patients to cover the excised wound bed when there is insufficient autograft skin

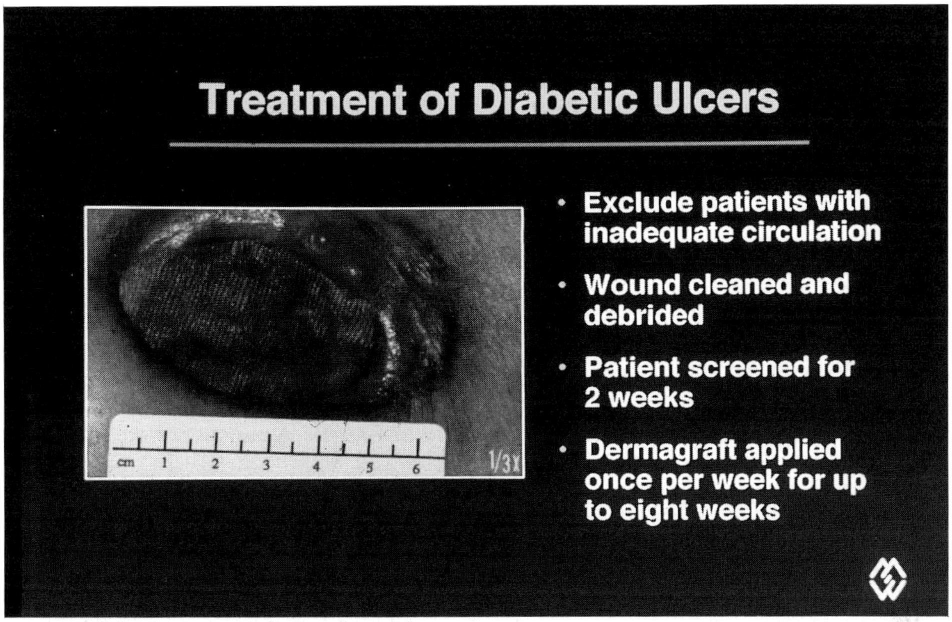

Fig. 7. Application of Dermagraft to a debrided diabetic ulcer wound.

Table 1.

Dermagraft™ Cell Line Testing

- Donor/family history obtained
- Mother's sera tested for:

– HIV 1 & 2	– ALT, Non A & Non B
– HTLV I & II	– Hepatitis B & C
– CMV IgM	– RPR

- Donor cells screened for:

– Adenovirus	– HIV 1 & 2
– HSV 1 & 2	– HVLT I & II
– CMV	– Mycoplasma

- Other tests include growth characteristics, USP sterility and karyology

[8, 9, 31, 43]. As additional autograft skin is available by repeated cycles of harvest and regrowth from unburned areas of the patient, the allograft is replaced. There are several problems with this procedure. The cadaveric allograft consists of both dermal and epidermal skin layers and is generally antigenic, resulting in graft rejection in days or weeks. In this situation additional surgery is required to replace the failed allograft and the process must be repeated until sufficient autograft skin is available to completely cover the burned area. The availability of cadaveric allograft skin and its potential to transmit disease has been recognized as an increasing problem since the FDA issued an interim rule on December 14, 1993, 21 CFR Part 1270, providing requirements for "banked human tissue" to prevent the transmission of AIDS and hepatitis through human tissue used in transplantation [44].

Dermagraft-Transitional Covering (Dermagraft-TC) is designed to meet this medical need. Human fibroblast cells prepared as described above are seeded onto a nylon mesh attached to a thin silicon rubber membrane (Biobrane™, Dow B. Hickam, Sugarland, TX). The nylon forms the three dimensional scaffolding for growth of the Dermagraft tissue and the silicon membrane forms an artificial (non-immunogenic) epidermis (Fig. 5). As the cells grow in culture, they secrete proteins and growth factors in a balanced mixture that is natural for these types of cells, thereby generating a three dimensional tissue matrix [34]. At the conclusion of the growth period, the closed production system is separated so that the individual bioreactors or cassettes used to grow the product are sealed and thus become the product's primary package. The sealed cassettes are frozen and stored at ultra cold temperatures (below $-70°C$) until use. The bioreactor/final product package system ensures delivery to the physician in an easily accessible form, providing the highest level of safety (Fig. 6). Sterility assays support this conclusion – assays are run on final production medium aliquots that have been exposed to every product unit and destructive final product release testing is performed on a statistical appropriate number of units per lot.

Ten patients in total were enrolled at three burn centers and each patient was treated both with Dermagraft-TC and a control of cryopreserved cadaveric allograft [35]. Preliminary clinical experience suggests that Dermagraft-TC provides a temporary wound coverage as effective as cryopreserved allograft and that it may remain in place for as long as approximately six weeks prior to autografting, while allograft is generally rejected in two to three weeks. A pivotal clinical trial is underway to obtain further supportive data regarding the safety and efficacy of Dermagraft-TC.

Chronic skin ulcers also represent a severe medical problem that may be addressed with this tissue engineering technology. There are over two million cases of chronic, slow healing or non healing skin ulcers treated in the U.S. each year. In these wounds the skin breaks down as a result of disruption of blood flow due to prolonged pressure over a localized area or due to chronic diseases that affect the circulatory or peripheral nervous systems. In the case of chronic ulcers associated with diabetes, more that a half million patients are treated annually, but healing is variable with high recurrence of the ulcers at the same site. Costs for treating these chronically open wounds ranges from $9,000 to $60,000 per year [45] and current treatments are often inadequate, resulting in approximately 50,000 amputations per year among diabetic patients [46]. Approximately 50% of all amputees die (all outcomes) within three years of surgery [47].

We have developed a similar neonatal human fibroblast product, Dermagraft-Ulcer, to address this medical need. In this application fibroblast cells are inoculated onto a dissolvable polymer mesh, Vicryl™ produced by Ethicon, for scaffolding and allowed to grow into the three dimensional tissue. This product is also frozen and stored long term until needed by physicians. Fig. 7 shows the application of this product to a diabetic foot ulcer. In a 50 patient pilot clinical trial, patients receiving the eight-dose regimen showed statistically significant improvement ($p < 0.05$) with complete wound closure in 50% of Dermagraft patients *versus* 8% in the control group. Of the Dermagraft treated ulcers that healed, there was no recurrence during long-term follow-up from four months to one year. Pivotal clinical trials are now underway in both the U.S. and France to further substantiate these results. In a separate clinical trial of patients with venous ulcers, recurrence of healed ulcers was 6.3% for treated patients *vs.* 19.7% for control patients ($p < 0.05$). We have also demonstrated success in treating the most prevalent type of these chronic wounds in a pilot trial of 50 patients with pressure ulcers. Complete healing was achieved in 46% of patients receiving the eight-dose regimen as compared to 25% of controls, with wound closure evaluated over a period of 12 weeks following the treatment period.

Summary

Over the last several decades, technologies have been developed making possible the *in vitro* cultivation of many different human cell types. At the same time the art and science of medicine has been expanded to include new biopharmaceutical and recombinant genetic therapies. Organ transplantation has become common place, limited not by surgical technique, but by donor availability. Tissue engineering holds the promise of combining the advances in culture technology with progress in medical and surgical intervention to provide new solutions by implantation of normal human tissue constructs that are best able to perform the natural functions for each tissue type. The therapeutic applications reported here for the use of dermal fibroblast based products for management of accidental or disease produced skin damage, demonstrates the principal that medical intervention by implantation of *in vitro* engineered human tissue is a feasible and effective treatment.

Tissue engineering has also succeeded in development of an *in vitro* skin model that may replace many toxicological tests currently done on animals and humans, and this procedure has received its initial regulatory approval with extensive additional regulatory work on-going. It has the potential to significantly help protect the individual from the increasing array of new chemicals entering his environment by providing a more economical and effective method to pre-screen for harmful properties of these materials.

References

1. Food and Drug Administration, Department of Health and Human Services (1993) Points to Consider in the Characterization of Cell Lines Used to Produce Biologicals. May.
2. De Wever B and Rheins LA (1994) Skin^{2TM}: An *in vitro* human skin analog. In: Rougier A, Goldberg AM and Maibach HI (eds) Alternative Methods in Toxicology, Vol. 10, pp 121–131. New York: Mary Ann Leibert Inc.
3. Landeen LK, Zeigler FC, Halberstadt C, Cohen R and Slivka SR (1992) Characterization of a human dermal replacement. Wounds 5: 167–175.
4. Rheinwald JG (1989) Methods for clonal growth and serial cultivation of normal human epidermal keratinocytes and mesothelial cells. In: Baserga R (ed) Cell Growth and Division, A Practical Approach, pp 81–94. New York: Oxford Univ. Press.
5. Lucas-Clerc C, Massart C, Campion JP, Launois B and Nicol M (1993) Long-term culture of human pancreatic islets in an extracellular matrix: morphological and metabolic effects. Mol Cell Endocrinol 94: 9–20.
6. Contard P, Bartel RL, Jacobs L, Perlish JS, MacDonald ED, Handler L, Cone D and Fleischmajer R (1993) Culturing keratinocytes and fibroblasts in a three-dimensional mesh results in epidermal differentiation and formation of a basal lamina-anchoring zone. J Invest Dermat 100: 35–39.
7. Langer R and Vacanti JP (1993) Tissue engineering. Science 260: 920–926.
8. Hansbrough JF, Morgan J and Greenleaf G (1993) Advances in wound coverage using cultured cell technology. Wounds 5: 174–194.
9. Hansbrough JF (in press) Status of cultured skin replacements. Wounds. July/August 1995; 7,IV: 130–136.
10. Mayer FL, Whalen EA and Rheins LA (1994) A regulatory overview of alternatives to animal testing: United States, Europe, and Japan. J Toxico-Cut & Ocular Toxicol 13: 3–22.
11. Perkins MA and Osborne R (1993) Development of an *in vitro* method for skin corrosion testing. J Invest Dermat 100: 535.
12. Rheins LA, Edwards SM, Miao O and Donnelly TA (1994) Skin^{2TM}: An *in vitro* model to assess cutaneous immunotoxicity. Toxic in Vitro 8: 1007–1014.
13. Leary T, Jones PL, Appleby M, Blight A, Parkinson K and Stanley M (1992) Epidermal keratinocyte self-renewal is dependent upon dermal integrity. J Invest Dermat 99: 422–430.
14. Zeigler FC, Landeen L, Naughton GK and Slivaka SR (1993) Tissue-engineered, three-dimensional human dermis to study extracellular matrix formation in wound healing. J Toxico-Cut & Ocular Toxicol 12: 303–312.
15. Fleischmajer R, Contard P, Schwartz E, MacDonald ED, Jocobs L and Sakai LY (1991) Elastin-associated microfibrils (10 nm) in a three-dimensional fibroblast culture. J Invest Dermatol 97: 638–643.
16. Fleischmajer R, MacDonald ED, Contard P and Perlish JS (1993) Immunochemistry of a keratinocyte-fibroblast co-culture model for reconstruction of human skin. J Histochem Cytochem 41: 1359–1366.
17. Koschier FJ, Roth RN, Stephens TJ, Spence ET and Duke MA (1994) *In vitro* skin irritation testing of petroleum-based compounds in reconstituted human skin models. J Toxico-Cut & Ocular Toxicol 13: 23–37.
18. Casterton PL, Potts LF and Klein BD (1994) Use of *in vitro* methods to rank surfactants for irritation potential in support of new product development. Toxic in Vitro 8: 835–836.
19. Edwards SM, Donnelly TA, Sayre RM and Rheins LA (1994) Quantitative in vitro assessment of phototoxicity using a human skin model, Skin^{2TM}. Photodermatol Photoimmunol Photomed 10: 111–117.
20. Kruse PF Jr and Patterson MK Jr (eds) (1973) Tissue Culture Methods and Applications. New York: Academic Press.
21. Petricciani JC, Hopps HE and Chapple PJ (eds) (1979) Cell Substrates. New York: Plenum.
22. Petricciani JC and Hennessen W (eds) (1979) Cells, products, safety: background papers for the WHO study group on biologicals. Developments in Biological Standardization 68.
23. Hansbrough JF (1990) Current status of skin replacements for coverage of extensive burn wounds. J Trauma 30: S155–S162.
24. Cuono C, Langdon R and McGuire J (1986) Use of cultured autografts and dermal allografts as skin replacement after burn injury. Lancet i: 1123.
25. Cuono CB, Langdon R and Birchall N (1987) Composite autologous-allogenic skin replacement: development and clinical application. Plast Reconstr Surg 80: 626.
26. Demarchez M, Hartmann DJ, Regnier M and Asselineau D (1992) The role of fibroblasts in dermal vascularization and remodeling of reconstructed human skin after transplantation onto the nude mouse. Transplantation 54: 317–326.
27. Hansbrough JF, Cooper ML, Cohen R, Spielvogel R, Greenleaf G and Naughton G (1992) Evaluation of a biodegradable matrix containing cultured human fibroblasts as a dermal replacement beneath meshed skin grafts on athymic mice. Surgery 4: 438–446.
28. Economou TP, Rosenquist MD, Lewis RW II, PA-C and Kealey GP (1995) An experimental study to determine the effects of Dermagraft on skin graft viability in the presence of bacterial wound contamination. J Burn Care Rehabil 16: 27–30.
29. Hansbrough JF, Dore C and Hansbrough WB (1992) Clinical trials of a living dermal tissue replacement placed beneath meshed, split-thickness skin grafts on excised burn wounds. J Burn Care Rehabil 13: 519–529.
30. Takeda N, Abe S, Tsukiyama H, Hata K, Saitou T, Komatsu K and Ariga A (1995) The clinical experience of artificial skin contained allogenic fibroblasts. Japanese Society of Burn Injury, Nagoya, Japan.
31. Hansbrough JF, Morgan J, Greenleaf B and Underwood J (1994) Development of a temporary living skin replacement composed of human neonatal fibroblasts cultured in Biobrane, a synthetic dressing material. Surgery 115: 633–644.
32. Allen LA, Landeen KK, Pieters R, Bice-Bentley K and Harman RJ (1994) Development of a porcine full-thickness burn model for the evaluation of human cadaveric skin and Dermagraft Transitional Covering, a biosynthetic skin substrate. 9th Congress of the International Society for Burn Injuries, Paris, France.
33. Ilten-Kirby B, Pinney E, Pavalec R and Allen L (1995) Human allograft vs. Dermagraft Transitional Covering, a tissue engineered skin substitute for use as a full-thickness burn. 2nd Annual Conference on Cellular Engineering, International Federation for Medical and Biological Engineering, San Diego.
34. Mansbridge M, Gould T, Pinney E, Snyder D, Fung M and Rose P (1995) Control of growth and matrix deposition in three-dimensional culture of human fibroblasts. 2nd Annual Conference on Cellular Engineering, International Federation for Medical and Biological Engineering, San Diego.
35. Davis MM, Hansbrough JF, Mozingo DW, Kealey GP, Gidner AF and Gentzkow GD (1995) Dermagraft-TC in the treatment of burns requiring temporary coverings. Symposium

on Advanced Wound Care and Medical Research Forum on Wound Repair.
36. Black K, Gentzkow G, Purchio T, Mansbridge J, Landeen L and Allen L (1994) The future for temporary and permanent wound coverage. 3rd International Congress on The Immune Consequences of Trauma, Shock and Sepsis, Munich, Germany.
37. Gentzkow G, Iwasaki S, Gupta S, Hershon K, Lipkin S, Steed D, Mengel M, Prendergast JJ and Ricotta J (1994) Cultured human dermal replacement tissue for the treatment of diabetic foot ulcers. European Tissue Repair Society, Oxford, England.
38. Jakoby WB and Pastan IH (eds) (1979) "Cell Culture". Methods in Enzymology 58: 641 p.
39. Committee for Proprietary Medicinal Products: Ad Hoc Working Party on Biotechnology/Pharmacy (1989). Notes to applicants for marketing authorizations on the production and quality control of monoclonal antibodies of murine origin intended for use in man. J Biol Stand 17: 213.
40. Blood S, Heck E and Baxter CR (1979) The importance of the bacterial flora in cadaver homograft burn donor skin. Proceedings of the 11th Meeting of the American Burn Association, New Orleans: 79–80.
41. White MJ, Whalen JD, Gould JA, Brown GL and Polk HC (1991) Procurement and transplantation of colonized cadaver skin. Am Surg 57: 402–407.
42. Bale JF, Kealey GP, Ebelhack CL, Platz CE and Goeken JA (1992) Cyotmegalovirus infection in a cyclosporine-treated burn patient: a case report. J Trauma 32: 263–267.
43. Hansbrough JF (1987) Biological dressings. In: Boswick J (ed) The Art and Science of Burn Care, pp 57–63. Rockville, MD: Aspen Publishers Inc.
44. Banked Human Tissue, Food and Drug Administration, Department of Health and Human Services (1993) Inspectional Guidance.
45. Wound healing: focus on growth factors, skin substitutes (1992) MedPRO Month 2: 91.
46. Centers for Disease Control (1991) The prevention and treatment of complications of diabetes: A guide for primary care practitioners. NIH Publication No. 93–3464.
47. Bild DE, Selby JV, Sinnock P, Browner WS, Braveman P and Showstack JA (1989) Lower-extremity amputation in people with diabetes: epidemiology and prevention. Diabetes Care 12: 24.

R. P. Lanza and W. L. Chick (eds.), Yearbook of Cell and Tissue Transplantation 1996/1997, 275–282.

Chapter 27

Tissue engineering: Tubular tissues

David J. Mooney[1], Joseph P. Vacanti[2] and Robert Langer[3]

[1] Depts. Biologic & Materials Sciences and Chemical Engineering, University of Michigan, Ann Arbor, MI 48109, U.S.A.; [2] Dept. Surgery, Harvard Medical School, Boston, MA 02115, U.S.A.; [3] Dept. Chemical Engineering, Massachusetts Institute of Technology, Cambridge, MA 02319, U.S.A.

A wide range of tissues within the body share a common tubular geometry. This chapter will focus on the approaches and challenges to engineer tubular tissues which contain a central lumen. These tissues (e.g., intestine, trachea) share the basic function of allowing flow, or convective transport of some material through the tissue. A variety of materials are transported through the central lumen of these tissues, including solids, liquids, and gases. A number of forces drive transport, including pressure gradients and smooth muscle contraction. The tissue may be relatively rigid (e.g., trachea), or capable of large strain (e.g., intestine). In all cases, these tissues are comprised of multiple cell types with a defined organization that is critical to the tissue's function. This chapter will not review engineered tubular tissues which do not contain a central lumen (e.g., nerves, ligaments, bone), or attempts to use tubular devices to transplant cells that do not normally exist in a tubular geometry (e.g., hollow fiber encapsulation of hepatocytes).

All tissues contain cells and a cellular-derived matrix, and multiple cell types are typically present in tubular tissues. Specialized epithelial cells (e.g., enterocytes) typically line the lumen of tubular tissues, and control directed transport of luminal contents into the tissue, or along the luminal surface of the tissue. Cells which directly or indirectly provide mechanical support or generation of transport driving forces are also present. The presence and organization of these varying cell types is critical to the function of the entire tissue. The mechanical properties of tissues are often dictated by the matrix surrounding the cells. Synthetic matrices utilized to engineer tissues must deliver transplanted cells to a specific anatomic location, serve as a template for the proper organization of the transplanted cells and cells from the host, and provide mechanical support during the process of tissue development.

This chapter will review the different types of materials which are being utilized to fabricate tissue engineering matrices. Specific examples of the cells, matrices, and strategies utilized to engineer tracheal [1–2], intestinal [3–5], and urological tissues [6–7] will then be presented. A brief discussion of the challenges which must be surmounted to engineer functional tubular tissues will conclude the chapter.

Tissue engineering matrices

The matrices utilized in tissue engineering have been fabricated from both naturally-derived and synthetic materials, and serve multiple functions. These matrices either transplant cells to a specific anatomic location and/or induce the ingrowth of a desired cell population from the tissue surrounding the implant. The matrix guides the development of the appropriate tissue structure from the transplanted cells and host tissue, and mechanically stabilizes the tissue during this time.

A variety of naturally-derived materials have been utilized as tissue engineering scaffolds, including type I collagen [8–12], laminin [13], and alginate [14–15]. These materials, in a pure state, typically exhibit excellent biocompatibility, and are amenable to cellular remodeling during the process of tissue formation [9]. In addition, extracellular matrix molecules (e.g., type I collagen) specifically interact with cells via cell membrane receptors [16]. This allows cell adhesion and function to be controlled by the presentation of the matrix to cells [17]. Recent work in the tissue engineering field has focused on developing processing techniques and improving the mechanical properties of type I collagen, the most widely utilized naturally-derived matrix material. Type I collagen can be processed into a variety of forms, including porous sponges and fabrics woven from small diameter fibers [8–12]. These fabrics can be formed into a variety of structures, including hollow tubes (Fig. 1). The mechanical properties of matrices formed from type I collagen can be greatly improved by chemical cross-linking of the collagen fibrils, although this may detrimentally effect the matrices' biocompatibility [10–11]. In addition, cells within collagen gels can be preferentially aligned in a desired direction [12]. This is achieved by magnetically aligning the individual collagen fibrils within the gels to which the cells adhere [12]. However, naturally-derived materials in general still suffer from the limitations of high cost and large batch-to-batch variations which result from their isolation from human, animal, or plant tissue.

A variety of synthetic polymers, both degradable and non-degradable, have also been utilized to fabricate tissue engineering matrices [18–19]. Polymers can be synthesized reproducibly and processed with a variety of techniques, and are typically cheaper than naturally-derived materials. In addition, the physical properties (e.g., mechanical strength) of synthetic polymers can be easily varied with manipulations that effect the polymer structure, molecular weight, and/or presence of pendant chains. Biodegradable polymers are especially attractive for tissue engineering applications because a completely natural tissue will result following polymer erosion. Polymers of lactic and glycolic acid have been extensively utilized to fabricate tissue engineering matrices [1–4, 6–7, 20–24]. These polymers can be formed into a variety of structures, including small diameter fibers (Fig. 2a), and porous films (Fig. 2b) that can be formed into hollow tubes (Fig. 2c). A number of cell types have been shown to adhere and proliferate on these polymers [1–4, 6–7, 20–24], and cell adhesion is likely mediated by extracellular matrix proteins randomly adsorbed onto the polymer surface.

Tissue engineering matrices, regardless of the material from which they are formed, guide the formation of a new tissue structure from the transplanted cells and host cells which surround and/or invade the device. Tubular devices with central lumens have been utilized to engineer tubular tissues. It is assumed that this matrix geometry will promote the development of a final tubular tissue structure [20, 24]. The pore size distribution and continuity dictate the interaction of the device and transplanted cells with the host tissue. Fibrovascular tissue will invade a device if the pores are larger than approximately 10 μm, and the rate of invasion will increase with the pore size and total porosity of a device [25–27]. This process results in the formation of a capillary network in the developing tissue [27–28]. Vascularization of the engineered tissue is required to meet the metabolic requirements of the tissue and integrate it with the surrounding host. A central stent may be temporarily placed in the lumen of tubular tissue engineering matrices to maintain the patency of the central lumen by preventing tissue ingrowth from filling the lumen [1–4].

Tubular tissues which exist to transport materials through their lumen are typically subjected to large mechanical forces. For example, large positive and negative pressures are present in the trachea during inhalation and exhalation. The large negative pressures (greater than 200 mm Hg) will collapse the airway and prevent airflow if the airway is not sufficiently rigid. The synthetic matrices utilized in tissue engineering applications must provide mechanical support during the process of tissue formation and prevent large distortions of tissue structure which hinder or prevent transport. Tubular cell delivery vehicles fabricated from polymers of lactic and glycolic acid (Fig. 2) can both withstand large compressive forces, and exhibit a viscoelastic response to mechanical loading [20–21] (Fig. 3a). The compressive stresses used to test these matrices (50–200 mm Hg) are in the same range as the pressures typically found in the trachea and large blood vessels [29]. These structures are also capable of resisting compressive forces from surrounding tissues following implantation in animal models [20–21]. Implantation of these devices leads to the formation of a vascularized, tubular tissue following fibrovascular tissue ingrowth (Fig. 3b). It is important to realize that the mechanical support provided by these matrices will decrease as the polymer degrades. The erosion rate of these devices can be varied from weeks to years to suit a specific application by fabricating the devices from copolymers of lactic and glycolic acid which contain varying ratios of the two monomers [21].

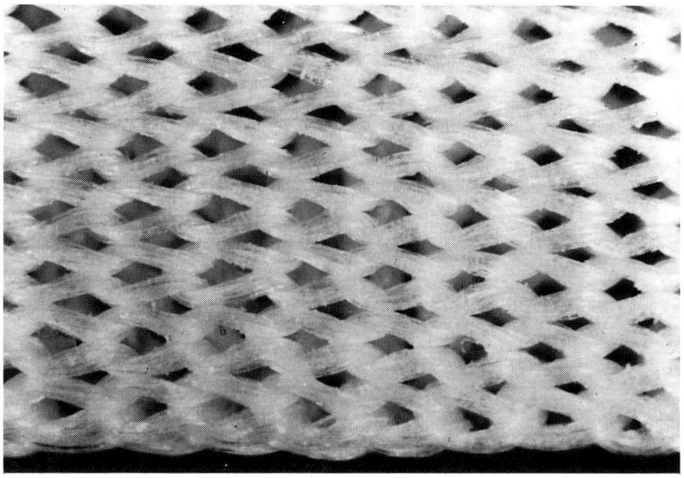

Fig. 1. A photomicrograph of type I collagen fibers knitted into a fabric [8]. Fiber diameters can be as small as 25 μm, and the fabrics can be formed into a variety of shapes, including tubes. Used with permission of John Wiley and Sons, Inc.

Engineering specific tubular tissues

Trachea

Reconstruction of the trachea following the management of trachea pathology is successful when up to 1/2 or 1/3 of the trachea is resected in adults and children, respectively [30]. However, there is a clinical need for a trachea equivalent tissue when a greater amount of the trachea must be resected. This tissue should contain cartilaginous rings to provide mechanical support and prevent collapse of the trachea during breathing. In addition, a layer of tracheal epithelial will be required to provide directional transport of mucus.

Cell transplantation has recently been utilized to engineer tracheal tissue [1–2]. Chondrocyte transplantation has been utilized to engineer the cartilaginous rings. Chondrocytes were isolated, expanded in culture, and transplanted into nude mice utilizing polyglycolic acid fiber-based tubular matrices [1]. The hollow, tubular cartilage tissues which formed were subsequently excised and cut into rings with approximately the same dimensions of tracheal rings. These cartilage rings were next sutured inside the lumen of harvested bowel to test their mechanical stability. They resisted a negative pressure of 200 mm Hg without collapse [1]. Experimental animals which underwent tracheal resection and transplantation of these engineered tracheal tissues survived for longer time periods than control animals. This approach was subsequently expanded by lining the engineered cartilage structures with isolated tracheal epithelial cells [2]. This led to the formation of a more complete tissue which contained the cartilaginous rings, and an epithelial cell lined lumen covering a vascularized submucosal tissue. This engineered trachea tissue may be capable of replacing the function of large sections of resected trachea.

Intestine

Short bowel syndrome is characterized by malabsorption due to resection of a large section of small intestine [31]. A variety of reconstructive procedures can be performed to manage short bowel syndrome, but their application is limited. Alternatively, total parenteral nutrition is often utilized to maintain patients suffering from short bowel syndrome, but this therapy often leads to long-term complications [31]. An engineered intestinal tissue could potentially cure these patients, and eliminate the complication of the established therapies. However, the intestine is a very complex organ. Intestinal epithelial cells (enterocytes) provide the absorptive function that is characteristic of intestine, and are present in a complex villi and crypt structure. Two layers of smooth muscle cells provide the locomotive force for transport of the contents of the intestine, and these two muscle layers are aligned perpendicular to each other. In addition, the intestine is highly vascularized and enervated. The multiple cell types and complex organization of these cell types clearly provide significant challenges to engineering a functional intestinal tissue.

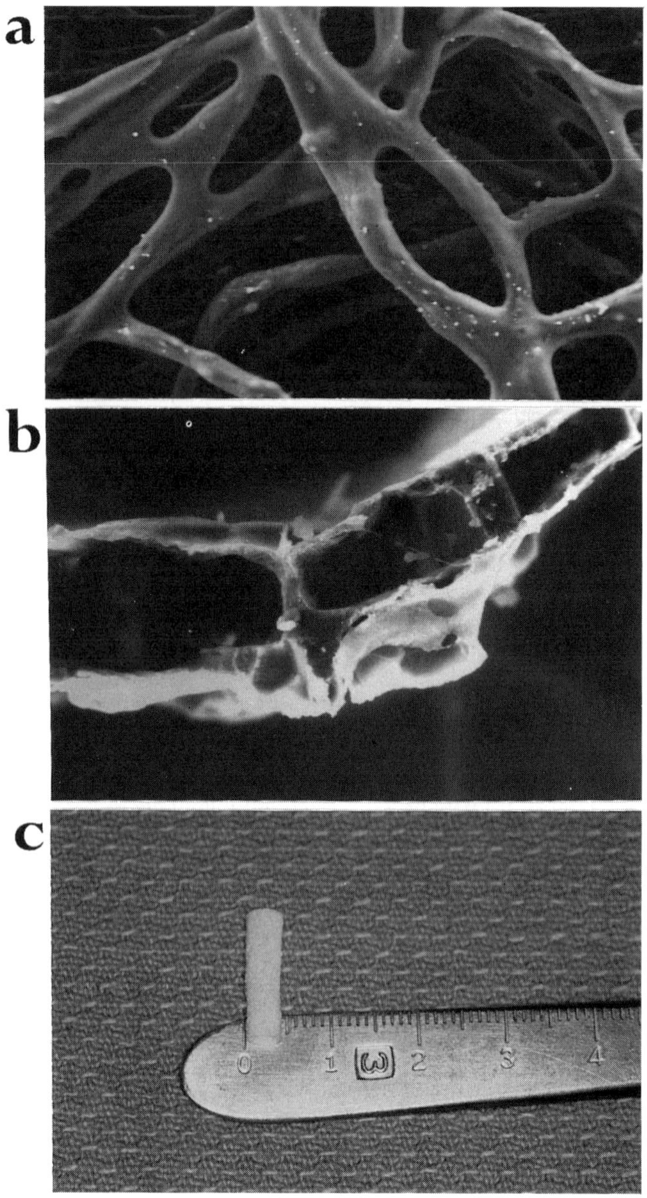

Fig. 2. Photomicrographs of synthetic polymer devices utilized as tissue engineering matrices. (a) A fabric formed from polyglycolic acid fibers after the fibers were physically bonded with polylactic acid to mechanically stabilize the structure [20]. (b) A cross-section of a thin, porous film fabricated from a 50/50 copolymer of lactic and glycolic acid that has been formed into a tube [24]. (c) A photograph of a tubular device fabricated from a porous film of 50/50 PLGA [21].

Initial efforts to engineer intestinal tissue have focused on creating an enterocyte-lined tubular tissue. Techniques have been developed to isolate, culture, and expand these cells [32]. Enterocytes will adhere to and organize in vitro on tubular devices fabricated from polyglycolic acid or poly (lactic-co-glycolic acid) (Fig. 4a). Transplantation of enterocytes isolated from the intestinal crypts of adult rats into the mesenteric tissue of syngeneic animals using these devices leads to the formation of a vascularized tubular tissue lined by enterocytes (Fig. 4b) [3–4]. These cells proliferate in vivo, indicating that the normally high growth rate of these cells is at least partially maintained. However, the formation of the more complex crypt and

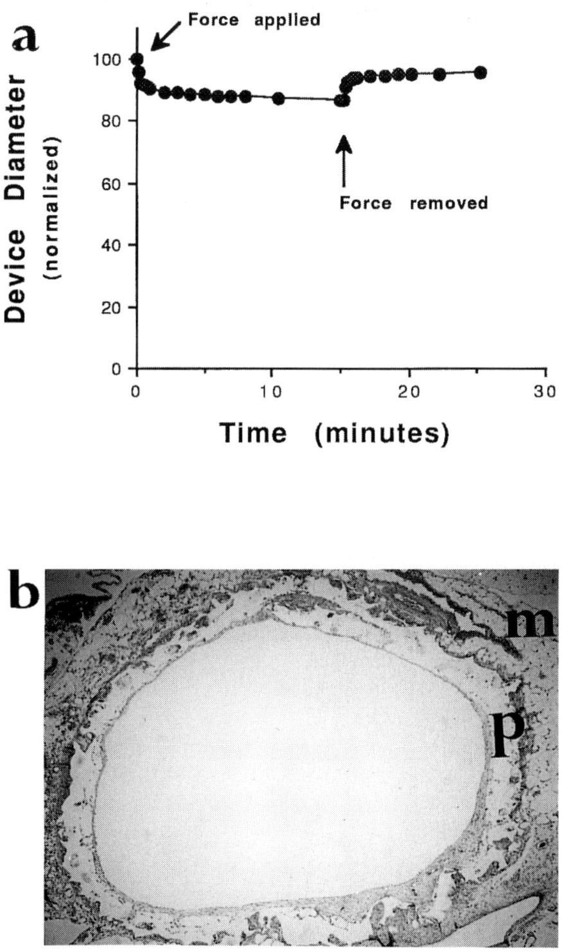

Fig. 3. Compression resistance of polymeric tissue engineering matrices in vitro and in vivo. (a) Representative strain diagram of a device (Fig. 2a) subjected to a compressive force of 200 mN applied in a direction perpendicular to the axis of the device lumen starting at 0 minutes. The force was removed at 10 minutes, and the change in the diameter of the tube (parallel to the direction of force application) was monitored both during and after the time of force application, and normalized to the initial diameter. (b) A photomicrograph of a histological section from a device implanted for 7 days in the mesentery of a Lewis rat [20]. This cross-section of the implanted device was cut perpendicular to the axis of the tube's lumen. The central lumen is visible, along with the polymer (p), the host mesenteric tissue (m), and the ingrown fibroblasts and fibrous tissue they deposited.

villi organization was not noted in these studies. In contrast, enterocytes isolated from fetal intestine and transplanted into the renal subcapsular space developed into simple tubular structures after one week, and rudimentary crypts and villi were observed by fourteen days [5]. The success of these studies suggests that it is possible to engineer an intestinal-like tissue, but no attempts to engineer an intestinal tissue containing the smooth muscle element have yet been made.

Urologic tissues

Ureters and other urothelial tissues are often required in reconstructive procedures to treat a variety of urological conditions. While natural tissues (e.g., bowel) and synthetic prosthesis (e.g., Teflon) have been utilized to reconstruct urologic tissues, they have either not performed satisfactorily in replacing function or led to numerous complications [33]. The limitations of current therapies have again led investigators to engineer urothelial structures using cell transplantation.

Urothelial structures have been engineered using transplantation of cells disassociated from bladder and cultured in vitro [6–7]. Cultured primary cells adhered to polymeric devices fabricated from polyglycolic acid, and formed urothelial lined structures following transplantation into experimental animals [6]. The transplanted cells successfully achieved correct spatial orientation. Further studies indicated that cotransplantation of urothelial and smooth muscle cells led to the formation of new urologic structures comprised of both cell types [7]. The smooth muscle cells segregated from the epithelial cells, forming a well defined sheet lining the multilayered epithelial cell structure. Extensive proliferation of the transplanted cells was noted during the formation of the new tissue. These data suggest that cell transplantation may be utilized to fabricate urologic tissues, and that smooth muscle and epithelial cells maintain the ability to organize appropriately following co-transplantation. This latter finding suggests that it may be possible to engineer tissues which contain multiple cells types present in a specific organization.

Other tubular tissues

A wide variety of other tubular tissues are also being engineered using the same concepts outlined above. These tissues include esophagus [34], kidney [35], and blood vessels [20, 36–38]. The failure of synthetic small diameter vascular grafts has driven the great number of studies devoted to engineering blood vessels. The biocompatibility of synthetic grafts has been enhanced by surface modification intended to promote endothelialization by pre-seeded or migrating endothelial cells [38]. Engineered vessels comprised of endothelial cells and smooth muscle cells have also been developed utilizing matrices fabricated from type

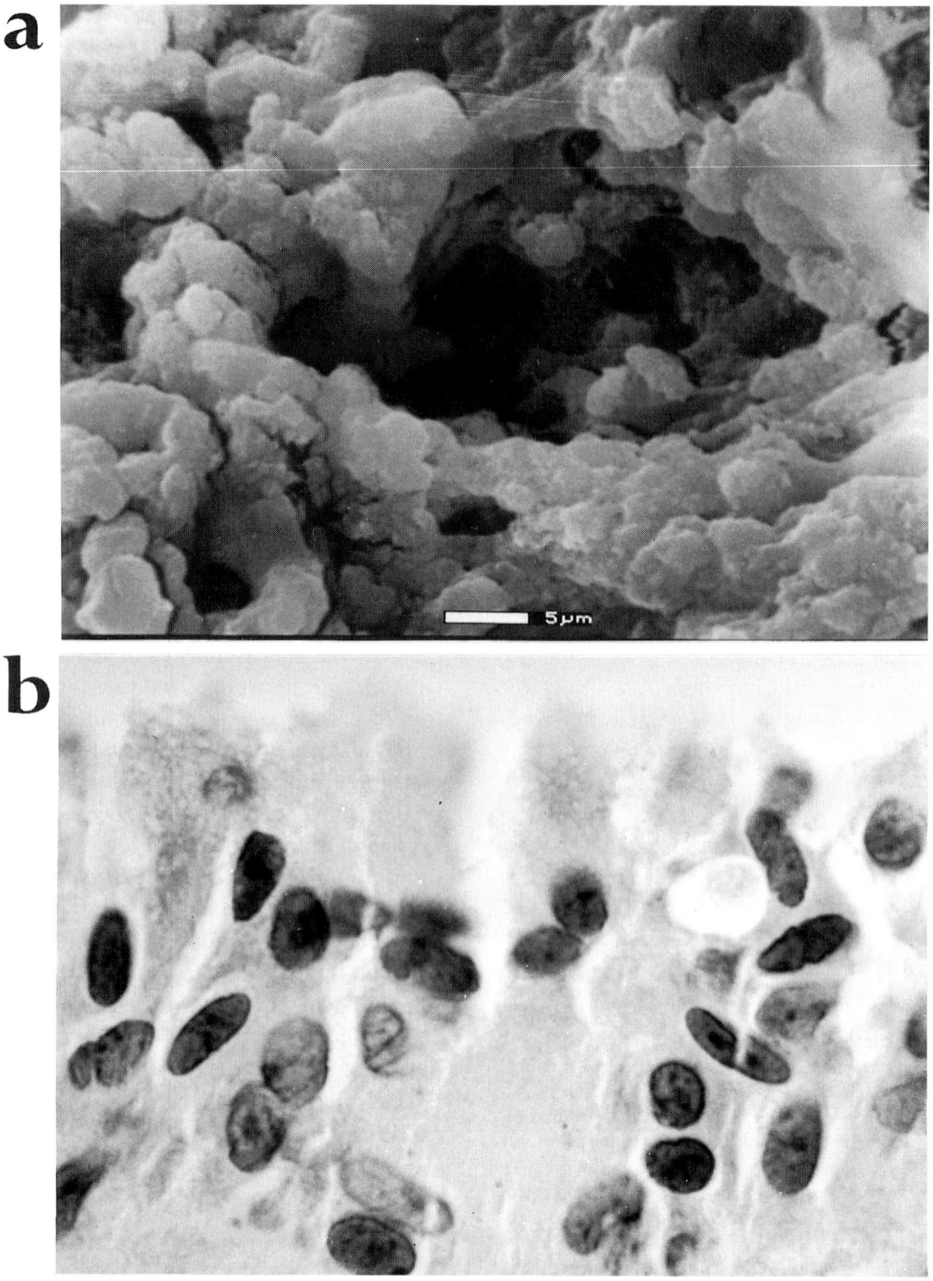

Fig. 4. Enterocytes *in vitro* and *in vivo*. (a) Photomicrograph of cultured enterocytes adherent to a tubular device fabricated from poly (lactic-co-glycolic acid) [Fig. 2]. (b) Photomicrograph of a histological section of an enterocyte seeded device implanted in the mesentery of a Lewis rat for 14 days. A layer of enterocytes, 3–5 cells thick, line the lumen of the engineered tubular tissue.

I collagen [36–37] or synthetic, degradable polymers [20]. More details on engineering blood vessels can be found in the chapter of this book focused on this specific tissue.

Summary and future challenges

Great strides have been made in engineering tubular tissues. Matrices and strategies have been developed to engineer tubular, vascularized tissues lined with the desired epithelium. Tissues containing multiple cell types have shown promise in a variety of animal models. However, several challenges remain in the engineering of functional tubular tissues.

An exciting area of research involves developing new biomaterials which combine the advantages of synthetic materials with the biological specificity and cell adhesion activity of naturally occurring extracellular matrix molecules. This can be achieved by covalently binding or physically adsorbing ECM molecules or small synthetic peptides which contain the biologically active amino acid sequences onto matrices fabricated from synthetic polymers [38–40]. In addition, matrices which mimic the mechanical and cell binding properties of naturally occurring extracellular matrix molecules are being synthesized using recombinant gene techniques [41].

A large number of tubular tissues are enervated, and this is critical to their function. Nerve ingrowth has not been noted up to this point in engineered tubular tissues, and special efforts may be required to induce this process. Specific extracellular matrices have been shown to promote nerve regeneration [42]. Tissue engineering matrices may be designed to incorporate the signals which are required to promote nerve ingrowth.

A critical feature of many tubular tissues is the specific organization of the varying cell types present in the tissue. The studies described in this chapter indicate that certain transplanted cells appear to maintain an inherent ability to reorganize into appropriate structures following implantation. However, as yet undefined signals from the tissue engineering matrices may be required to promote appropriate organization of some tissue structures (e.g., the two smooth muscle layers present in intestinal tissue).

Acknowledgements

The authors thank the University of Michigan, the National Science Foundation (BCS-9202311), and Advanced Tissue Sciences for financial support.

References

1. Vacanti CA, Paige KT, Kim WS *et al.* (1994) Experimental tracheal replacement using tissue-engineered cartilage. J Ped Surg 29: 201–205.
2. Sakata J, Vacanti CA, Schloo B *et al.* (1994) Tracheal composites tissue engineered from chondrocytes, tracheal epithelial cells, and synthetic degradable scaffolding. Transplant Proc 26: 3309–3310.
3. Organ GM, Mooney DJ, Hansen LK *et al.* (1992) Transplantation of enterocytes utilizing polymer-cell constructs to produce a neointestine. Transplant Proc 24: 3009-3011.
4. Organ GM, Mooney DJ, Hansen LK *et al.* (1993) Enterocyte transplantation using cell-polymer devices to create intestinal epithelial-lined tubes. Transplant Proc 25: 998–1001.
5. Tait IS, Evans GS, Kedinger M *et al.* (1994) Progressive morphogenesis in vivo after transplantation of cultured small bowel epithelium. Cell Transplant 3: 33–40.
6. Atala A, Vacanti JP, Peters CA *et al.* (1992) Formation of urothelial structures in vivo from dissociated cells attached to biodegradable polymer scaffolds in vitro. J Urology 148: 658–662.
7. Atala A, Freeman MR, Vacanti JP *et al.* (1993) Implantation in vivo and retrieval of artificial structures consisting of rabbit and human urothelium and human bladder muscle. J Urology 150: 608–612.
8. Cavallaro JF, Kemp PD and Kraus KH (1994) Collagen fabrics as biomaterials. Biotech Bioeng 43: 781–791.
9. Yannas IV, Lee E, Orgill DP *et al.* (1989) Synthesis and characterization of a model of extracellular matrix that induces partial regeneration of adult mammalian skin. Proc Natl Acad Sci (USA) 86: 933–937.
10. Anselme K, Petite H and Herbage D (1992) Inhibition of calcification in vivo by acyl azide cross-linking of a collagen-glycolsaminoglycan sponge. Matrix 12: 264–273.
11. Koide M, Osaki K, Oyamada K *et al.* (1993) A new type of biomaterial for artificial skin: dehydrothermally cross-linked composites of fibrillar and denatured collagens. J Biomed Mat Res 27: 79–87.
12. Dickinson RB, Guido S and Tranquillo RT (1994) Biased cell migration of fibroblasts exhibiting contact guidance in oriented collagen gels. Ann Biomed Eng 22: 342–356.
13. Dixit V (1994) Development of a bioartificial liver using isolated hepatocytes. Artificial Organs 18: 371–384.
14. Lim F and Sun AM (1980) Microencapsulated islets as bioartificial endocrine pancreas. Science 210: 908–910.
15. Atala A, Kim W, Paige KT *et al.* (1994) Endoscopic treatmen of vesicoureteral reflux with a chondrocyte-alginate suspension. J Urology 152: 641–643.
16. Hynes RO 1987) Integrins: a family of cell surface receptors. Cell 48: 549–554.
17. Stoker AW, Streuli CH, Martins-Green M *et al.* (1990) Designer microenvironments for the analysis of cell and tissue function. Curr Opin Cell Biol 2: 864–874.
18. Langer R and Vacanti JP (1993) Tissue engineering. Science 260: 920–926.
19. Peppas NA and Langer R (1994) New challenges in biomaterials. Science 263: 1715–1720.
20. Mooney DJ, Mazzoni CL, Breuer C *et al.* (in press) Stabilized polyglycolic acid fiber-based devices for tissue engineering. Biomaterials.
21. Mooney DJ, Breuer C, McNamara K *et al.* (1995) Fabricating tubular tissues with devices of poly (D,L-lactic-co-glycolic acid). Tissue Eng. 1: 107–118.
22. Hansbrough JF, Cooper ML, Cohen R *et al.* (1992) Evaluation of a biodegradable matrix containing cultured human fibroblasts as a dermal replacement beneath meshed skin grafts on athymic mice. Surgery 4: 438–446.
23. Mooney DJ, Park S, Kaufmann PM *et al.* (1995) Biodegradable sponges for hepatocyte transplantation. J Biomed Mat Res. 29: 959–966.

24. Mooney DJ, Organ GM, Vacanti JP *et al.* (1994) Design and fabrication of biodegradable polymer devices to engineer tubular tissues. Cell Transplant 3: 203–210.
25. Wesolowski SA, Fries CC, Karlson KE *et al.* (1961) Porosity: primary determinant of ultimate fate of synthetic vascular grafts. Surgery 50: 91–96.
26. White RA, Hirose FM, Sproat RW *et al.* (1981) Histopathologic observations after short-term implantation of two porous elastomers in dogs. Biomaterials 2: 171–176.
27. Mikos AG, Sarakinos G, Lyman MD *et al.* (1993) Prevascularization of porous biodegradable polymers. Biotech Bioeng 42: 716–723.
28. Mooney DJ, Kaufmann PM, Sano K *et al.* (1994) Transplantation of hepatocytes using porous biodegradable sponges. Transplant Proc 26: 3425–3426.
29. Caro CG, Pedley TJ, Schroter RC *et al.* (1978) The mechanics of circulation. New York: Oxford Univ. Press.
30. Grillo HC (1990) Tracheal replacement. Ann Thorac Surg 49: 864–865.
31. Vanderhoof JA, Langnas AN, Pinch LW *et al.* (1992) Short bowel syndrome. J Ped Gastroenter Nutrition 14: 359–370.
32. Weiser MM (1973) Intestinal epithelial cell surface membrane glycoprotein synthesis. J Biol Chem 248: 2536–2541.
33. Atala A and Retik A (in press) Pediatric urology – future perspectives. In: Krane RJ, Siroky MB and Fitzpatrick JM (eds) Clinical urology. Philadelphia: J.B. Lippincott.
34. Sato M, Ando N, Ozawa S *et al.* (1993) A hybrid artificial esophagus using cultured human esophageal epithelial cells. ASAIO J: M554–M557.
35. Humes HD (in press) Tissue engineering of the kidney. In: Bronzino JD (ed) CRC Handbook of Biomedical Engineering, Boca Raton, FL: CRC Press.
36. Weinberg CB and Bell E (1986) A blood vessel model constructed from collagen and cultured vascular cells. Science 231: 397–400.
37. Ziegler T and Nerem RM (1994) Tissue engineering a blood vessel: regulation of vascular biology by mechanical stresses. J Cell Biochem 56: 204–209.
38. Hubbell JA, Massia SP, Desai NP *et al.* (1991) Endothelial cell-selective materials for tissue engineering in the vascular graft via a new receptor. Bio/Technology 9: 568-572.
39. Hubbell JA (1994) Chemical modification of polymer surfaces to improve biocompatibility. TRIP 2: 20–25.
40. Barrera DA, Zylstra E, Lansbury PT *et al.* (1993) Synthesis and RGD peptide modification of a new biodegradable copolymer: poly(lactic acid-co-lysine). J Am Chem Soc 115: 11010–11011.
41. Anderson JP, Cappello J and Martin DC (1994) Morphology and primary crystal structure of a silk-like protein polymer synthesized by genetically engineered Escherichia coli bacterial. Biopolymers 34: 1049–1058.
42. Guenard V, Kleitman N, Morrisey TK *et al.* (1992) Syngeneic Schwann cells derived from adult nerves seeded in semipermeable guidance channels enhance peripheral nerve regeneration. J Neuro 12: 3310–3320.

R. P. Lanza and W. L. Chick (eds.), Yearbook of Cell and Tissue Transplantation 1996/1997, 283–285.

Chapter 28

Tissue engineering: Vascular system

Bruce E. Jarrell and Stuart K. Williams
Department of Surgery, University of Arizona, Health Sciences Center, Tucson, AZ 85724, U.S.A.

While major advances have been made in the field of clinical vascular surgery, a confounding problem remains that small diameter (< 6 mm) vascular grafts continue to exhibit poor, clinically unacceptable long term patency. The conspicuous complications which lead to this poor performance are the occurrence of intimal hyperplasia and thrombosis. The basic inherent problem with these prosthetic grafts, that is grafts constructed of synthetic polymeric materials (e.g. ePTFE and Dacron), is the relative lack of biocompatibility of these polymeric materials. Numerous solutions have been hypothesized, developed and tested toward improvement in polymeric graft function. Some of these solutions are provided in Table 1.

In addition to these device modifications, numerous pharmacologic agents have been evaluated to improve the function of synthetic and native vessel bypass grafts. This pharmacologic approach has attacked both thrombosis as well as intimal hyperplasia as the major modes of graft failure.

The first three general methods to improve blood vessel function have been directed toward the lumenal surface of polymeric grafts with the premise that a non thrombogenic surface could be achieved. Short term improvements in polymer thrombogenicity have been achieved, however, the long term performance of these polymers measured in terms of years has generally not been altered by these modifications. The last category of methods, defined using the broad term tissue engineering holds promise as a means to improve polymer function through the creation of an active antithrombogenic cellular lining on vascular replacement devices. The focus of the remainder of this chapter will be on the recent developments in tissue engineering of vascular devices.

Table 1. Methods to improve the biocompatibility of prosthetic grafts

Method	Example
New polymers [1]	– Polyurethanes – Silicone composites
Polymer modification [2]	– Radio frequency glow discharge – Pyrolytic carbon coating – Protein passivation
Anticoagulant bindings [3]	– Heparin – Hirudin – Plasminogen activator – Prostaglandins
Tissue engineering [4]	– Polymers which support spontaneous healing – Cell/Matrix composites – Cell transplantation

Tissue engineering

The term tissue engineering defined in its broadest sense denotes the formation or synthesis of a tissue replacement from cellular and/or polymeric components. In vascular tissue engineering the device to be engineered will ultimately serve as a replacement to an element in the vascular system. For this review, arterial blood vessel replacements will be the focus. Design parameters for a tissue engineered device are rather simple necessitating 1. a tube of appropriate diameter, 2. able to withstand arterial pressure, 3. of selective porosity capable of appropriately entrapping blood elements and plasma components, 4. exhibiting an antithrombogenic lumenal cellular lining, 5. providing clinically acceptable long term patency. Numerous investigators have endeavored to achieve tissue engineered blood vessel replacements using a variety of techniques combining molecular biology, cell biology, materials sciences and biochemistry.

Synthetic polymer tissue engineered vessels

The structure of tissue engineered polymeric blood vessels replicates the medial components of a normal artery using polymeric material. Thus the polymer will perform the function of withstanding arterial blood pressure. Two types of design configurations of polymers have been evaluated. First polymers with compliance simulating the intact vessel (i.e. intima, media and adventitia) have been constructed from compliant materials such as polyurethanes [5, 6]. Alternatively investigators have used clinically available polymers such as ePTFE and Dacron for tissue engineering [7, 8]. These materials have limited compliance resulting in compliance mismatch at the anastomotic interface. The importance of this compliance mismatch toward accelerated intimal hyperplasia has yet to be definitively established.

The formation of a cellular intima on these polymeric devices has been engineered using either cell transplantation prior to vessel implantation or through spontaneous intima formation after the polymer is implanted. While the formation of a cellular intima is the final goal of both these methods, the first method relies on cellular engineering prior to graft implantation. On the other hand the second method involves a more complex design of polymer chemistry and polymer porosity since cell lining formation occurs in vivo. Cell transplantation onto polymeric vascular grafts represents one of the earliest forms of blood vessel cell engineering. Methods for this form of cell engineering have undergone significant modification since originally described by Dr. Malcolm Herring in 1978 [9]. Endothelial cell transplantation has progressed to the stage of human clinical trials and reviews of this technology appear elsewhere in this yearbook as well as in several recent reviews [4, 10].

A more recent development in tissue engineering of blood vessels has been the synthesis and modification of polymeric devices which accelerate the formation of an antithrombogenic cellular lining on the lumenal surface of polymeric grafts after the graft has been implanted as an arterial bypass. The major cellular mechanism which is hypothesized to lead to the cellular lumenal lining is the ability of vascular cells to migrate through the interstices of the polymer onto the lumenal surface of the graft, and subsequently cells which have successfully reached the lumenal surface divide to a degree sufficient to create a complete monolayer [11]. In comparison to cell transplantation, monolayer development is retarded by the need for cells to penetrate the interstices of the graft and proliferate on the graft surface. Nevertheless this method is highly attractive since cell transplantation methods need not be employed.

The design of polymers to enhance spontaneous transmural cell migration has initially focussed on the synthesis of polymers with higher porosity [11]. Commercially available clinically approved polymers have a limited porosity and thus may inhibit cell penetration by acting simply as a physical barrier. The earliest use of higher porosity polymers has resulted in significant preclinical success in animal models [11]. Unfortunately these animal results have not been followed by any corresponding human clinical successes toward formation of a lumenal cell lining. Further developments in this technology have focussed on the incorporation of chemical agents into polymer grafts to further stimulate spontaneous cell linings [12]. Again, a significant amount of engineering is being performed on these polymers before they are implanted, followed by cellular engineering as the implant heals. These chemical modifications have included the incorporation of known endothelial cell growth promoting factors into the graft interstices with the hypotheses that these factors will stimulate the angiogenic process [13]. To date all of the factors used have been known promoters of the angiogenic process.

Tissue engineered blood vessels from cell and matrix components

An alternate method to the use of synthetic polymeric materials has been the reconstruction of blood vessel components using individual cellular components. This technology has been advanced by the development of methods to stimulate the growth of human endothelial cells in culture in sufficient quantity to permit high density seeding during tube construction. van Buul-Wortelboer *et al.* [14] and Weinberg and Bell [15] took advantage of newly developed vascular cell culturing techniques to construct artificial blood vessels. The vessels are composed of three layers consisting of

1. an outer layer of fibroblasts and collagen, 2. an intermediate middle layer of smooth muscle cells in collagen, 3. an inner layer of endothelial cells [16]. These layers thus represent the native blood vessel adventitia, media and intima respectively. The in vitro longevity of these artificial vessel constructs is limited to less then twenty days. However, these studies represent a novel tissue engineering approach to the construction of vascular replacement devices. Although significant technical difficulties must be surmounted before this vessel construct could achieve clinical acceptability, the concept remains a fascinating example of tissue engineering.

The discussion of tissue engineered vascular devices has been limited in this chapter to arterial replacements. The methods and concepts described also have great application possibilities for use in other vascular replacements including venous replacements, valves, ventricular assist devices and the total artificial heart. Microvascular tissue replacements are also being studied wherein vascular elements are engineered within three-dimensional matrices for subsequent use as vascularized tissue replacements [12]. This field has immense application potential including the cellular basis for somatic cell gene therapy. The field of tissue engineering, while practiced by a limited number of investigators for several decades, is now being actively pursued by a wide variety of research groups. The use of this ever emerging technology in tissue engineered vascular devices will continue to be the focus of active research activity in the future.

References

1. Helmus MN (1991) Overview of biomedical materials. Material Research Society Bulletin 16: 33–38.
2. Kim SW, Feijen J (1985) Surface modification of polymers for improved blood compatibility. CRC Critical Reviews in Biocompatibility 1: 229–260.
3. Salzman EW, Merrill EW, Binder A, Wolf CFW, Ashford TP and Austen WG (1969) Protein-platelet interactions on heparinized surfaces. J Biomedical Materials Research 3: 69–81.
4. Williams SK (1995) Endothelial cell transplantation. Cell Transplantation 4: 401–410.
5. Lelah MD and Cooper SL (1986) Polyurethanes in Medicine. Boca Rotan, FL: CRC Press.
6. Williams SK, Carter T, Park PK, Rose DG and Jarrell BE (1992) Formation of a multilayer cellular lining on a polyurethane vascular graft following endothelial cell sodding. J Biomedical Materials Research 26: 103–117.
7. Graham LM, Vinter DW and Ford JW (1980) Immediate seeding of enzymatically derived endothelium in Dacron vascular grafts. Early studies with autologous canine cells. Arch Surgery 115: 1289–1294.
8. Williams SK, Rose DG and Jarrell BE (1994) Microvascular endothelial cell sodding of ePTFE vascular grafts: Improved patency and stability of the cellular lining. J Biomedical Materials Research 28: 203–212.
9. Herring M, Compton RS, Gardner AL and LeGrand DR (1987) Clinical experiences with endothelial seeding in Indianapolis. In: Zilla P, Fasol R and Deutsch M (eds) Endothelialization of Vascular Grafts, pp 218–224. Basel: Karger.
10. Williams SK (1996) The transplantation of endothelial cells. In: Lanza RP and Chick WL (eds) 1996 Yearbook of Cell and Tissue Transplantation. Kluwer.
11. Clowes AW (1991) Graft endothelialization: the role of angiogenic mechanisms. J Vascular Surgery 13: 734–736.
12. Thompson JA, Hadenschild CC and Anderson KD (1989) Heparin binding growth factor-1 induces the formation of organoid neovascular structures in vivo. Proceeding National Academy of Sciences (USA) 86: 7928–7932.
13. Greisler HP, Cziperle DJ, Kim DU, Garfield JD, Petsikas D, Murchan PM, Applegren EO, Drohan W and Burgess WH (1992) Enhanced endothelialization of expanded polytetrafluoroethytlene grafts by fibroblast growth factor type 1 pretreatment. Surgery 112: 244–255.
14. Van Buul-Wortelboer MF, Brinkman HJM, Dingemans KP, DeGroot PG, Van Aken WG and Van Mourick JA (1986) Reconstruction of the vascular wall in vitro: a novel model to study interactions between endothelial and smooth muscle cells. Experimental Cell Research 162: 151–158.
15. Weinberg CB and Bell E (1986) A blood vessel model constructed from collagen and cultured vascular cells. Science 231: 397–399.
16. L'Heureux N, Germain L, Labbe R and Auger FA (1993) In vitro construction of a human blood vessel from cultured vascular cells: a morphological study. J Vascular Surgery 17: 499–509.

Section XIV:

Chimerism and Tolerance

R. P. Lanza and W. L. Chick (eds.), Yearbook of Cell and Tissue Transplantation 1996/1997, 287–290.

Chapter 29

Hematopoietic cell transplantation for tolerance induction

Norma S. Kenyon and Camillo Ricordi
Diabetes Research Institute, University of Miami School of Medicine, Miami, FL 33136, U.S.A.

A major focus of the entire transplant community has been to delineate anti-rejection strategies which, alone or via multifaceted approaches, result in the development of donor specific tolerance, thus obviating the need for chronic, generalized immunosuppression of the recipient. Many approaches which have proven successful in rodent models of solid organ or cellular transplantation have not translated well to large animals or humans. This may be due, in part, to a need for multifaceted immunointerventions when transplanting cells or organs between members of outbred, large animal populations.

Historically, both therapeutic manipulation of the recipient and immunomodulation (or immunoisolation) of the graft itself have been explored as potential avenues to acceptance of transplanted cells and/or organs. Intrathymic administration of donor cells or antigens, monoclonal antibody therapy and treatment of cellular grafts to deplete or modulate the function of antigen presenting cells [1] are among the many approaches which have resulted in consistent prolongation of graft survival in rodent models. Although many monoclonal antibody strategies which have shown great promise for the induction of donor specific tolerance in rodents have not been effective in large animals, such studies have led to the widespread use of OKT3 and ATG (anti-thymocyte globulin), in combination with immunosuppressive agents (CSA, FK506, steroids, etc.) for the purposes of induction and/or anti-rejection therapy in humans [2]. With regards to immunomodulation, utilization of this approach for islet cell transplantation has resulted in modest prolongation of allograft survival in dogs treated with decreased levels of CSA [3, 4], although the results have been inconsistent, and this type of approach has not been widely studied for human transplantation.

Hematopoietic reconstitution of cytoablated mice, via transplantation of bone marrow from an allogeneic donor, results in the production of a chimeric animal which, depending on the irradiation/transplant protocol, posesses varying degrees of both host and donor derived hematopoietic cells. Even prior to the introduction of organ transplantation in humans, it was demonstrated that the induction of specific immunological tolerance could be achieved in inbred rodent strains by infusing allogeneic bone marrow or bone marrow derived cell suspsensions, with subsequent lifelong acceptance of donor specific skin grafts [5, 6]. With the development of techniques for vascularized anastomoses, organ transplantation in rodents became possible [7, 8], and it was demonstrated that specific immunological tolerance to organ allografts could also be achieved via infusion of bone marrow derived cells [9, 10]. The perceived requirement for cytoablation of the recipient, however, to provide a niche for donor bone marrow cells, has severely limited the application of such an approach to the fields of solid organ and cellular transplantation. Chemotherapy and irradiation are generally not indicated nor desirable for the types of diseases which result in a patient being placed on an organ waiting list.

Infusion of bone marrow derived cells, in the form of donor specific blood transfusions, has been shown to have a favorable effect on human allograft survival [11], although the mechanism responsible for the beneficial effect remains controversial. It has been proposed that enhancement of allograft survival, with either blood transfusion or bone marrow infusion, is a result of the establishment of low grade chimerism [12]. Sharing of Class II antigens between donor and recipient appears to augment the effect, possibly due to the sparing of stem cells from immune attack, with subsequent engraftment and survival of the donor derived cells [12].

Thomas and her colleagues have developed a primate model in which infusion of donor bone marrow (into non-cytoablated recipients), which has been depleted of the Class II bright cells, results in the prolongation of renal allograft survival [13, 14]. A veto mechanism has been proposed to account for the observed effect [15], and the critical cell has been reported to be Class II dim/negative and CD8 positive [15]. Interestingly, this cell shares phenotypic characteristics with the recently described facilitating cell of Ildstad and colleagues [16].

Attempts to tolerize humans with donor bone marrow infusions, however, have been skeptically viewed by clinicians. Using a protocol similar to those which he had described for mice and dogs [17, 18], which employed anti-thymocyte globulin as induction therapy, Monaco and his colleagues performed simultaneous kidney/bone marrow transplants in humans, with mixed results [19]. In the discussion of his findings, Monaco emphasized the critical importance of cell dose, as well as the number and timing of infusions, on allograft survival. Barber and his colleagues utilized PCR to detect the presence of microchimerism in patients who received donor bone marrow infusions in conjunction with a kidney transplant [20]. A significant correlation ($p = 0.01$) between the presence of chimerism (as detected by PCR) and the absence of rejection episodes was observed [20]. The majority of patients (91.3%) who received donor bone marrow were also rejection free; in contrast, 85.7% of patients with undetectable chimerism experienced at least one rejection episode [20].

The recent demonstration that chimerism can be achieved in non-cytoablated animals via multiple infusions of donor bone marrow [21, 22], however, has led to renewed interest in the potential of this approach, especially in light of results from studies of long term human transplant recipients who have discontinued their immunosuppressive medications, yet have maintained excellent graft function 27 to 29 years posttransplant [23]. Analysis of recipient tissues has revealed the presence of microchimerism, in the form of donor derived dendritic cells (DC), thus leading to the concept that the establishment of chimerism may be a prerequisite for the induction and maintenance of donor specific tolerance. Since DC are the most potent antigen presenting cells in the body, and have long been thought to be responsible for direct stimulation of the rejection response, it was postulated that non-antigen presenting, progenitor DC may somehow be critical to the tolerogenic process [24–26]. These findings led to a series of clinical trials at the University of Pittsburgh, designed to enhance the potential for the establishment of chimerism via infusion of donor vertebral body marrow.

Initially, two recipients of simultaneous liver – bone marrow transplants were conditioned with total lymphoid irradiation. One of the patients experienced a rejection episode, and the second one developed severe graft versus host disease (GVHD), which was resolved by infusion of cryopreserved autologous marrow [27, 28]. All subsequent patients were transplanted in the absence of radiation conditioning, using FK506 and steroids as immunosuppressive agents. Whole bone marrow was infused, in the absence of antibody based induction therapy, into recipients of kidneys, livers, hearts, and intestinal allografts [29]. The key findings of this study were that donor bone marrow infusion was safe, and enhancement of chimerism was achieved [29]. It was not possible to establish any clear correlation between the degree of chimerism and the presence of rejection and/or GVHD [29]. It therefore appeared that one donor bone marrow infusion, at the time of transplant, was not of substantial benefit to patients. The dose and timing of the bone marrow infusion, coupled with the type and timing of immunosuppression, may not have been optimal for the establishment of donor specific tolerance.

At the University of Miami, we have begun large animal trials of donor vertebral body marrow cell (DBMC) infusion in the dog. Our results indicated that, similar to the results obtained in the Pittsburgh trial, peri-transplant DBMC infusion with standard immunosuppression did not produce any beneficial effect. In fact, an increase in the incidence of rejection episodes was observed, as compared to control dogs which did not receive DBMC. A short course of induction immunosuppression (days -3 to -1) with a monoclonal antibody specific for canine T cells, however, allowed for allograft survival in all animals that received DBMC and CSA (20 mg/kg, days -5 to $+30$) treatment [30].

These observations, together with recent reports regarding the importance of DBMC dose and timing on the ability of DBMC to engraft [31], prompted us to reconsider clinical application of this approach. We therefore initiated a trial to test the effect of dose, timing, and inductive immunosuppression on human liver and kidney allograft survival. Since July of 1994, we have performed 63 liver allografts (32 with DBMC) and 30 kidney allografts (8 with DBMC).

We now have results which clearly indicate that 2 infusions of donor vertebral body marrow (5×10^8/kg at day 0 and at day 11) can improve liver allograft survival [31]. The incidence of rejection was reduced following 2 infusions of DBMC. In contrast, patients who received only one DBMC infusion, with or without OKT3 induction therapy, had a higher incidence of rejection than the control group. Despite the presence of rejection, allograft survival at four month follow-up was significantly better in recipients of either one DBMC (89%) or two DBMC (100%) infusions, as compared to controls (74%) [31]. No significant difference between the groups was observed in either the dose of steroids, the dose and level of FK506, or the degree of HLA matching. As was the case in Pittsburgh, we were unable to establish a correlation between the presence or absence of rejection and the degree of chimerism. Longer follow-up time will be required in order to assess the stability, lineage specificity, and effect of chimerism on the incidence of chronic rejection. Reduction in the levels of immunosuppression will be initiated one year posttransplant, in order to determine if donor specific tolerance has been achieved.

While we have not yet attained the donor specific tolerance which has been observed in rodent models, the encouraging results emerging from clinical trials of donor bone marrow infusion provide an impetus to continue to explore this approach in humans.

References

1. Charlton B, Aunchicloss Jr H, Fathman CG (1994) Mechanisms of transplantation tolerance. Annu Rev Immunol: 707–734.
2. Waldmann TA (1992) Immune receptors: Targets for therapy of leukemia/lymphoma, autoimmune diseases and for the prevention of allograft rejection. Annu Rev Immunol 10: 675–704.
3. Kenyon NS, Strasser S and Alejandro R (1990) Ultraviolet light immunomodulation of canine islets for prolongation of allograft survival. Diabetes 39 (3): 305–311.
4. Alejandro R, Latif Z, Noel J *et al.* (1987) Effect of anti-Ia antibodies, culture and cyclosporin on prolongation of canine islet allograft survival. Diabetes 36 (3): 269–273.
5. Billingham RE, Brent L and Medawar PB (1953) Actively acquired tolerance of foreign cells. Nature 172: 603–606.
6. Billingham RE and Brent L (1954) Quantitative studies of tissue transplantation immunity. II. The origins, strength and duration of actively and adoptively acquired immunity. Proc R Soc B 143: 58–80.
7. Lee S and Fisher B (1961) Portocaval shunt in the rat. Surgery 50: 668–673.
8. Fabre JW, Lim SH and Morris PJ (1971) Renal transplantation in the rat: Details of a technique. Aust NZ J Surgery 41: 69–75.
9. Stuart FP, Saitoh T and Fitch FW (1968) Rejection of renal allografts: Specific immunologic suppression. Science 160: 1463–1465.
10. French ME and Batchelor JR (1969) Immunological enhancement of rat kidney grafts. Lancet 2: 1103–1106.
11. Opelz G, Sengar DPS, Mickey MR *et al.* (1973) Effect of blood transfusions on subsequent kidney transplants. Transplant Proc 5: 253–259.
12. De Waal LP and Van Twuyver E (1991) Blood transfusions and allograft survival: Is mixed chimerism the solution for tolerance induction in clinical transplantation? Crit Rev Immunol 10: 417–425.
13. Thomas JM, Carver FM, Cunningham PR *et al.* (1991) Kidney allograft tolerance in primates without chronic immunosuppression – the role of veto cells. Transplantation 51: 198–207.
14. Thomas JM, Carver FM, Kasten-Jolly J *et al.* (1994) Further studies of veto activity in rhesus monkey bone marrow in relation to allograft tolerance and chimerism. Transplantation 57: 101–115.
15. Thomas JM, Verbanac KM and Thomas FT (1991) The veto mechanism in transplant tolerance. Transplantation Rev 5: 209–229.
16. Kaufman CL, Colson YL, Wren SM *et al.* (1994) Phenotypic characterization of a novel bone marrow-derived cell that facilitates engraftment of allogeneic bone marrow stem cells. Blood 84 (8): 2436–2446.
17. Monaco AP, Wood ML and Russel PS (1966) Studies on heterologous anti-lymphocyte serum in mice: III. Immunologic tolerance and chimerism produced across the H-2 locus with adult thymectomy and anti-lymphocyte serum. Ann NY Acad Sci 129: 190–206.
18. Hartner WC, De Fazio SR, Maki T *et al.* (1986) Prolongation of renal allograft survival in antilymphocyte-serum-treated dogs by postoperative injection of density-gradient-fractionated donor bone marrow. Transplantation 42 (6): 593–597.
19. Monaco AP, Clark Aw, Wood ML *et al.* (1976) Possible active enhancement of a human cadaver renal allograft with antilymphocytic serum (ALS) and donor bone marrow: Case report of an initial attempt. Surgery 79: 384–392.
20. McDaniel DO, Naftilan J, Hulvey K *et al.* (1994) Peripheral blood chimerism in renal allograft recipients transfused with donor bone marrow. Transplantation 57: 852–856.
21. Stewart FM, Crittenden RB, Lowry PA *et al.* (1993) Long-term engraftment of normal and post-5-fluorouracil murine marrow into normal mice. Blood 81: 2566–2571.
22. Harrison DE (1993) Competitive repopulation in unirradiated normal recipients. Blood 81: 2473–2474.
23. Starzl TE, Demetris AJ, Trucco M *et al.* (1993) Chimerism and donor specific nonreactivity 27 to 29 years after kidney allotransplantation. Transplantation 55: 1272–1277.
24. Starzl TE, Demetris AJ, Murase N *et al.* (1993) Donor cell chimerism permitted by immunosuppressive drugs: a probable basis of organ transplant acceptance and tolerance. Immunology Today 14: 326–332.
25. Ricordi C, Ildstad ST, Demetris AJ *et al.* (1992) Donor dendritic cells repopulation in recipients after rat-to-mouse bone marrow transplantation. Lancet 339: 1610–1611.

26. Ricordi C, Ildstad ST and Starzl Te (1992) Induction of pancreatic islet graft acceptance: The role of antigen presenting cells. Transplantation Science 2: 34–38.
27. Ricordi C, Tzakis AG, Demetris AJ *et al.* (1993) Reversal of graft-versus-host disease with infusion of stored autologous bone marrow cells following combined liver-bone marrow allotransplantation in man. Transplantation Science 3: 76–77.
28. Ricordi C, Tzakis Ag, Zeevi A *et al.* (1994) Reversal of graft-versus-host disease with infusion of autologous bone marrow. Cell Transplant 3: 187–192.
29. Fontes P, Rao A, Demetris AJ *et al.* (1994) Augmentation with bone marrow of donor leukocyte migration for kidney, liver, heart, and pancreas islet transplantation. Lancet 344: 151–155.
30. Brendel M, Kong SS, Schachner RD *et al.* (1995) The effect of donor specific vertebral body derived bone marrow cell infusion on islet allograft survival in the dog without cytoablative conditioning of the recipient. Presented at the International Symposium on Tolerance Induction, Breckenridge, Colorado, Jan 22–25.
31. Ricordi C, Karatzas T, Selvaggi G *et al.* (1995) Multiple bone marrow infusions to enhance acceptance of allografts from the same donor. Ann NY Acad Sci: in press.

List of contributors

Patrick Aebischer, M.D.
Professor of Surgery
Division of Surgical Research & Gene Therapy Center
Centre Hopitalier Universitaire Vaudois (CHUV)
Lausanne, Switzerland

Robert B. Aramant, Ph.D.
Associate Professor
Department of Ophthalmology and Visual Sciences
University of Louisville Medical School
Louisville, Kentucky, USA

James O. Armitage, M.D.
Professor and Chairman
Department of Internal Medicine
University of Nebraska Medical Center
Omaha, Nebraska, USA

Michael R. Bishop, M.D.
Assistant Professor
Department of Internal Medicine
University of Nebraska Medical Center
Omaha, Nebraska, USA

Susan Bonner-Weir, Ph.D.
Investigator
Section on Islet Transplantation and Cell Biology
Joslin Diabetes Center
Boston, Massachusetts, USA

Reinhard G. Bretzel, M.D., Ph.D.
Professor of Medicine
Third Medical Department
Justus-Liebig-University
Giessen, Germany

William L. Chick, M.D.
President & Scientific Director
BioHybrid Technologies Inc
and Sensor Technologies Inc
Shrewsbury, Massachusetts, USA

Bruce M. Carlson, M.D., Ph.D.
Professor and Chairman
Department of Anatomy and Cell Biology
The University of Michigan
Ann Arbor, Michigan, USA

Steve Chen, M.D.
Fellow, Liver Support Unit
Cedar-Sinai Medical Center
Los Angeles, California, USA

Zhong-Ping Chen, M.D., Ph.D.
Division of Neurosurgery
Lady Davis Institute for Medical Research
Jewish General Hospital
Montréal, Québec, Canada

Achilles A. Demetriou, M.D., Ph.D.
Director of Surgical Research,
Director, Liver Support Unit
Vice-Chairman, Academic Affairs
Cedar-Sinai Medical Center
and Professor of Surgery
University of California
Los Angeles, California, USA

Ghenima Dirami, Ph.D.
Research Assistant Professor
Department of Cell Biology
Georgetown University Medical Center
Washington, DC, USA

Martin Dym, Ph.D.
Professor and Chairman
Department of Cell Biology
Georgetown University Medical Center
Washington, DC, USA

Susumu Eguchi, M.D.
Fellow, Liver Support Unit
Cedar-Sinai Medical Center
Los Angeles, California, USA

Konrad F. Federlin, M.D.
Professor and Head
Third Medical Department
Justus-Liebig-University
Giessen, Germany

William J. Freed, Ph.D.
Chief, Section on Preclinical Neurosciences
NIMH Neuroscience Center
at Saint Elizabeths
National Institute of Mental Health
Washington, D.C., USA

Robert P. Gale, M.D., Ph.D.
Director of Stem Cell and
Bone Marrow Transplantation
Salick Healthcare Inc
Los Angeles, California, USA

Christoph Geier, M.D.
Third Medical Department
Justus-Liebig-University
Giessen, Germany

John F. Hansbrough, M.D.
Professor of Surgery
Director, Regional Burn Center
University of California
San Diego, California, USA

Bernhard J. Hering, M.D.
Registry Coordinator
International Islet Transplant Registry
Department of Medicine
Justus Liebig University
Giessen, Germany

Mary M. Horowitz, M.D., M.S.
Professor of Medicine
Scientific Director,
Intl. Bone Marrow Transplant Registry
Medical College of Wisconsin
Milwaukee, Wisconsin, USA

Celmente Ibarra, M.D.
Surgical Research
The Children's Hospital
Boston, Massachusetts, USA

Bruce E. Jarrell, M.D.
Professor and Head
Department of Surgery
University of Arizona
Tucson, Arizona, USA

Meng-Chun Jia, M.D.
Research Fellow
Department of Cell Biology
Georgetown University Medical Center
Washington, DC, USA

Norma Kenyon, Ph.D.
Research Assistant Professor
Diabetes Research Institute
University of Miami
Miami, Florida, USA

Anne Kessinger, M.D.
Professor of Medicine
Chief, Section of Oncology/Hematology
University of Nebraska Medical Center
Omaha, Nebraska, USA

John P. Klein, Ph.D.
Director of Biostatistics
Medical College of Wisconsin
Milwaukee, Wisconsin, USA

Robert Langer, Sc.D.
Kenneth J. Germeshausen Professor
of Chemical & Biomedical Engineering
Massachusetts Institute of Technology
Cambridge, Massachusetts, USA

Robert P. Lanza, M.D.
Director, Transplantation Biology
BioHybrid Technologies Inc
Shrewsbury, Massachusetts, USA

David F. Lawlor, M.D.
U.S. Army Institute
of Surgical Research
Fort Sam Houston, Texas, USA

Thomas E. Mandel, M.D.
Head, Transplantation Unit
The Walter and Eliza Institute of
Medical Research
University of Melbourne and
The Royal Melbourne Hospital
Victoria, Australia

Henry J. Mankin, M.D.
Chief, Orthopedic Surgery
Edith M. Ashley Professor of
Orthopedic Surgery
Harvard Medical School,
and Director, Orthopedics
Massachusetts General Hospital
Boston, Massachusetts, USA

Michio Mito, M.D.
Vice-President, Asahikawa Medical College
Director, Asahikawa Medical College Hospital
Professor of Surgery
Second Department of Surgery
Asahikawa Medical College
Asahikawa, Japan

David J. Mooney, Ph.D.
Assistant Professor of Chemical
Engineering and Dentistry
University of Michigan
Ann Arbor, Michigan, USA

Stanislaw Moskalewski, Ph.D.
Professor of Histology
Department of Histology
and Embryology
Institute of Biostructure
Medical school
Warsaw, Poland

Marion Murray, Ph.D.
Professor of Anatomy
and Neurobiology
Medical College of Pennsylvania
and Hahnemann University
Philadelphia, Pennsylvania, USA

Gail Naughton, Ph.D.
Executive Vice-President
and Chief Operating Officer
Advanced Tissue Sciences
La Jolla, California, USA

Terence A. Partridge, Ph.D.
Head of Muscle Cell Biology Group
MRC Clinical Sciences Centre
Royal Postgraduate Medical School
Hammersmith Hospital
London, UK

Jacob R. Passweg, M.D.
Clinical and Research Fellow
International Bone Marrow Transplant Registry
Medical College of Wisconsin
Milwaukee, Wisconsin, USA

Basil A. Pruitt, Jr., M.D., F.A.C.S.
Colonel, MC
Commander and Director
U.S. Army Institute of Surgical Research
Fort Sam Houston, Texas, USA

Neelakanta Ravindranath, Ph.D.
Research Assistant Professor
Department of Cell Biology
Georgetown University Medical Center
Washington, DC, USA

Camillo Ricordi, M.D.
Professor of Surgery
Director, Division of Cellular
Transplantation
Co-Director, Diabetes Research Institute
University of Miami School of Medicine
Miami, Florida, USA

Philip A. Rowlings, M.D.
Assistant Professor
Intl. Bone Marrow Transplant Registry
Medical College of Wisconsin
Milwaukee, Wisconsin, USA

Jacek Rozga, M.D., Ph.D.
Director, Liver Support Unit
Research Laboratory
Cedar-Sinai Medical Center
Los Angeles, California, USA

Jacqueline Sagen, Ph.D.
Associate Director,
Pharmacology and Behavioral Research
Cytotherapeutics Providence,
Rhode Island, USA

Masayuki Sawa, M.D.
Second Department of Surgery
Asahikawa Medical College
Asahikawa, Japan

Andreas O. Schultz, M.I.M.
Third Medical Department
Justus-Liebig-University
Giessen, Germany

Magdalene Seiler, Ph.D.
Assistant Professor
Department of Ophthalmology
and Visual Sciences
University of Louisville
Medical School
Louisville, Kentucky, USA

Kathleen A. Sobocinski, M.S.
Associate Director, International
Bone Marrow Transplant Registry
Medical College of Wisconsin
Milwaukee, Wisconsin, USA

Thomas E. Starzl, M.D., Ph.D.
Professor of Surgery
Director, Pittsburgh Transplantation Institute
University of Pittsburgh School of Medicine
Pittsburgh, Pennsylvania, USA

David E.R. Sutherland, M.D., Ph.D.
Professor of Surgery
University of Minnesota Medical School
Minneapolis, Minnesota, USA

Alan Tessler, M.D.
Research Associate Professor
Department of Anatomy
and Neurobiology
Medical College of Pennsylvania
and Hahnemann University
Philadelphia, Pennsylvania, USA

William R. Tolbert, PhD
President
W.R. Tolbert and Associates
San Diego, California, USA

William W. Tomford, M.D.
Associate Professor
Orthopaedic Surgery
Harvard Medical School and
Massachusetts General Hospital
Boston, Massachusetts, USA

Joseph P. Vacanti, M.D.
Associate Professor of Surgery
Harvard Medical School and
Director of Organ Transplantation
The Children's Hospital
Boston, Massachusetts, USA

David C. Wahoff, M.D.
Department of Surgery
University of Minnesota
Minneapolis, Minnesota, USA

Gordon C. Weir, M.D.
Professor of Medicine
Harvard Medical School and
Head, Section on Islet Transplantation
and Cell Biology
Joslin Diabetes Center
Boston, Massachusetts, USA

Keryn A. Williams, Ph.D.
Department of Ophthalmology
Flinders Medical Centre
Bedford Park, Australia

Stuart K. Williams, Ph.D.
Professor of Surgery and Physiology
Chief, Section of Surgical Research
Department of Surgery
University of Arizona
Health Sciences Center
Tucson, Arizona, USA

Mei-Jie Zhang, Ph.D.
Assistant Professor
International Bone Marrow Transplant Registry
Medical College of Wisconsin
Milwaukee, Wisconsin, USA